The Fiber Society's Spring 2015 Conference

Functional Fibers and Textiles

In Conjunction with the 2015 International Conference on Advanced Fibers and Polymer Materials

Shanghai, China
24 - 27 May 2015

ISBN: 978-1-7138-0624-0

Printed from e-media with permission by:

Curran Associates, Inc.
57 Morehouse Lane
Red Hook, NY 12571

Some format issues inherent in the e-media version may also appear in this print version.

Printed with permission by Curran Associates, Inc. (2025)

For permission requests, please contact Donghua University
at the address below.

Donghua University – College of Materials Science & Engineering
2999 North Renmin Road
Songjiang, Shanghai
China, 201600

Phone: 0086-21-67792362
Fax: 0086-21-67792855

http://cmse.dhu.edu.c

Additional copies of this publication are available from:

Curran Associates, Inc.
57 Morehouse Lane
Red Hook, NY 12571 USA
Phone: 845-758-0400
Fax: 845-758-2633
Email: curran@proceedings.com
Web: www.proceedings.com

THE FIBER SOCIETY

Advancing Scientific Knowledge
Pertaining to Fibers and Fibrous Materials

ICAFPM

The Fiber Society's Spring 2015 Conference

in conjunction with the

2015 International Conference on Advanced Fibers and Polymer Materials

Functional Fibers and Textiles

May 24–27, 2015

Program

Conference Chair

Dr. Meifang Zhu, Donghua University, China

Venue

Songjiang Campus, Donghua University, Shanghai, China

Organizers

State Key Laboratory for Modification of Chemical Fibers and Polymer Materials
College of Materials Science and Engineering, Donghua University
The Fiber Society

SKLFPM
DONGHUA UNIVERSITY

Sunday, May 24

12:00 PM–6:00 PM	Registration: No. 2 Teaching Bldg., Entrance Hall, Songjiang Campus, Donghua University
3:00 PM–5:00 PM	111 Project Annual Meeting (Jing-Gang-Shan Room, 12F, Neo-Sunshine Hotel)
6:30 PM–9:30 PM	Governing Council Meeting (Jing-Gang-Shan Room, 12F, Neo-Sunshine Hotel)

****

Donghua University Songjiang Campus

Monday, May 25

Morning

8:00 Registration Desk Opens, No. 2 Teaching Building, Entrance Hall

Opening Ceremony

No. 2 Teaching Building, Room 2137

8:30 Welcoming Remarks, Business, and Announcements

Plenary Lectures

No. 2 Teaching Building, Room 2137

Session Chair: *Ton Peijs*

9:00 *The Development of High-Performance Fibers at Donghua University* – Yimin Wang, Donghua University...2

9:30 *Future Textiles: From Functional to Smart* – Jinlian Hu, Hong Kong Polytechnic University...3

10:00 Photo Session

10:10 Coffee Break

Session Chair: *Meifang Zhu*

10:30 *Application of High-Speed Melt Spinning Process for Enhanced Fiber Structure Formation of PHBH* – Takeshi Kikutani, Tokyo Institute of Technology...5

11:00 *3D Scaffolds of Short Electrospun Nanofibers* – Andreas Greiner, University of Bayreuth...7

11:30 *Carbon Fibre Microelectrodes for Neuroscience Applications* – Xungai Wang, Deakin University...8

12:00–1:00 Lunch: Donghua Restaurant II (3F)

Afternoon

Poster Session

No. 2 Teaching Building, Hallway

1:00–3:30 Poster Presentations and Break

Oral Sessions

No. 2 Teaching Building, Room 2118
Session: Nanofibrous Materials
Session Chair: René Rossi

3:30 *Electrospun Nanofiber-Assembled Cellular Aerogels with Superelasticity and Multifunctionality* – Bin Ding, Donghua University...16
4:00 *Artificial Muscles from Carbon Nanotube and Nylon Fibres* – Javad Foroughi, University of Wollongong...18
4:20 *Functional Nonwovens by Green Electrospinning* – Seema Agarwal, University of Bayreuth...20
4:40 *Nanofibrous Membranes with One-Way Oil Transport Ability and Advanced Oil-Water Separation Performance* – Tong Lin, Deakin University...21
5:00 *Controlled Release of Active Agents from Electrospun Fibers* – René Rossi, Empa...23

No. 2 Teaching Building, Room 2109
Session: Functional Textiles and Clothing
Session Chair: Dominique Adolphe

3:30 *Functional Textiles for Biomedical Engineering* – Yi Li, Hong Kong Polytechnic University...26
4:00 *Textile Heart Valve: Fabric Construction Parameter and In Vivo Results* – Antoine Vaesken, ENSISA...28
4:20 *Thermal Comfort of Professional Sportswear with Physiological and Psychophysical Measurements During High-Intensity Exercise* – Tsung-lin Liu, Taiwan Textile Research Institute...29
4:40 *Parameters Impact on Pressure Comfort in System "Female Body-Dress"* – Victor Kuzmichev, Ivanovo State Polytechnic University (presented by Dominique Adolphe)...31
5:00 *Open*

No. 2 Teaching Building, Room 2105
Session: Fiber Applications in Medical, Transportation, and Energy
Session Chair: Jintu Fan

3:30 *Regenerated Keratin Fibers and Their Potential Applications in Textile and Biomedical Industries* – Yiqi Yang, University of Nebraska-Lincoln...34
4:00 *Geometry Modeling of a Tubular Weft-Knitted Intestinal Stent* – Gang Li, Soochow University...36
4:20 *Fabrication of Multiscale Elastic Electrospun Fibrous Scaffolds for Repairing Augmentation of Rotator Cuff* – Wenguo Cui, Soochow University...38
4:40 *Hyaluronic Acid-Functionalized Electrospun Random and Aligned PLA Nanofibers for Cancer Cell Capture* – Yili Zhao, Donghua University...39
5:00 *Cellulose/Chitosan Composites Prepared in Ethylene Diamine/Potassium Thiocyanate for Adsorption of* Cu^{2+} – Min Xiao, Guangzhou Testing and Research Institute...41

No. 2 Teaching Building, Room 2106
Session: Nonwovens; Green Textile Manufacturing
Session Chair: Richard Kotek

3:30 *Solvent-Free, Solid State Processed Tapes of UHMWPE: Influence of Molar Mass and Molar Mass Distribution on Tensile Properties* – Sanjay Rastogi, Loughborough University...44
4:00 *Evolution of Chain Entanglements of Ultrahigh Molecular Weight Polyethylene in Dilute Solution Blends During Thermal Annealing and Its Influence on Crystallization* – Wei Li, Ningbo University...46
4:20 *Polypropylene (PP) Polymer and Its Nonwovens in Medical Applications* – Jamil Siddiqui, TASNE-NIPRAS Plastics Research Center...47
4:40 *Investigation of Hydrostatic Pressure and Blood Penetration of a Nonwovens Material* – Peter Tsai, University of Tennessee-Knoxville...48
5:00 *Open*

6:00–8:30 Banquet: Neo-Sunshine Hotel Shanghai

Tuesday, May 26

Morning

8:00 Registration Desk Opens, No. 2 Teaching Building, Entrance Hall

Oral Sessions

No. 2 Teaching Building, Room 2118

Session: Nanofibrous Materials; High-Performance Fibers and Composites

Session Chair: Sanjay Rastogi

8:30 *Chitosan Nanofibers for Cr (VI) Adsorption* – Chuanfang Yang, Chinese Academy of Sciences...51

9:00 *Gel Spinning of Nylon 6 and PVA* – Richard Kotek, North Carolina State University...52

9:20 *Modeling PA-6 Nanofiber Properties Produced by Needleless Electrospinning: Effect of Solution Concentration and Voltage* – Nabyl Khenoussi, ENSISA...54

9:40 *Effect of Reaction Solvents on Morphology and Properties of Interfacial Polymerized Polyaniline Nanofibers* – Tao Li, Donghua University...56

10:00 Break

Session Chair: Yingfeng Tu

10:20 *Production of Ultra-Strong Polymer Fibers from Extended Chains* – Donggang Yao, Georgia Institute of Technology...58

10:50 *Morphology and Structure Development of Aromatic Polysulfonamide Fibers During Spinning Process* – Yumei Zhang, Donghua University...60

11:20 *Study on the Silane Modification of PPTA Fibers* – Junrong Yu, Donghua University...62

11:50 Lunch: Donghua Restaurant II (3F)

No. 2 Teaching Building, Room 2109

Session: Functional Textiles and Clothing

Session Chair: Laurence Schacher

8:30 *Flow Asymmetricity in Fibrous Constructs* – Jintu Fan, Cornell University...64

9:00 *Change the Fiber Inner Structure by Twisting Monofilament Under a Hot Condition* – Zhigang Xia, Wuhan Textile University...66

9:20 *Flame-Resistant Protective Clothing in Industries* – Zina Zhu, Milliken & Company...68

9:40 *Fully-Printed Flexible Temperature Sensors on Textile Substrates for Textronic Applications* – Zbigniew Stempién, Lodz University of Technology...70

10:00 Break

Session Chair: Yimin Wang

10:20 *Scalable and Controllable Fabrication of Conductive Fibers and Fabrics Through Facile Dip-Dying Approach: New Design and Potential Applications* – Zong-Yi Qin, Donghua University...72

10:50 *Synthesis of Triarylmethane Reversible Thermochromic Colorant Using Bromcresol Purple for Smart Fabrics* – Wan Zhang, Jiangnan University...74

11:10 *Simulation of Stoll's Empirical Burn Criterion Using a Mathematical Model for Evaluation of Thermal Protective Clothing* – Li-Na Zhai, Donghua University...76

11:30 *3D Clothes Shaping by Means of the Sewing Edges Configuration and Fabric Properties* – Dominique Adolphe, ENSISA...78

11:50 Lunch: Donghua Restaurant II (3F)

No. 2 Teaching Building, Room 2105

Session: Fiber Applications in Medical, Transportation, and Energy

Session Chair: Janice Gerde

8:30 *Functional Polymer Nanofibers for Water Purification* – Xuefen Wang, Donghua University...81

9:00 *Melt-Spun Nanofibrous Filtration Media for Smog Control for Personal Respiratory Protection and Environmental Air Quality* – Tao Huang, DuPont...83

9:20 *Nanofibrous Scaffolds from Polycaprolactone as Wound Dressings* – Kateřina Pilařová, Technical University of Liberec...85

9:40 *Development of Poly(L-lactide-co-caprolactone) Multichannel Nerve Conduit with Aligned Electrospun Nanofibers for Schwann Cell Proliferation* – Muhammad Aqeel Bhutto, Donghua University...87

10:00 Break

Session Chair: Xuming Xie

10:20 *Development of Biocompatible Hydrogels for Biomedical Applications* – Bhuvanesh Gupta, Indian Institute of Technology...88

10:50 *Electrospun Porous Carbon Nanofiber@MoS_2 Core/Sheath Fiber Membranes as Highly Flexible and Binder-Free Anode for Lithium-Ion Batteries* – Yue-E Miao, Fudan University...89

11:10 *Hierarchical $ZnCo_2O_4$@$NiCo_2O_4$ Core-Sheath Nanowires: Bifunctionality Toward High-Performance Supercapacitors and Oxygen Reduction Reaction* – Yunpeng Huang, Fudan University...90

11:30 *Controllable Preparation of Multidimensional Hybrid Materials of Nickel-Cobalt Layered, Double Hydroxide Nanorods/Nanosheets on Electrospun Carbon Nanofibers for High-Performance Supercapacitors* – Feili Lai, Fudan University...91

11:50 Lunch: Donghua Restaurant II (3F)

No. 2 Teaching Building, Room 2106

Session: Functional and Smart Fibers

Session Chair: Yusuf Ulcay

8:30 *Fascinating Electroactive Hydrogels-Based Bacterial Cellulose* – Guang Yang, Huazhong University of Science and Technology...93

9:00 *Shape Memory Effect of Biopolymer Fibers Responsive to Humidity* – Xueling Xiao, Hong Kong Polytechnic University (presented by Jinlian Hu)...95

9:20 *Nanocellulose Hierarchical Structures and Functional Materials* – You-Lo Hsieh, University of California-Davis...97

9:40 *Molecular Orientation in Melt-Spun P3HB Fibers* – Rudolf Hufenus, Empa...99

10:00 Break

Session Chair: Jinlian Hu

10:20 *Changes in Microclimate Temperature by Temperature-Adaptable Materials Under Transient Conditions* – Eunae Kim, Yonsei University...101

10:50 *Effect of Hot Drawing on the Mechanical Properties, Moisture Absorption, and Dyeability of Novel Nylon 5,6 Fibers* – Yassir Eltahir, Donghua University...103

11:10 *Thermal-Stored Energy and Performance of Protective Clothing Upon Hot Water Exposure Using Instrumented Spray Manikin* – Farzan Gholamreza, University of Alberta...105

11:30 *Centrifugally-Spun Energy-Storage Nanofibers* – Xiangwu Zhang, North Carolina State University

11:50 Lunch: Donghua Restaurant II (3F)...107

Afternoon

Oral Sessions

No. 2 Teaching Building, Room 2118

Session: Nanofibrous Materials; High-Performance Fibers and Composites

Session Chair: Frederic Heim

1:00 *Electrospun Nanoyarn Scaffold for Tissue Engineering* – Xiumei Mo, Donghua University...110

1:30 *Air-Impedance Electrospinning Technique for Cartilage Regeneration* – Gary Bowlin, University of Memphis...111

1:50 *Polydopamine-Derived Porous Carbon Fiber Embedded with CoO Nanoparticles for Efficient Oxygen Reduction Reaction* – Jiajie Yan, Fudan University...112

2:10 *Spherical Nanocellulose Formates as Green Fillers to Improve Crystallization and Mechanical Property of Poly(lactic acid)* – Houyong Yu, Zhejiang Sci-Tech University...113

2:30 *Construction Nanomaterials for Modification of General Thermal Plastic Polymer Resin and Fiber* – Bin Sun, Donghua University...115

2:50 Break

Session Chair: Yumei Zhang

3:10 *Silk Fibroin Fibers and Scaffolds from Aqueous Solutions* – Yaopeng Zhang, Donghua University...116
3:40 *Green Graphene-Silk Nanocomposite Materials* – Xiao Hu, Rowan University...118
4:00 *In Situ Self-Assembly Nanocomposites from Electrospun Poly(lactic acid)/Lipid and Bacterial Cellulose Nanofibers* – Chunhui Xiang, Iowa State University...120
4:20 *Poly(ether ester) Alternative Multiblock Copolymers: Synthesis and High Resilience Properties* – Yingfeng Tu, Soochow University...122
4:40 *ZnS/Bacterial Cellulose/Epoxy Resin (ZnS/BC/E56) Nanocomposites with Good Transparency and Flexibility* – Shiyan Chen, Donghua University...123

No. 2 Teaching Building, Room 2109

Session: Functional Textiles and Clothing; Functional Finishing and Dyeing

Session Chair: Xiangwu Zhang

1:00 *Research on the Nanogenerators Based on Electrospun Polymeric Nanofiber Mats* – Hao Yu, Donghua University...126
1:30 *Characterizing Thermal Protective Fabrics of Firefighters' Clothing in Hot Surface Contact* – Guowen Song, University of Alberta (presented by Farzan Gholamreza)...127
1:50 *Fabrication of ZnO & SiO_2 with Eliminated Photocatalytic Activity for Preventing Wool Photoyellowing* – Lu Sun, Deakin University...129
2:10 *Dynamical Wettability Switch via Photoinduced F/TiO_2 for Cellulose Substrates* – Yunjie Yin, Jiangnan University...131
2:30 *Open*
2:50 Break

Session Chair: Guang Li

3:10 *Methodologies for High-Performance Electrospun Fibers* – Ton Peijs, Queen Mary University of London...133
3:40 *Role of Relative Humidity and Ca^{2+} Concentration on Molecular Assembly of Silk Fibroin* – Qingfa Peng, Donghua University...135
4:00 *Open*
4:20 *Open*
4:40 *Open*

No. 2 Teaching Building, Room 2105

Session: Fiber Applications in Medical, Transportation, and Energy; Natural Fiber Spinning, Weaving, and Knitting

Session Chair: Huili Shao

1:00 *Hydrogen-Bonding Polymer Complexation for Low-Dimensional Material Preparation* – Shuguang Yang, Donghua University...138
1:30 *Directly Deposited Si NP/CNT/PVA Nanofiber Anodes for Li-Ion Battery Application* – Yong Lak Joo, Cornell University...139
1:50 *Membranes of Manganese Oxide Beaded Carbon Nanofibers as High-Performance Flexible Anodes for Lithium Ion Batteries* – Xin Zhao, Donghua University...141
2:10 *Carbon Nanotube Sheet Electrodes for Anisotropic Actuation of Dielectric Elastomers* – Tushar Ghosh, North Carolina State University...143
2:30 *Open*
2:50 Break

Session Chair: Xungai Wang

3:10 *Design of Electrospun Hybrid Nanofibrous Materials for Biomedical Applications* – Xiangyang Shi, Donghua University...145
3:40 *Efficient Dissolution of Cellulose in Novel Complex Solvent and Its Spinning* – Yue Zhang, Donghua University...146
4:00 *Cellulose Multifilament Prepared from Cellulose Carbamates in NaOH/ZnO Aqueous Solutions on a Pilot Scale* – Jinping Zhou, Wuhan University...148
4:20 *Lower Modulus Research on Cellulose Diacetate Fiber* – Jiajun Wu, Donghua University...149
4:40 *Open*

No. 2 Teaching Building, Room 2106

Session: Synthesis, Structure, and Properties of Fiber-Related Materials

Session Chair: Donggang Yao

1:00 *Self-Healable, Super-Tough GO/PAA Nanocomposite Hydrogels Facilitated by Dual Cross-Linking Effects* – Xu-Ming Xie, Tsinghua University...152

1:30 *Strengths of Fiber Products: Current and Historical Perspectives on Statistical Size Effects* – Moon Suh, North Carolina State University...153

1:50 *Prediction of the Tensile Strength of a Two-Ply Polyester Filament Yarn* – Huang Wei, Donghua University...155

2:10 *Improvement of the Testing Instrument of Fabric-Skin Adhesion Properties* – Feng Ji, Donghua University...157

2:30 *Open*

2:50 Break

Session Chair: Dajun Chen

3:10 *Development of Novel Poly(p-phenylene terephthalamide)s* – Xinhua Wan, Peking University...159

3:40 *Solubility, Discoloration and ^{13}C NMR Spectra of Stereoregular Poly(vinyl chloride) Prepared by Urea Clathtrate Polymerization at Low Temperature* – Masatomo Minagawa, NPO, Dream-Create-Laboratories...160

4:00 *Hyperbranched Linear Block Co-Polybenzimidazole for Non-Covalent Functionalization of Carbon Nanomaterials and the Effective Reinforcement of Their Composites* – Yan Wang, Donghua University...162

4:20 *Synthesis and Characterization of Melt Processable Acrylonitrile-N-Vinylimidazole Copolymers* – Na Han, Tianjin Polytechnic University...164

4:40 *Open*

5:00–6:30 Dinner: Donghua Restaurant II (3F)

Wednesday, May 27

Morning

Plenary Lectures

No. 2 Teaching Building, Room 2137

Session Chair: Yiqi Yang

9:00 *Chemical Modifications of Synthetic Fibers* – Gang Sun, University of California-Davis...10

9:30 *Crystal Networks in Fibrous Soft Materials: From Hierarchical Structure to Ultra-Performance* – Xiang Yang Liu, National University of Singapore...12

10:00 Break

Session Chair: Rudolf Hufenus

10:20 *Wearable Fiber-Shaped Energy Harvesting and Storage Devices* – Huisheng Peng, Fudan University...13

10:50 *Exploring the Nature of Cellulose Microfibrils* – Benjamin Hsiao, Stony Brook University...14

11:20–11:50 Closing Ceremony

12:00–1:00 Lunch: Donghua Restaurant II (3F)

****

Poster Presentations

Plenary Speakers

The Development of High-Performance Fibers at Donghua University and the Preparation of TLCP Polyarylate Fibers

Yimin Wang, Yanping Wang, Yumin Xia
State Key Laboratory for Modification of Chemical Fibers and Polymer Materials, College of Materials Science and Engineering, Donghua University, Shanghai, China

ymw@dhu.edu.cn

ABSTRACT
The studies on the preparations of aromatic polyamide and fiber, flexible chain PE, PP fiber, rayon based carbon, polyimide polymerization and fiber by dry spinning, extrude reaction polymerization and spinning of PBO were introduced and the preparation of polyarylate and its melt spinning of fibers were described in detail.

The results on the preparation of polyarylate and its melt spinning of fibers showed that the melt spinning process could be effectively improved by optimizing spinning attachments and conditions and fiber with high mechanical properties was obtained by low molecular weight polymer spinning and post heat treatment. The results showed that the tensile strength of the fiber spun by low molecular weight and post treated by 32 hours at vacuum increased from 5.67 to 16.42 cN/dtex and Young's modulus from 295.23 to 389.32 cN/dtex, however for the higher molecular weight polymer fiber, tensile strength from 8.52 to 19.33 cN/dtex and Young's modulus from 382.41 to 410.42 cN/dtex respectively.

Future Textiles: From Functional to Smart

Jinlian Hu
The Hong Kong Polytechnic University
jin-lian.hu@polyu.edu.hk

STATEMENT OF PURPOSE
Smart textiles can also be thought as the higher stage of functional textiles. Shape memory polymers applied in smart textiles in fiber, foam, solution, gel, etc forms; and varieties of functions such as shape memory, conductive, color change, etc. to achieve ventilation, water vapor management, liquid protective, biological protective mechanical protective, thermal protective and medical protective functions. The market demand and future trend of smart textiles are also describes in this article.

INTRODUCTION
Functional textiles are specifically engineered to deliver a pre-defined performance or functionality to the end-user, over and above its normal and basic functions. Smart is a further step of functional textiles or an intelligentized function which can be defined as these materials and structures can sense and react to environmental conditions or stimuli. Miniaturisation of electronic components and the use of conductive materials and other structures have enabled the rapid development of smart textiles. Growing demand for connectivity and enhancement in wireless technologies is expected to further market growth over the next few years. According to a new report, the global demand for smart and interactive textiles was valued at USD 1.53 billion in 2013 and is expected to reach USD 3.82 billion in 2020[1]. However, emerging regions such as Asia Pacific are likely to witness the fastest growth over the next few years. Currently, the market for smart textiles in Asia Pacific is sluggish as compared to that in Europe and North America.

SMART MATERIALS
Smart materials can response to controllable external stimulus significantly reverses. Thermal storage materials, conductive materials, and shape memory materials will be introduced.

Stimuli-responsive polymers (SRPs) rapidly change their configuration, dimension or physical properties with small changes in the appropriate stimuli such as heat [2], moisture/water [3], pH value [4], electricity [5], light [6], magnetic field [7], and solvent [8]. Stimuli responsive materials include thermochromic organic materials, photochromic organic, materials, photoluminescent organic materials and shape memory materials. Shape memory alloys (SMAs) and shape memory polymers [9] (SMPs) are the main shape memory materials used in smart textiles. The large amount of strain by more than several hundred percent can be recovered in shape memory polymers. Because of the easy process ability, soft handle and better compatibility, the present form of shape memory polymer fiber is most preferred for textile applications and some engineering applications compared to shape memory alloys (SMAs), shape memory foam and shape memory film.

Inherently conductive polymers have the ability to emulate the biological muscles due to their both sensing and actuation properties. Some commonly known inherently conductive polymers are polyacetylene, polypyrrole, polyaniline, and polythiophene. Polypyrrole (PPy) which has high mechanical strength with high elasticity, relatively stable in air and electroactive in both organic and aqueous solutions.

Latent heat-storage materials are also called phase change materials (PCM). PCM can absorb or release heat with a slight temperature change and may be repeatedly converted between solid and liquid phases to utilize their latent heat of fusion to absorb, store and release heat or cold during such phase conversions.

Other materials like plastic optical fibers can serve a number of functions in a smart garment - transmit data signals, transmit light for optical sensing, detect deformations in fabrics due to stress and strain and perform chemical sensing.

APPLICATIONS OF SHAPE MEMORY POLYMERS FOR SMART TEXTILES
Shape memory polymers have tremendous affinity to smart textiles and clothing. SMPs have been applied to make various protective textiles with smart functions and have tremendous potential for wide applications to smart textiles. Quite significant progress had been made in making textiles using SMPs in the shape memory textile center.

Self-adaptability of SMP textiles
Self-adaptable shape memory textiles include high pressure garments with easy wear&down property, comfort garments with low pressure changing with the body figures; shape retention garments with flat appearance, crease retention bagging recovery and wrinkle recovery properties; fashion design with style change and three dimension effects.

SMPs for ventilation textiles
Multi-layered fabrics made of shape memory polymers can automatically open when wearer starts to sweat and flap close again when skin returns to normal humidity. This kind of fabric mimics the pine cone structure which reacts differently to humidity to control the open and close the layers.

SMPs for liquid management textiles
Fabrics coated by a thin film of liquid-crystal shape memory polymers to get a micrometer and nanometer scale

roughness surface. The water droplets rolled off at 23℃ but were pinned fast at 75℃ even when the surface was held upside –down. A hydrophobic fabric with shape memory polymer coating can be scratched and self-healing after more than 100 cycles to keep the good hydrophobic effect.

SMPs for biological protective textiles
The nanofibers with core-shell structure or bead on string structure can be electro-spun successfully. The excellent shape memory effect and excellent antibacterial activity against bacteria are achieved in the shape memory core-shell nanofiber.

SMPs for medical textiles
Compress garments made of shape memory fibers can supply enough pressure on body parts and easy to wear and down by heating the garments to get a larger temporary size. The pressure force can be controlled by periodically heating. When heating above the switch temperature of the materials, the pressure increases. When the temperature is below the switch temperature, the pressure force decreases too.

SMPs for mechanical protective textiles
Shape memory materials can show a good damping property for mechanical protection. For example, an automotives seatbelt fabric using SMP fibers can absorb energy from the body's forward motion.

SMPs for thermal protective clothing
Shape memory fibers or thread is added between two layers. When temperature is higher than the switch temperature, the gap becomes thicker automatically to insulate the heat danger.

APPROACH
Smart materials can be incorporated into the textile structure by different technologies such as embroidering, sewing, non-woven textile, knitting, weaving, braiding, coating/laminating and printing. The methods are quite different and each one has very specific features. The advantage of coatings is suitable for many fibers types textiles and they produce good functions without significantly altering existing key substrate properties such as density, flexibility and handle. Nanoparticles are being widely used to improve the performance and functionality of textiles by coating method. Embroidery allows precisely specifying the functional thread layout and stitching pattern, from which any number of articles can be sewn under machine control. [11]. Bio-inspired method or structure is successful for smart textiles development. The notion of bio-inspired smart textiles is intended to perform artificial mechanosensing and actuation phenomenon for different benefits. Camouflage phenomenon, like in chameleon is put a natural mechanism to design the dream apparel for the military and defense services to hide or make an entity invisible. Light sensing mechanism in algae eyespot inspire designing light sensitive apparel which can be worn as optical protective device filtering harmful light rays from electromagnetic spectrum. It is possible to design artificial touch sensitive actuators incorporated into apparels learning from sensing actuation mechanism of touch sensitive mimosa plant.

FUTURE WORK
Wearable electronics are now a huge trend. Fashion and wearable will likely meet in more obvious ways too, with colors, designs and trims controlled dynamically while a garment is being worn. Rising demand for smart and interactive textiles from the transportation industry is likely to be one of the major factors for the market. Furthermore, growth in medical and healthcare industries is anticipated to lead to higher demand for smart textiles during the forecast period. However, market growth is likely to be sluggish due to high prices of finished products of smart textiles as compared to conventional textiles.

KEYWORDS
Functional textiles, stimuli responsive, shape memory, thermal storage, and wearable electronics

REFERENCES
[1]http://www.transparencymarketresearch.com/pressrelease/smart-fabrics-and-interactive-textiles-market.htm
[2] Hu, J.L. *Shape Memory Polymers and Textiles.* Cambridge: Woodhead Publishing, 2007.
[3] Yang, B, Huang, W.M, Li, C., Li, L. "Effects of moisture on the thermomechanical properties of a polyurethane shape memory polymer." *Polymer,* 47, 2006: 1348-1356.
[4] Feil, H., Bae, Y.H., Feijen, T., Kim, S.W. "Mutual influence of pH and temperature on the swelling of ionizable and thermosensitive hydrogels." *Macromolecules*, 25, 1992: 5228–5230.
[5] Asaka, K., Oguro, K. "Bending of polyelectrolytemembrane platinum composites by electric stimuli." *J. Electroanal. Chem.*, 48, 2000: 186-198.
[6] Jiang, H.Y, Kelch, S., Lendlein, A. "Polymers move in response to light." *Adv. Mater.*, 18, 2006: 1471-1475.
[7] Makhosaxana, X.P., Filipcsei. G., Zrınyi, M. "Preparation and responsive properties of magnetically softpoly (N-isopropylacrylamide) gels." *Macromolecules*, 33, 2003: 1716-1719.
[8] Siegal, R.A., Firestone, B.A. "pH-dependentequilibrium swelling properties of hydrophobicpolyelectrolyte copolymer gels." *Macromolecules*, 21, 1988: 3254-3259.
[9] Hu, Jinlian, Zhu, Yong, Huang, Huahua, Lu, Jing. "Recent advances in shape-memory polymers: Structure, mechanism, functionality, modeling and applications." *Progress in Polymer Science*, 37 (12), 2012: 1720-1763.
[11] Guptaq, Deept. "Functional clothing: definition and classification." *Indian Journal of Fibre & Textile Research,* 2011: 321-326.

Application of High-Speed Melt Spinning Process for Enhanced Fiber Structure Formation of PHBH

Qing Qin, Wataru Takarada, Takeshi Kikutani
Department of Organic and Polymeric Materials,
Graduate School of Science and Engineering, Tokyo Institute of Technology, Japan
kikutani.t.aa@m.titech.ac.jp

ABSTRACT

High-speed melt spinning of poly(3-hydroxybutyrate-*co*-3-hydroxyhexanoate) (PHBH) with the 3-hydroxyhexanoate composition of 5.4 mol% was carried out. Even though it has been reported that PHBH fibers of good appearance can be prepared through the melt spinning process only when extrusion temperature is lower than the melting temperature of pure PHB (176° C), it was revealed in this study that the crystallization of PHBH proceeds at high take-up velocities even when the extrusion temperature is higher than 180° C. Crystalline orientation of α-form crystal and fraction of β-form crystals increased with an increase in the take-up velocity and decrease in the extrusion temperature, whereas the effect of through-put rate was not straight forward because there was a significant effect of thermal decomposition during the extrusion process.

INTRODUCTION

Recently, bacterial poly(3-hydroxybutyrate) (PHB) attracts a great deal of attention not only for its biodegradability and biocompatibility but also for its nature of bio-based polymer. PHB is an aliphatic polyester, however, its melting temperature is close to the temperature where thermal degradation starts to occur. To overcome such drawback, various PHB based co-polymers have been developed. Poly(3-hydroxybutyrate-*co*-3-hydroxyhexanoate) (PHBH) is one of those polymers, which has been known as a melt-processable aliphatic co-polyester. However, processing of this polymer also is not easy because the glass transition temperature of PHBH is below the room temperature whereas the crystallization rate of this polymer is quite low. Accordingly, it is difficult to form fibers, films and other products from PHBH through the melt processing because the polymer can be still in a rubbery state after cooled down to the room temperature. In this research, attempts were made to overcome the low crystallizability of PHBH in the melt processing by applying the high-speed melt spinning process. When take-up velocity is increased, high tensile stress can be applied to the spinning line. Molecular orientation and orientation-induced crystallization are expected to proceed under such high tensile stress.

EXPERIMENTAL

PHBH with the 3-hydroxyhexanoate content of 5.4 mol% supplied by Kaneka Co. Ltd. was used in this study. Molecular weight of the polymer was 5.3×10^5.

Melt spinning experiment was carried out using an extrusion system consisting of a single-screw extruder and a gear pump. A spinneret with four nozzles of 0.5 mm diameter was attached to the spinning head. The polymer was extruded at temperatures of 160, 170, 180 and 190° C. The total throughput rate was controlled to 5, 10, and 20 g/min. To evaluate the spinability, take-up velocity was increased from 0.5 km/min to the attainable highest speed. Molecular weight of the as-spun fibers were examined by GPC. Molecular orientation and crystalline structure of the as-spun fibers were investigated through the measurements of birefringence and two dimensional wide-angle X-ray diffraction (WAXD) patterns, respectively. Mechanical properties of the as-spun fibers were analyzed using a tensile testing machine with the gage length of 20 mm and tensile speed of 100%/min.

RESULTS AND DISCUSSION

Figure 1 shows the variation of the molecular weight (M_w) of as-spun fibers prepared by changing the total throughput rate and extrusion temperatures. As can be seen in the figure, there was a significant reduction of M_w at higher extrusion temperatures presumably because of the thermal decomposition. On the other hand, M_w increased with the increase of throughput rate from 5 to 20 g/min. The higher throughput rate corresponds to the shorter residence time of the molten polymer at elevated temperatures, which lead to the minimization of the effect of thermal decomposition. Rate constant for the thermal decomposition estimated from the M_w of as-spun fibers and residence time of polymer melt in the spinning head was approximately 100 times higher than that for poly(ethylene terephthalate) at ordinary extrusion temperature.

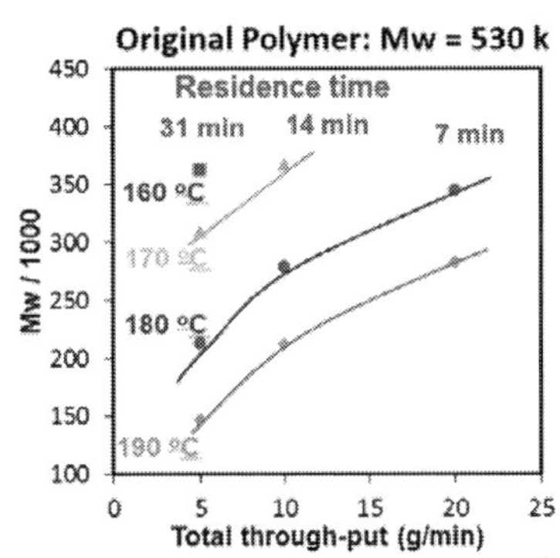

Fig. 1. Effect of throughput rate and extrusion temperature on molecular weight of as-spun PHBH fibers.

In general, spinnability was improved when the extrusion temperature was raised. This was speculated to be due to the lower viscosity of polymer melt originated not only from the temperature effect but also from the effect of M_w reduction by the thermal decomposition. There was a reduction of spinnability at an extremely high extrusion temperature of 190° C, indicating that the spinnability might be deteriorated also if the effect of thermal decomposition was too severe. Appearacne of the as-spun fibers prepared at the take-up velocities of 0.5 and 6 km/min are compared in Figure 2. Extrusion temperature and total through-put rate were 190° C and 20 g/min, respectively. It can be seen that the fibers prepared at lower take-up velocities stuck together whereas the well-separated fibers were obtained at higher take-up velocities.

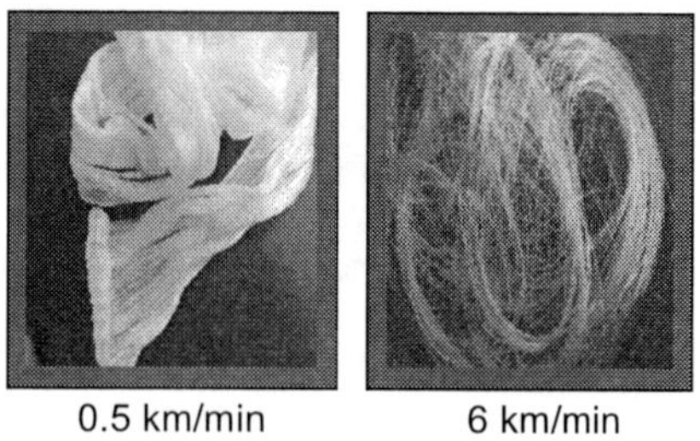

Fig. 2. Appearance of as-spun fibers prepared at take-up velocities of 0.5 and 6 km/min. Extrusion temperature and total through-put rate were 190° C and 20 g/min.

WAXD equatorial profiles of as-spun PHBH fibers prepared at extrusion temperature of 190° C and total throughput rate of 20 g/min are shown in Figure 3. The crystalline reflections from (020) and (110) of α-form crystal were clearly detected for all the fibers [1]. The reflections from (021), (101) and (111) at around 2θ = 22° were observed as overlapped peaks on the equator at lower take-up velocities. This result indicates that the fibers have low crystalline orientation. The peak disappeared with an increase in the take-up velocity and improvement of crystalline orientation. In addition, a small diffraction peak was detected at around 2θ=19.7° on the equator for the WAXD patterns of the fibers prepared at high take-up velocities. This reflection corresponds to the highly oriented β-form crystals, indicating that the fiber structure was developed under high tensile stress.

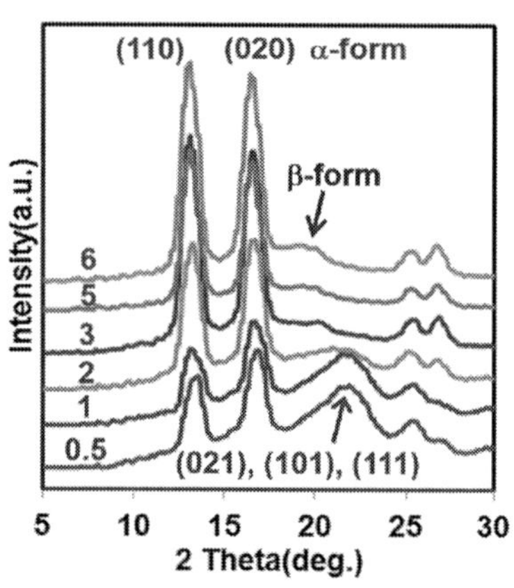

Fig. 3. WAXD equatorial scans of as-spun PHBH fibers prepared at the extrusion temperature of 19° C and total throughput rate of 20 g/min.

In order to evaluate the relative proportions of two crystalline forms, the changes in the integrated intensity of β-form reflection with the take-up velocity and the total throughput rate were analyzed. In the WAXD equatorial scans, the β-form could not be noted at low take-up velocities, while the intensity of β-form reflection increased with the increment of take-up velocity. On the other hand, the take-up velocity at which β-form crystals started to appear increased with the increase of extrusion temperature. In addition, the intensity of β-form reflection was higher at the extrusion condition of higher throughput rate. These results suggested that the development of the β-form crystal was promoted under the effect of higher tensile stress applied in the spin-line.

Tensile strength of as-spun fibers was plotted against the β-form fraction obtained through the WAXD analysis as shown in Figure 4. There was a positive correlation between the two parameters, and the highest tensile strength and β-form fraction were achieved for the fibers prepared at the extrusion temperature of 180° C, total throughput rate of 10 g/min and take-up velocity of 6 km/min.

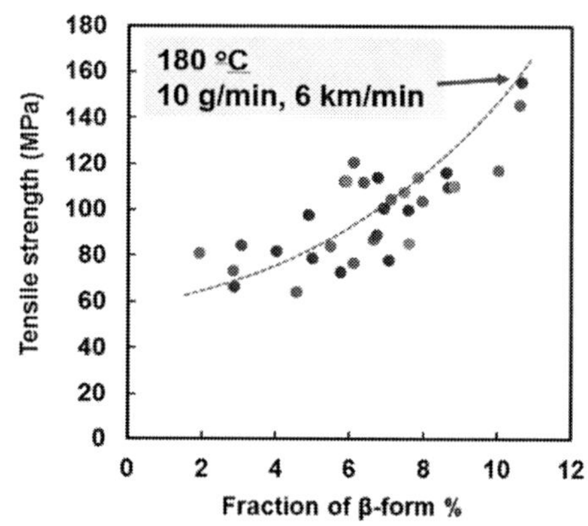

Fig. 4. Relation between tensile strength and β-form fraction of high-speed spun PHBH fibers prepared under various take-up velocities, extrusion temperatures and through-put rates.

CONCLUSION

Crystallization proceeded at high take-up velocities and the β-form crystal formation was promoted in the melt spinning of PHBH fibers even at the extrusion temperature of 190° C. Because there was a significant effect of thermal decomposition during the extrusion process, effect of the extrusion conditions on the structure development behavior was not straight forward.

REFERENCES

1. Yokouchi, M., Chatani, Y., Tadokoro, H., Teranishi, K., Tanii, H. *Polyme.*, 14, 1973: 267.
2. Lee, J.C., Furuhashi, Y., Kawahara, Y., Yamane, H. *Sen'i Ga.*, 62, 2006: 115.

3D Scaffolds of Short Electrospun Nanofibers

Andreas Greiner

University of Bayreuth, Macromolecular Chemistry II and Bayreuth Center for Colloids and Interfaces
Bayreuth, Germany

andreas.greiner@uni-bayreuth.de

Electrospinning is the state-of-the-art technique for the preparation of polymer nanofiber-based breathable nonwovens. Fibers of these nonwovens could have a large variety of chemical composition and could represent many different fiber shapes and dimensions which give access to a near endless number of new materials.[1] Therefore electrospinning is so attractive. Typically these nonwovens are two-dimensional and find applications in filtration, textile, medicine, pharmacy, agriculture, composites, and sensors.[1] The electrospun nanofibers are continuously long and entangled that limits post-spinning modifications and their use as filler additives. Post electrospinning alterations are limited mostly to chemical modifications. A large variety of new options would be given if electrospun fibers could be applied in standard materials processing like film casting, melt extrusion, freeze-drying – just to name a few. The basic concept for new chances with electrospun fibers is the separation of the electrospinning process itself from the final processing step.

In this presentation the preparation, formulation, and processing of short electrospun fibers is presented as a new concept for getting novel materials based on electrospun fibers. Following this novel concept, short electrospun fibers can be processed from suspensions and form stable nonwovens on nearly any substrate in a way similar to the direct electrospinning. Suspension or powders of electrospun fibers can be also easily melt extruded with thermoplasts or mixed with thermosets for the preparation of novel nanofiber reinforced nanocomposites[2] which is otherwise with electrospun nonwovens very difficult.

Very exciting novel three-dimensional nanofiber scaffolds were obtained by freeze-drying of short nanofiber suspensions. Depending on the nanofiber diameters and their concentration in suspension ultralight nanofiber sponges were obtained (Figure 1). These sponges are similar in architecture to natural sponges but display additionally pore-pore architecture which contributes to soft compressibility and high mechanical integrity. These sponges show other exciting properties which will be presented in detail.

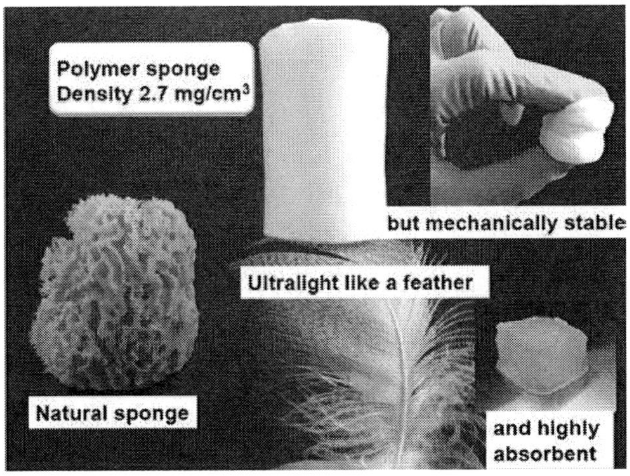

Figure 1. Infographic of polymer sponges versus natural sponges

KEYWORDS
Electrospinning, nanofibers, ultralight

ACKNOWLEDGMENT
Financial support by Deutsche Forschungsgemeinschaft (SFB 840) is kindly acknowledged.

REFERENCES

1) Agarwal, S., et al. *Prog. Polym. Sci.,* 38, 2013: 963-991.
2) Jiang, S., et al. *Compos. Sci. Techn.,* 88, 2013: 57-61.
3) Duan, et al. *Adv. Funct. Mater.*, accepted.

Carbon Fibre Microelectrodes for Neuroscience Applications

Jian Fang[1,2], Zhigang Xie[1,2], Gordon Wallace[1,3], Xungai Wang[1,2]
[1]ARC Centre of Excellence for Electromaterials Science (ACES)
[2]Australian Future Fibres Research and Innovation Centre (AFFRIC), Institute for Frontier Materials, Deakin University, VIC, Australia
[3]Intelligent Polymer Research Institute (IPRI), University of Wollongong, NSW, Australia
xwang@deakin.edu.au

INTRODUCTION

Microelectrodes have shown outstanding performance in neural signal recording, neural stimulation and electrochemical sensing[1,2]. Compared with their micro-sized counterparts, microelectrodes normally exhibit improved signal-to-noise ratio, fast response time and can work with limited sample volumes. Microelectrodes are required to have good biocompatibility, low electrical impedance and long-term stability in many biomedical applications. Carbon fibres are manufactured from polymeric precursor fibres through carbonization, and high carbon content makes carbon fibres electrically conductive, corrosion resistant, biologically safe and inert[3]. Therefore, carbon fibre has been considered as an ideal candidate for making microelectrodes.

In this work, single carbon fibres were loaded into capillary tubes and fabricated into microelectrodes. The surface of the microelectrodes was functionalized with electrochemically reduced graphene oxide (rGO) sheets, decorated with carbon dots (CDs). It has been demonstrated that this surface modified microelectrode could achieve a high sensitivity and selectivity in detecting dopamine (DA), in the presence of ascorbic acid (AA) and uric acid (UA).

APPROACH

Unsized carbon fibres without any surface treatment were used in this work. Graphene oxide was prepared via the Hummer[4] method and water-soluble carbon dots with strong blue fluorescence were synthesized through a hydrothermal process using sodium citrate and ammonium bicarbonate as the starting materials.

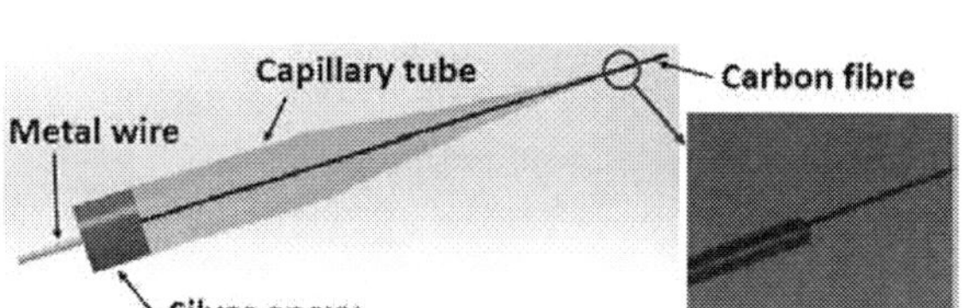

Figure 1. Illustrated structure of carbon fibre microelectrode

Three kinds of carbon fibres were used during the microelectrode preparation in this work: pure carbon fibres (CF), carbon fibres with electrochemically reduced GO on the surface (rGO/CF) and carbon fibres modified with electrochemically reduced GO and carbon dots (rGO+CD/CF). The structure of the fabricated carbon fibre microelectrode is illustrated in Figure 1. One end of the carbon fibre was sealed in a pulled capillary tube while the other end was bonded through conductive epoxy to a gold wire for electrical connection.

RESULTS AND DISCUSSION

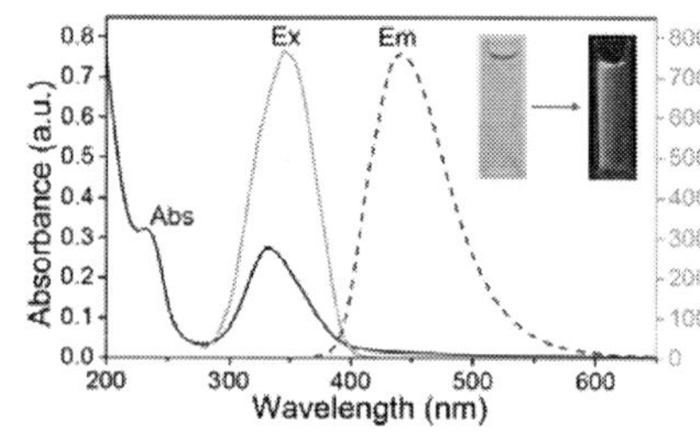

Figure 2. UV absorption (Abs), fluorescence excitation (Ex) and emission (Em) spectra of aqueous CD solution, the inset are the pictures of the solution before and after 365 nm UV radiation.

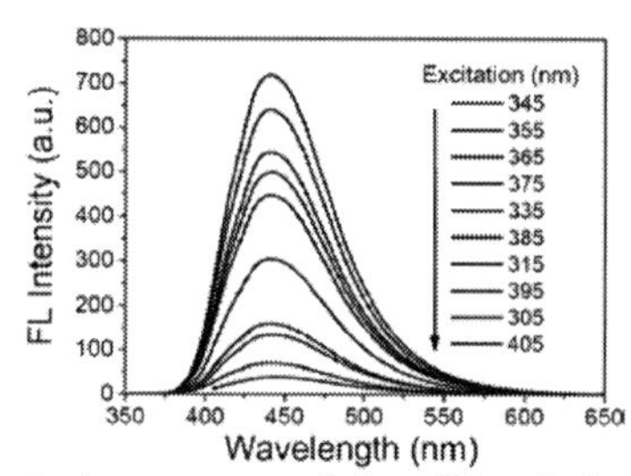

Figure 3. Emission spectra of the CD solution at different excitation wavelengths

Carbon dots are a new type of fluorescent carbon nanomaterials with versatile surface chemistry and non-toxic nature. They have been intensively investigated in many biomedical applications, especially biosensing and bioimaging[5]. The UV absorption, fluorescent excitation and emission spectra of the CDs are shown in Figure 2.

It can be seen that the positions of UV absorption maximum peak (332 nm) and fluorescent excitation peak (345 nm) are very close. The inset image clearly shows the bright blue light displayed from the CD solution under UV light with excitation at 365 nm, which is corresponding to its emission spectra, also shown in Figure 2. It has also been observed that the

CDs exhibited the highest fluorescent intensity when they were excited at 345 nm (Figure 3).

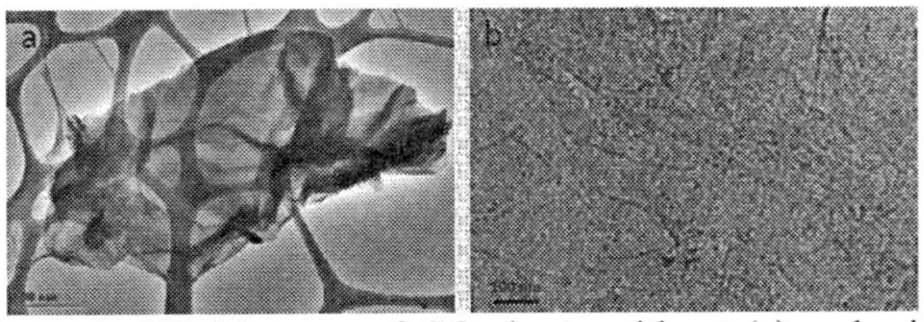

Figure 4. TEM images of GO sheets without (a) and with (b) carbon dot decoration.

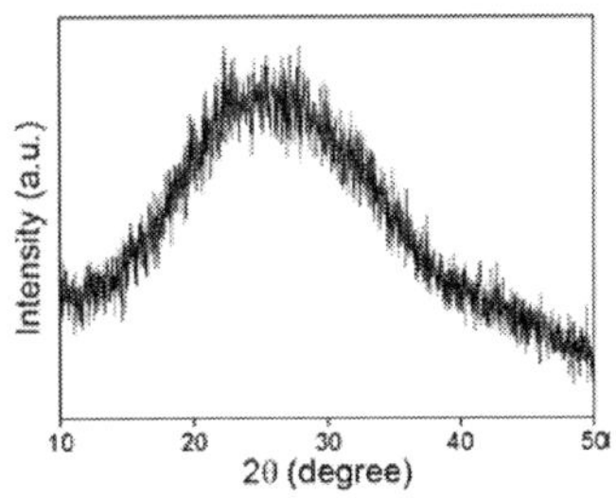

Figure 5. XRD pattern of the CD

Through the strong covalent bonding, CDs can be easily decorated onto GO sheets. The difference between pristine and decorated GO sheets is clearly seen in the TEM images in Figure 4. The broad diffraction peak in the XRD result has also confirmed the amorphous nature of the CDs (Figure 5).

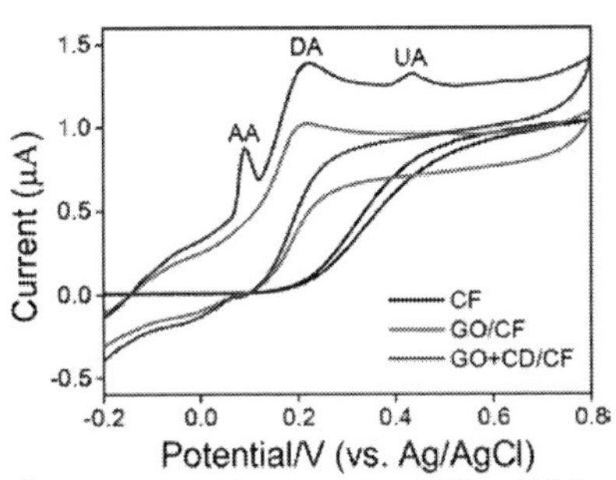

Figure 6. CV responses (scan rate: 100 mV/s) of a 0.1 M PBS solution (pH 7.0) containing 100 μM DA, 1 mM AA and 100 μM UA on three different carbon fibre microelectrodes

Cyclic voltammetry (CV) of a 100 μM DA solution (with 1 mM AA and 100 μM UA) was carried out for the three kinds of microelectrodes prepared in this work. As shown in Figure 6, due to limited surface area of pure carbon fibres, none of the three neurotransmitters in the solution was identified by the CF microelectrode. After GO sheets were electrochemically deposited and reduced onto pure carbon fibres, an obvious oxidation peak at 0.197 V was shown on the rGO/CF microelectrode, however it's still not sensitive enough to detect AA and UA in the same solution.

By decorating the GO sheets with carbon dots, oxidation peak of all three neurotransmitters in the solution can be found on the CV curve of the rGO+CD/CF microelectrode. This significant improvement has probably been contributed by increased surface area and improved electrostatic interaction.

CONCLUSION

Graphene oxide sheets have been decorated with carbon dots, electrochemically reduced and deposited onto carbon fibre surface for detecting low concentration dopamine (DA), ascorbic acid (AA) and uric acid (UA) in aqueous solution. The decoration of carbon dots leads to significantly improved sensitivity and selectivity of the surface modified carbon fibre microelectrodes, which can be attributed to increased accessible surface area and enhanced electrostatic interaction with DA. The results suggest that surface modified carbon fibre maybe an ideal microelectrode material for various neuroscience applications, such as biosensing and neuro signal recording.

KEYWORDS

Carbon fibre, Microelectrode, Neurotransmitter, Neuroelectrode

ACKNOWLEDGMENT

This work is financially support by Australian Research Council (ARC) through ARC Centre of Excellence for Electromaterials Science (ACES). The authors also would like to thank Carbon Nexus for kindly providing unsized carbon fibres.

REFERENCES

1. Cogan, Stuart. "Neural Stimulation and Recording Electrodes." *Annual Review of Biomedical Engineering* 10 (2008): 275-309.
2. Bond, Alan. "Past, Present and Future Contributions of Microelectrodes to Analytical Studies Employing Voltammetric Detection: A Review." *Analyst* 11 (1994): 1R-21R.
3. Frank, Eric, Steudle, Lisa, Ingildeev, Denis, Sporl, Johanna, Buchmeiser, Michael. "Carbon Fibers: Precursor Systems, Processing, Structure, and Properties." *Angewandte Chemie International Edition* 53, 2014: 5262-5298.
4. Hummers, William, Offeman, Richard. "Preparation of Graphite Oxide." *Journal of the American Chemical Society,* 80, 1958: 1339.
5. Lim, Shi Ying, Shen, Wei, Gao, Zhiqiang. "Carbon quantum dots and their applications." *Chemical Society Review,* 44, 2015: 362-381.

Chemical Modifications of Synthetic Fibers

Gang Sun
University of California, Davis, CA, USA
Donghua Research Institute, Donghua University, Shanghai, China
gysun@ucdavis.edu

INTRODUCTION

Synthetic fibers are advantageous in mechanical, thermal, and chemical properties and have gained applications in textile apparels, protective clothing, and engineering composite. However, most commodity synthetic fibers are hydrophobic, chemically inert, and uncomfortable to wear when used as clothing materials. Functional modifications of synthetic fibers to overcome the drawback properties have been the efforts of many fiber and polymer researchers for decades. Chemical modifications of polymers are one of the major approaches in functionalizing polymers and fibers in order to achieve durable and permanent novel properties. But, due to the differences in chemical structures and manufacturing processes, there are several methods to implement the chemical modifications of synthetic fibers, including copolymerization during synthesis, graft copolymerization during processing, and graft polymerization on surfaces of fibers. Chemistries and properties of all three processes in producing refreshable biocidal synthetic fibers will be presented here.

REACTION APPROACHES

Copolymerization Halamine precursor monomers such as 3-allyl-5,5-dimethy hydantoin (ADMH), can be copolymerized with acrylonitrile to form a copolymer, which can then be extruded to fibers, and the fibers can be chlorinated to biocidal halamine [1].

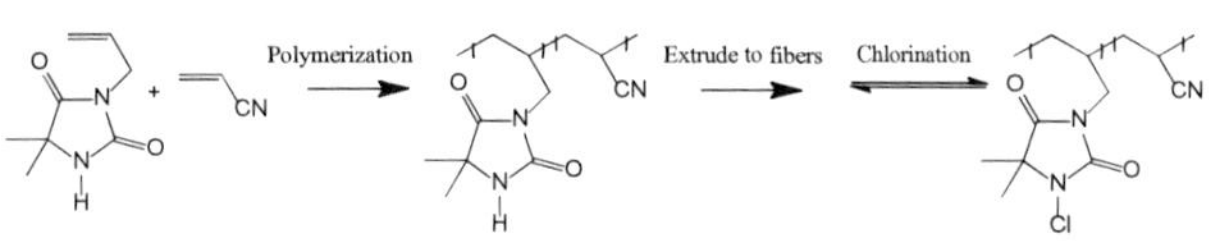

Scheme 1 Copolymerization ADMH with acrylonitrile [1]

Reactive Extrusion The same halamine precursor, ADMH and several other precursors (Figure 1) were coextruded with olefin polymers and peroxide initiators to have a radical graft polymerization in a twin-screw extruder, which can incorporate the halamine precursor structures and also simultaneously extrude them into fibers, such as in a melt blowing process, or into pellets for additional melt spinning into fibers. The fibers can be chlorinated to become biocidal halamines. The overall reaction scheme of ADMH reacting with polypropylene is shown in Scheme 2 [2].

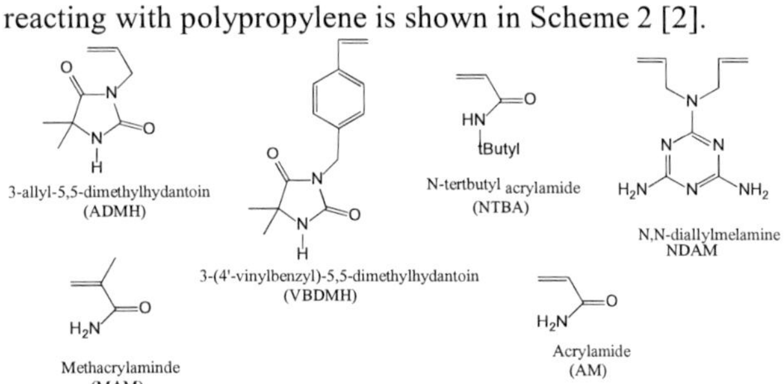

Figure 1 Structures of halamine precursors [2]

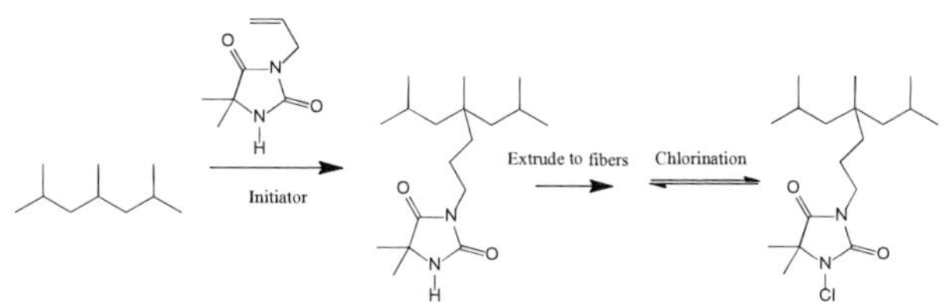

Scheme 2 Radical graft polymerization

Surface Graft Modification ADMH can be grafted onto surfaces of fibers such as polyester by a radical graft polymerization method, shown in Scheme 3. This process could work on fibers or fabrics [3].

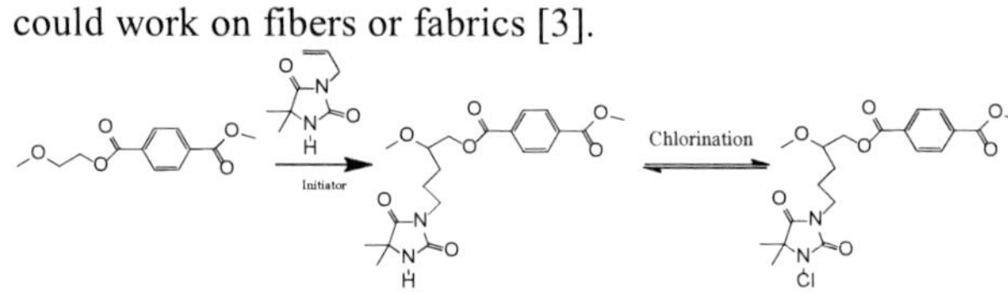

Scheme 3 Surface graft polymerization

RESULTS AND DISCUSSIONS

The three methods can all effectively incorporate the intended functional moieties onto the polymers or fibers, respectively [1-3]. However, the differences of polymer structures and the methods in modifying the materials also resulted in interesting products.

The copolymer of ADMH-AN (Cop7-1) has to be blended with polyacrylonitrile commercial sample to prepare spinning solutions. It seems the produced copolymer blend fibers demonstrated suitable antibacterial functions when the content of the copolymer was above 10% in the blends. (Figure 1)

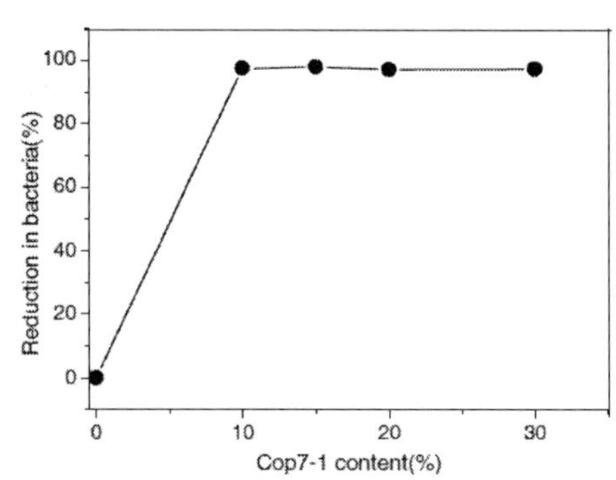

Figure 1 Antibacterial results of ADMH-AN copolymers

The graft polymerization reactions within a twin-screw extruder occurred effectively, and the incorporated halamine precursors could be converted to biocidal halamine and demonstrated different levels of antibacterial properties, shown in Table 1. Due to the nature of hydrophobicity of polypropylene, the copolymer fibers all revealed relatively slow functions, requiring longer contact times. However, several halamine structures could provide complete kills.

Table 1 Antibacterial functions of the polypropylene copolymer fibers (average size of fibers <0.6 um)

	Bacterial Reduction (%)				
Contact time	1h	2h	4h	8h	16h
PP-g-ADMH	30%	40%	45%	68%	80%
PP-g-NTBA	90%	100%	100%	100%	100%
PP-g-AM	40%	42%	40%	70%	80%
PP-g-MAM	40%	80%	100%	100%	100%
PP-g-NDAM	100%	100%	100%	100%	100%
PP-g-VBDMH	25%	50%	80%	82%	99%

The surface graft polymerization reaction was conducted on poly (ethylene terephthalate) PET fibers, and the grafted fibers could contain over 200 ppm active chlorine, which should be able to provide effective biocidal functions. But the products also revealed slow reaction due to the hydrophobic nature of PET fibers (Table 2). Interestingly the surface graft polymerization reactions occurred at the defected areas or mostly on the ends of the fibers (Figure 2).

Table 2 Results of surface grafted PET fibers [3]

	Chlorine (ppm)	Percentage Reduction of *E. coli* at Different Contact Time (%)		
		30min	60min	120min
Poly(ADMH)-g-PET	288	50%	96%	>99.999

Figure 2 SEM of surfaces of PET fibers [3]

KEYWORDS: Chemical modifications; functional fibers; graft polymerizations

ACKNOWLEDGMENTS
Research was funded by National Science Foundation (CTS 0424716); DTRA HDTRA1-08-10005). Drs. Mohammad Badrossamay and Song Liu worked on the areas.

REFERENCES

1. Wang, L.Y., Xie, J., Gu, L.X., Sun, G. "Preparation of antimicrobial polyacrylonitrile fibers: Blending with polyacrylonitrile-co-3-allyl-5,5-imethylhydantoin." *Polymer Bulletin*, 2006, 56 (2-3): 247-256.
2. Badrossamay, M.R., Sun, G. "Durable and Rechargeable Biocidal Polypropylene Polymers and Fibers Prepared by Using Reactive Extrusion." *Journal of Biomedical Materials Research: Part B-Applied Biomaterials,* 2009, V. 89B: 93-101.
3. Liu, S., Sun, G. "Functional Modification of Poly(ethylene terephthalate) with an Allyl Monomer: Chemistry and Structure Characterization." *Polymer*, 2009, V.49:5225-5232.

Crystal Networks in Fibrous Soft Materials: From Hierarchical Structure to Ultra-Performance

Xiang Yang Liu
Department of Physics and Department of Chemistry
National University of Singapore, Singapore, China
Research Institution for Biomimetics and Soft Matter, Xiamen University, Xiamen, China

phyliuxy@nus.esu.sg

A talk will be delivered on the structural characteristics of crystal networks of fibrous soft materials in correlation with the macroscopic properties/performance, and the network formation mechanisms. The correlation between the hierarchical mesoscopic structures and the mechanical properties of silk fibrous soft materials including silk fibroin hydrogels and naturally spun silk fibers are addressed based on the hierarchical crystal network models. Namely, two types of hierarchical networks are identified: the weak nanofibril-nanofibril interaction case (i.e. silk fibroin hydrogels), and the strong nanofibril-nanofibril interaction case (i.e. silk fbers), respectively. The macroscopic properties, i.e. the rheological/mechanical properties, can be controlled in terms of tuning different levels of hierarchical network structures by ultrasonication-induced gelation, introducing the initial nucleation centers, etc. Such controls take effect by different meso-scale assembly pathways, which are found to occur via different routes of the nucleation-growth processes. Furthermore, the hierarchical network model of soft materials can be applied to explain the superior mechanical properties and the unique strain-hardening behaviors of spider silk fibers within the framework of hierarchical breaking mechanism. Obviously, the knowledge on crystal networks will allow the prediction of the performance of soft materials in generals and engineering strategy.

References

1. Li, J.-L., Liu, X.-Y. *Advanced Functional Materials,* 2010, 20 (19): 3196-3216.
2. Li, J.-L., Liu, X.Y. "Molecular Gels for Controlled Formation of Micro-/Nano-Structures." *Soft Fibrillar Materials*. Wiley-VCH Verlag GmbH & Co. KGaA, 2013: 163-181.
3. Tang, S., Liu, X.Y., Strom, C.S. *Advanced Functional Materials,* 2009, 19 (14): 2252-2259.
4. Du, N., Liu, X.Y.; Narayanan, J., Li, L., Lim, M.L.M., Li, D., *Biophysical Journal,* 2006, 91 (12): 4528-4535.
5. Du, N., Yang, Z., Liu, X.Y., Li, Y., Xu, H.Y. *Advanced Functional Materials,* 2011, 21 (4): 772-778.
6. Xu, G., Gong, L., Yang, Z., Liu, X.Y. *Soft Matter,* 2014, 10 (13): 2116-2123.
7. Liu, X.Y., Sawant, P.D. *Angewandte Chemie-International Edition,* 2002, 41 (19): 3641-3645.

Wearable Fiber-Shaped Energy Harvesting and Storage Devices

Huisheng Peng

State Key Laboratory of Molecular Engineering of Polymers, Collaborative Innovation Center of Polymers and Polymer Composite Materials, Department of Macromolecular Science, and Laboratory of Advanced Materials, Fudan University, Shanghai, China

penghs@fudan.edu.cn

It is critically important to develop miniature energy harvesting and storage devices in modern electronics, e.g., for portable and foldable electronic facilities. Here novel miniature fiber-shaped solar cells, electrochemical capacitors and lithium-ion batteries as well as their integrated devices are carefully discussed with unique and promising advantages such as lightweight and weaveable compared with the conventional planar architecture.[1-5]

REFERENCES

(1) Chen, T., Qiu, L., Yang, Z., Peng, H. *Chem. Soc. Rev.*, 42, 2013: 5031-5041.

(2) Pan, S., Yang, Z., Li, H., Qiu, L., Sun, H., Peng, H. *J. Am. Chem. Soc.,* 135, 2013: 10622-10625.

(3) Yang, Z., Sun, H., Chen, T., Qiu, L., Luo, Y., Peng, H. *Angew. Chem. Int. Ed.*, 52, 2013: 7545-7548.

(4) Ren, J., Li, L., Chen, C., Chen, X., Cai, Z., Qiu, L., Wang, Y., Peng, H. *Adv. Mater.*, 25, 2013: 1155-1159.

(5) Chen, T., Qiu, L., Yang, Z., Cai, Z., Ren, J., Li, H., Lin, H., Sun, X., Peng, H. *Angew. Chem. Int. Ed.*, 51, 2012: 11977-11980.

Exploring the Nature of Cellulose Microfibrils

Ying Su, Benjamin Chu, Benjamin S. Hsiao
Department of Chemistry, Stony Brook University, Stony Brook, New York, United States
benjamin.hsiao@stonybrook.edu

ABSTRACT

Ultra-thin cellulose microfibril fractions were extracted from spruce wood powder using combined delignification, TEMPO-catalyzed oxidation and sonication. Small-angle X-ray scattering analysis [1] of these microfibril fractions in a 'dilute' aqueous suspension (concentration 0.077 wt%) revealed that their shape was in the form of nanostrip with 4 nm width and only about 0.5 nm thicknesses. These dimensions were further confirmed by TEM and AFM measurements. The 0.5 nm thickness implied that the nanostrip could contain only a single layer of cellulose chains. At a higher concentration (0.15 wt%), SAXS analysis indicated that these nanostrips aggregated into a layered structure. The X-ray diffraction of samples collected at different preparation stages suggested that microfibrils were delaminated along the $(1\bar{1}0)$ planes from the I_β cellulose crystals. The degree of oxidation and solid-state ^{13}C NMR characterizations indicated that in addition to the surface molecules, some inner molecules of microfibrils were also oxidized, facilitating the delamination into cellulose nanostrips (Figure 1).

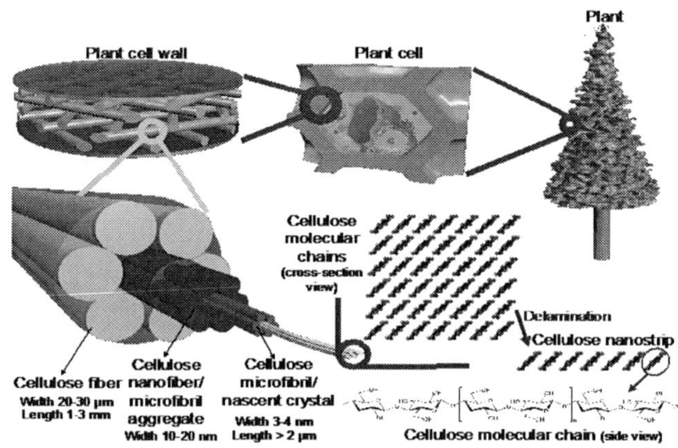

Figure 1 Hierarchical structure of plant cellulose

ACKNOWLEDGMENT

The financial support for this work was provided by a grant from the SusChEM Program of the National Science Foundation (DMR-1409507). The authors thank Drs. Christian Burger and Hongyang Ma for technical assistance.

REFERENCE

1. Su, Ying, Burger, Christian, Hsiao, Benjamin S., Chu, Benjamin. "Characterization of TEMPO-oxidized cellulose nanofibers in aqueous suspension by small-angle X-ray scattering." *Journal of Applied Crystallography*, 47(2), 2014: 788-798.

Nanofibrous Materials

Electrospun Nanofiber-Assembled Cellular Aerogels with Superelasticity and Multifunctionality

Bin Ding
College of Textiles, Donghua University, Shanghai, China
binding@dhu.edu.cn

ABSTRACT
Three-dimensional (3D) nanofibrous aerogels (NFAs) that are both highly compressible and resilient would have broad technological implications; however, creating such NFAs has proven extremely challenging. Here, we report a novel strategy to create fibrous, isotropically-bonded elastic reconstructed (FIBER) NFAs with a hierarchical cellular structure and superelasticity by combining electrospun nanofibers and the fibrous freeze-shaping technique. Our approach causes the intrinsically lamellar deposited electrospun nanofibers to assemble into elastic bulk aerogels with tunable densities and desirable shapes. The resulting FIBER NFAs exhibit densities of > 0.12 mg cm^{-3}, rapid recovery from deformation, and multifunctionality in terms of the thermal insulation, sound absorption, emulsion separation, and elasticity-responsive electric conduction. The successful synthesis of such fascinating materials may provide new insights into the design and development of multifunctional NFAs for various applications.

INTRODUCTION
Ultralow-density materials are attractive for their wide range of applications in the fields of thermal insulation, tissue engineering, and shock energy damping.[1] The effective properties of the ultralight materials are defined by their cell geometry and properties of the solid constituents.[2] Thus, creating 3D NFAs with high continuity and an open-cell cellular structure could be another strategy for achieving promising performance for widespread applications. Several cellulosic materials (such as bacterial cellulose) have recently been assembled into NFAs.[3] However, the inherent limits on the diversity of bulk materials, combined with the lack of precise control of the physicochemical and mechanical properties have limited their practical applications.

Electrospun nanofibers, which are at the forefront of advanced fibrous materials, combine fine flexibility, extremely high aspect ratio, and ease of scalable synthesis from various materials.[4] Despite their outstanding potential for constructing macroscopic NFAs, the major problem associated is their anisotropic lamellar deposition character, with the resultant nanofibers usually assembling into close-packed membranes (with thicknesses < 100 μm).[5] Current knowledge on the 3D shaping of nanofibers suggests that the NFAs should be prepared by stacking layers of nanofiber membranes together or by using specially designed fiber collectors.[6] However, the anisotropic lamellar deposition problem has not been solved. Moreover, most of these products are not real 3D structured aerogels but stacks of membranes or floccose nanofiber deposits, which exhibit poor mechanical strength with no elasticity.[7] The challenge, therefore, is to construct homogeneous, continuous, and mechanically robust NFAs capable of forming a 3D isotropic cellular architecture, without compromising the functionality of nanofibers.

In this work, we present a robust methodology for creating superelastic NFAs with a hierarchical cellular structure that consists of bonded nanofibers, which we call "fibrous, isotropically-bonded elastic reconstructed" (FIBER) NFAs. The premise of the design is that the lamellar deposited nanofibers are reconstructed into 3D elastic bulk aerogels with tunable densities and desirable shapes. We demonstrate that our FIBER NFAs exhibit integrated properties of extremely low density (> 0.12 mg cm^{-3}), superelasticity, and multifunctionality, all originating from the hierarchical cellular fibrous networks and well-bonded nanofibers.

RESULTS AND DISCUSSION
Formation of hierarchical cellular-structured FIBER NFAs
Polyacrylonitrile (PAN) nanofibers were selected as the major precursors to construct elastic fibrous networks, and the bifunctional benzoxazine (BA-a) was blended into PAN nanofibers as a novel in situ crosslinking agent. Rigid SiO_2 nanofibers were introduced into aerogels to enhance the structural stability. The as-prepared nanofiber membranes were first homogenized to form well-dispersed nanofiber dispersions and then freeze-dried into an uncrosslinked NFA. Subsequently, the uncrosslinked NFAs were heated at 240°C for 1 h to form crosslinked 3D fibrous networks, endowing FIBER NFAs with elastic resilience.

Due to the simplicity of the assembly process in our methodology, great versatility in controlling the shapes of aerogels was possible. As illustrated in Fig. 1a, integrated FIBER NFAs with desired shapes, such as pentagonal prisms, cones, cubes, triangular prisms, and intricate ear-like shapes, were readily prepared. It is noteworthy that fibrous freeze-shaping also simultaneously induces the nanofibers to be organized in a unique cellular architecture. Fig. 1b present typical scanning electron microscope (SEM) images of the FIBER NFAs with a density of 9.6 mg cm^{-3}. The cellular architecture exhibited an obvious open-cell geometry with a major cellular pore size of 10–30 μm, and these cells were highly interconnected with each other through triangular junctions (Fig. 1b).

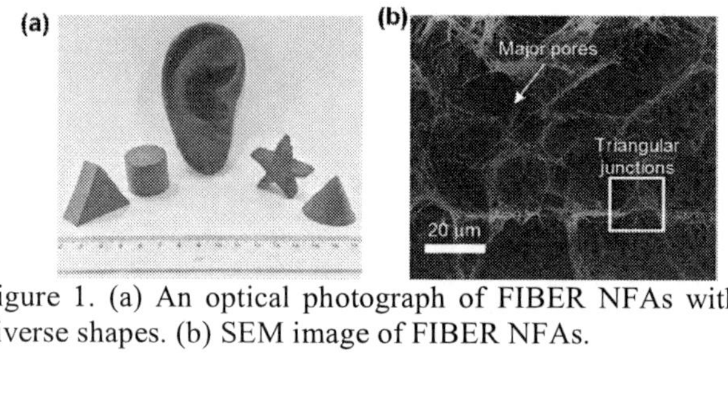

Figure 1. (a) An optical photograph of FIBER NFAs with diverse shapes. (b) SEM image of FIBER NFAs.

Characterization of superelasticity

In dramatic contrast to the brittle nature of traditional colloidal aerogels, the FIBER NFAs can bear a compressive strain (ε) as high as 80% and can recover their original volume after the release of the stress (σ), as shown in insets of Fig. 2. The plots of the compressive σ-ε curves or the FIBER NFAs at set ε maxima of 40, 60, and 80%, revealing two characteristic deformation regimes typically observed in open-cell foams[3]: a linear elastic regime for ε<60% with a stable tangent modulus ($d\sigma/d\varepsilon$) of ~30 kPa and a densification regime for ε>60% with σ and the $d\sigma/d\varepsilon$ increasing steeply. The maximum σ was 10.6 kPa for the first elastic regime and 36.1 kPa at 80% strain; these values were significantly higher than those of other fibrous aerogels with similar densities.[2]

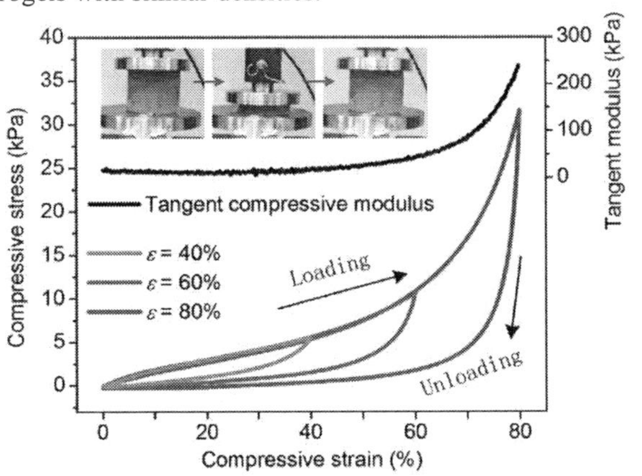

Figure 2. Compressive s and ds/de versus e curves for the FIBER NFAs. Insets: photographs of the FIBER NFAs under a compressing and releasing cycle (ε=80%).

Tunable densities of FIBER NFAs

The FIBER NFAs with various densities could be prepared simply by adjusting the concentrations of the nanofiber dispersions. Fantastically, the lowest density of FIBER NFAs we have obtained could reach 0.12 mg cm^{-3} (Fig. 3a), which is lower than that of previously reported ultralight aerogels.[8] In addition, the ultralight feature was further confirmed using a designed concept experiment, as illustrated in Fig. 3b. Compared with light feathers, the FIBER NFAs with a density of 0.12 mg cm^{-3} can easily float on cold air and behave as a boat, relying only on the minute density difference between the room and cold air.

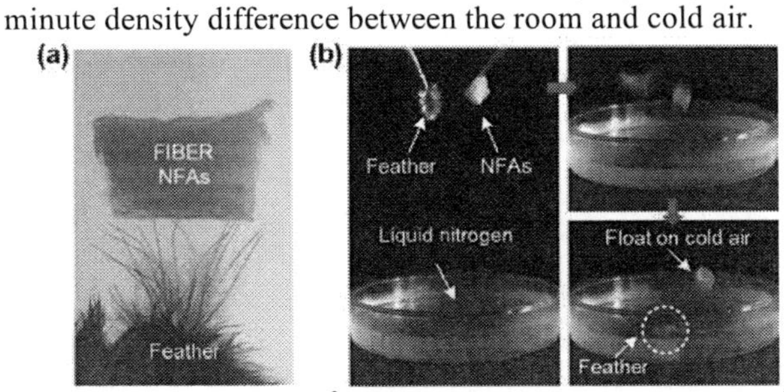

Figure 3. (a) A 20-cm^3 FIBER NFA (ρ=0.12 mg cm^{-3}) stands on the tip of dyed feathers. (c) Photographs show that the ultralight FIBER NFAs (ρ=0.12 mg cm^{-3}) float on cold air.

Multifunctionality of FIBER NFAs

With the ultralow density, superelasticity, selective wettability, and hierarchical cellular fibrous networks, the as-prepared FIBER NFAs could be used in areas such as thermal insulation, sound absorption, emulsion separation, etc. FIBER NFAs with a density of 0.12 mg cm^{-3} exhibited a thermal conductivity of 0.026 W m^{-1} K^{-1}, which is similar to that of air at ambient conditions (0.023 W m^{-1} K^{-1}). Compared with the well-known sound absorption materials—3M Thinsulate TAI2047 felts, which have a thickness of 10 mm and an areal density of 210 g m^{-2}, the FIBER NFAs with half of thickness (5 mm) and a quarter of areal density (48 g m^{-2}) exhibited a higher sound absorption coefficient over the entire testing frequency. Furthermore, the FIBER NFAs could separate surfactant-stabilized water-in-oil emulsions with a high flux (8100 L m^{-2} h^{-1}) driven by gravity, and could also collect pure oil from emulsions via a pumping method. Moreover, the FIBER NFAs can be facilely carbonized to yield conductive carbon NFAs, which exhibited a promising initial conductivity of 0.25 S cm^{-1}.

CONCLUSION

In summarize, we have demonstrated a gelation-free assembly strategy for fabricating hierarchical cellular structured FIBER NFAs with tunable densities and desirable shapes. With their ultralow density, super recyclable compressibility, and unique multifunctionality, we envision that such exceptional FIBER NFAs will open up numerous opportunities for a range of applications in heat insulators, separators, and tissue engineering scaffolds.

ACKNOWLEDGMENT

This work is supported by the National Natural Science Foundation of China (No. 51322304 and 51173022), the National Basic Research Program of China (973 Program, 2012CB525005) and the "DHU Distinguished Young Professor Program."

REFERENCES

1. Pierre, A., Pajonk, G. "Chemistry of Aerogels and Their Applications." *Chem. Rev.*, 102, 2002: 4243-4265.
2. Schaedler, T., et al. "Ultralight Metallic Microlattices." *Science*, 334, 2011: 962-965.
3. Liang, H., et al. "Highly Conductive and Stretchable Conductors Fabricated from Bacterial Cellulose." *NPG Asia Mater.*, 4, 2012: e19.
4. Wang, X., et al. "Electro-spinning/Netting: A Strategy for the Fabrication of Three-dimensional Polymer Nano-fiber/Nets." *Prog. Mater. Sci.*, 58, 2013: 1173-1243.
5. Greiner, A., Wendorff, J. *Angew. Chem. Int. Ed.*, 46, 2007: 5670-5703.
6. Ramakrishna, S. *An Introduction to Electrospinning and Nanofibers*. World Scientific, 2005.
7. Teo, W., Inai, R., Ramakrishna. "Technological Advances in Electrospinning of Nanofibers." *Sci. Technol. Adv. Mater.*, 12, 2011: 013002.
8. Mecklenburg, M., et al. "Aerographite: Ultra Lightweight, Flexible Nanowall, Carbon Microtube Material with Outstanding Mechanical Performance." *Adv. Mater.*, 24, 2012: 3486-3490.
9. Si Y., et al. "Ultralight Nanofibre-Assembled Cellular Aerogels with Superelasticity and Multifunctionality." *Nat. Commun.*, 5, 2014: 5802.

Artificial Muscles from Carbon Nanotube and Nylon Fibres

Javad Foroughi[1], Geoff Spinks[1], Gordon Wallace[1], Carter Haines[2], Ray Baughman[2]
[1]Centre of Excellence for Electromaterials Science, University of Wollongong, Wollongong NSW, Australia
[2]Alan G MacDiarmid NanoTech Institute, University of Texas at Dallas, USA
foroughi@uow.edu.au

ABSTRACT

The high cost of powerful, large-stroke, high-stress artificial muscle fibers and wires has combined with typical performance problems, like low-cycle-life under high load, hysteretic behavior, and low efficiencies, to restrict applications. We here demonstrate that inexpensive, high-strength polymer fibers used for fishing line and sewing thread can be easily transformed by non-conventional, extreme twist-insertion processes to provide fast, long-life tensile and torsional muscles that contract by over 49% and lift 100 times heavier loads than the same length and weight natural muscle. These muscles, which are normally thermally actuated, can be powered electrically using sub-volt voltages, photonically, chemically, or by harvesting chemical or thermal energy to provide over 2 times higher specific work than competing nickel-titanium actuator wires. A single muscle fiber weighing 600 mg, made by coiling an 860 μm diameter nylon fishing line, can reversibly lift a kilogram weight 3 cm, and accomplish this once per second when powered by a 75°C temperature change. By weaving or braiding these fibers, we demonstrate textiles that open and close in response to temperature, which might lead to clothing that changes porosity to provide protection and comfort. Use of these muscles for harvesting thermal energy and for window shutters that help conserve energy is demonstrated.

INTRODUCTION

Artificial muscles have been of practical interest for over 60 years, but few types have found important applications. Slow response, low strain or force generation, short cycle life, creep in actuator length, and low energy efficiency are typical problems. Electrochemically powered carbon nanotube yarn muscles[1] can generate over a thousand times larger rotation per muscle length than previous torsional muscles based on shape memory alloys, ferroelectric ceramics, and conducting polymers. The actuating yarn electrode can accelerate an 1800 times heavier paddle up to 590 revolutions/min in 1.2 s and provide similar gravimetric torque and mechanical power generation per yarn weight as for large electric motors. In addition, infiltrated twist-spun carbon nanotube yarns[2] can be used as a single-electrode, electrolyte-free muscle to provide fast, high-force, high-strain torsional and tensile actuation that is reversible for over a million cycles. A power density of 10 kW/kg resulted during muscle contraction, which is 200 times that for natural skeletal muscle. Actuation is by thermally, chemically, and photochemically powered dimensional changes of an infiltrated guest, which generates torsional rotation and contraction of the helical yarn host. Here we will present the inexpensive, high-strength polymer fibers into artificial muscles that undergo millions of reversible strokes while generating over 20% tensile contraction [3]. These muscles should approach or exceed the power generation capabilities of shape memory metal wires, while providing the hysteresis-free cycle performance needed for convenient control. Additionally, these new muscles should be weavable into textiles and braids that actuate to either accomplish amplified mechanical work or change porosity. The ability to provide fast, high-stroke torsional actuation is another targeted benefit of these new muscles.

RESULTS AND DISCUSSION

We start from commercially available, high-strength fibers (either single filament or multifilament) – most often those used as fishing line or sewing thread. Commercially produced polyethylene, nylon 6, and nylon 6,6 fibers are especially important muscle precursors, since they have the combination of properties that we find enhance muscle stroke: reversible fiber direction thermal contraction when heated, and large anisotropy in thermal expansion. Like for previous carbon nanotube yarn muscles, twist is inserted into these high strength polymer fibers to make large stroke muscles. This twist makes the fibers chiral, which enables them to function as torsional artificial muscles. Most importantly, in order to maximize actuator tensile stroke, we insert such a large amount of twist (measured in turns per meter of initial fiber length) that some of this twist is converted to fiber coiling, called writhe. These new muscles exploit hierarchical structures that cascade upwards in scale from the molecular and nanoscale morphological levels of the muscle precursor fiber, to the fiber-diameter-scale structures introduced by coiling and plying, and to centimeter and larger scale structures obtained when individual muscles are assembled in textiles and braids (Fig. 1).

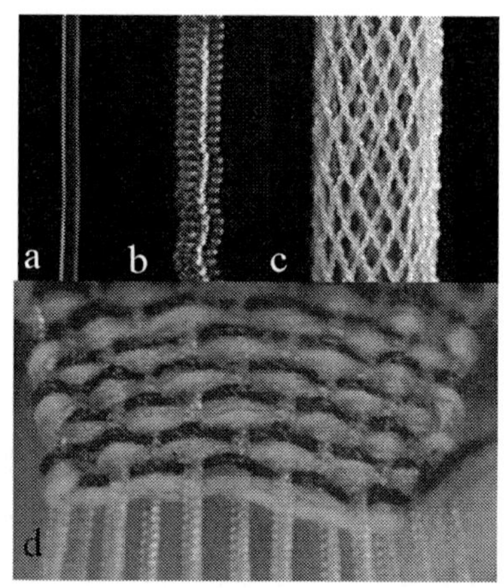

Fig. 1: Muscle and precursor structures using nylon 6,6 monofilament . Optical images of (a) a nontwisted 300-mm-diameter fiber; (b) the fiber of (a) after coiling by twist insertion; (c) a braid formed from 32 two-ply, coiled, and (d) an actuating textile woven from conventional polyester, cotton, and silver-plated nylon (to drive electrothermal actuation) yarn in the weft direction and coiled nylon monofilament muscle fibers in the warp direction.

The reversible thermal contraction between 20 and 240°C for nylon 6,6 monofilament fiber increased from 4% to 34% as a result of fiber coiling, providing nearly 9 times amplification. While the maximum temperature of useful dimensional stability of the gel-spun Berkley FireLine (Dyneema) polyethylene was much lower (~130°C), the higher modulus and strength of these fibers is especially useful as a thermally powered artificial muscle that lifts heavy loads, and provides increased energy efficiency. Before coiling, Berkley FireLine polyethylene exhibits less than 0.3% contraction between 20 and 130°C, whereas when coiled it provided contraction of 16% for the same temperature range.

The tensile stroke and stress capabilities of coiled muscles can be tuned across a wide range by simply adjusting the spring index of the coils. Figure 2 shown the dependence of tensile stroke versus applied load for 127 μm nylon 6,6 monofilaments coiled under stresses of 9.9, 16 and 35 MPa. These stresses produced coils with spring indices of C = 1.1, 1.4 and 1.7, respectively. For each coil, the point of maximum stroke corresponds to the lowest stress that prevented coil contact over the 20 to 120°C temperature change used for actuation. For the largest diameter coil (C=1.7), this occurs at the lowest stress, 22 MPa, to provide the highest stroke of 21%. When tightly coiled (C=1.1) the maximum stroke decreases to 9.3%, but occurs at a higher 50 MPa stress not attainable by the low stiffness C=1.7 coil. Similarly, coils formed by annealing around a large-diameter mandrel can yield even greater strokes at lower loads, such as the 49% contraction provided at 1 MPa for nylon 6 fibre having C = 5.5.

A coiled nylon 6,6 muscle delivered over 1 million cycles during periodic actuation at 1 Hz, raising and lowering a 10-g weight producing 22 MPa of nominal stress. This actuation was powered by applying a 30 V/cm squarewave potential (normalized to coil length) at a 20% duty cycle. Although the coiled fiber did experience creep, this creep was below 2% over the 1.2 million investigated cycles, stroke amplitude was negligibly affected, and the creep rate decreased with cycling.

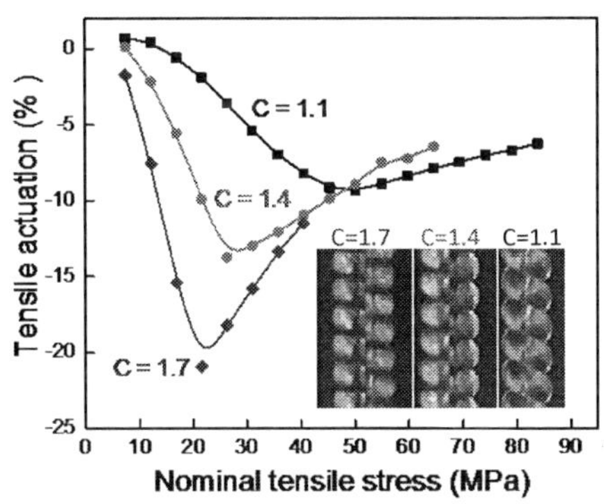

Fig. 2: Tensile stroke versus load, as a percent of the loaded muscle length, for a 127 μm nylon 6,6 monofilament fiber that was coiled by twist insertion under loads of 10, 16, and 35MPa, which resulted in spring indices of 1.7, 1.4, and 1.1, respectively. Optical images of the coils are inset.

CONCLUSION

We demonstrated that inexpensive high-strength polymer fibers used for fishing line and sewing thread can be easily transformed by twist insertion to provide fast, scalable, nonhysteretic, long-life tensile and torsional muscles. Extreme twisting produces coiled muscles that can contract by 49%, lift loads over 100 times heavier than can human muscle of the same length and weight, and generate 5.3 kilowatts of mechanical work per kilogram of muscle weight, similar to that produced by a jet engine. Woven textiles that change porosity in response to temperature and actuating window shutters that could help conserve energy were also demonstrated. Large-stroke tensile actuation was theoretically and experimentally shown to result from torsional actuation.

ACKNOWLEDGMENT

This work has been supported by The Australia Research Council under Discovery Early career researcher award (Javad Foroughi DE12010517), through the Centres of Excellence (ACES) scheme for financial support and Australian National Fabrication Facility (ANFF).

REFERENCES

1. Foroughi, J., et al. "Torsional Carbon Nanotube Artificial Muscles." *Science,* 334 (6055), 2011: 494-497.
2. Lima, M.D., et al. "Electrically, Chemically, and Photonically Powered Torsional and Tensile Actuation of Hybrid Carbon Nanotube Yarn Muscles." *Science,* 338 (6109), 2012: 928-932.
3. Haines, C.S., et al. "Artificial Muscles from Fishing Line and Sewing Thread." *Science,* 343 (6173), 2014: 868-872.

Functional Nonwovens by Green Electrospinning

Seema Agarwal, Andreas Greiner
University of Bayreuth, Macromolecular Chemistry II
Bayreuth, Germany
agarwal@uni-bayreuth.de

Solution electrospinning is a well-established method to process homogeneous fibers from different polymers with a huge range of properties by using a wide variety of organic solvents. However, the use of highly toxic and flammable solvents for electrospinning limits its use for many different application areas. The problem is more severe when electrospinning has to be carried out in large production scale. It is desirable to have water based electrospinning formulations for applications like biomedical field, food packaging and agricultural field. From the point-of-view of the work-safety aspects also it is desirable to use water as solvent for electrospinning for every application.

Water soluble polymers like poly(vinyl alcohol) and poly(ethyleneoxide) can be easily spun from water but the fibers are not water stable. The real challenge is to use water as a solvent for electrospinning of otherwise water insoluble polymers providing water stable fibers. The solution to this challenge is the suspension electrospinning which will be discussed in the present talk. Besides tackling toxicity and safety issues, the suspension electrospinning is highly productive as solid contents up to 50% and more can be spun. Moreover, it allows additional functionality by simple addition of additives in spinning formulations.

Some water-based spinnable formulations providing functional non-wovens by electrospinning like water-repelling textiles, antibacterial nonwovens, etc. will be presented.

Figure 1: Functional nonwovens by green water-based electrospinning

KEYWORDS
Green electrospinning, water-based formulations, functional nonwovens

ACKNOWLEDGMENT
Deutsche Forschungsgemeinschaft (DFG) and BMBF are acknowledged for financial support.

Nanofibrous Membranes with One-Way Oil Transport Ability and Advanced Oil-Water Separation Performance

Hongxia Wang, Hua Zhou, Tong Lin
Australian Future Fibers Research and Innovation Centre, Deakin University, Geelong, VIC, Australia
tong.lin@deakin.edu.au

INTRODUCTION

One-way liquid-transport through fibrous materials driven by the surface property of solid substrate has received great attraction because of the ability to transfer liquid automatically toward a specific location without costing extra energy.[1-3] Such a "smart" function is promising for many applications ranging from water harvesting and microfluidics to membrane separation, functional clothing, and defense. However, research on surface-driven one-way liquid-transport have mainly been focused on transport of water. One-way transport of oil fluids has received less attention until very recently.[4]

In this work, we have prepared a fibrous membrane that shows novel one-way oil-transport ability and incredible oil-water separation behavior.

APPROACH

Single-layer nanofiber membranes from poly(vinylidene fluoride-hexafl uoropropylene) (PVDF-HFP) and PVDP-HFP containing well-dispersed FD-POSS (fluorinated decyl polyhedral oligomeric silsesquioxanes, also referred to as "PVDF-HFP/POSS" in this abstract) were prepared by electrospinning the respective solutions using a needle electrospinning setup. [5] One-way transport nanofiber membrane was prepared by layered electrospinning of the PVDF-HFP and PVDF-HFP/POSS solutions in sequence.

RESULTS AND DISCUSSION

Fig. 1a shows the chemical structure of the materials used for making nanofiber membranes. The electrospinning setup is schematically shown in Fig. 1b. A metal drum was used to collect nanofibers during electrospinning.

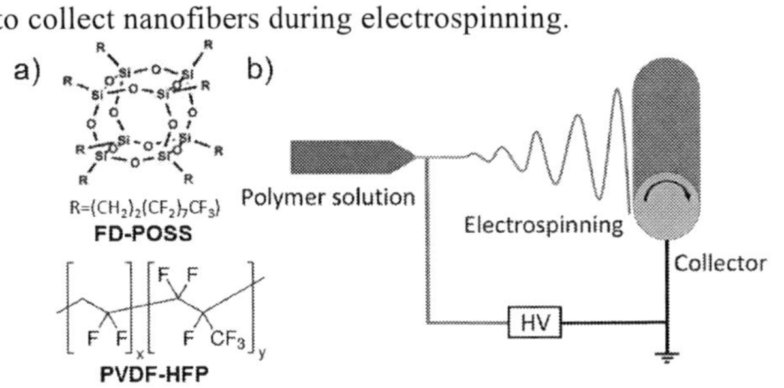

Figure 1. (a) Chemical structures of the materials used, (b) schematic illustration of electrospinning setup.

Before making one-way transport nanofibrous membranes, we first prepared PVDF-HFP and PVDF-HFP/POSS nanofibrous membranes and studied their wetting features. Fig. 2a shows the SEM images of the electrospun nanofibers. All nanofibers looked uniform without containing bead. Upon dropping liquids onto nanofiber membranes, the two nanofiber membranes showed different wetting behaviors, as shown in Fig. 2b. Contact angle (CA) measurement indicated that the PVDP-HFP nanofiber membrane had a CA of 153° and 0° respectively to water and diesel. For the PVDP-HFP/POSS membrane, the water and diesel CAs were 168° and 146°, respectively. Our further experiments indicated that the PVDP-HFP nanofiber membrane was superhydrophobic and oleophobic, whereas the PVDP-HFP/POSS membrane was superamphiphobic (i.e., being both superhydrophobic and superoleophobic).

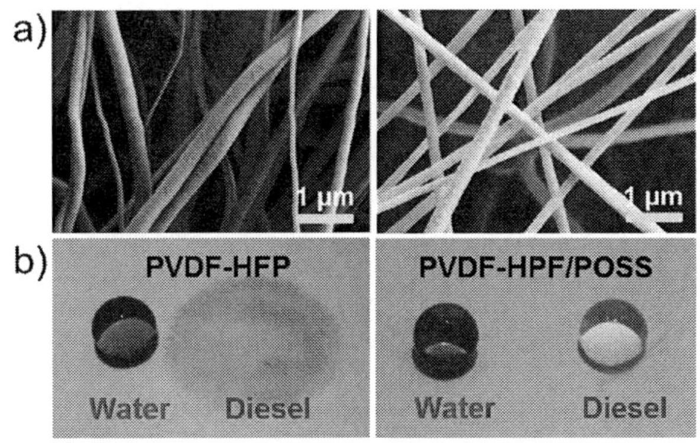

Figure 2. (a) SEM images of PVDF-HFP (left) and PVDFHFP/POSS (right) nanofiber membrane, (b) photos of water and diesel droplets on nanofiber membrane.

When a layer of PVDF-HFP nanofibers and a layer of PVDF-HFP/POSS were electrospun in sequence to form a dual-layer membrane, the resulting membrane showed a novel wettability. Fig. 3 shows dropping water and diesel on the dual-layer fibrous membrane. When water (35 μL) was dropped on the membrane surface, no matter onto which side, the droplet can always stay stably in a spherical shape. This can be explained by the superhydrophobic feature of both PVDF-HFP and PVDF-HFP/POSS membranes.

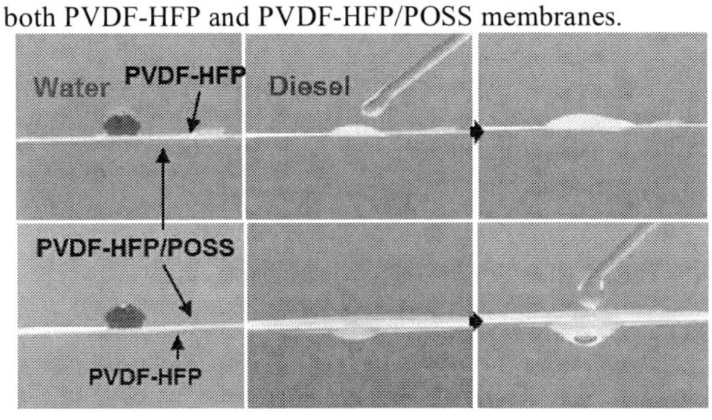

Figure 3. Still frames taken from digital videos to show red water and clear diesel drops on different sides of a dual-layer nanofiber membrane.

When diesel was dropped onto the dual-layer nanofibrous membrane, however, the wetting behavior was completely different to the single-layer fibrous membranes. On the

PVDF-HFP side, diesel just spread into the surface layer without penetrating through the membrane. When dropping diesel onto the PVDF-HFP/POSS side, it penetrated through the fibrous membrane automatically, and spread in the PVDF-HFP surface. This indicates that the dual-layer nanofibrous membrane has a one-way diesel-transport property.

Breakthrough pressure (*P*) was measured to understand the fluid resistance in the fibrous membranes. Water breakthrough pressure on the single-layer PVDF-HFP nanofiber membrane (thickness 200 μm) was 37 kPa (Fig. 4a), which was much higher than that of diesel (2 kPa), because of the superhydrophobic oleophilic feature of the fibrous membrane.

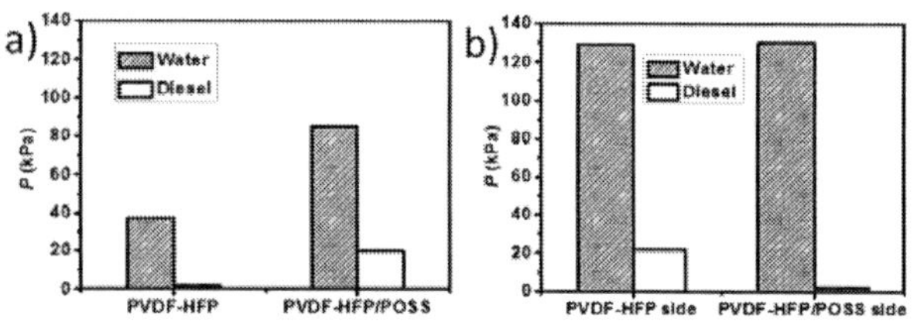

Figure 4. Breakthrough pressure on nanofiber membranes, a) single-layer nanofi brous membranes, and b) dual-layer nanofibrous membrane (overall thickness 450 μm, PVDF-HFP/POSS layer thickness 200 μm).

For the single-layer PVDF-HFP/POSS nanofibrous membrane at the same thickness and size, the breakthrough pressure to water and diesel was 85 kPa and 20 kPa, respectively, which was higher than that of the single layer PVDF-HFP nanofiber membrane. There was no effect of liquid feeding side on the breakthrough pressure value for both fibrous membranes. The difference in the liquid breakthrough pressure between the two single-layer nanofibrous membranes was attributed to the different liquid repellency.

It was interesting to note that liquid breakthrough pressure on the dual-layer fibrous membrane was considerably different to that on the single-layer ones. Here, we used a dual-layer nanofiber membrane with an overall thickness of 450 μm (PVDF-HFP/POSS layer thickness, 200 μm). The breakthrough pressure was 129 kPa for water and 22 kPa for diesel when the liquids were fed from the PVDF-HFP side. When the liquids were fed from the opposite side, however, the respective breakthrough pressure changed to 130 kPa and 2.3 kPa (Fig. 4b). The low diesel breakthrough pressure on the PVDF-HFP/POSS side originates from the novel one-way oil transport function.

The highly water-repellent and one-way oil-transport ability suggests strong water-rejection and oil-transport efficiency in oil-water separation. To prove the advanced oil-water separation performance, we used the one-way nanofiber membrane to separate a diesel-water mixture (20 mL + 15 mL). For comparison, single layer PVDF-HFP nanofiber membrane (the same overall thickness) was used for the same test.

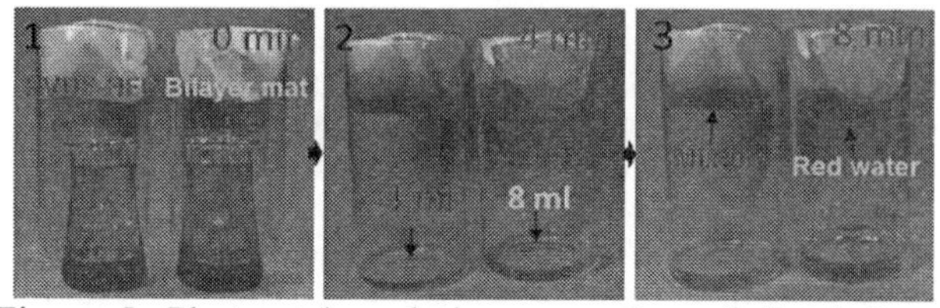

Figure 5. Photos taken during separation of diesel from water using nanofibrous membranes. (thickness 450 μm for both single-layer PVDF-HFP membrane and dual-layer membrane, thickness of the PVDF-HFP/POSS layer in the dual-layer membrane is 200 μm).

When diesel-water mixture was poured on the one-way nanofibrous membrane, 8 mL diesel was separated in 4 minutes as show in Fig. 5 In the same time period, a single-layer PVDF-HFP nanofiber membrane (the same thickness and size) just separated about 4 mL diesel. The dual-layer membrane took about 10 minutes to complete the separation of the entire diesel from water. For the single-layer PVDF-HFP membrane under the same condition, it took 15 minutes to complete the separation. Obviously, oil-water separation on the dual-layer nanofibrous membrane was faster than that on the single-layer PVDF-HFP nanofibrous membrane.

CONCLUSION

We have proved that a dual-layer nanofibrous membrane comprising a layer of superamphiphobic nanofibers and a layer of superhydrophobic oleophilic nanofibers has an interesting one-way oil-transport property. This novel fibrous membrane shows advanced oil-water separation capacity. These may lead to development of new, intelligent membranes for oil-water separation, fuel-recovery and other applications in industry.

REFERENCES

[1] Parker, A.R., Lawrence, C.R. *Nature*, 414, 2001: 33.

[2] Zheng, Y., Bai, H., Huang, Z., Tian, X., Nie, F.-Q., Zhao, Y., Zhai, J., Jiang, L. *Nature*, 463, 2010: 640.

[3] Wang, H., Ding, J., Dai, L., Wang, X., Lin, T. *J. Mater. Chem.*, 20, 2010: 7938.

[4] Zhou, H., Wang, H., Niu, H., Lin, T. *Sci. Rep.*, 3, 2013: 2964.

[5] Wang, H., Ding, J., Lee, B., Wang, X., Lin, T. *J. Membr. Sci.*, 303, 2007: 119.

Controlled Release of Active Agents from Electrospun Fibers

Gökçe Yazgan, Rene M. Rossi, Katharina Maniura-Weber, Josep Puigmarti-Luis, Giuseppino Fortunato
Laboratory for Protection and Physiology, Empa, Swiss Federal Laboratories for Materials Science and Technology
St. Gallen, Switzerland
giuseppino.fortunato@empa.ch; rene.rossi@empa.ch

OBJECTIVE

Our aim is to study and control the release kinetics of a model substance from emulsion and solution electrospun fibers in function of their morphology and composition. Thereby, both dyes as well as spinning environmental conditions are selectively varied.

INTRODUCTION

Electrospinning is a promising technique to obtain enhanced delivery systems due to high surface to volume ratio, adjustable fiber morphologies as well as tunable inter-fiber porosities [1] [2].

Composite fiber meshes can be developed by conventional electrospinning of a mixture of polymers and active substances either in solution or suspension. However, the active compounds are often not well incorporated within the polymer fiber matrix or the efficiency is low due to the contact with harmful organic solvents. Therefore emulsion electrospinning is a good alternative to better couple incompatible polymer and drug pairs[3].

Not only the inter fiber porosity but also single fiber morphology is expected to play an important role regarding controlled release kinetics. Hence, specified drug delivery systems can be tailored by adjustment of individual fibers morphologies. Up to now, pores on electrospun fibers have been induced either by fiber post-treatment [4] or phase separation processes during fiber formation [5].

Here we show that appropriate adjustment of spinning dyes as well as the control of spinning parameters (e.g. relative humidity, temperature, etc.) leads to hierarchically structured fiber and membrane morphologies exhibiting controlled release kinetics of the encapsulated species.

APPROACH

The fabrication of nanostructured fibers with a water-soluble core (dispersed phase polymer), which can act as drug reservoir, and a porous, water insoluble shell (continuous phase polymer) by emulsion electrospinning was investigated in different relative humidity conditions for selected emulsion formulations.

Table I: PS solutions and emulsions to investigate the influence of surfactant on pore formation (PS: Polystyrene, PVP: Polyvinylpyrrolidone, FLU: fluorescein sodium salt)

Name	10.7 w/w% PS in $CHCl_3$:DMF (g)	SPAN80(w/w% wrt continuous phase)	1% FLU in 50w/w% aq PVP (mL)
s1	2	0	0
s2	2	2.5	0
s3	2	7.5	0
em1	2	2.5	0.05
em2	2	7.5	0.05
em3	2	2.5	0.1

s: solution; em: emulsion

RESULTS AND DISCUSSION

It has been identified that the most influencing factors of pore formation are surfactant (SPAN80) concentration as a solution/emulsion parameter and relative humidity (RH) as a process environmental parameter.

Hence, the effect of surfactant concentration on pore formation was studied for both solution (Figure 1) and emulsion (Figure 2) systems. The formulations are listed in Table I.

The surfactant free solution was processed into porous fibers (Figure 1a), while the electrospun fibers from the solution with the highest SPAN80 amount yielded no pores (Figure 1c). It has been hypothesized that the surfactant acts as a plasticizer, which in turn decreased the glass transition temperature (Tg) of PS. Tg of atactic PS pellet used for fiber spinning as well as solution spun fibers were determined with DSC measurements. While the Tg of neat PS was around 90°C, it decreased to the range of 70-75°C with the addition of surfactant. Additionally the increase of surfactant might change the solvent evaporation rate. This can postpone the solidification of the fiber that phase separates into polymer rich and solvent rich regions as proposed by Dayal and Ku [6].

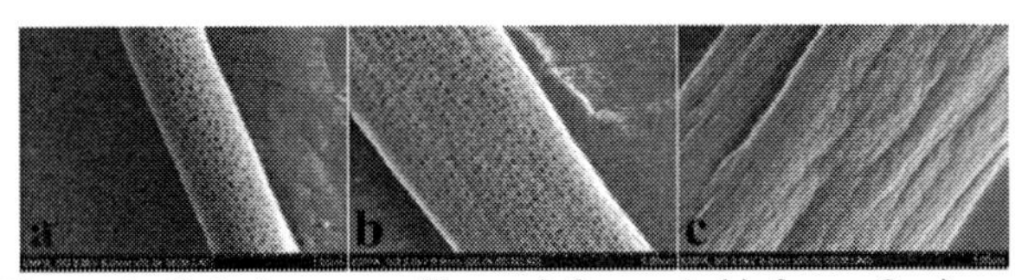

Figure 1: Solution spun fibers a) from s1, b) from s2, c) from s3.

For emulsion spun fibers, an increase of surfactant concentration caused loss of pore formation (Figure 2a, d). An additional effect of the surfactant is the distribution of the model compound along the fiber direction. The CLSM images show that the SPAN80 amount had distinct influences regarding the distribution of FLU within the fiber (Figure 2c, f).

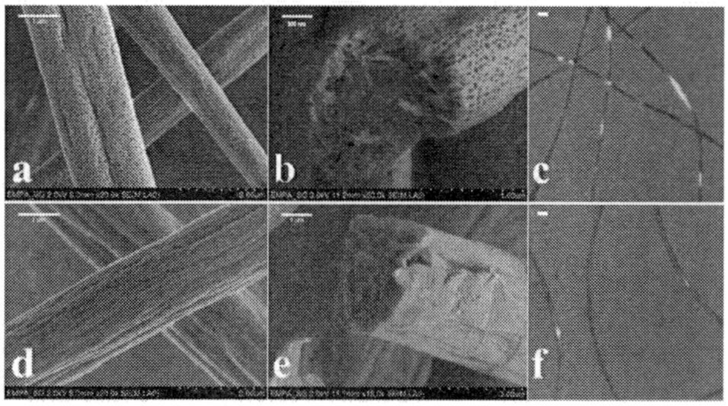
Figure 2: a) and b) SEM micrographs of electrospun fibers for em1; d) and e) for em2. c) and f) show CLSM images of electrospun fibers for em1 and em2, respectively. In c) and f) the scale bar is 5μm

Additionally, as an environmental factor, the influence of RH % was studied for em1. When the RH was decreased below 30%, the porous morphology of the electrospun fibers was replaced by smooth fiber surface (Figure 3).

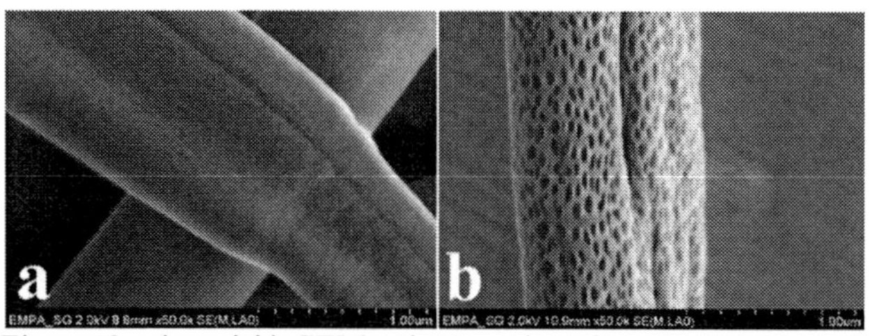
Figure 3: a) and b) SEM micrographs of em1 electrospun fibers synthesized in relative humidity below 30% and above 30%, respectively.

To further investigate the influence of surface pores on the release behaviour of incorporated compound, em3 with low surfactant concentration was electrospun both at 10% and 60% relative humidity conditions. The release of FLU from membranes fabricated at 10% RH was less in comparison to the one fabricated at 60% RH after 120 min in PBS at 37°C.

CONCLUSION
Surfactant concentrations and relative humidity were investigated in term of yielded fiber properties in relation to the release kinetics of a model compound, fluorescein sodium salt (FLU). The resulting fiber structure and surface morphology were prominently influenced by the relative humidity. Consequently, the release kinetics of the encapsulated compound was varied based on such fiber properties.

On the other hand, when surfactant concentration was increased, a transition from porous flattened to circular and smooth fiber forms was observed.

KEYWORDS
Emulsion electrospinning, drug release

ACKNOWLEDGMENT
The authors acknowledge financial support from the Swiss Science Foundation (NRP 62 Grant, IRENA: Intelligent responsive nanofibres (406240_126128)).

REFERENCES
1. Guex, A.G., Hegemann, D., Giraud, M.N., Tevaearai, H.T., Popa, A.M., Rossi, R.M., Fortunato, G. *Colloids and Surfaces B: Biointerfaces*, 123 (0), 2014: 724-733.
2. Camerlo, A., Bühlmann-Popa, A.M., Vebert-Nardin, C., Rossi, R., Fortunato, G. *Journal of Materials Science*, 49 (23), 2014: 8154-8162.
3. Qi, H., Hu, P., Xu, J., Wang, A. *Biomacromolecules*, 7 (8), 2006: 2327-2330.
4. Friedemann, K., Corrales, T., Kappl, M., Landfester, K., Crespy, D. *Small*, 8 (1), 2012: 144-153.
5. Megelski, S., Stephens, J.S., Chase, D.B., Rabolt, J.F. *Macromolecules*, 35 (22), 2002: 8456-8466.
6. Dayal, P., Kyu, T. *Journal of Applied Physics*, 100 (4), 2006.

Functional Textiles and Clothing

Functional Textiles for Biomedical Engineering

Yi Li[1,2]
[1]Institute of Textiles and Clothing, The Hong Kong Polytechnic University, Hong Kong, China
[2]School of Materials, the University of Manchester, UK
henrylydr@gmail.com; yi.li@polyu.edu.hk

ABSTRACT
With the increasing of aging population in the world, functional textiles become more and more important for healthcare and biomedical applications. To meet the increasing demand in the society, textile biomedical engineering (TBE) aims to develop a holistic and integrative approach of designing and engineering textile devices to meet the biomedical and healthcare needs by providing various functions such as drug delivery, tissue engineering and artificial implants, temperature and pressure therapy, bioelectric/magnetic detection and stimulation for the health and recovery of human body. TBE is experimental in nature and needs extensive multi-disciplinary collaborations among the experts in medical professionals, physiologists and designers, scientists and engineers in textile and clothing industry. Design and engineer of functional textiles for biomedical applications involve the integration of the fundamental research in physical, chemical, mathematical, and computational sciences with the development of engineering principles with sound understanding on the relationship between textiles and human biology, medicine, behavior and health. Fundamental concepts need to be advanced together with creating new knowledge from molecules to body-textile systems, and developing innovative materials, devices, and apparel products for healthy lifestyles with the functions of comfort, protection, prevention, diagnosis, and treatment of disease and healthcare. A systematic, quantitative and integrative way needs being adopted for approaching the solutions of how fibers and textiles can be engineered to the benefits of biology, physiology, medicine, behavior and health of human populations. This paper reviews the theoretical foundations of functional textiles for biomedical engineering.

KEYWORDS: Biomedical, Healthcare, Functional textiles, Engineering design, Biomaterial and Computational models

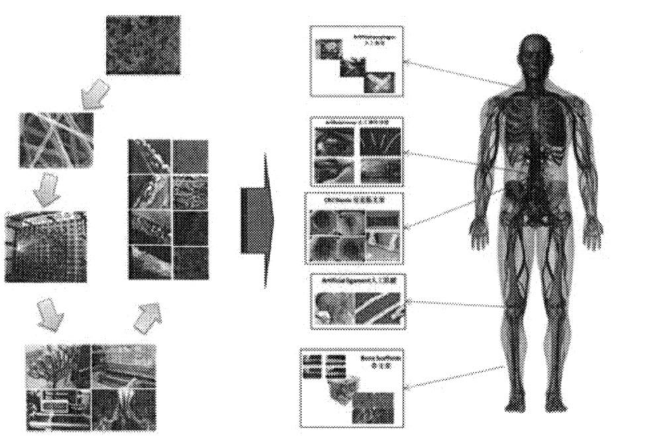

Figure 1. Textile biomedical engineering

REFERENCES

1. Angnew, B., *Science*, 280, (5369), 1998: 1516.
2. Chen, Aizheng, Zhao, Zheng, Wang, Shibin, Li, Yi, Zhao, Chen, Liu, Yuangang. "A Continuous RESS Process to Prepare PLA-PEG-PLA Microparticles." *Journal of Supercritical Fluids*, SUPFLU-D-11-00204R1.
3. Aihua, Mao, Jie, Luo; Yi, Li, Xiaonan, Luo, Ruomei, Wang. "A multi-disciplinary strategy for computer-aided clothing thermal engineering design." *Journal of Computer-aided Design*, CAD-D-10-00261R3.
4. Guo, Y.P., Wong, P.M., Li, Y., Or, Peggy P.L. "Is double-gloving really protective? A comparison between the glove perforation rate among perioperative nurses with single and double glove during operations." *The American Journal of Surgery*, DOI:10.1016/j.amjsurg.2011.08.017.
5. Lai, Joanna Y.F., Guo, Y.P., Or, Peggy P.L., Li, Y. "A Comparison Between Hand Contamination Rates and the Levels of Environmental Contamination when Glove Removal Follows Two Different Methods and Distances." *Am. J. Infect. Control*, 39, 2011: 104-111.
6. Chen, Dong, Tan, Longfei, Liu, Huiyu, Tang, Fangqiong, Hu, Junyan, Li, Yi. "Fabrication of Fast Absorbing and Quick Drying Wool Fabrics with Good Washing Durability." *Chem. Sus. Chem.* (SCI, IF: 4.767).
7. Yao, Lei, Gohel, Mayur D.I., Li, Yi, Chung, Waiyee J. "Investigation of Pyjama Properties in Skin Under Mild Cold Conditions–Exploring the Interaction Between Skin and Clothing." *International Journal of Dermatology*. Vol. 50, 7, 2011: 819-826. DOI:10.1111/j.1365-4632.2010.04798.x.
8. Yao, Lei, Li, Yi, Gohel, Mayur Danny Indulal, Chung, Waiyee J. "The effects of pajama fabrics' water absorption properties on the stratum corneum under mildly cold conditions." *J. Am. Acad. Dermatol.* 10.1016/j.jaad.2009.12.035.
9. Hu, J.Y., Li, Y., Ding, X., Hu, Junyan. "Neuromechanical representation of fabric-evoked prickliness: a fiber-skin-neuron model." *Cognitive Neurodynamics*, Vol. 5, 2, 2011: 161-170.
10. Linmm, Yinglei, Choi, Kai-Fi, Luximon, Ameersing, Yao, Lei, Hu, J.Y., Li, Y. "Finite element modeling of male leg and sportswear: contact pressure and clothing deformation." *Text. Res. Journal* (Accepted).
11. Chen, Dong, Tan, Longfei, Liu, Huiyum, Hu, Junyan, Li, Yi, Tang, Fangqiong. "Fabricating Superhydrophilic Wool Fabrics." *Langmuir*, 26 (7), 2010: 4675-4679.
12. Chen, Dong, Tan, Longfei, Liu, Huiyu, Tang, Fangqiong, Hu, Junyan, Li, Yi. "Fabrication of Fast-Absorbing and Quick-Drying Wool Fabrics with Good Washing Durability." *Chem. Sus. Chem.* 3, 2010: 1031-1035.
13. Liu, Xuqing, Chang, Haixin, Li, Yi, Huck, Wilhelm T.S., Zheng, Zijian Zheng. "Polyelectrolyte-Bridged Metal/Cotton Hierarchical Structures for Highly Durable Conductive Yarns." *ACS Applied Materials & Interfaces*, Vol. 2, 2, 2010: 529-535, 2010.
14. Liu, Xuqing, Li, Yi, Zhen, Zheng, Zijian. "Programming nanostructures of polymer brushes by dip-pen nanodisplacement lithography (DNL)." Nanoscale, Vol. 2, 12, 2010: 614-2618.
15. Liu, Z.B., Zhang, Y., Yu, J.J., Mak, A.F.T., Li, Y., Yang, M. "A microfluidic chip with poly(ethylene glycol) hydrogel microarray on nanoporous alumina membrane for cell patterning and drug testing." *Sensors and Actuators B-Chemical*, 143, 2, 2010: 776-783.
16. Li J.S., Chen, Y., Mak, A.F.T., Tuan, R.S., Li, L., Li, Y. "A one-step method to fabricate PLLA scaffolds with deposition of

bioactive hydroxyapatite and collagen using ice-based microporogens." *Acta Biomateriali*, 6, 6, 2010: 2013-2019.
17. Liu, R, Kwok, Y.L, Li, Y., et al. "Fabric Mechanical-Surface Properties of Compression Hosiery and Their Effects on Skin Pressure Magnitudes When Worn." *Fibres & Textiles in Eastern Europe*, 18, 2, 2010: 91-97.
18. Liu, R., Lao, T.T., Kwok, Y.L., Li, Yi, et al. "Effects of Compression Legwear on Body Temperature, Heart Rate, and Blood Pressure Following Prolonged Standing and Sitting in Women." *Fibers and Polymers*, 11, 1, 2010: 128-135.
19. Zhu, Q.Y., Li, Y. "Numerical simulation of the transient heat and liquid moisture transfer through porous textiles with consideration of electric double layers." *International Journal of Heat and Mass Transfer*, 53, 7-8, 2010: 1417-1425.
20. Li, L., Au, W.M., Li, Y., et al. "Design of Intelligent Garment with Transcutaneous Electrical Nerve Stimulation Function Based on the Intarsia Knitting Technique." *Textile Research Journal*, 80. 3, 2010: 279-286.
21. Zhou, Liya, Feng, Xunwei, Li, Yi. "Influences of the Fibre Hygroscopicity of Connecting Yarn on the Liquid Water Transfer Property of Knitted Double-Layer Fabric." *Fibres & Textiles in Eastern Europe*, Vol. 18, 6, 2010: 72-75.
22. Zhou, Jinyun, Li, Yi, Lam, Jimmy, Cao Xuyong. "The Poisson Ratio and Modulus of Elastic Knitted Fabrics." *Textile Research Journal*, Vol. 80, 18, 2010: 1965-1969.
23. Hu, Jiyong, Li, Yi, Hu, Junyan. "Neuromechanical Representation of Fabric-evoked Prickle: Spatial and Probability Integration." *Fibers and Polymers*, Vol. 11, 5, 2010: 790-797.
24. Zhu, Q.Y., Xie, M.H., Yang, J., Li, Y. "Investigation of the 3D model of coupled heat and liquid moisture transfer in hygroscopic porous fibrous media." *International Journal of Heat and Mass Transfer*, Vol. 53, 19-20, 2010: 3914-3927.
25. Yu, Bo, Zheng, Zijian, Li, Zhou, Feng. "Surface Grafted Polymer Brushes: Potential Applications in Textile Engineering." *Journal of Fiber Bioengineering and Informatics*, Vol.1, 4, 2009: 249-260.
26. Chen, Ai-Zheng, Li, Yi, Chau, Foo-Tim, Lau, Tsui-Yan, Hu, Jun-Yan, Zhao, Zheng, Mok, Daniel Kam-wah. "Application of Organic Nonsolvent in the Process of Solution-enhanced Dispersion by Supercritical CO^2 to Prepare Puerarin Fine Particles." *The Journal of Supercritical Fluids*, 49, 3, 2009: 394-402.
27. Chen, Ai-Zheng, Li, Yi, Chau, Foo-Tim, Lau, Tsui-Yan, Hu, Jun-Yan, Zhao, Zheng, Mok, Daniel Kam-wah. "Microencapsulation of puerarin nanoparticles by poly(L-lactide) in a supercritical CO^2 process." *Acta Biomaterialia*, 5, 2009: 2913-2919. DOI: 10.1016/j.actbio.2009.04.032.
28. Chen, Ai-Zheng, Li, Yi, Chen, Dong, Hu, Jun-Yan. "Development of core-shell microcapsules by a novel supercritical CO2 process." *Journal of Materials Science: Materials in Medicine*, 20, 3, 2009: 751-758. on-line publication DOI: 10.1007/s10856-008-3633-9.
29. Chen, Ai-Zheng, Kang, Yun-Qing, Pu, Xi-Ming, Yin, Guang-Fu, Li, Yi, Hu, Jun-Yan. "Development of Fe3O4-Poly(l-lactide) Magnetic Microparticles in Supercritical CO2." *Journal of Colloid and Interface Science*, 330, 2, 2009: 317-322.
30. Li, Linlin, Li, Hongbo, Chen, Dong, Liu, Huiyu, Tang, Fangqiong, Zhang, Yanqi, Ren, Jun, Li, Yi. "Preparation and Characterization of Quantum Dots Coated Magnetic Hollow Spheres for Magnetic Fluorescent Multimodal Imaging and Drug Delivery." *Journal of Nanoscience and Nanotechnology*, Vol. 9, 4, 2009: 2540-2545.
31. Li, Jiasen, Li, Yi, Li, Lin, Mak, A., Ko, F., Qing, L. "Fabrication of poly(L-latic acid)/keratin scaffold for osteoblast cultivation." *Composites Part B: Engineering*, 2009.
32. Li, Lin, Li, Y., Li, J.S., Yao, L., Mak, A.F.T., Ko, F, Qin, L. "Antibacterial Properties of Nano Silver PLLA fibrous Membranes." *Journal of Nanomaterials*, Article No. 168041, 2009. DOI: 10.1155/2009/168041.
33. Li, Jiashen, Li, Yi, Li, Lin, Mak, Arthur F.T., Ko, Frank, Qin, Ling. "Preparation and Degradation of PLLA/Keratin Electrospun Membranes." *Polymer Degradation and Stability*,DOI:10.1016/j.polymdegradstab.200.06.004, on-line publication.
34. Aihua, Mao, Li, Yi. "Heat transfer coupled with multi-dimensional liquid moisture diffusion in porous textiles with a measurable-parameterized model." *Numerical Heat Transfer, Part A: Applications*, 56, 3, 2009: 246-268.
35. Chen, Ai-Zheng, Li, Yi, Chau, Foo-Tim, Lau, Tsui-Yan, Hu, Jun-Yan, Zhao, Zheng, Mok, Daniel Kam-Wah. "Effect of Operating Parameters on Yield of Puerarin in Supercritical Process." *Journal of Fiber Bioengineering and Informatics*, 1, 4, 2009.
36. Ng, Sun-pui, Yu, Winnie, Li, Yi. "Photogrammetric prediction of girdle pressure." *Measurement Science & Technology*, Vol. 20, 1, 2009.
37. Ho, Simone S. M, Yu, Winnie W.M, Lao, Terence T., Chow, Daniel H.K, Chung, Joanne W.Y, Li, Yi. "Garment needs of pregnant women based on content analysis of in-depth interviews." *Journal of Clinical Nursing,* Vol. 18, 17, 2009: 2426-2435.
38. Ho, Simone S.M, Yu, Winnie W. M, Lao, Terence T, Chow, Daniel H.K, Chung, Joanne W.Y, Li, Yi. "Effectiveness of maternity support belts in reducing low back pain during pregnancy: a review." *Journal of Clinical Nursing*, Vol. 18, 11, 2009: 1523-1532.
39. Liu, Xuqing, Zhou, Feng, Li, Yi, Zheng, Zi-Jian, Hu, Junyan. "Selective Attachment of Gold Nanoparticles to Ionic Liquids Adsorbed Multiwalled Carbon Nanotube." *Journal of Fiber Bioengineering and Informatics*. Vol. 2, 1, June 2009.
40. Mao, Ai-Hua, Li, Yi Li, Luo, Xiao-Nan, Wang, Ruo-Mei. "Multi-structural computational scheme for textile thermal bioengineering design." *Journal of Fiber Bioengineering and Informatics*, Vol. 2, 1, June 2009.
41. Ying, Bo-an, Kwok, Yi-lin, Li, Yi, Li, Feng-zhi, Yick, Kit-lun, Wong, Kar-Yin. "An improved mathematical model of thermal physiological response of naked infants." *Journal of Fiber Bioengineering and Informatics*, Vol. 2, 2, September 2009.
42. Guo, Yue-Ping, Li, Yi, Wong, Thomson G.C., Chung, Joanne H.Y., Wong, Anthony S.W., Gohel, Mayur D.I., Leung, Polly H.M., Luximon, Ameersing. "Evaluation of Different Kinds of PPE Ensembles in Terms of Sensation, Usability, Satisfaction and Preference." *Journal of Fiber Bioengineering and Informatics*, Vol. 2, 2, September 2009.
43. Zhou, Dangxia, Li, Yi, Li, Quanhai Li. "Study of the relationship between point resistances and temperatures of 12 channels and body health status." *Journal of Fiber Bioengineering and Informatics*, Vol. 2, 2, September 2009.
44. Guo, Yueping, Li, Yi, Tokura, Hiromi, Wong, Thomas, Chung, Joanne, Wong, Anthony S. W., Gohel, Mayur Danny Indulat, Leung, Polly Hang Mei. "Impact of Fabric Moisture Transport Properties on Physiological Responses When Wearing Protective Clothing." *Textile Research Journal*, Vol. 78, 12, 2008: 1057-1069.
45. Liu, Huiyu, Chen, Dong, Tang, Fangqiong, Du, Gangjun, Li, Linlin, Meng, Xianwei, Liang, Wei, Zhang, Yangde, Teng, Xu, Li, Yi. "Photothermal therapy of Lewis lung carcinoma in mice using gold nanoshells on carboxylated polystyrene spheres." *Nanotechnology*, Vol. 19, 45, 2008.
46. Li, Y., Guo, Y.P., Wong, T., Chung, J., Gohel, D., Leung, P. "Transmission of Communicable Respiratory Infections and Facemasks. *Journal of Multidisciplinary Healthcare*, 2008: 117–27
47. Chen, Binjiao, Wang, Xin, Zeng, Ruolang, Zhang, Yinping, Wang, Xichun, Niu, Jianlei, Li, Yi, Di, Hongfa. "An experimental study of convective heat transfer with microencapsulated phase change material suspension: Laminar flow in a circular tube under constant heat flux." *Experimental Thermal and Fluid Science*, Vol. 32, 8, 2008: 1638-1646.
48. Li, Y., Dai, Xiao-qun. *Biomechanical Engineering of Textiles and Clothing*. Cambridge, UK: Woodshead Publishing Limited, 2006. ISBN 1 84569 0524, 1-412.

Textile Heart Valve: Fabric Construction Parameters and *In Vivo* Results

Frederic Heim[1], Antoine Vaesken[1], Nabil Chakfe[2]
[1]Laboratoire de Physique et Mécanique Textiles (LPMT), ENSISA, Université de Haute Alsace, Mulhouse, France
[2]Service de Chirurgie Vasculaire - Geprovas, Hôpitaux Universitaires de Strasbourg, Strasbourg, France
frederic.heim@uha.fr

INTRODUCTION

Trans-catheter (non-invasive) aortic valve implantation has become an alternative technique to surgical valve replacement for an increasing number of patients [1]. Biological tissue, currently used in clinical practice, remains, however, expensive to harvest and fragile. Heim et al. showed already in vitro that textile polyester could be an alternative solution to replace valve leaflets [2]. However, only limited results have been published about the in vivo behavior of textile as valve leaflet material. The goal of the present work is to assess in vivo the influence of the fibrous construction used as valve leaflet material on the interaction with living tissues.

APPROACH

Textile valve prototypes were obtained from shaping tubular textile membranes of various constructions (C1, C2, C3, C4). Obtained valves were implanted in sheep models in mitral position and harvested after 6 months implantation for histology study and calcium dosage.

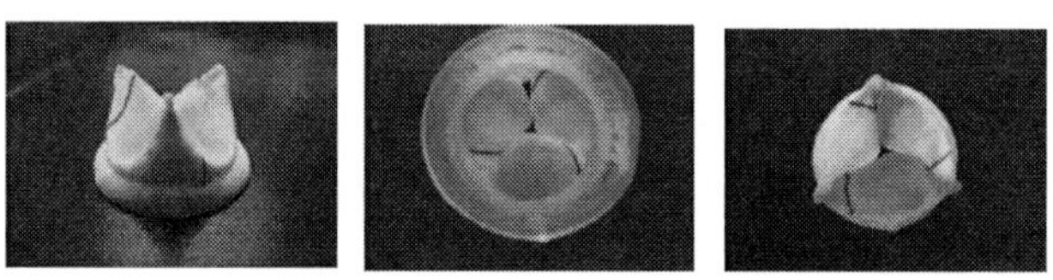

Figure 1: Valve prototype

RESULTS AND DISCUSSION

Results showed significant differences between the tested constructions in terms of interaction with living tissues. Calcium deposit seems to vary with the fiber size, the roughness and the porosity of the material (Fig. 1).

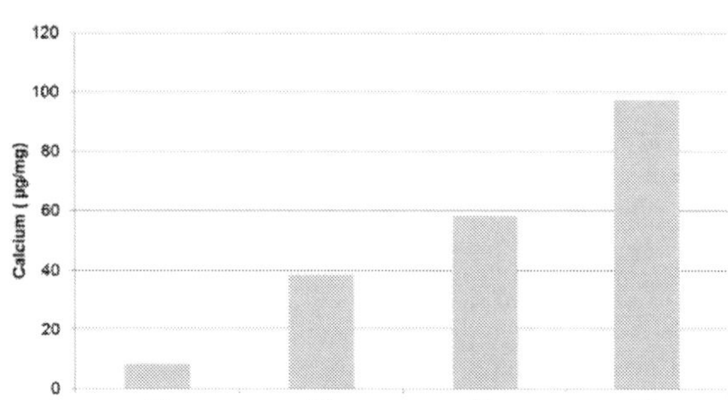

Figure 2: Calcium dosage

With regard to the fibrin ingrowth, the thickness of the fibrous layer seems to depend largely on the construction as well. Figure 2 brings out significant differences among the tested samples.

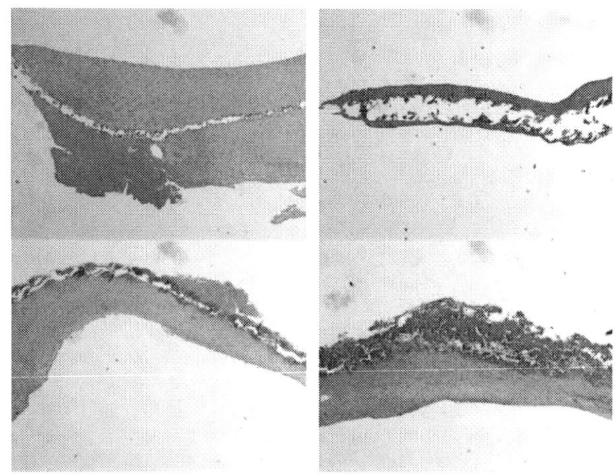

Figure 3: Fibrin deposition models

The in vivo observations show that it is possible to limit the fibrin ingrowth as well as the calcium deposit on the valve leaflets just by adjusting the fabric construction in a first approach. Surface functionalization of fabric leaflets (plasma treatments) can then be additionally performed in order to enhance the in vivo performances of the textile valve.

CONCLUSION

According to these early results, the textile construction appears to influence the valve interaction with living tissues but further tests are necessary to identify which parameter is the most critical.

REFERENCES

[1] Kidane, A., et al. "Current Developments and Future Prospects for Heart Valve Replacement Therapy." *Journal of Biomedical Materials Research Part B: Applied Biomaterials*, 88B (1): 2009: 290-303.
[2] Heim, F., Durand, B., Chakfe, N. "Textile Heart Valve Prosthesis: Manufacturing Process and First in Vitro Performances." *Textile Res. J.*, 78: 2008: 1124-113.

KEYWORDS
heart valves, trans-catheter, textile, non-invasive

Thermal Comfort of Professional Sportswear with Physiological and Psychophysical Measurements During High-Intensity Exercise

Tsung-lin Liu[1], Yu-sian Ciou[1], Ku-hsin Chang[1,2], Sheng-fu Chiu[1]
[1]Department of Testing and Certification, Taiwan Textile Research Institute (TTRI), Taipei, Taiwan
[2]Department of Material and Fiber, Oritental Institute of Technology, Taipei, Taiwan
tlliu.1242@ttri.org.tw; sfchiu.0222@ttri.org.tw

OBJECTIVE
The aim of this study was to investigate the thermal comfort of professional sportswear in a neural-warm environment (23 ± 1.5°C, 65 ± 3% rh, no wind) during high-intensity exercise. Eight healthy males ran on the treadmill for 40 minutes at 75% VO_{2max} exercise intensity. Three types of sport fabrics were used then developed to four different wearing types. For the objective assessments, the measurements of mean skin temperature and sweating rates (corresponding to ISO 9886), the microclimate temperature and humidity of different subjects were analyzed. The psychophysical sensations like thermal, wettedness, and comfort sensations during exercise were also recorded. The results suggested that the compression suits are more suitable for activity on high-intensity exercise and high-sweating rates condition.

INTRODUCTION
Recently, with the growing interest in sports activities, the demand for high performance sportswear is also increase. To achieve this goal, both the thermal comfort and functionality of the sportswear should be enhanced. Therefore, compression garments have become more and more popular due to the possibility decrease of the energy cost of locomotion [1] and the improved recovery and health benefits [2-3]. However, a compression garment always contains the elastic yarn to increase its elasticity and tension. The increased percentage of elastic yarn of compression fabric could reduce the water absorption ability and reduce the thermal-physiological wear comfort [4].
The main goal of this study was to study the comfort sensation of different types of professional sportswear in high-intensity running exercise. During the tests, physiological and psychological measurements are recorded. An effective assessment process for the thermal comfort of professional sportswear has also been developed.

APPROACH
Material of sportswear. Table 1 shows the composition material of the fabrics and their basic weights.

Table 1:Fabric properties.

Code	Compositions	Basic weight
A (blue)	70% nylon / spandex 30%	255 g/m^2
B (black)	74% nylon / spandex 26%	280 g/m^2
C (green)	65% polyester / Texcare-Smart dry 35%	160 g/m^2

Those fabrics are used to make four wearing styles, as shown in figure 1. D-type is a body-mapping compression garment that is combined with fabric code (A) and (B). It uses the seams of clothing and body-mapped fabric design for focused support of major muscle groups. G-type is a compression suit made by fabric (B) without body-mapping design. T-type is dressed with upper loose-fitting sportswear made with fabric (C). The control group is N-type, which is set as nude upper body for subjects. To be noted that, all the G, T, and N-type have the same compression long-tight made with fabric (B).

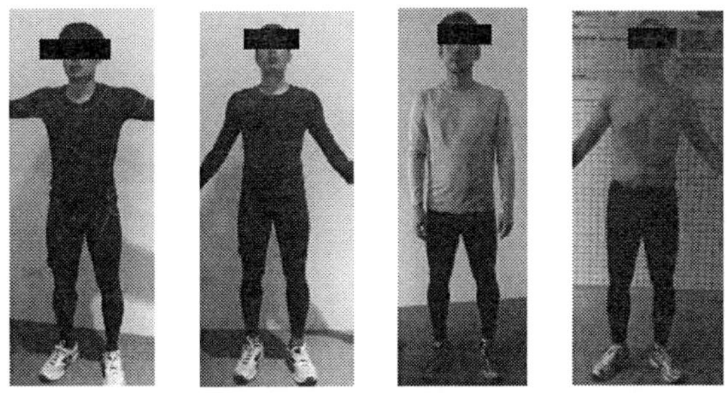

(a) D-Type (b) G-Type (c) T-Type (d) N-Type

Figure 1: Four wearing types of this study.

Experiment protocol. Eight healthy man (age: 26.8±2.3 yr; height: 1.75±0.03m; body mass: 69.2±5.8 kg; VO_{2max}: 49±3 kgl/min) participated in this study. All the volunteers have recreational activities two or three times per week. The VO_{2max} of subjects were tested on a treadmill (H/P/cosmos) with a cardiopulmonary system (CORTEX MetaLyzer®3B). Body mass loss was determined pre- and post-exercise. All the temperature and humidity measurements are used i-bottom sensors (DS1923, MAXIM). The environment temperature and humidity were set to 23±1.5°C and 65± 3% rh, no wind condition. The subject's nude body mass was measured first. The protocol starts with 3 minutes rest, and then running with 75% VO_{2max} of each subjects for 40 minutes.

RESULTS AND DISCUSSION
Mean skin temperature. The measurement of mean skin temperature in the present study is corresponding to the ISO 9886. According to this standard, four point measurement in neck (T0), right scapula (T1), left hand (T2), and right shin (T3) are recorded. The formula can be showed as :

$$T_{skin} = 0.28\times T_0 + 0.28\times T_1 + 0.16\times T_2 + 0.28\times T_3$$

Figure 2 shows the mean skin temperature comparison between the four different wearing types. During the running test, the skin temperature of the compression suit (type D&G) are mean higher than the other ones. N-type has the lowest skin temperature comparing to others, which is consistent with the experience.

Mean micro-humidity. Figure 3 shows the comparisons of mean micro-humidity with different wearing types. Interestingly, T-type with higher micro- humidity in the early to mid- exercise period. It suggests the loose-fit garment may meet higher demand of vapor permeability to disperse the sweat vapor.

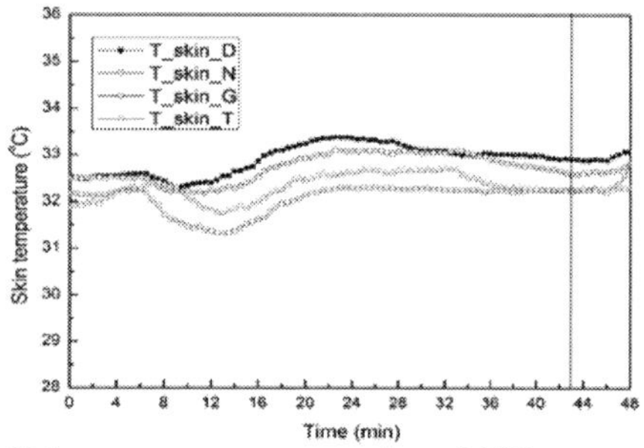

Figure 2. Skin temperature comparsion of different wearing types.

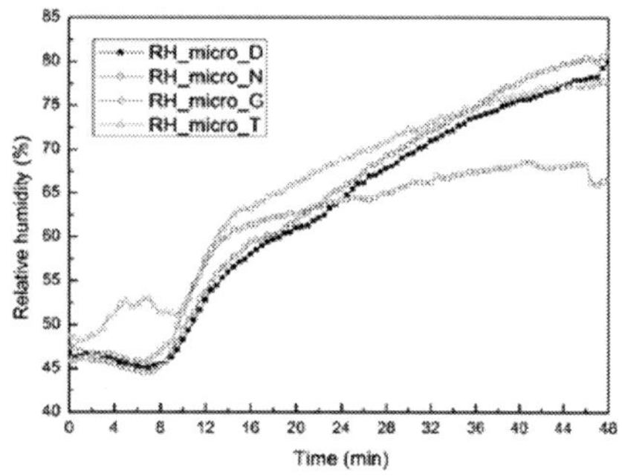

Figure 3. Micro-humdity comparison of different wearing types.

Body mass loss. Three items of body mass loss are estimated; they are respiration loss, mass loss due to the difference molecular weight between CO_2 and O_2, and body sweating loss. Figure 4 shows the body mass loss comparison between differnet wearing types. It suggested that no significant differnce between respiration loss and O_2/CO_2 loss. The lowest mean body sweatig loss apears in N-types, and relativly highest in the D-type wearing style. However, the standard deviation in the body sweating loss is large, which could exceed 400 g after end of test.

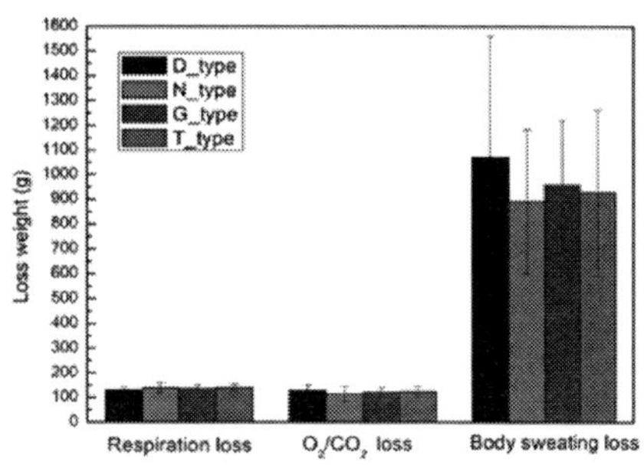

Figure 4. Body mass loss comparison of different wearing types.

Psychophysical sensations. Figure 5 shows the ratings of thermal sensaton, wettedness sensation, and total comfort sensation of whole body during exercise. The range of thermal sensation is from -3 (very cold) to 3 (very hot), and from 0 to 3 of wettedness and comfort sensations. The higher values indicate higher wet feeling and uncomfortable. The results of the thermal sensation between different clothing types seems no significant different exception of T-type in the early-stage of exercise. The thicker air lawyer and less cloth ventilation of the loose-fit long sleeve sportswear in no wind condition caused that result. The wettedness sensation of T-type can be found relatively higher than the compression suit types during the excercise period, which is consistant with the physiological mesurment of microclimate humdity (see fig. 3).

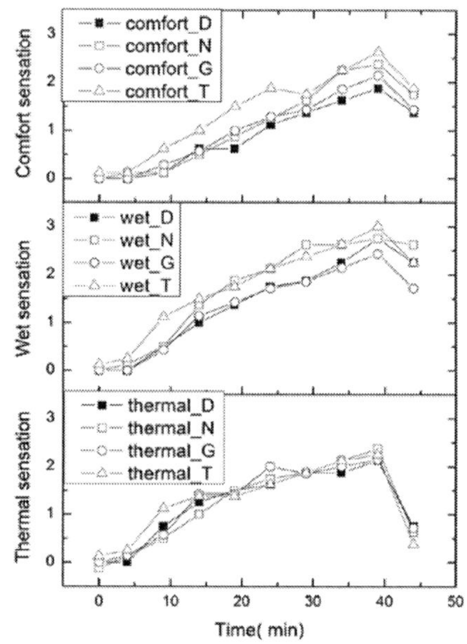

Figure 5. Psychophysical sensation comparison between different wearing types during exercise.

CONCLUSION

The thermal comfort of different professional sportswear during high-intensity exercise with human trials was studied. The results shows that the compression garment with thinner air layer between skin and cloth could let subjects with lower hot feeling and wet sensation, which can be proved by micro-humidity measurement and psychological sensations evaluated of the present study.

ACKNOWLEDGMENT

The authors would like to acknowledge the financial support for this study by Ministry of Economic Affairs, Taiwan, R.O.C.

REFERENCES

[1] Bringard, A., Perrey, S., Belluye, N. "Aerobic Energy Cost and Sensation Responses During Submaximal Running Exercise–Positive Effects of Wearing Compression Tights." *International Journal of Sports Medicine*, 27, 2006: 373-378.

[2] Chatard, J.C., Atlaoui, D., Farjanel, J., Louisy, F., Rastel, D., Guézennec, C.Y. "Elastic stockings, performance and leg pain recovery in 63-year-old sportsmen." *Eur. J. Appl. Physiol.*, 93, 2004: 347-352.

[3] Doan, B., Kwon, Y.-H., Newton, R., Shim, J., Popper, E.V.A., Rogers, R., Bolt, L., Robertson, M., Kraemer, W. "Evaluation of a lower-body compression garment." *Journal of Sports Sciences*, 21, 2003: 601-610.

[4] Manshahia, M., Das, A. "Thermo-physiological comfort of compression athletic wear." *Indian Journal of Fibre & Textile Research*, 39, 2014: 139-146.

Parameters Impact on Pressure Comfort in System “Female Body-Dress”

V. Kuzmichev, Meng Na Guo
Ivanovo State Polytechnic University
amyidaguo@gmail.com; vwkd37@list.ru

KEYWORDS
Woman dress, Pattern design, Ease allowance, Dress structure, Sleeve, Pressure, Comfort

INTRODUCTION
Ease allowance which designing in 2D pattern block is the core issue. They are influenced on the clothing structures, detailed indexes of pattern in the longitudinal and transverse directions, the diversification of clothing shape and the pressure comfort perception when wearing. [1]

In this research we studied how the pattern indexes influenced on the comfort in the system "body-clothes" considering 3 aspects of structure: 1) Parameters designing on front and back pattern; 2) Different sleeve lengths; 3) The position of armhole lines to the body.

APPROACH
Experiment was designed to find the relations between pattern block parameters and pressure comfort in different sleeve lengths:

• Five female dresses patterns were prepared in the same size: Ease on bust/waist/hip = (6/4/6, cm); but they were varied for ease distribution on front and back pattern.

Table I. Parameters distribution on front and back pattern

	Ease distribution, %						
	Bust level			Waist level		Hip level	
	Ebf	*Eba*	*Ebb*	*Ewf*	*Ewb*	*Ehf*	*Ehb*
S1	0	30	70	15	85	15	85
S2	10	40	50	30	70	30	70
S3	25	50	25	50	50	50	50
S4	50	40	10	70	30	70	30
S5	70	30	0	85	15	85	15

Ebf - ease to front bust width; Eba - ease to armhole width; Eb - ease to back bust width; Ewf - ease to front waist width; Ewb - ease to back waist width; Ehf - ease to front hip width; Ehb - ease to back hip width. S1,S2...S5 - sample of dress

• These samples were prepared in three groups with different sleeve lengths: long sleeves /short sleeves/without sleeves. (Fig. 1)

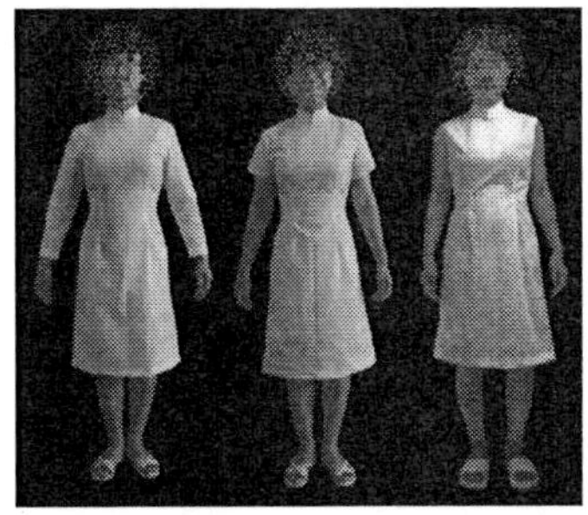

Figure 1. Samples of dresses with different sleeves

• Thirteen anthropometric points as same as in our earlier research [2] were selected out to test pressure comfort.

• Five expects volunteers participated in the experiment to test objective pressure indexes with FlexiForce sensor and described their subjective perceptions in different dynamic postures.

• The objective pressure indexes and subjective pressure comfort levels were compared with dresses which were designed by different parameters in details. In this way we find out the better way to plan the ease allowance distribution in a proper way, which would increase the pressure comfort with the same size. In addition, influence of different sleeve length was concerned.

RESULTS

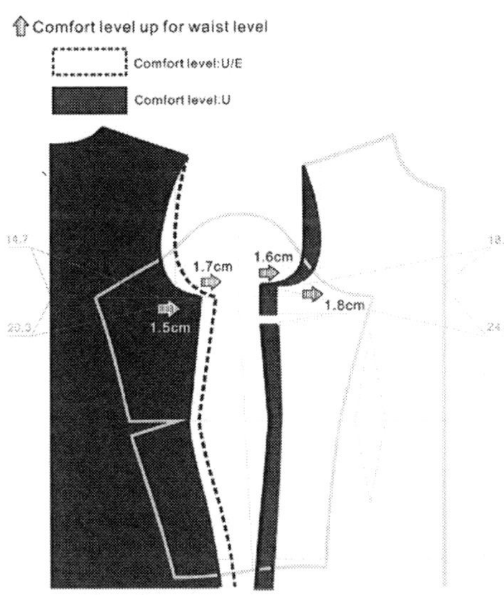

Figure 2. Pressure comfort improved by adjusting ease allowance parameters in detail (with long sleeves)

As proved in our earlier research [3], the change of sleeve length will also influence the pressure comfort. In this work, we find out how to improve comfort perception by adjusting parameters in detail with different sleeve length. As shown in the above picture, the pressure comfort perception could be improved from level U (uncomfortable) to U/E (uncomfortable/endurable) by distributing more ease allowance on back pattern for dresses with long sleeves. (Fig. 2)

In addition, the armhole plane angle (APA) measured between two vertical planes (first - across armhole arc of dress, second - across torso and upper arm of body) (Table II) have been calculated. Ease allowance on bust level helped us to find this angle with the pressure indexes.

Table II. APA of dresses

Sample	Distribution of ease allowance (ease on half bust = 3 cm)			Angle of the armhole plane (°)
	Ebwf	*Eaw*	*Ebwb*	
1	0	0.9	2.1	33.8
2	0.3	1.2	1.5	29.7
3	0.75	1.5	0.75	23.9
4	1.5	1.2	0.3	18.2
5	2.1	0.9	0	13.7

In this way, we compared how APA would influence the objective pressure measurements in different group of sleeve length.

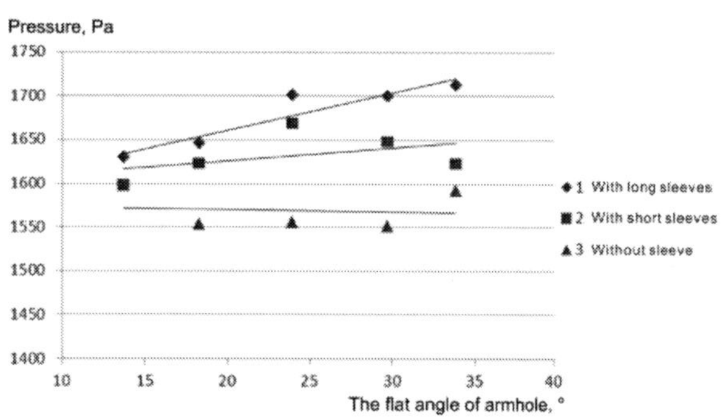

Figure 3. Relations between APA and the pressure on bust level for dresses with different sleeve length

From the Fig. 3, we could observe that APA plays an obvious effect on the pressure measurements. The increasing of APA leads to the increasing of pressure indexes, only the sleeveless dress is exception.

CONCLUSION

The results of this study demonstrate and complement our previous achievements:

• The relations between ease allowance distribution and pressure comfort were studied. The more reasonable way for designing ease allowance that would improve the pressure comfort level was suggested to pattern designers. At the same time, the changing of sleeve length was concerned to be an important factor which impact on comfort level.

• A new index which presents APA was proved effective for predicting pressure comfort of bust level. This value differs from the traditional ones as it was calculated from the major indexes on bust level; it helped us to describe the angle shaping of armhole in section's view. The relations between APA and the pressure measurements had shown that the parameters design in detail strongly influenced the pressure comfort.

FUTURE WORK

In the next step, we will continue to study how pattern parameters design in other details (e.g. the change of armhole depth) will influence outside shaping and inside pressure comfort in the "Female body-dress" system.

REFERENCES

[1] Guo, Mengna, Kuzmichev, Victor E. "Pressure and comfort perception in the system "female body-dress." *AUTEX Research Journal*, Vol. 13, No. 3, 2013: 71-78.

[2] Kuzmichev, V., et al. "Complex Approach to Designing of Comfortable System "Body-Dress." 14th AUTEX World Textile Conference, May 26-28, 2014, Bursa, Turkey.

[3] Го, М., Кузьмичев, В.Е. "Прогнозирование давления платьев на поверхность женских фигур." *Швейная промышленность*. No. 2, 2014: 37-40 (Russian).

Applications in Medical, Transportation, and Energy

Regenerated Keratin Fibers and Their Potential Applications in Textile and Biomedical Industries

Helan Xu[1], Xiang Mi[2], Yiqi Yang[1,3,4]

[1]Department of Textiles, Merchandising and Fashion Design, 234, University of Nebraska-Lincoln, Lincoln, NE, United States; [2]Key Laboratory of Science and Technology of Eco-Textiles, Ministry of Education, Donghua University, Shanghai, China; [3]Department of Biological Systems Engineering, University of Nebraska-Lincoln, Lincoln, NE, United States; [4]Nebraska Center for Materials and Nanoscience, University of Nebraska-Lincoln, Lincoln, NE, United States

yyang2@unl.edu

ABSTRACT

We report regeneration of protein fibers from poultry feathers and waste wool using green processes after more than two decades of research. There is potential to apply the technologies on highly crosslinked proteins, which are or will be largely available from agricultural and biofuel industries.

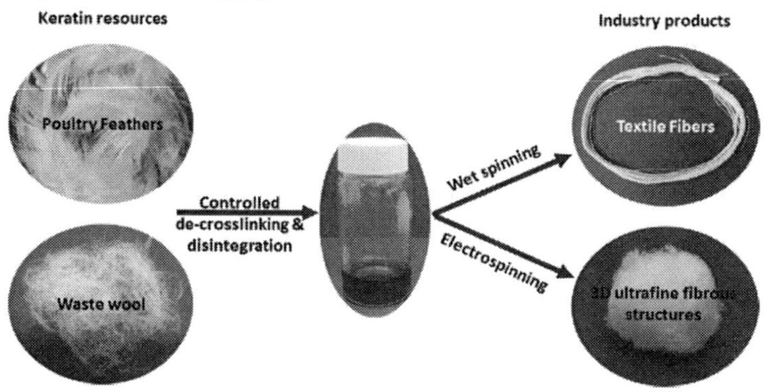

BACKGROUND

Protein fibers, mainly wool and silk, are relatively expensive, with unparallel performance properties, such as hand, moisture transmission, thermal insulation, etc., comparing to cellulosic and synthetic fibers. All the protein fibers are expensive, as breeding of animals/insects and fiber collection need extensive care, and are highly labor intensive. Developing fibers from bio-based non-textile proteins could be an alternative approach to provide new resources to satisfy customer demands and compensate potential fiber shortage due to decreased production of cotton.

However, till now, all the production approaches for regenerated protein fibers had environmental issues. In 1940s, regenerated protein fibers from a variety of proteins, such as zein, soy protein, peanut protein, etc., did not yield good fiber properties even after using toxic crosslinkers, such as formaldehyde. Current commercial protein fibers are actually polyvinyl alcohol (PVA) fibers blended with less than 50% of hydrolysates from soy or milk protein. In addition, textile usage of these proteins formed competition with food industries.

Feathers and discarded wool are waste non-food protein resources with annual availability, abundance, low cost, and thus, could be resources for protein fiber production. More importantly, these keratins contain 80-90% of keratins that meet the basic technical requirements for fiber spinning. However, to the best of our knowledge, no efficacious method has been developed to produce regenerated keratin fibers, despite global efforts in the last two decades.

In addition to textiles, keratin fibers could also be good substrates in biomedical applications. Current protein biomaterials lacked water stability. However, keratins could be produced into water-stable scaffolds due to their high content of cysteine to form inter- or intra-molecular crosslink. Three-dimensional (3D) ultrafine fibrous tissue engineering scaffolds from keratins could be advantageous due to their structural and molecular similarities, comparing to native extracellular matrices (ECMs) of most soft tissues. Also due to technical difficulty, 3D ultrafine fibrous tissue engineering scaffolds from pure keratin have not been reported.

KEY SCIENTIFIC PROBLEMS

Currently, by achieving de-crosslinking via alkaline hydrolysis, the backbones of keratin were substantially destroyed. The small keratin hydrolysates could not ensure good spinnability, and thus required incorporation of PVA.

To produce 100% keratin fibers, the highly crosslinked networks of feathers and wool should be disrupted controlledly to preserve the backbones of keratins, and controlled disentanglement and alignment of keratin was required to endow spinnability.

APPROACH

We effectively extracted keratin macromolecules for fiber regeneration under mild conditions, endowed them with satisfactory spinnability and produced keratin fibers in aqueous systems. Urea and cysteine were used to disintegrate the keratin crosslinking network, while sodium dodecyl sulfate (SDS) was used to controlledly disentangle and align keratin macromolecules, and endowed them with spinnability. Normal fibers were wet spun from SDS-keratin solution, while 3D ultrafine fibrous scaffolds were electrospun with substantially higher content of SDS in solution to ensure high stretchability. SDS was removed after fiber solidification. The whole processes were green as all the chemicals could be either produced renewable resources or could be reused.

In terms of biomedical applications, adipose derived mesenchymal stem cells (ADMSCs) were cultured on the 3D keratin scaffolds, and induced chondrogenic differentiation. Biological assays and differentiation assays were used.

RESULTS AND DISCUSSION

The hypothesis of fiber spinning was verified, as retaining of macromolecules during extraction and effective control on dis-entanglement of de-crosslinked keratin molecules were proved using technologies of X-ray diffraction, viscosity, and circular dichroism. The keratin dopes demonstrated exceptional spinnability.

The 100% keratin fibers targeting textile use showed fineness of around 20 µm, and dry mechanical properties close to wool without external crosslinking. The wet properties were low but still higher than most regenerated protein fibers from other reports.

The 3D keratin tissue engineering scaffolds with fiber diameter of 1-5 µm, demonstrated exceptional water stability without external crosslinking, as well as satisfactory support for the proliferation and cartilaginous differentiation of ADMSCs, comparing to plant protein-based 3D scaffolds with similar morphologies. The existence of cell binding tripeptides, Arg-Gly-Asp (RGD) and Leu-Asp-Val (LDV) in the amino acid composition and high dimensional stability.

CONCLUSIONS

Keratins in poultry feathers and waste wool were used to develop 100% protein fibers, which demonstrated potential for textile and biomedical applications. The "green" processes developed will have broad impact on developing value-added industrial products from largely-available highly crosslinked proteins, e.g., sorghum proteins, from biofuel and agricultural industries.

Geometry Modeling of a Tubular Weft-Knitted Intestinal Stent

Gang Li[1,2], Lei Liu[1], Yi Li[3], Ping Lan[4]

[1]National Engineering Laboratory for Modern Silk, College of Textile and Clothing Engineering, Soochow University, China
[2]State Key Laboratory for Modification of Chemical Fibers and Polymer Materials, Donghua University, China
[3]Institute of Textiles and Clothing, The Hong Kong Polytechnic University, Hung Hom, Kowloon, Hong Kong
[4]Department of Colorectal Surgery, Sun Yat-Sen University, Guangzhou 510655, Guangdong Province, China

tcligang@suda.edu.cn

ABSTRACT
The paper describes a geometry modeling for a weft-knitted stents for colorectal cancer. Polydioxanone monofilament with 150 tex was used to produce tubular stents using a small weft-knitting machine. A geometrical model for single jersey weft-knitted structures is suggested here based on the plain knitted fabric model of Perice and Kurbak. The main objective is to model tubular weft-knitted stent so as to investigate its biomechanical functions by optimizing different stent structures and weft-knitting parameters of the stent can be carried out using Finite element modeling (FEM).

KEYWORDS: stents; geometry modeling; mechanical properties.

STATEMENT OF PURPOSE
The aim is to model tubular weft-knitted stent so as to study its biomechanical functions of the stent.

INTRODUCTION
Intestinal cancer is one of the most common cancers in terms of both incidence and mortality in the world[1,2]. Many systematic reviews revealed that stents can prevent acute intestinal obstruction in patients[3,4]. Saito et al[5] demonstrated that using a Polydioxanone (PDO) stent can provide time for systematic support and obviate the need for faeces diversion or on-table lavage. The stents mainly give mechanical support of obstructed intestines which can restore intestinal lumen and waste flow smoothly[6]. Design and selection of stents having proper mechanical functions are very important. Although much progress was reported, few literatures were focused on the effect of weft-knitting structures and parameters on the mechanical properties of intestinal stents[6]. Sirry et al. (2010) model is focus on solid model and static connection[7]. In fact, the loops are not static connection but small slipping movement. Hence, mathematical modeling for stent is necessary and more accurate. That is required to achieve good interaction between intestines and stents. With this model, it can provide a theoretical reference for other textile-forming structural biomedical stents.

APPROACH
PDO monofilament with linear density at 150 tex were knitted into a tubular form using a small weft-knitting machine with 22 needles. Referring to the metal stents in market, the diameter and length of the samples are fixed at 20 and 60mm, respectively.

RESULTS AND DISCUSSION
The heat-setting treatment with temperature of 80°C and time of 5 minutes was used for the stents to restore a tubular shape. Images of the tubular stent after heat-setting, weft-knitted structural loops and simulated single loop using Matlab 8.0 are shown in Figures 1a, 1b and 1d. The front view and A-A cross section of the loop of the single jersey weft-knitted stent are shown in Figure 1c. The whole loop can be divided into seven curves in which each curve has its different equation. As this assumption, the equations of the curves between the points O, A, B, C, D, E, F and Q of the loop were denoted by the following formulas:

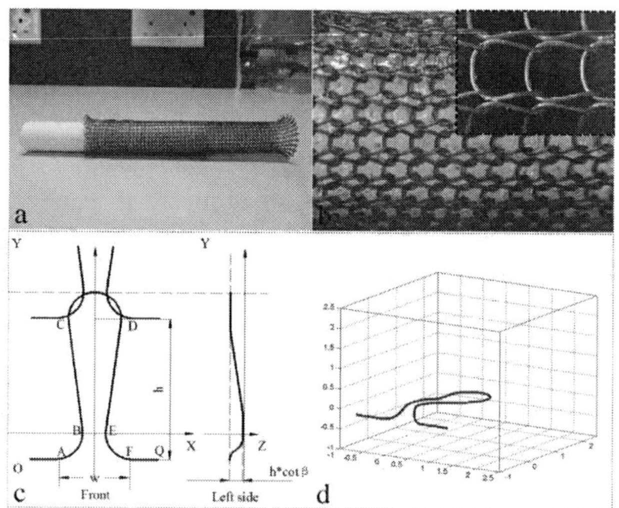

Figure 1. Images of the tubular stent after heat-setting, weft-knitted loops (a and b); the front view, A - A cross section of the loop (a) and a single simulated loop using Matlab 8.0 (d).

$$\text{OA}\quad \begin{cases} -1 \le x \le -0.5 \\ y = -0.4 \\ z = -2\cot\beta \end{cases}$$

$$\text{AB}\quad \begin{cases} x = \dfrac{w-d}{2}\cos\theta - \dfrac{w}{2} \\ y = -\dfrac{w-d}{2}\sin\theta \\ z = \dfrac{h\cot\beta}{2}\sin\left|\dfrac{1}{2} + \sin\theta\right|\pi - \dfrac{h\cot\beta}{2} \end{cases},\quad 0 \le \theta \le \dfrac{\pi}{2}$$

$$\text{BC}\quad \begin{cases} -\dfrac{w-d}{2} \le x \le -\dfrac{d}{2} \\ y = \dfrac{2h}{w-2d}\left|x+\dfrac{d}{2}\right| \\ z = \dfrac{h\cot\beta}{2}\sin\left|-\dfrac{2x+d}{w-2d}+\dfrac{1}{2}\right|\pi - \dfrac{h\cot\beta}{2} \end{cases}$$

$$\textbf{CD}\quad \begin{cases} x = \dfrac{w-d}{2}\cos\theta \\ y = \dfrac{w-d}{2}\sin\theta + h \qquad 0 \le \theta \le \pi \\ z = -M_1M_5 = -h\cot\beta \end{cases}$$

$$\textbf{DE}\quad \begin{cases} \dfrac{d}{2} \le x \le \dfrac{w-d}{2} \\ y = \dfrac{2h}{w-2d}\left|x-\dfrac{d}{2}\right| \\ z = \dfrac{h\cot\beta}{2}\sin\left|\dfrac{2x-d}{w-2d}+\dfrac{1}{2}\right|\pi - \dfrac{h\cot\beta}{2} \\ 0 \le \theta \le \dfrac{\pi}{2} \end{cases}$$

$$\textbf{EF}\quad \begin{cases} x = -\dfrac{w-d}{2}\cos\theta + \dfrac{w}{2} \\ y = -\dfrac{w-d}{2}\sin\theta \\ z = \dfrac{h\cot\beta}{2}\sin\left|\dfrac{1}{2}+\sin\theta\right|\pi - \dfrac{h\cot\beta}{2} \end{cases}$$

$$\textbf{FQ}\quad \begin{cases} 0.5 \le x \le 1 \\ y = -0.4 \\ z = -2\cot\beta \end{cases}$$

Where: d represents the diameter of a three dimensional yarn; h is the loop height; w is the loop distance; β impacts the thickness of the loop; θ is the viable angle.

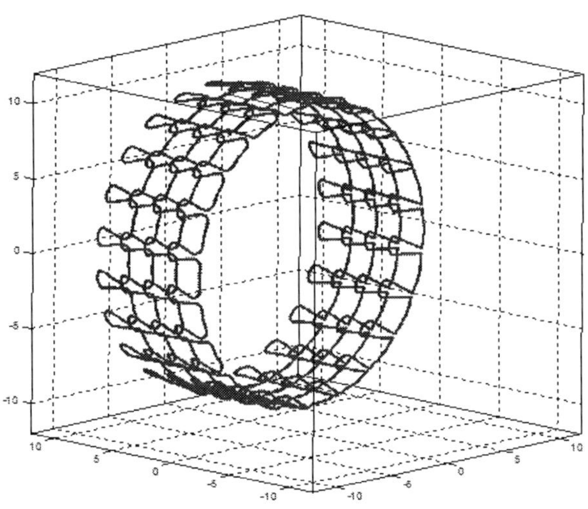

Figure 2. Geometry structural modeling of weft-knitted stent using Matlab 8.0.

The single knitting loop was obtained by connecting these different lines. Then all data points were generated using single loop model (d=0.2 mm, w=1 mm; h=2 mm; β=0.45pi). The method of transposed matrix will be used to generate other 21 knitted loops left. Finally, the geometry model of full stent was set up (Figure 2).

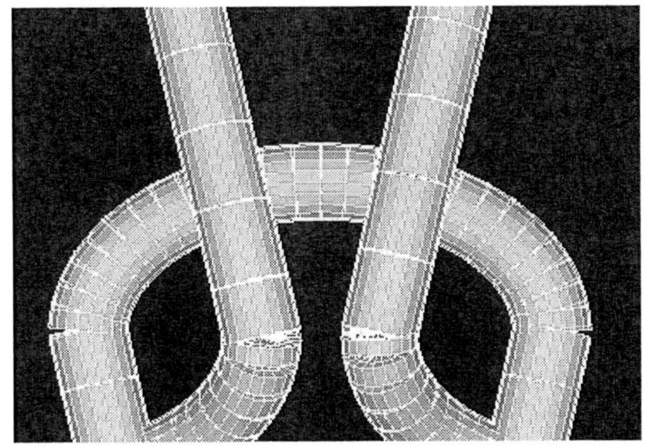

Figure 3. Stent loop simulation using ANSYS 13.0 software. Still some cracks can be seen at the connection sites. More data points should be input.

CONCLUSIONS

The geometry modeling for the stent was successfully conducted. With this mathematical model, it can provide a theoretical reference for other textile-forming structural biomedical stents.

FUTURE WORK

The investigation on biomechanical functions of stent by optimizing different stent structures and weft-knitting parameters of the stent will carried out using FEM.

ACKNOWLEDGMENT

The project was supported State Key Laboratory for Modification of Chemical Fibers and Polymer Materials, Donghua University with code LK1421, Guangdong Provincial Department of Science and Technology with project codes 2012B050800002 and Natural Science Foundation for Young Scholars of Soochow University.

REFERENCES

[1] André, T., Boni, C., Mounedji-Boudiaf, L., Navarro, M., Tabernero, J. *New England Journal of Medicine*, 350, 2004: 2343.

[2] Li, G., Lan, P., Li, Y., Hu, H. *J. Fiber. Bioeng. Inform.*, 3, 2011: 187.

[3] Li, G., Chen, Y., Hu, J., Wu, X., Hu, J., He, X., Li, J., Zhao, Z,. Chen, Z., Li, Y. *Biomaterials,* 34, 2013: 9451.

[4] Li, G., Li, Y., Lan, P., Li, J., Zhao, Z., He, X., Zhang, J., Hu, H. *J. Biomed. Mater. Res. A.*, 102, 2014: 982.

[5] Saito, Y., Tanaka, T., Andoh, A., Minematsu, H., Hata, K., Tsujikawa, T., Nitta, N., Murata, K., Fujiyama, Y. *Dig. Dis. Sci.*, 53, 2008: 330.

[6] Li, G., Li, Y., Lan, P., He, X., Hu, H. *Text. Res. J.*, 83, 2013: 2129.

[7] Sirry, M.S., Zilla, P., Franz, T. *Cardiovasc. Eng. Technol.,* 1, 2010: 269.

Fabrication of Multiscales Elastic Electrospun Fibrous Scaffolds for Repairing Augmentation of Rotator Cuff

Song Zhao, Wenguo Cui
Orthopedic Institute, Soochow University, Suzhou, Jiangsu, China
wgcui80@hotmail.com

Rotator cuff tear (RCT) is a common cause of shoulder pain and disability in middle and older age. Despite advances in imaging and surgery, the etiology is still undefined, and the high failure rate after repair is still problematic [1]. In humans, the rotator cuff often ruptures at the tendon-bone insertion site, which is transitioned by a well-defined zone composed of fibrocartilage. The application of scaffolds to promote flexible fibrocartilage regeneration at the tendon-bone insertion site may be an attractive strategy for minimizing stress concentration and improving the outcome of RC repairs.Fibrous scaffolds are commonly used in tissue engineering due to their characteristic high porosities and interconnected pores, which have been shown to facilitate cellular infiltration and homogenous tissue regeneration [2]. Electrospun fibrous scaffolds from a wide variety of synthetic or natural polymers are considered to be extremely promising candidates for the preparation of engineered tissue scaffolds, as they can constitute a temporary supportive network on which cells can regenerate new extracellular matrix (ECM) structures.Previous reports have indicated that the size/scale of the fibers in such a scaffold can have a significant effect on tissue development. Specifically, nanoscale features are desired because they more closely resemble native components of the extracellular matrix (ECM), such as collagen fibers [3]. It is also fairly well accepted that high concentrations of nanoscale fibers often lead to increased cell spreading and promote new tissue regeneration. However, nanofibers are difficult to mimic the bio-mechanic and micro-structure of tendon-bone insertion because of single fibers structure. For this reason, the logical progression is to fabricate scaffolds that combine microfibers with nanofibers in an effort to maintain larger pore sizes, while simultaneously harnessing the positive effects of nanofibers, such as increased cellular attachment and ECM production as well as improved differentiation.

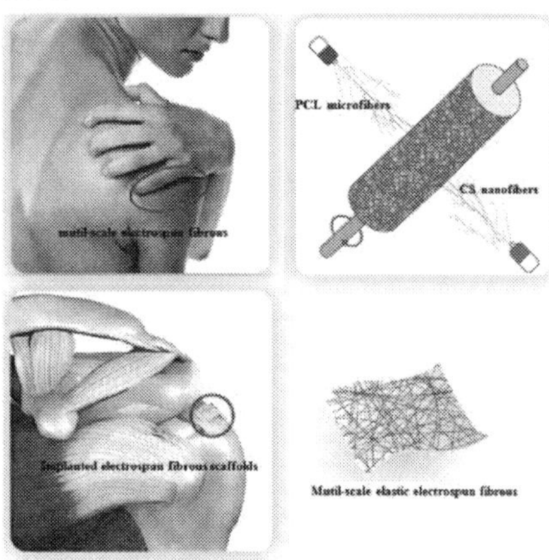

Fig. 1. Illustrations of multiscale fibers fabrication of PCL and CS fibrous scaffolds using a dual electrospinning process with a rotating metal mandrel and repairing augmentation of RCT.

Poly (ε-caprolactone) (PCL) has been used for both soft and hard tissue regeneration. PCL has various advantages, including mechanical flexibility, low antigencity, easy processability, and low degrees of chronic persistence[4]. Chitosan (CS), the N-deacetylated derivative of chitin which forms the exoskeleton of crustacean shells, has been widely used in biomedical applications [5]. The objectives of the present study were to demonstrate the feasibility of fabricating electrospun scaffolds consisting of two differently scaled fibers interspersed evenly throughout an entire construct, to fabricate scaffolds containing PCL and CS fibers, and to evaluate the extent to which augmentation of chronic RCTs can be repaired (Fig 1). The multi-scale electrospun fibrous scaffolds were characterized and evaluated by in vitro and in vivo results of cell proliferation and repair outcomes of chronic RCTs.

This study facilely fabricated multi-scale micro- and nano-fibrous elastic scaffolds using dual electrospinning for repairing rotator cuff tears (RCTs). Multi-scale fibrous scaffolds consisted of 1.16±0.27 μm Poly (ε-caprolactone) (PCL) fibers and 0.36±0.23 μm Chitosan (CS) fibers, were moderately hydrophilicity, had high water absorption and accelerated degradation, with controlling tensile strength. Local application of multi-scale fibers to the healing tendon-bone interface in a chronic RCT rat model was found to strengthen the healing enthesis, increase fibrocartilage area, and improve new bone formation and collagen organization compared with repair alone. RC regeneration using multi-scale PCL-CS is a promising treatment for RCT.

KEYWORDS: Rotator cuff tear, multi-scales fiber, electrospinning, micro and nano.

ACKNOWLEDGMENT
This work was supported by the National Natural Science Foundation of China (51373112 and 31370945).

REFERENCES

[1] Domb, B.G., Glousman, R.E., Brooks, A., Hansen, M., Lee, T.Q., El Attrache, N.S. *J. Bone Joint Surg. Am.*, 90, 4, 2008: 35-9.
[2] Hu, C., Cui, W. *Adv. Health Mater.*, 1, 2012: 809-14.
[3] Levorson, E.J., Raman Sreerekha, P., Chennazhi, K.P., Kasper, F.K., Nair, S.V., Mikos, A.G. *Biomed Mater.*, 8, 2013: 014103.
[4] Zhu, Y., Hu, C., Li, B., Yang, H., Cheng, Y., Cui, W. *Acta Biomater.*, 9, 2013: 8328-36.
[5] Sun, X., Cheng, L., Zhu, W., Hu, C., Jin, R., Sun, B., et al. *Colloids and Surfaces B, Biointerfaces*, 115C, 2013: 61-70.

Hyaluronic Acid-Functionalized Electrospun Random and Aligned PLA Nanofibers for Cancer Cell Capture

Yili Zhao[1], Karen Cheung[2], Frank K. Ko[3], Xiangyang Shi[1,4]
[1]Key Laboratory of Textile Science & Technology, Ministry of Education, College of Textiles, Donghua University, Shanghai, China
[2]Department of Electrical and Computer Engineering, University of British Columbia, British Columbia, Canada
[3]Department of Materials Engineering, University of British Columbia, British Columbia, Canada
[4]College of Chemistry, Chemical Engineering and Biotechnology, Donghua University, Shanghai, China
frank.ko@ubc.ca; xshi@dhu.edu.cn

INTRODUCTION

Electrospun nanofibers offer the advantage of strong interaction with cells by inducing topographical interactions with nanoscale cell components (Hou et al.; L. Zhao et al.). Hyaluronic acid (HA) shows high affinity to cancer cells over-expressing CD44 receptors (Li et al.). In this work, HA was conjugated to the surface of the electrospun random and aligned poly lactic acid (PLA) nanofibers for the specific capture of cancer cells.

EXPERIMENTAL

Formation of HA-modified random and aligned PLA nanofibers

PLA was dissolved into a mixed solvent of Chloroform/DMF (v/v = 7:3) at the concentration of 10 wt%. Random PLA nanofibers were fabricated at 0.5 mL/h, 16 kV and 18 cm. Aligned fibers were conducted using a drum electrospinning unit, Nanon (MECC CO., LTD) by the following parameters: 2000 rpm, 1.0 mL/h, 21 kV and 10 cm. The produced PLA nanofibrous mats were first covalently conjugated with PEI via EDC/NHS coupling reaction to introduce amine groups on their surfaces. Thereafter, HA was further covalently conjugated on the surface of PEI, followed by acetylation to neutralize the remaining PEI surface amines to form the R-HA-PLA-Ac (R = random) and A-HA-PLA-Ac (A = aligned) fibrous mats.

Characterization

SEM, FTIR, TGA, water contact angle, FBS uptake assay, hemocompatibility assay, static and dynamic cancer cells capture assay were conducted according to the procedure reported in our previous work (Y. Zhao et al.).

RESULTS AND DISCUSSION

Formation of HA-modified random and aligned PLA nanofibers

Fig. 1 shows that smooth and continuous PLA nanofibers with random and aligned orientation are able to be produced with a mean diameter of 773.1 and 711.0 nm, respectively. After the HA modification, the 3D porous structure does not appreciably change except that the diameters increase to 1014.4 and 1093.1 nm, respectively. The increased fiber diameter should be stemmed from the surface-grafted HA molecules.

As shown in Fig. 2, the major peaks at 1750 cm^{-1} in the spectra of both the pristine and HA-modified PLA nanofibers are attributed to the ester stretching of PLA. Both HA-modified nanofibers and HA have a peak at 1616 cm^{-1}, which can be assigned to the amide stretching bands of HA, confirming the successful modification of HA onto the fiber surfaces.

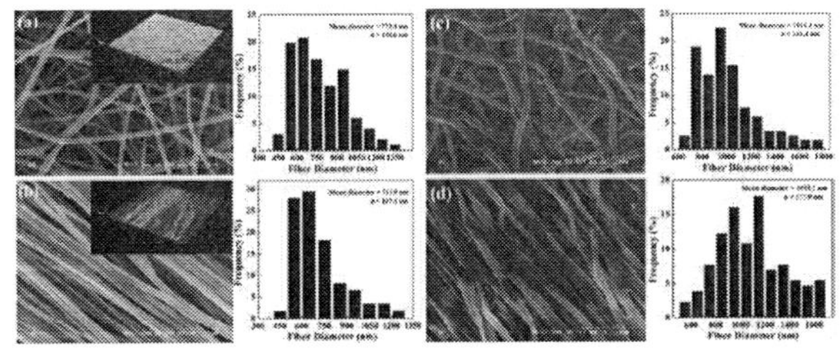

Fig. 1. SEM micrographs and diameter distribution histograms of (a) random, (b) aligned PLA nanofibers, (c) R-HA-PLA-Ac and (d)A-HA-PLA-Ac nanofibers.

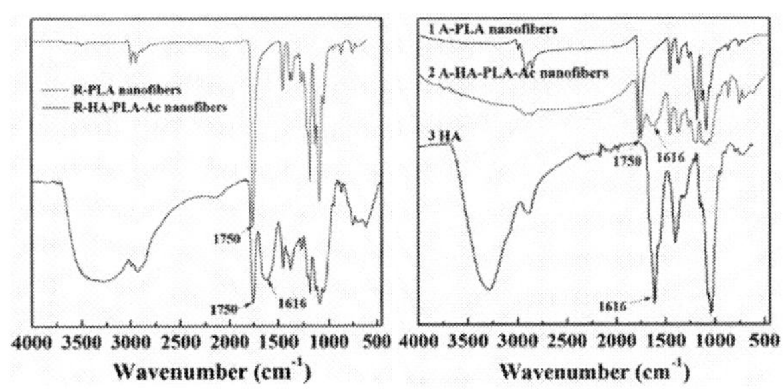

Fig. 2. FTIR spectra of pristine and HA-modified random and aligned PLA nanofibers.

TGA was used to characterize the fiber composition. HA shows an initial weight loss from 60°C to 150°C due to the evaporation of residually bound water, and at 220°C with a residual weight of 32.8% at 550°C. For PLA nanofibers, the weight loss started at approximately 320°C. However, the weight loss of R-HA-PLA-Ac and A-HA-PLA-Ac started at 220°C. The higher residual mass of R-HA-PLA-Ac (6.6 wt%) and A-HA-PLA-Ac (5.0 wt%) compared with that of PLA nanofibers (0.5 wt%) at 550°C could be correlated with the grafted HA.

For the surface property assay, the pristine random and aligned PLA nanofibers showed a contact angle of 128° and 116°, respectively, whereas the HA-modified nanofibers have decreased contact angles of 114° and 57°, respectively. Undoubtedly, this reduction was due to the presence of hydrophilic HA molecules on the PLA nanofiber surface.

FBS uptake assay

FBS assay shows that all the fibers exhibit higher FBS uptake ability than TCP due to their porous structure. The aligned nanofibers exhibited relatively high FBS sorption probably due to the wicking effect of their aligned structure. After modification, the FBS uptake of the PLA nanofibers increased 0.36-fold for random fiber from 336 % to 460 %

and 0.39-fold for aligned fiber from 497 % to 693 % after 1 h with the grafting of highly hydrophilic HA molecules. After 4 h, the FBS uptake ability increased and similar trend can be seen for all the samples.

Hemocompatibility assay

It is shown that no obvious hemolytic phenomenon can be observed except the positive control of water. Furthermore, hemolysis percentages of all the samples were calculated to be less than the threshold value of 5%, indicating that all the samples do not display appreciable hemolytic effect.

Static cancer cell capture assay

The cell capture efficiency of HA-modified nanofibrous mats at different time points was investigated (Fig. 3). It can be seen that with the capture time, the efficiency to capture MCF-7 cells and HEK293 cells increased. At a given time point, all the fiber samples exhibited significant higher efficiency than TCP control, which is due to their ECM mimicking structure. What's the most interesting is that, HA-modified aligned fiber mats first displays significantly higher MCF-7 cells capture ability than the unmodified random PLA fibers at 60 and 120 min, this should be due to the high FBS absorb ability of aligned nanofibers combined with ligand-receptor interaction. After 240 min, both R-HA-PLA-Ac and A-HA-PLA-Ac displayed significantly higher MCF-7 cells capture efficiency than non-targeted control group, this probably due to the fact that HA-mediated targeting began to play a main role. However, the efficiencies to capture HEK293 cells using HA-targeted PLA mat is almost equivalent to those using the non-targeted PLA mat at a given time point, indicating that the HA-modified nanofibrous mat in this case does not show specificity to HEK293 cells.

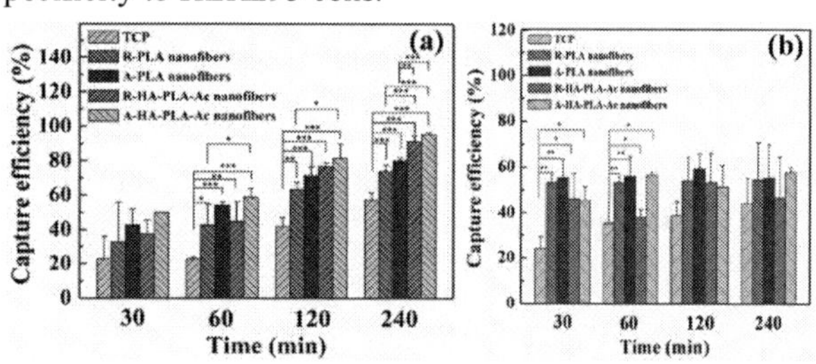

Fig. 3. The capture efficiency of (a) MCF-7 cells and (b) HEK293 cells using TCP, random, aligned PLA nanofibrous mats, R-HA-PLA-Ac and A-HA-PLA-Ac at different time points.

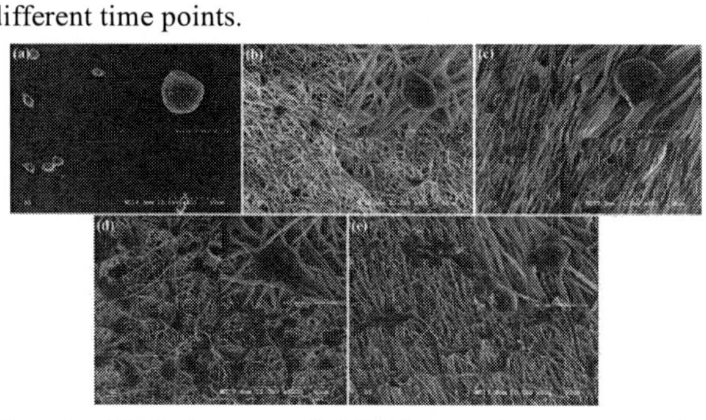

Fig. 4. SEM images of CD44 receptor-overexpressing MCF-7 cells captured on (a) TCP, (b) random PLA fibers, (c) aligned PLA fibers, (d) R-HA-PLA-Ac and (e) A-HA-PLA-Ac nanofibers after 240 min.

The specific cancer cell capture ability of the HA-modified nanofibers was also confirmed by SEM observation (Fig. 4). Coinciding to the quantitative cell counting assay data after 240 min, much more MCF-7 cells can be captured onto the HA-modified PLA nanofibrous mat than onto unmodified PLA mat.

Dynamic cancer cell capture assay

Similar to the results shown in static assay, the HA-modified PLA nanofibrous mat shows significant higher capture efficiency towards the CD44 receptor-overexpressing MCF-7 cells rather than HEK293 cells without CD44 receptor expression and peripheral blood mononuclear cells (Fig. 5).

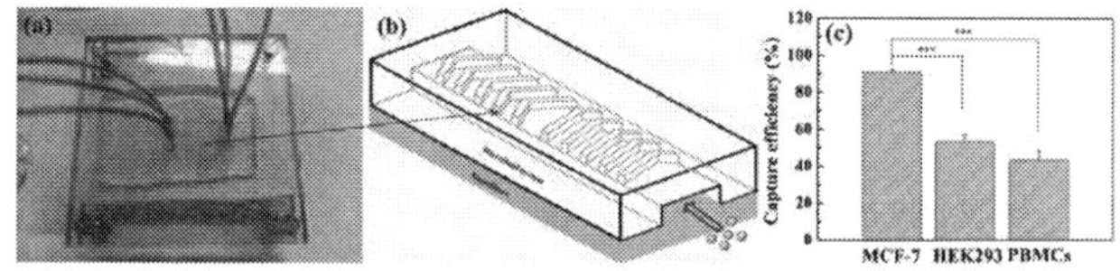

Fig. 5. (a) The chip consists of three inlets and exits, (b) the inner structure of the channels, (c) cells capture efficiency using R-HA-PLA-Ac nanofibers.

CONCLUSION

In summary, we developed a convenient approach to fabricating HA-modified electrospun PLA nanofibrous mats for targeted cancer cell capture applications. The formed HA modified aligned PLA nanofibers firstly displayed significant higher MCF-7 capture ability at the time point of 60 min and 120 min. At the time point of 240 min, both HA-modified random and aligned PLA nanofibers demonstrated higher MCF-7 capture efficiency. Therefore, the HA-modified PLA nanofibers may be applied for capturing circulating tumor cells for cancer diagnosis and therapy.

ACKNOWLEDGMENT

This research is financially supported by the Program for Professor of Special Appointment (Eastern Scholar) at Shanghai Institutions of Higher Learning, and the Innovation Funds of Donghua University Doctorate Dissertation of Excellence (101-06-0019014).

REFERENCES

1. Hou, Shuang, et al. "Polymer Nanofiber Embedded Microchips for Detection, Isolation, and Molecular Analysis of Single Circulating Melanoma Cells." *Angewandte Chemie, International Edition,* 52.12, 2013: 3379-83.

2. Zhao, Libo, et al. "High-Purity Prostate Circulating Tumor Cell Isolation by a Polymer Nanofiber-Embedded Microchip for Whole Exome Sequencing." *Advanced Materials,* 25.21, 2013: 2897-902.

3. Li, Jingchao, et al. "Hyaluronic Acid-Modified Fe 3 O 4@ Au Core/Shell Nanostars for Multimodal Imaging and Photothermal Therapy of Tumors." *Biomaterials,* 38, 2015: 10-21.

4. Zhao, Yili, et al. "Dendrimer-Functionalized Electrospun Cellulose Acetate Nanofibers for Targeted Cancer Cell Capture Applications." *Journal of Materials Chemistry B.*, 2.42, 2014: 7384-93.

Cellulose/Chitosan Composites Prepared in Ethylene Diamine/Potassium Thiocyanate for Adsorption of Cu^{2+}

Min Xiao, Jiancan Hu
Guangzhou Testing and Research Institute for Fibrous and Textile Products
gttxiaom@163.com

INTRODUCTION

Water pollution has become one of serious environmental concerns since heavy metals are widely used in many industrial fields and discharged in industrial effluents. Heavy metals tend to accumulate in the aquatic and terrestrial environments and eventually enter the food chains. Over the years a variety of techniques have been developed to remove toxic heavy metals in industrial wastewater, such as chemical precipitation, carbon adsorption, ion exchange, membrane filtration, reverse osmosis and biosorption. Among those techniques, biosorption based on biopolymers has recently gained increasing interest due to its low cost, biodegradability and high efficiency. It is known that chitosan has a natural affinity for heavy metal ions and has been applied as biosorbent materials in wastewater treatment[1]. However, chitosan is soluble in dilute mineral and organic acids and shows low mechanical strength and poor resistance to acid media[2]. Although chemical modification is often used to improve the chemical stability of chitosan under acidic conditions, the modification process involves the reaction of amino groups in chitosan molecules, which will decrease the number of available binding sites and lead to a decrease in metal adsorption capacity of chitosan[3]. An alternative approach is through formation of chitosan-based composites, such as utilization of cellulose as a blending polymer to reinforce the mechanical properties, since cellulose has high abundance, low cost, desirable mechanical strength as well as similar chemical structure with chitosan[4]. However, cellulose has extensive hydrogen bonding and high crystallinity that make it difficult to dissolve in most organic solvents. The dissolving solvents for chitosan are also very limited. In this paper, we explored a novel, cost-effective and recyclable solvent system, ethylene diamine/potassium thiocyanate (EDA/KSCN), in the formation of cellulose/chitosan composites. The composite materials were directly prepared in 70/30 (w/w) EDA/KSCN solvent system using a freeze-thaw method followed by coagulation in methanol. A series of batch adsorption experiments were conducted to investigate the effects of pH and metal ion concentration on the adsorption capacity of Cu^{2+} in acidic aqueous solutions together with the analyses of adsorption mechanisms and isotherms.

APPROACH

Preparation of cellulose/chitosan composites

100 grams of 70/30 (w/w) EDA/KSCN solution was chilled in a freezer at ca. -20°C for approximately 15 minutes. 5 grams of cellulose powder was immediately added into the chilled solvent system at room temperature to form 5wt% cellulose solution. Similarly, 5wt% chitosan solution was prepared using this method. The resultant cellulose and chitosan solutions were then mixed at a predetermined ratio by weight to form a mixture. The cellulose/chitosan composites were obtained by injecting the mixture through a syringe with a needle into a methanol coagulant bath. Removal of the residual solvent and salt was achieved by thoroughly washing the samples in distilled water until the pH of the residual water was neutral. The samples were freeze-dried under vacuum for subsequent characterization and adsorption studies.

Adsorption measurements

Batch adsorption measurements were conducted by introducing the prepared composites into Cu^{2+} solution and then soaking for 24h to reach adsorption equilibrium. The adsorption capacity at equilibrium was calculated by equation (1)

$$Q_e = (C_0 - C_e)V / W_d \quad (1)$$

where Q_e is the adsorption capacity of the composite at equilibrium (mmol/g); C_0 is the initial concentration of Cu^{2+} (mmol/L); C_e is the concentration of Cu^{2+} at equilibrium (mmol/L); V is the volume of the Cu^{2+} solution (L); W_d is the dry weight of the composite (g). Each adsorption measurement was taken in triplicate and the mean of three measurements was reported in this paper.

RESULTS AND DISCUSSION

Characterization of the composites

The cellulose/chitosan composites were obtained by cutting the as-spun fibers to small pieces ca. 1-3mm in length. Figure 1 shows that the composites were capable of removing Cu^{2+} ions from the acid solution.

Figure 1. Images of the composites (a) before and (b) after adsorption of Cu^{2+} ions.

Compared with the mixture of cellulose and chitosan, the absorption peak in the –OH and –NH stretching region (3200-3600 cm^{-1}) for the cellulose/chitosan composite in Figure 2 was broadened and shifted to a lower frequency at 3320cm^{-1}. This suggests that intermolecular interactions existed in the blends of cellulose and chitosan by forming hydrogen bonds between -OH groups of cellulose and –OH/-NH_2 groups of chitosan, which ensures the good compatibility of the components in the composite[5].

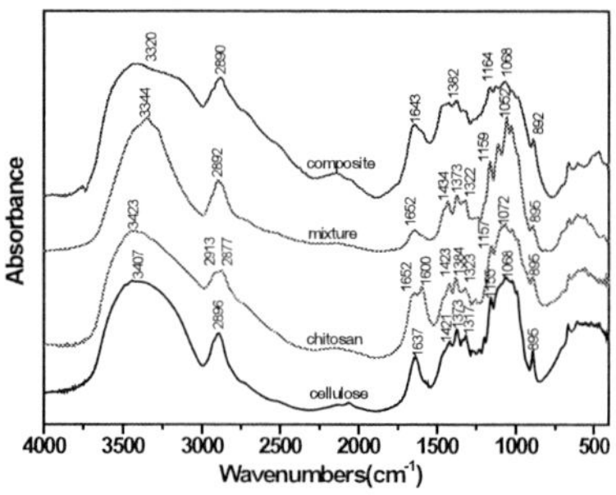

Figure 2. FT-IR spectra of cellulose, chitosan, mixture of cellulose and chitoan, and cellulose/chitosan composite.

X-ray diffraction studies reveal that the crystallinities of cellulose and chitosan were both reduced after dissolution, indicating that the EDA/KSCN solvent system is able to break the hydrogen bonding network in both cellulose and chitosan. The increase in the amorphous regions of the composite facilitated the diffusion of metal ion into the cellulose/chitosan composite.

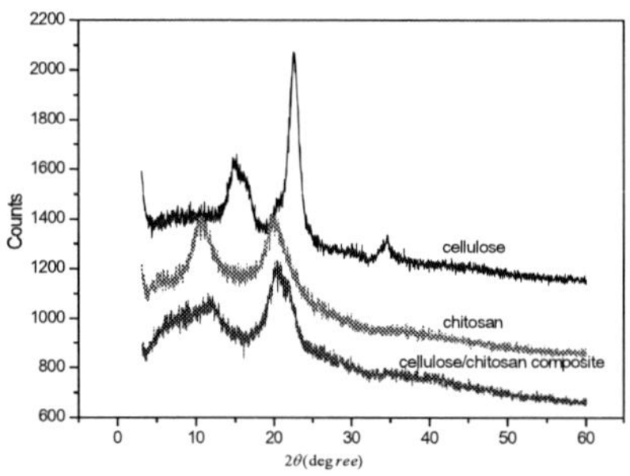

Figure 3. X-ray diffraction graphs of cellulose, chitosan, and cellulose/chitosan composite.

Effect of pH on adsorption

Figure 4 shows that the amount of Cu^{2+} adsorbed onto the composite increased with an increase in pH and the maximum metal uptake was achieved at pH 5.0. This could be attributed to the fact that the amino groups at the surface of the composite are readily protonated at low pH values (NH_3^+), thus inducing electrostatic repulsion effect between the composite surface and positively charged Cu^{2+}. When the pH is increased to weak acidity or close to neutrality, the induced electrostatic repulsion becomes less dominant and hence Cu^{2+} ions can closely approach the composite surface to complete metal fixation process.

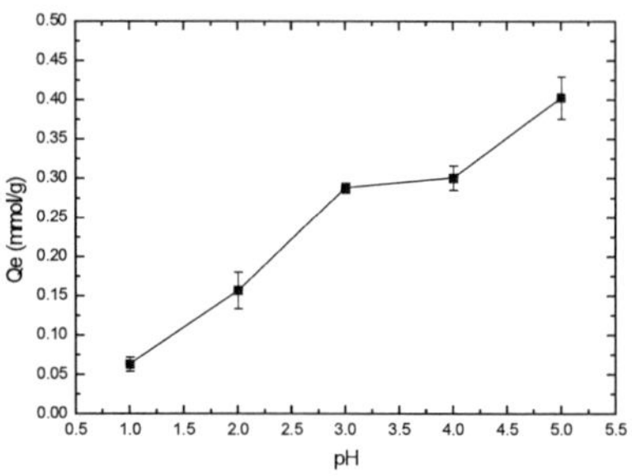

Figure 4. Effect of pH on the adsorption of Cu^{2+} onto the cellulose/chitosan composite. $C_0(Cu^{2+})$=3.15mmol/L.

Effect of metal ion concentration on adsorption

Figure 5 shows that the amount of Cu^{2+} adsorbed was significantly increased with an increase in the metal ion concentration. The driving force resulting from concentration gradient enhances the diffusion of Cu^{2+} ions from the bulk solution to the surface of the composite, hence giving rise to the high level of metal uptake. When the Cu^{2+} concentration is increased to a point where no more functional groups are available to bind Cu^{2+}, the adsorption capacity reaches a plateau.

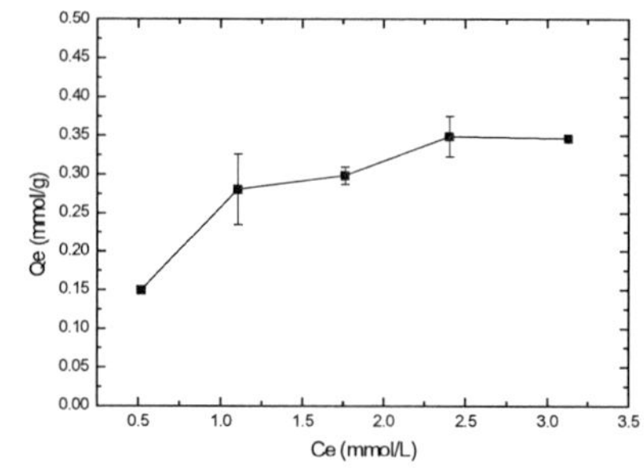

Figure 5. Effect of metal ion concentration on the adsorption of Cu^{2+} onto the cellulose/chitosan composite.

CONCLUSIONS

Cellulose/chitosan composites were successfully prepared in the EDA/KSCN solvent system. pH and metal ion concentration affected the adsorption capacity of Cu^{2+} onto the composite. The adsorption mechanisms were likely due to electrostatic interactions and chelation between the functional groups of the composite and the metal ions.

KEYWORDS

Cellulose, Chitosan, Composite, Adsorption, Heavy Metals

ACKNOWLEDGMENT

This work was financially supported by Guangzhou Testing and Research Institute for Fibrous and Textile Products and Administration of Quality and Technology Supervision of Guangdong Province (project number 2014ZZ09).

REFERENCES

1. Ravi Kumar, M.N. "A review of chitin and chitosan applications." *Reactive and Functional Polymers*, 46, 2000: 1-27.

2. Liu, Z., Wang, H., Liu, C., Jiang,Y., Yu, G., Mu, X., Wang, X. "Magnetic cellulose–chitosan hydrogels prepared from ionic liquids as reusable adsorbent for removal of heavy metal ions." *Chemical Communications*, 48, 2012: 7350-7352.

3. Guibal, E. "Interactions of metal ions with chitosan-based sorbents: a review." *Separation and Purification Technology*, 38, 2004: 3-74.

4. Guibal, E., Cambe, S., Bayle, S., Taulemesse, J.M., Vincent, T. "Silver/chitosan/cellulose fibers foam composites: From synthesis to antibacterial properties." *Journal of colloid and interface science*, 393, 2013: 411-420.

5. Miya, M., Iwamoto, R., Mima, S. "FT -IR stu intermolecular interactions in polymer blends." *Journal of Polymer Science: Polymer Physics Ed.*, 22, 1984: 1149-1151.

Nonwovens

Green Textile Manufacturing

Solvent-Free, Solid State Processed Tapes of UHMWPE: Influence of Molar Mass and Molar Mass Distribution on Tensile Properties

Sanjay Rastogi[1,2,3], Dario Romano[1], Sara Ronca[1], Giuseppe Forte[1], Yefeng Yao[4], Niek Tops[2]
[1]Department of Materials, Loughborough University, Leicestershire, England (UK); [2]Research Institute, Teijin Aramid B.V., Arnhem, The Netherlands; [3]Materials Science, Maastricht University, Maastricht, The Netherlands; [4]Shanghai Key Laboratory of Magnetic Resonance, East China Normal University, Shanghai, China
sanjay.rastogi@teijinaramid.com; s.rastogi@lboro.ac.uk; sanjay.rastogi@maastrichtuniversity.nl

SUMMARY

In normal practice, ease in processing comes at the expense of reduction in mechanical properties. Here we show that by controlled synthesis, it is possible to synthesize a wide range of linear Ultra High Molecular Weight Polyethylenes that can be stretched uniaxially into tapes. In these uniaxially drawn tapes the bundles of fibres align themselves in a preferred crystal plane orientation accounting for an extraordinary high tensile modulus. The stretching process is accomplished in the solid state, without any solvent requirement. The ease in processing combined with controlled synthesis provides a unique opportunity to follow the influence of molar mass on mechanical properties.

INTRODUCTION

Recently, we have shown that making use of a single-site catalytic system it is possible to control entangled state in a semi-crystalline UHMWPE, having number average molar mass greater than a million g/mol, which can be drawn more than 200 times in solid state below the equilibrium melting temperature (~142 °C) of linear polyethylene[1-10]. The high deformation ratio of the compressed nascent reactor powder results into tapes having tensile modulus approaching 200 GPa and tensile strength 4.0 GPa.[3] For the synthesis, in our laboratory, two single site catalytic systems, Fujita FI catalyst having titanium as the metallic centre (Figure 1a)[6] and Enders catalyst having chromium as the metallic centre (Figure 1b)[13], have been identified. The synthesis predominantly done in the homogeneous condition, in the presence of a solvent or solvent mixture, can be translated to heterogenous support where the support used can vary from nanoparticles to elongated wires for example carbon nanotubes[4]. This process not only provides homogeneous mixing of nanoparticles (or functionalities) in the intractable polymer matrix, but also offers possibilities of upscaling requirements for commercial purposes. The principles adopted for controlling the entanglement density, during synthesis, are mainly control of the polymerization and crystallization kinetics – earlier proposed by Smith et al while using a Vanadium catalyst[11].

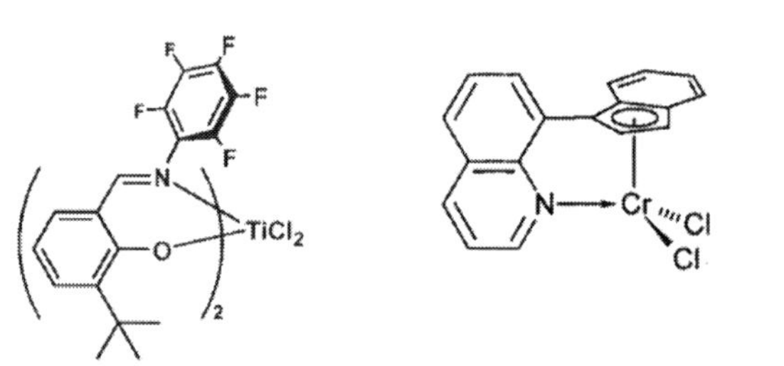

Figure 1: (a) Fujita catalyst with Titanium metallic center and (b) Enders catalyst with Chromium metallic center.

Characterization of Entangled State

To follow differences in the topological constraints, in the amorphous region of the semi-crystalline polymer and there influence on the resultant melt, solid state NMR, DSC and rheological studies have been performed. What follows is a brief overview of salient findings that have been of relevance in molecular understanding of the synthesised polymer.

Influence of Entanglements on Segmental Mobility of the Noncrystalline Region: To have insight on the entangled state of the polymer solid state NMR studies have been performed. Differences in the entangled and disentangled state become evident by following the chain diffusion of methylene segments from amorphous to crystalline regions. For the purpose use of ^{13}C NMR is made where the pulse sequence is tailored in a way to differentiate between the mobile and rigid components of the amorphous (or non-crystalline) and crystalline regions. For details the readers are requested to follow the references 5,7. Figure 2 highlights differences in the NMR spectra of the entangled and disentangled UHMWPE, synthesised using a conventional Z-N and single-site catalytic system, respectively.

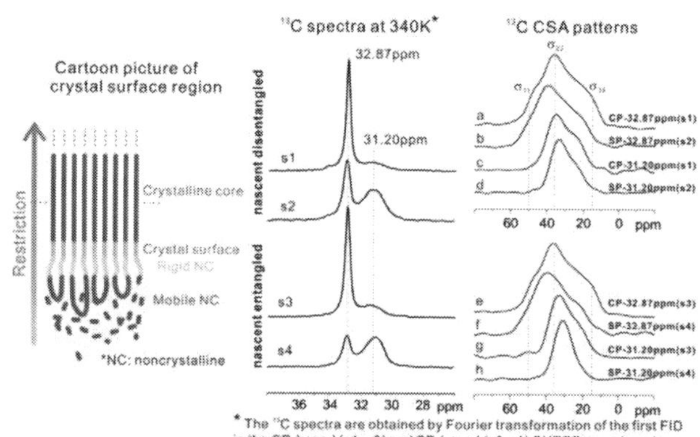

Figure 2: Left: schematic of crystal describing gradual transition of the geometrical restriction present in the sample; Middle: the ^{13}C spectra of the nascent disentangled and entangled samples obtained by Fourier transformation experiments: s1 and s3 are from the CP-based SUPER experiments of the disentangled and entangled samples; s2 and s4 are from the SP-based SUPER experiments of the disentangled and entangled samples, respectively. Right: the ^{13}C chemical shift anisotropy (CSA) powder patterns of the two nascent UHMWPEs

Melting Kinetics: In a series of publications we have also shown that melting behaviour of crystalline component strongly depends on entangled state of the amorphous region.[2,6,10] To recall, polymer melting that involves detachment of chains from crystal surface and their reeling in melt can occur either by consecutive detachment or in cluster, the latter involving more than seven chain segments. During slow heating or annealing at the onset of the melting temperature, the cluster melting is observed in the crystals having entangled amorphous region whereas consecutive detachment of chains is observed in disentangled polyethylene. The two different melting processes of the crystalline zones that are dependent on the chain topology in the amorphous region, become evident by following the kinetics involved in the change in the ratio of the low and high melting temperature peaks, where the former is associated with the melt crystallised component arising on crystallization of the molten part whereas the high melting temperature peak is related with the remainder of the nascent component that did not melt on annealing. The rate at which the high temperature peak transforms into the low temperature peak is related with the entanglement density and molar mass. Figure 3, reproduced from reference 6 shows an example of melting kinetics related to the entangled state of the polymer.

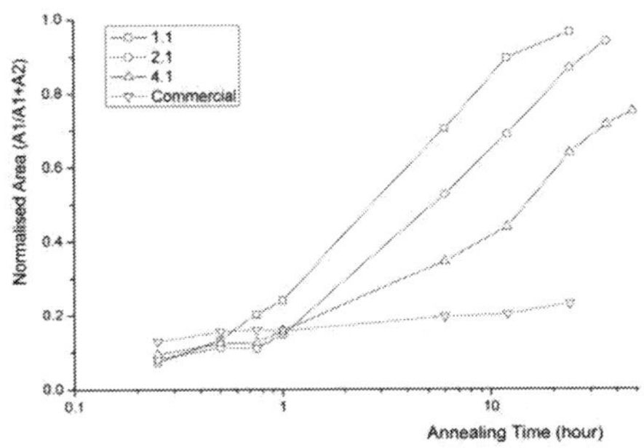

Figure 3: Evolution of the normalised area as a function of time for the synthesised polymers at three different pressures; unfilled squares, 1.1 atm (Mw ~9.9 x 10^6 g/mol; Mw/Mn ~ 2.6); unfilled circles, 2.1 atm (Mw ~15.3 x 10^6 g/mol; Mw/Mn ~ 3.0); unfilled up-triangles, 4.1 atm (Mw ~34.0 x 10^6 g/mol; Mw/Mn ~ 7.1); unfilled down-triangles, refers to a commercial Sigma-Aldrich UHMWPE for comparison.

Solid state deformation (biaxial): The disentangled UHMWPE provides opportunity to deform the compressed nascent powder not only uniaxially but also biaxillay, where the former can be used for applications in hard ballistics while the latter in membranes required for Lithium ion batteries or water purification or gas permeability and other demanding applications that require high isotropic high tensile and high modulus flexible films having thickness in the order of several microns. An example of the same is given in Figure 4.

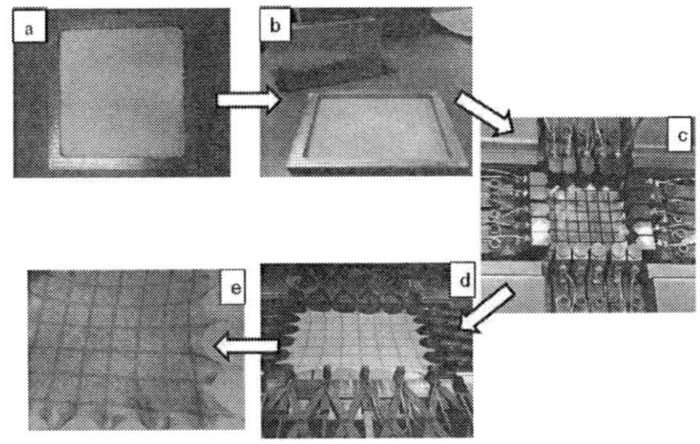

Figure 4: The figure shows biaxial deformation of compressed nascent powder. The powder depicted in figure a is compressed to make a film having thickness 225 microns shown in figure b. The film is mounted on Brukner biaxial machine and stretched 10 x 10 times, below the equilibrium melting point to get a film having metallic sound of several microns thickness. The film thus drawn, from the sample having molar mass 8 million g/mol, shows isotropic tensile strength in the region of 0.5 – 0.7 GPa. Permeability of the film can be varied by adding functionality such as hydroxyapatite or carbon nanotube or ethylene copolymer or graphene in the polymerization medium prior to the synthesis of disentangled UHMWPE. This route provides homogeneous dispersion of functionality in the intractable matrix of UHMWPE. Such a possibility is shown in the following section.

REFERENCES

[1] Rastogi, S., Lippits, D., Peters, G.W.M., Graf, R., Yao, Y., Spiess; H.W. *Nature Materials*, 4, 2005: 635-641.

[2] Lippits, D.R., Rastogi, S., Hohne, G.W., *Physical Review Letters*, 96, 2006: 218303-218307.

[3] Rastogi, S., Yao, Y., Ronca, S., Bos, J., Van der Eeem, *J. Macromolecules*, 44, 2011: 5558-5568.

[4] Andablo-Reyes, E.A., De Boer, E., Romano, D., Rastogi, S. *Journal of Rheology*, 58 (6), 2014: 1981-1986.

[5] Yao, Y., Jiang, S., Rastogi, S. *Macromolecules*, 47 (4), 2014: 1371-1382.

[6] Romano, D., Tops, N., Andablo-Reyes, E., Ronca, S., Rastogi, S. *Macromolecules*; 47 (14), 2014: 4750-4760.

[7] Yao, Y., Rastogi, S., Xue, H.J., Chen, Q., Graf, R., Verhoef, R. *Polymer*, 54 (1), 2013: 411-422.

[8] Ronca, S., Forte, G., Tjaden, H., Yao, Y., Rastogi, S. *Polymer*, 53, 2012: 2897-2907.

[9] Pandey, A., Champouret, Y., Rastogi, S. *Macromolecules*, 44 (12), 2011: 4952-4960.

[10] Pandey, A., Toda, A., Rastogi, S. *Macromolecules*, 44 (20), 2011: 8042-8055.

[11] Smith, P., Chanzy, H.D., Rotzinger, B.P. *Journal of Material Science*, 22, 1987: 523-531.

[12] Porter, R.S., Kanamoto, T., Zachariades, A.E. *Polymer*, 35, 23, 1994: 4979-4984.

[13] Romano, D., Ronca, S., Rastogi, S. *Macromolecular Rapid Communications*, 2014. DOI: 0.1002/marc.201400514.

Evolution of Chain Entanglements of Ultrahigh Molecular Weight Polyethylene in Dilute Solution Blends During Thermal Annealing and Its Influence on Crystallization

Wei Li, Jie Xu, Peng Chen, Chuanzhuang Zhao, Zhong-Ren Chen
Ningbo Key Laboratory of Specialty Polymers, Department of Polymer Science and Engineering, School of Material Science and Chemical Engineering, Ningbo University, Ningbo, Zhejiang, China
liwei@nbu.edu.cn

ABSTRACT
A new method to investigate the evolution of chain entanglement of ultrahigh molecular weight polyethylene (UHMWPE) in dilute solution blends was reported. The systems prepared from blending polyethylene wax and UHMWPE. Component one had a molecular weight (M_w=1000 g/mol) slightly less than the critical entanglement molecular weight (M_e=4ρRT/5G_N). Component two had an ultrahigh molecular weight (M_w =4500 000 g/mol). The fraction of component two was 2.5 wt%. Such a blend was referred as the dilute solution blend. The evolvement of chain entanglement during annealing was monitored by rheology. The morphology of the crystallized blends after different stage of annealing was investigated through scanning electron microscope, differential scanning calorimeter, small angle X-ray scattering and rheology, respectively. It was showed that the blends would transfer from heterogeneous melt to homogenous melt with the increasing of annealing time. The entanglement density first decreased and then increased to thermodynamic stable state upon annealing. Finally, the mechanism for the transition of melts was discussed.

KEYWORDS
UHMWPE; dilute solution; blends; entanglements

ACKNOWLEDGMENT
Prof. Dr. Julie A. Kornfield is gratefully acknowledged for her helpful discussion. Funding from the Project of Natural Science Foundation of China (No. 21206078), the Natural Science Foundation of Zhejiang Province (LQ12B06003), Zhejiang Province Department of Education Fund (Y201223797), the Key Innovation Team of Zhejiang Province (2011R50001), Ningbo Key Laboratory of Specialty Polymers 2014A22001, K. C. Wong Education Foundation, and K. C. Wong Magna Fund, Hong Kong, are gratefully acknowledged. We thank Shanghai Synchrotron Radiation Facility (SSRF) for supporting SAXS tests.

Polypropylene (PP) Polymer and Its Nonwovens in Medical Applications

Jamil A. Siddiqui
TASNEE–NIPRAS Plastics Research Center, Industrial Support Area, Jubail–Kingdom of Saudi Arabia
j.siddiqui@tasnee.com

ABSTRACT

Polypropylene (PP) is the first stereo regular polymer to have achieved industrial importance and it is presently, the fastest growing fiber for technical end-uses where high tensile strength fastened with low-cost are essential features. The fibers from polypropylene (PP) polymer were introduced to the textile industries in late 50s and become an important member of the rapidly growing family among the synthetic fibers. Today, polypropylene (PP) possesses fourth position behind the big three fiber classes i.e. polyester, rayon and acrylics. A large volume of PP utilized in the market segment known as fibers and fabrics. Polypropylene (PP) fiber is utilized in a host of applications including raffia/slit-film, tape, strapping, (BCF) bulk continuous filament, (SF) staple fibers, continuous filament and spun bond as in nonwovens fabric.

Polypropylene (PP) polymer is also the material of choice in medical applications as well not only because of it is light weight, low cost or both but also due to high chemical and bacterial resistance. For polypropylene (PP) polymer, it is possible to formulate in PP compounding to enable them to undergo steam and radiation sterilization. The most notable such application is the disposable syringes. Polypropylene polymer fulfills the essential need of a medical syringe extremely well. It can be precision-molded to yield the proper dimensions; it is clear enough to view the contents; it can be printed if necessary and can be irradiation-sterilized.

Medical vials, diagnostic devices, petri dishes, intravenous bottles, specimen bottles, food trays, pans, and amber-colored small containers that hold pills and capsules are also of polypropylene (PP) polymer products. 'Sharp containers' used to collect used syringes and other contaminated medical wastes are also the product of PP.

The other unique feature of polypropylene (PP) polymer makes it ideal for others medical use is its ability to be processed into nonwovens fabrics. Polypropylene (PP) has been used in Hernia and Pelvic organ prolapse repair operations to protect the body from new hernias in the same location. A small patch of the material is placed over the spot of the hernia, below the skin, and is painless and rarely, if ever, rejected by the body.The use of nonwovens in medical application is large and growing.

Nonwovens fabrics account for more polypropylene (PP) usage than any other single fiber application. Nonwovens fabrics are used in several areas like, liners in disposable diapers, feminine hygiene products, and medical applications. Polypropylene monofilament used in making ropes and twines, PP rope and twine are very strong and moisture resistant very suitable for marine applications.

In this paper, an overview of main nonwovens processes e.g. spunbond, spunlace, needlepunch, carded, airlaid and wetlaid along with applications of nonwovens in medical would be generally discussed.

KEYWORDS

Nonwovens, medical application, polypropylene

Investigation of Hydrostatic Pressure and Blood Penetration of a Nonwovens Material

Peter P. Tsai[1,2], Yurong Yan[2]
[1]The University of Tennessee, Knoxville, USA; [2]South China University of Technology, China
ptsai@utk.edu

ABSTRACT
Disposable materials receive an annually steady growth in the field of medical applications as surgical gowns, isolation gowns, and medical drapes to avoid the cross contamination from the repeated use of a durable protective apparel in a hospital. There are four levels of liquid barrier performance according to ANSI/AAMI PB70:2012[1] in terms of liquid impact penetration, hydrostatic pressure, and blood penetration. SMS (spunbond, meltblown, spunbond) materials and porous membranes are widely used as medical protective clothing. SMS materials are usually made to achieve up to Level 3, which requires a hydrostatic pressure of greater than 50 cm. Above that, Level 4 requires the liquid barrier property that resists to the penetration by blood, in which a porous membrane is employed to laminate with a substrate. The objective of this paper is to use the pore size to estimate the hydrostatic pressure, and further to correlate the hydrostatic pressure with the resistance to blood penetration. Theories were successfully applied to calculate the pore size and the hydrostatic pressure of a fibrous material and to correlate the hydrostatic pressure with the blood penetration. The requirement of the fiber fineness capable of resisting to the penetration of blood was also calculated.

RESEARCH MATERIALS AND EXPERIMENTS
Five materials - a SMMMS fabric, three MB (MB1, MB2, and MB3) fabrics, an electrospun (ES) fabric, an ePTFE porous membrane, and a PE film - that are potentially to be or have been used for protecting apparel were chosen for this research. The SMMMS is a 35g commercial medical protective clothing, which has a total weight of 8g in the MMM layers. The MB1 and MB2 have basis weight of 8g each while MB3 has a basis weight of 20 gsm, all made from the Reifenhauser's 24" MB pilot line at TANDEC. They were made to have different fiber diameters for this research purpose. MB1 had a fiber diameter of 0.98 μm, MB2 1.2 μm, and MB3 1.82 μm calculated from the equations and the master curves derived by Tsai[2]. The ES fabric is a PTFE material made by Zeus. The porous ePTFE was supplied by Zeus as well. The PE film is a commercial material used as the back sheet of a diaper.

RESULTS AND DISCUSSION
Listed in Table I are the results of air permeability, the hydrostatic pressure, and the blood penetration of the materials used in this research.

Table I. Results of Materials Intended for Protective Clothing.

Sample	Air Perm (ft3/ft2/min) @1/2")	Hydrostatic (cm)	Blood Pene. (Pass/Fail)
ePTFE	1.67	805	Pass
SMMMS	65	52	Fail
MB1	43	110	Fail
PE Film	0.0072 (0)	474	Pass
ES Fabrics	9.96	395	pass

The materials for this research are all highly hydrophobic and their blotter weight gain of the water penetration by impaction was zero.

The air pressure used to test the blood penetration was 2 psig, which is 143 cm of H_2O. The synthetic blood had a surface tension of 42 dynes/cm, which is lower than that of pure H_2O at 72 dynes/cm. Theoretically, in order to pass the blood penetration test, the hydrostatic pressure needs to be greater than 245 cm calculated from the hydrostatic pressure of 143 cm and the viscosity difference between water and the synthetic blood. The SMMMS and the MBs failed the blood penetration test as shown in Figure 1. However, they successfully passed Levels 1, 2, and 3. The microfiber MB1 fabric could theoretically pass the blood penetration test if its fiber diameter was reduced to 0.45 μm, which can be barely achieved by an MB process but can be achieved by an ES process.

Figure 1. A "Fail" test of blood penetration.

It was reported[3] that pore size (D) is proportional to the fiber diameter (d_f) and inversely proportional to a function of packing density (α),

$$D = \frac{d_f}{g(\alpha)} \quad (1)$$

where

$$g(\alpha) = 1/\left(\frac{1}{2(1-\alpha)^2 \alpha^{1.5}(1+56\alpha^3)}\right)^{0.5} \quad (2)$$

It is known[3] that the resistance to a fluid pressure through a pore is inversely proportional to the pore size,

$$P = \frac{4\gamma}{D} \quad (3)$$

where (γ) is the surface tension of the fluid. By combining Eqs. (1) and (3), the resistance to a fluid pressure through a pore is

$$P = \frac{4\gamma}{d_f} g(\alpha) \quad (4)$$

The hydrostatic pressure of MB1 calculated from its fiber diameter of 0.98 μm was 113 cm while its measured hydrostatic pressure was 110 cm. Shown in Table II, for a makeup sample, it requires a fiber diameter of 0.45 μm or lower for equal packing density to achieve a hydrostatic pressure of 245 cm that passes the blood penetration test. Maximum pore, which is usually twice the size of a mean pore size[1], was used to calculate the resistance to penetration of a liquid at a certain pressure.

The ePTFE, the ES fabric, and the porous PE film had measured hydrostatic pressures of 395, 805, and 474 cm, respectively, as shown in Table I, which were higher than the hydrostatic pressure of 245 cm required to pass the blood penetration test. Therefore, they all passed the blood penetration test as shown in the same table.

The e-spun PTFE passed the blood penetration test and had much higher air permeability than those of ePTFE and the PE film. The PE film had very low, nearly zero, air permeability.

Table II. Theoretical and Experimental Hydrostatis Values.

Sample	Fiber Dia. (μm)	Mean Pore (μm)	Max. Pore (μm)	Hydrostatic Pressure Theo(cm)	Ex[(cm)
MB1	0.98	12.7	25.4	113	110
MB2	1.2	15.6	31.2	92	90
SMMMS	2.1	27.7	55.4		52
Imaginary	0.45	5.9	11.8	245	

The MB2 fabric had a calculated fiber diameter of 1.2 μm, in which theoretical hydrostatic pressure was calculated to be 92 cm in comparison with its experimental hydrostatic pressure of 90 cm. The theoretical hydrostatic pressure values agreed very well with the experimental ones for MB1 and MB2 fabrics as listed in Table II. In the SMMMS sample, the fiber diameter was calculated as 2.1 μm from its measured hydrostatic pressure of 52 cm.

KEYWORDS: surgical gowns, surgical drapes, protective clothing, hydrostatic pressure, blood penetration, SMS, meltblown, membranes

REFERENCES

1. Liquid barrier performance and classification of protective apparel and drapes intended for use in health care and facilities, *Association for the Advancement of Medical Instrumentation*, ANSI/AAMI PB70: 2012.
2. Tsai, P.P. "Theoretical and Experimental Investigation on the Relationship Between the Nonwoven Web Structure and Web Properties." *International Nonwovens Journal,* Winter, 2002: 33-36.
3. Tsai, P.P., Yan, Y.Y. "The Influence of Fiber and Fabric Properties on Nonwoven Performance." *Applications of Nonwovens in Technical Textiles*, Woodshead Publishing Limited, ISBN978-1-84569-437-1, 2010:18-45.

Nanofibrous Materials

High-Performance Fibers and Composites

Chitosan Nanofibers for Cr (VI) Adsorption

Lei Li[1, 2], Yanxiang Li[1], Lixia Cao[1], Chuanfang Yang[1]

[1]Key Laboratory of Green Process and Engineering, Institute of Process Engineering, Chinese Academy of Sciences, Beijing, China

[2]University of Chinese Academy of Sciences, Beijing, China

cfyang@ipe.ac.cn

Chitosan is a natural polymer with abundant functional groups, which is nontoxic and biodegradable. Chitosan has been used as an efficient adsorbent for removal of heavy metals. However, chitosan powders and flakes have lower specific surface area which limits their application in some extent. Electrospinning is an efficient technique for the fabrication of nano and submicron-sized polymer fibers ranging from 5 to 500 nm, 10^2 to 10^4 times smaller than those prepared by traditional method of solution or melt spinning. By electrospinning technology, chitosan nanofibers with large surface area can be obtained for enhanced adsorption of chromium (VI) from water.

In this work, 90 wt% acetic acid was used as the solvent to prepare electrospun chitosan nanofibers with optimized conditions. The as-prepared chitosan nanofibers 75 nm in diameter were cross-linked by glutaraldehyde vapor to minimize the material's solubility in aqueous solutions. The cross-linked chitosan fibers were used to adsorb chromium (VI) from water directly. The effect of contact time, pH, and co-ions on the adsorption was studied, the adsorption kinetics and isotherms were also determined. Furthermore, the mechanism of adsorption was investigated with Fourier transform infrared spectroscopy (FT-IR) and X-ray photoelectron spectrophotometer (XPS). We found that the chitosan nanofibers indeed gave higher adsorption capacity than other forms of materials based on chitosan, and the adsorption was not affected by other anions and cations studied except for SO_4^{2-}. It was revealed that both the amino and the hydroxyl groups of chitosan were involved in the adsorption.

This work provided the foundation for further application of the chitosan nanofibers for tap water decontamination, where the concentration of Cr (VI) are relatively low, and where bacteria may still present a concern that can be handled by chitosan due to its anti-bacteria properties.

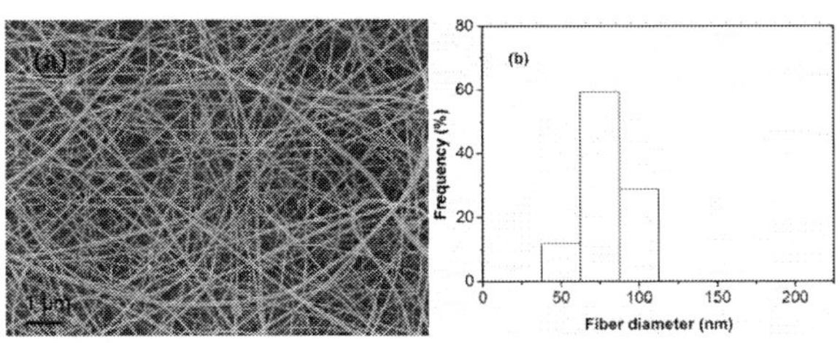

Fig. 1. SEM image (a) and fiber size distribution (b) of electrospun chitosan nanofibers.

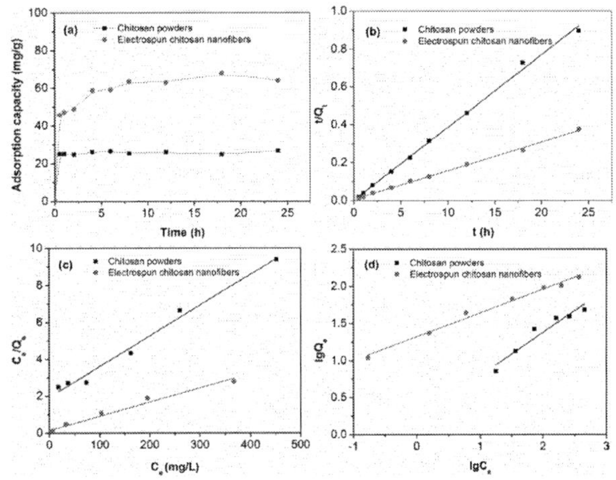

Fig. 2. (a) The sorption curves and (b) Pseudo-second order sorption kinetics for Cr (VI) sorption; (c) Langmuir plot and (d) Freundlich plot for Cr (VI) adsorption.

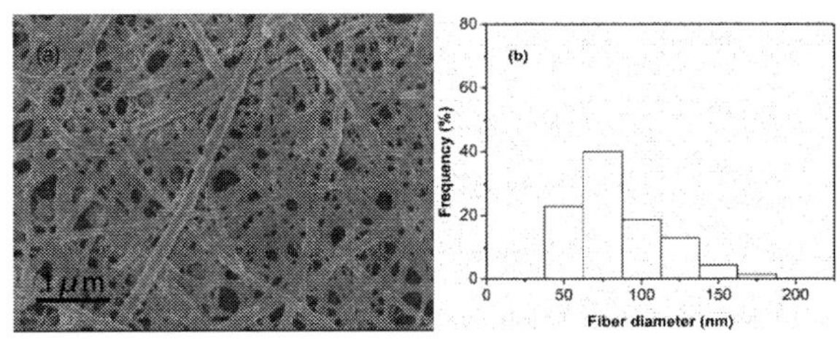

Fig. 3. SEM image (a) and fiber diameter distribution (b) of electrospun chitosan nanofibers after adsorption of Cr (VI).

REFERENCES

1. Desai, K., Kit, K., Li, J., Zivanovic, S. "Morphological and surface properties of electrospun chitosan nanofibers." *Biomacromolecules*, 9 (3), 2008: 1000-1006.
2. Desai, K., Kit, K., Li, J.J., Davidson, P.M., Zivanovic, S., Meyer, H. "Nanofibrous chitosan non-wovens for filtration applications." *Polymer*, 50 (15), 2009: 3661-3669.
3. Geng, X., Kwon, O. H., Jang, J. "Electrospinning of chitosan dissolved in concentrated acetic acid solution." *Biomaterials*, 26 (27), 2005: 5427-5432.
4. Haider, S., Park, S.-Y. "Preparation of the electrospun chitosan nanofibers and their applications to the adsorption of Cu (II) and Pb (II) ions from an aqueous solution." *Journal of Membrane Science*, 328 (1-2), 2009:90-96.

Gel Spinning of Nylon 6 and PVA

Richard Kotek[1], Amit Gupta[2]
[1]College of Textiles, North Carolina State University, Raleigh North Carolina, USA; [2]SABIC, Bangalore, India
rkotek@ncsu.edu

INTRODUCTION

Over the past many years, numerous advances have been made in the development of polymeric fibers for high performance engineering applications. To date, two major research directions have been taken to produce such fibers: 1) processing of rigid rod or liquid crystalline polymers like Zylon® PBO, Vectran®, Kevlar®, and M5 and, 2) processing of flexible chain polymers into highly oriented structures like polyesters, polyamides, and polyolefins. In the second class most of the techniques use multiple drawing stages in order to achieve high levels of orientation such as solid-state extrusion, solid-state co-extrusion, solution and gel spinning, and zone drawing and annealing techniques. It is required that a mechanism or means of relaxation must be sought which allows for drawing to occur in later stages.

In this respect gel spinning has proved to be an important technique as it allows the processing of an ultra-high molecular weight polymer into fibers. There is already ample evidence in the literature about the successful use of this technique to form gel spun ultra-high molecular weight polyethylene (UHMW-PE) fibers and for some other semi-crystalline polymers as well, such as polyvinyl alcohol and isotactic polypropylene. However, there is no evidence for the use of gel spinning of nylon 6 to produce fibers with improved mechanical properties. The inability to ultra-draw nylons is largely due to the presence of intermolecular hydrogen bonding between adjacent amide groups. Hydrogen bonding gives polyamides high melting points, but also limits the possibility of developing highly oriented structures with improved mechanical properties. Several attempts have been made till date to overcome this barrier with little or no success. In this research we attempt to control hydrogen bonding in Nylon 6 and PVA to develop high performance fibers via the gel-spinning approach.

MATERIALS

In the present work we have used two different types of nylon 6 polymers differing from each other in terms of their molecular weight. The first one is high viscosity (HV) nylon 6 purchased from BASF® Corp. (trade name ULTRAMID@ B40 01) and hereafter will be referred to as N6 HV. The second one is ultra-high molecular weight nylon 6 polymer prepared in our laboratory here at NC State University, College of Textiles by Dr. Anushree Mohan working alongside with us on this project. It was prepared via ring opening anionic polymerization and we will refer to it as N6 AP.

RESULTS AND DISCUSSION

In this study, the thermoreversible gelation of two hydrogen bond forming polymers namely, nylon 6 and atactic-polyvinyl alcohol (a-PVA) in N-methyl pyrrolidinone is investigated. Upon cooling down to room temperature, the hot solutions of nylon 6 and a-PVA in N-methyl pyrrolidinone form gels. This process is thermoreversible and clear solutions are obtained when the gels are heated again. From the phase diagrams of the two polymers it was evident that the gel melting and gel forming temperatures are concentration dependent and increased with increasing polymer concentration. Fibers were spun from these gels and subsequently hot drawn to different draw ratios. In this paper, we studied two different molecular weights for nylon 6 and one ultra-high molecular weight a-PVA polymer. The influence of draw ratio on the mechanical properties, thermal properties, overall and crystalline orientation of the fibers was investigated. Tensile strength, modulus, birefringence and crystalline orientation factor all increased with increasing draw ratio for both nylon 6 as well as PVA. For the high molecular weight nylon 6, we were able to produce fibers with tensile strength up to 1.23 GPa and modulus up to 25 GPa at a high draw ratio of 8.97. Similarly, PVA fibers with very high strength of up to 1.82 GPa and modulus up to 40 GPa have also been reported.

CONCLUSION

The thermoreversible gels of nylon 6 and PVA in N-methyl pyrrolidinone were extruded into homogeneous filaments and subsequently hot drawn after removal of NMP to yield fibers with improved mechanical properties. High molecular weight nylon 6 fibers could be drawn to a high draw ratio of 8.97 and gave a maximum strength and modulus of up to 1.23 and 25 GPa respectively. PVA fibers on the other hand had even better properties than nylon 6 fibers owing to lesser hydrogen bonding and more drawability. The maximum strength and modulus for PVA fibers were 1.82 and 40 GPa respectively. The results of DSC experiments indicated that a very low amount of crystallinity in the as spun fibers is beneficial for obtaining good mechanical properties upon hot drawing in later stages. Overall and crystalline orientation, crystal size and perfection all improved with increasing draw ratio for both nylon 6 as well as PVA fibers. Gel spinning provides a possible and convenient route to process such high molecular weight polymers to form fibers with improved mechanical properties.

KEYWORDS

Gel spinning, nylon 6, PVA, gelation

ACKNOWLEDGMENT

We appreciate NTC for funding this project. The authors of this paper would like to acknowledge Dr. Alan Tonelli and his PhD student Anushree Mohan (both in Fiber and Polymer Science, College of Textiles, NCSU, Raleigh, NC) for providing the high molecular weight nylon 6 polymer and for all the valuable suggestions. We thank Kuraray, Co. for a gift of high molecular weight PVA.

REFERENCES

1. Fukae, R., Maekawa, A., Sangen O. "Gel-spinning and drawing of gelatin." *Polymer*, 46, 2005: 11193–11194.
2. Cunniff, P.M., Auerbach, M.A., Vetter, E., Sikkema, D.J. "High-performance M5 fiber for ballistics structural composites." Online, March 7, 2015, at https://www.google.com/webhp?tab=ww&ei=BjflU7T9HMissQTss4HIAg&ved=0CBYQ1S4#q=HIGH+PERFORMANCE+%E2%80%9CM5%E2%80%9D+FIBER+FOR+BALLISTICS.
3. Kaplan, D.L., Adams, W.W., et al. *Silk Polymers: Materials Science and Biotechnology.* Washington, D.C.: American Chemical Society, 1994.
4. Vollrath, F., Knight, D.P. "Liquid crystalline spinning of spider silk." *Nature*, 410, 2001: 541-549.
5. Seidal, A., Liivak, O., et al. "Regenerated Spider Silk: Processing, Properties, and Structure." *Macromolecules*, 33, 2000: 775-780.
6. Predecki, P., Statton, W.O. "Extrusion of high density polyethylene below its melting point." *J. Polym. Sci., Polym. Lett. Ed.*, 10, 1972: 87-90.
7. Zachariades, A.E., Porter, R.S. "Ultradrawing of amorphous polymers by coextrusion illustrated with atactic polystyrene." *J. Appl. Polym. Sci.*, 24, 1979: 2137-2146.
8. Wijnenburg, A., Pennings, A.J. *J. Colloid Polym. Sci.*, 1975: 253, 452.
9. Smith, P., Lemstra, P.J. "Ultrahigh-strength polyethylene filaments by solution spinning/drawing" *and* "Influence of solvent on the drawability." *Makromol. Chem.*, 180, 1979: 2983-2986.
10. Kunugi, T., Suzuki, A., Akiyama, I., Hashimoto, M. *Polym Prepr.*, Am. Chem. Soc., 20, 1979: 778.
11. Hearle, J.W.S. *High-Performance Fibers.* Cambridge, 2001.
12. Smith P., Lemstra P.J. UK Patent 2051661, 1979.

Figure 1. Digital images for 20% N6 HV in NMP for cooling.

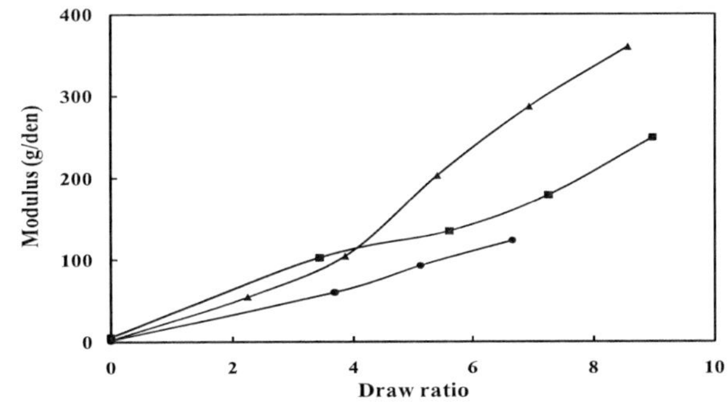

Figure 2. Fiber modulus vs. draw ratio: (●) Nylon 6 HV; (■) Nylon 6 AP; and (▲) PVA.

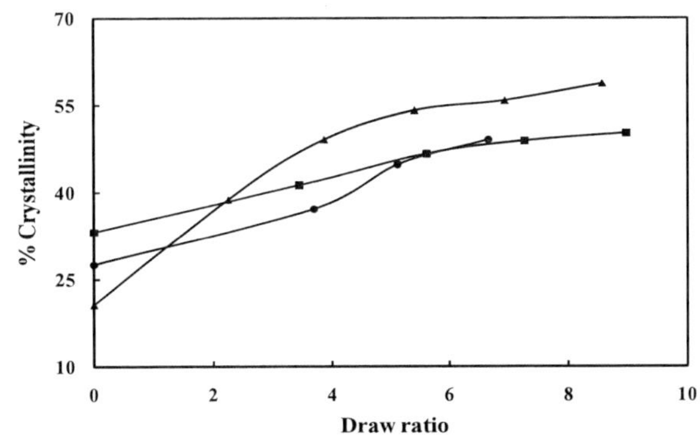

Figure 3. % Crystallinity vs. draw ratio: (●) Nylon 6 HV; (■) Nylon 6 AP; and (▲) PVA.

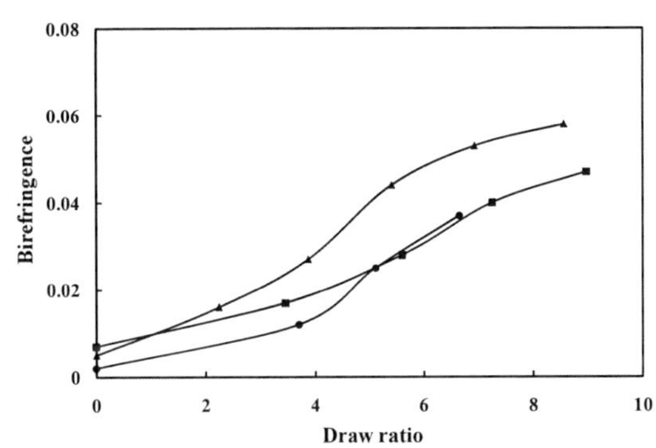

Figure 4. Dependence of birefringence on draw ratio: (●) Nylon 6 HV; (■) Nylon 6 AP; and (▲) PVA.

Modeling PA-6 Nanofiber Properties Produced by Needleless Electrospinning: Effect of Solution Concentration and Voltage

A. Nazir[1], L. Schacher[2], T. Hussain[1], D.C. Adolphe[2], N. Khenoussi[2]
[1]National Textile University, Faisalabad, Pakistan
[2]Université de Haute-Alsace–ENSISA–Laboratoire de Physique et Mécanique Textiles Mulhouse, France
nabyl.khenoussi@uha.fr

OBJECTIVE

Needleless electrospinning is increasingly considered as a possible solution for development of industrial nanofiber manufacturing facilities, where optimization of parameters for a desired output is always important. This paper discusses the modeling and optimization of nanofiber diameter as a function of two input factors i-e polymer solution concentration and voltage.

INTRODUCTION

Electrospinning received plenty of attention in recent years because of versatility and potential for applications in various fields [1]. Electrospinning has been carried out from last decade with needle based machines that use an electrical charge to draw fibers from a needle containing polymer [2-4].

Needleless electrospinning mechanisms have also been developed for getting more versatile products with higher production efficiencies [5, 6]. One of the most successful among needless electrospinning setups is wire electrode based machine. In this setup a polymer solution is coated on a wire electrode with lower voltage, from where it is drawn by electrostatic force from another wire electrode at higher voltage, allowing the deposition of nanofibrous web on a collector below the upper wire electrode. This technique allows the formation of multiple jets from free surface of a polymer solution. Ability to form multiple jets at the same time makes it a bulk scale production process for nanofibers [7, 8]. However, it is a relatively new method with little studies on process optimization and modeling, which is much important for a process developed for industrial scale production.

MATERIAL AND METHOD

The electrospinning solution was prepared by dissolving Polyamide-6 (PA-6) in formic acid (99%) with stirring on a magnetic stirrer at 40℃ for 24 hours. The solution was electrospun on Elmarco's Nanospider NS 1WS500 using the wire electrode module. The variables selected were concentration of polymer solution with three levels (i-e 18%, 19% and 20%) and the voltage of upper electrode, with three levels (i-e 30 KV, 40 KV and 50 KV). The experiments were designed according to full factorial methodology. The constant parameters were voltage of lower wire electrode (-30 KV), the air flow (80/130 m^3/h), electrode to collector distance of 290 mm, carriage speed of 290 mm/s and substrate speed of 15 mm/min. The orifice used for coating the wire electrode had a diameter of 0.6 mm. During electrospinning the temperature was maintained at 20℃ with relative humidity of 35%. After electrospinning the morphology of nanofibers was studied using a Scanning Electron Microscope (SEM). The data were analyzed using response surface methodology to develop a model to predict the diameter of nanofibers using concentration and voltage as input parameters. Moreover, the samples were also tested for their filtration capability to employ them as face mask after coating on a spunbond nonwoven substrate. This filtration system is expected to improve the filtration efficiency of spunbond filters when used as a protection media against different type of agents that may affect the human respiratory system. Moreover, the smarter protection offered by such a system is expected to allow for improvement of comfort properties such as thermal conductivity, moisture management and air permeability.

RESULTS AND DISCUSSION

Effect of PA-6 Concentration

The effect of polymer solution concentration on diameter of PA-6 nanofibers was found to be significant, as confirmed by analysis of variance (ANOVA) in table 1. P-value from analysis of variance was found to be 0.00 for (concentration) ^{2}, which shows that the concentration has a strong non-linear effect on nanofiber diameter. The diameter of fibers was found to decrease when concentration was changed from 17 wt% to 18 wt% and then it started to increase as shown in figure 1. This trend needs to be studied further.

Effect of Voltage

Voltage of upper electrode also posed a significant impact on fiber diameter, as confirmed by ANOVA for it (P-value = 0.058; table 1). Moreover, as shown in figure 1 the impact of voltage was found to be a linear one. The diameter was found to increase with increasing voltage. This could be attributed to stronger attractive forces, pulling more polymer solution at higher voltages.

The effects produced by PA-6 concentration and voltage are shown in figure 1, which gives overall and detailed view of nanoweb at optimized parameters.

Table1: Analysis of variance for input variables

Term	Concentration (C)		Voltage (V)		(Concentration)^2		(Voltage)^2	
	Coeff.	P-Value	Coeff.	P-Value	Coeff.	P-value	Coeff.	P-value
Diameter	16.2	0.21	30.3	0.05*	52.85	0.04*	-5.62	0.77
*Statistically significant term with 95% confidence								

Development of Model

The model between output variable i-e diameter and input variables i-e concentration and voltage was developed using response surface regression. The analysis of variance and respective coefficients in coded units are show in table 1.

The regression model developed from analysis of data was

Diameter = 16727.1 – 1885.0(C) + 7.54(V) + 52.85(C)^2 - 0.46(V)^2

R-sq for the above mentioned regression equation was found to be 83.66%, which shows that the model is strongly significant.

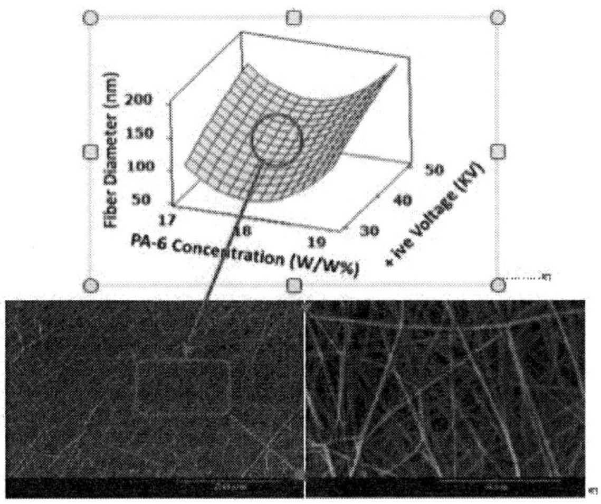

Figure 1: Effect of PA-6 concentration and voltage on diameter, overall morphology of optimized structure and detailed micrograph of optimized structure

Table 2: Validation of model from predicted and actual results.

Sr. No.	Concen-tration (%)	Upper Voltage (KV)	Diameter (nm)		
			Actual	Predicted with model	Error (%)
1	18	45	169.00	168.59	0.24
2	18	35	135.53	130.25	3.90
3	20	45	380.45	415.82	-9.30
4	19	45	230.34	239.34	-3.91

Validation of Model

The models were validated by electrospinning nanofibers with input parameters as mentioned in table 2 and comparing the resultant output variables with those predicted by the models. From the percentage (%) error it can be inferred that the models offer acceptable accuracy at lower concentration range.

CONCLUSION

This study discusses the modeling and optimization of nanofiber diameter prepared thanks to wire-based electrospinning setup as a function of two input factors i-e polymer solution concentration and electrospinning voltage. A model was developed to determine the diameter from required set of inputs. ANOVA was used to identify the statistically significant parameters. Both the input factors were found to produce a significant impact on nanofiber diameters. For a required fiber diameter an optimum electrospinning settings of the concentration and voltage must be selected. It was found that the concentration doesn't put a linear impact on diameter. Moreover, the proposed model could prove to be a helpful tool for getting the required set of parameters.

FUTURE WORK

Other variables may be varied to study and optimize the output/dependent variables: solution viscosity, conductivity of solution, solvents, distance between nozzle and collector, etc. The process variables will be selected by Taguchi's method and regression analysis will allow to obtain a quantitative relationship between selected electrospinning variables and average fiber diameter and set the optimal level for each parameter.

Finally, experiment will be conducted to check agreement between the predicted values of the response obtained from optimum level parameters and the obtained experimental data.

KEYWORDS: nanofibers, modelling, electrospinning

REFERENCES

[1] Bhardwaj, N., Kundu, S.C. "Electrospinning: a fascinating fiber fabrication technique." *Biotechnology Advances*, 28, 2010: 325-347.

[2] Sell, S.A., McClure, M.J., Garg, K., Wolfe, P.S., Bowlin, G.L. "Electrospinning of collagen/biopolymers for regenerative medicine and cardiovascular tissue engineering." *Advanced Drug Delivery Reviews*, 61, 2009: 007-1019.

[3] Lee, K.Y., Jeong, L., Kang, Y.O., Lee, S.J., Park, W.H. "Electrospinning of polysaccharides for regenerative medicine." *Advanced Drug Delivery Reviews*, 61, 2009: 1020-1032.

[4] Haghi, A., Akbari, M. "Trends in electrospinning of natural nanofibers." *Physica Status Solidi (a)*, 204, 2007: 1830-1834.

[5] Yarin, A., Zussman, E. "Upward needleless electrospinning of multiple nanofibers." *Polymer*, 45, 2004: 977-2980.

[6] Wang, X., Niu, H., Lin, T., Wang, X. "Needleless electrospinning of nanofibers with a conical wire coil." *Polymer Engineering & Science*, 49, 2009: 1582-1586.

[7] Elmarco. (06 January). *Nanospider™ electrospinning equipment.* Available: http://www.elmarco.com.

[8] Khenoussi, N., Schacher, L., Adolphe, D. "Nanofiber production: Study and development of electrospinning device." *Experimental Techniques*, 36, 2012: 32-39.

Effect of Organic Solvents on Morphology and Properties of Interfacial Polymerized Polyaniline Nanofibers

Tao Li[1], Zongyi Qin[1,2], Banglei Liang[1], Na Liu[2], Zhe Zhou[2], Meifang Zhu[1,2]
[1]State Key Laboratory for Modification of Chemical Fibers and Polymer Materials, Donghua University, Shanghai, China
[2]College of Material Science and Engineering, Donghua University, Shanghai, China
taoli_dhu@163.com; phqin@dhu.edu.cn

INTRODUCTION

Among the family of π-conjugated polymers, polyaniline (PANI) has curved a unique niche because of its wide applications in anti-corrosion coatings, wave-absorption materials, sensors, supercapacitors, solar cells, etc [1-3]. Recently, the heightened interest in one-dimensional electrical conductors and the need to further understand the formation mechanism have caused the research to focus on PANI nanofibers. Although a quality of approaches have been proposed to prepare PANI nanofibers, such as hard-template and soft-template methods, rapid-mixing and dilute solution routes, some disadvantages including boring post-treatment, hard control and low yield restrict the wide application of them. Meanwhile, interfacial polymerization of PANI nanofibers has been proposed and widely investigated in recent years [3-5], which mainly profits from the easy synthesis, convenient operation, well controllable repeatability, as well as no tedious post-treatment. On the other hand, the influence of solvent in interfacial system is scarcely researched. Therefore, interfacial polymerization was selected as the preparation approach and various organic solvents with different solution parameters were studied to understand the role of solvent in the formation of PANI nanofibers.

APPROACH

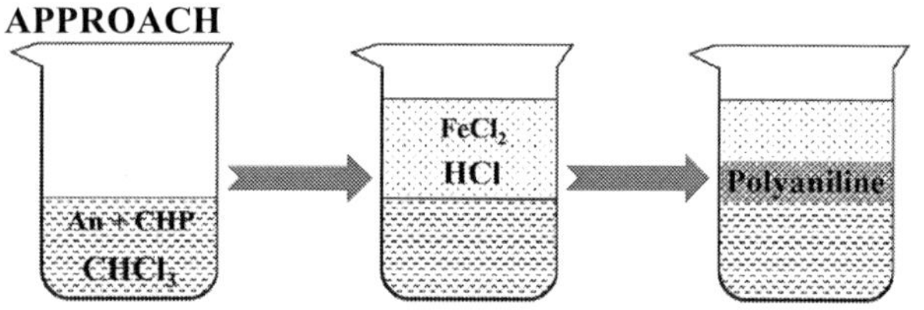

Fig. 1: A schematic illustration of interfacial polymerization of PANI nanofibers

The typical procedure for the interfacial polymerization of PANI can be carried out as follows: dissolve the aniline (An) monomers and cumene hydroperoxide (CHP) into the organic phase, the same as ferrous chloride ($FeCl_2$) into acidic aqueous phase. After complete stirring for 30 minutes, there formed uniform solution in each beaker, respectively. After standing for a period of time, pour the lighter aqueous phase into the beaker containing the organic solution alongside the wall to keep the interface formed between two phases still. During the course of polymerization, the system stood without any disturbance. After the period of 24 h, the product diffused into the aqueous phase was purified by filtering and washing for several times. Finally, the sample was dried under vacuum at 60℃ for 24 h to obtain the PANI powder.

RESULTS AND DISCUSSION

Morphology

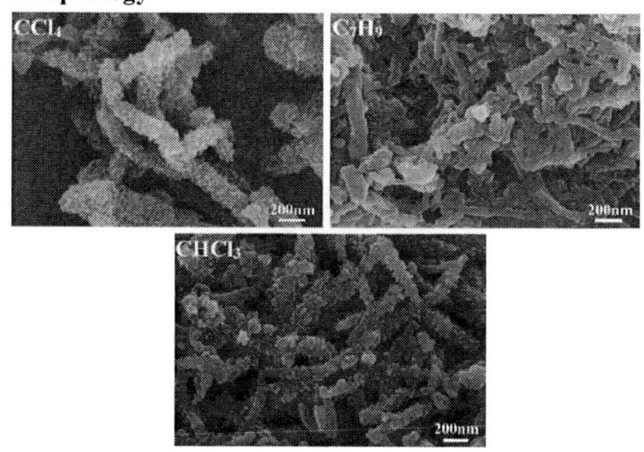

Fig. 2: SEM images of PANI nanofibers prepared with different organic solvents

As well known that the liquids incompatible with each other can form an interface between them, and their position of upper or lower is decided by the density correlative to water. In the interfacial polymerization of PANI, the products are generated at the interface and then disperse into the aqueous phase due to the hydrophility of polymer. Thus, the correlative density, as well as other parameters, has some influence on the morphology of products directly. As can be seen, Fig. 2 shows SEM images of PANI nanofibers obtained via interfacial polymerization in different organic solvents. Among the organic solvents, the denser tetrachloromethane (CCl_4) and lighter methylbenzene (C_7H_9) were selected to compare with chloroform ($CHCl_3$) referring to their different parameters as listed in Table Ⅰ. Obviously, PANI nanofibers synthesized in CCl_4 has a rough diameter and coarse surface, but a thinner diameter and smooth surface for that in C_7H_9, which can be mainly arisen from the density difference.

In the polymerization process, PANI experienced the nucleation and growth periods, the most common formation mechanism [2, 5]. Expect from the hydrophility of PANI, the effect of gravity cannot be ignored because of its consistent role when the water phase lie in the lower position.

Table I. Parameters of organic solvents used in the interfacial polymerization of PANI nanofibers

Organic solvent	Density	Solubility parameter	Polarity
CCl_4	1.60	8.6	1.6
C_7H_9	0.87	8.9	2.4
$CHCl_3$	1.50	9.3	4.4

Structural Analysis

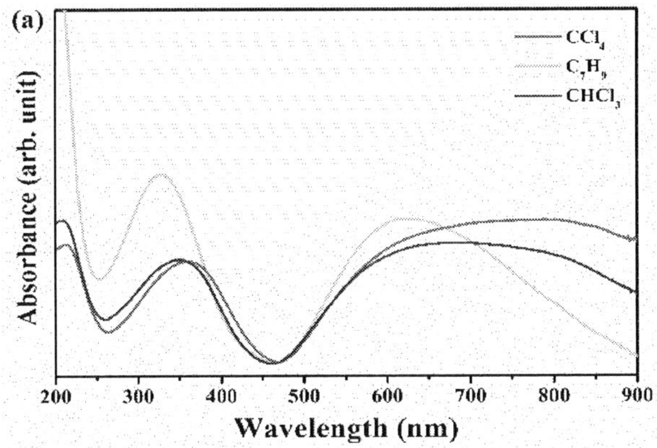

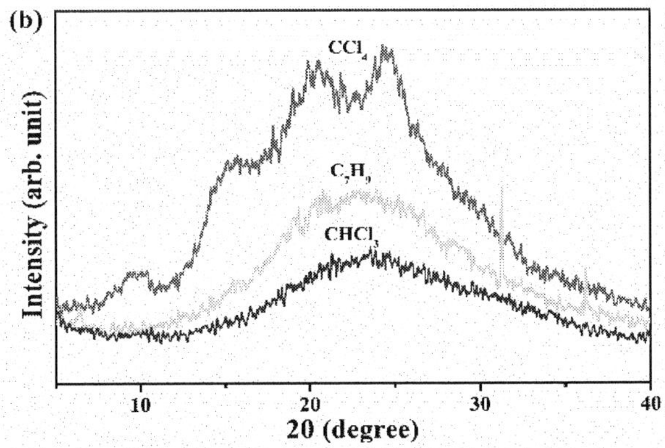

Fig. 3: UV-Vis spectra (a) and XRD patterns (b) of PANI nanofibers prepared with different organic solvents

Fig. 3(a) indicates the UV-Vis spectra of PANI nanofibers synthesized via interfacial polymerization in different organic solvents. From the point-view of PANI formation, there's no doubt that organic solvent used in interfacial polymerization plays an essential role in the morphology as well as the structure of product. For PANI prepared by interfacial polymerization containing $CHCl_3$ as organic solvent, the characteristic bands at about 350 and 620 nm along to 800 nm can be contributed to π-π^* and π-polaron transitions, respectively [6], which means the products are in the emeraldline state. However, the bands in PANI obtained with C_7H_9 shift to lower wavelength, indicating the polymer are in the peroxide state, where the quinone rings over the benzene rings. The result suggests that the diffusion rate of created product from the interface to the close aqueous phase differs among the organic solvents, and it's quite faster in the system with lighter organic solvent.

The crystalline property of PANI nanofibers was investigated by the XRD patterns as observed in Fig. 3(b). Obviously, PANI synthesized in CCl_4 appears many characteristic peaks at 9.8°, 15°, 20° and 24.7°, which are in the monoclinic space group $P2_1$. However, for PANI synthesized in the interfacial system composing of C_7H_9 and $CHCl_3$, only a broad peak around 23° can be observed. In comparison with each other, the absence of the peak at 2θ = 20° which is generally observed in bulk PANI may be due to the lack of ordering along the periodicity parallel to the polymer chains [4]. Therefore, the sample obtained at the presence of organic solutions with high polarity may be disadvantage to highly ordered products.

Electrochemical Property

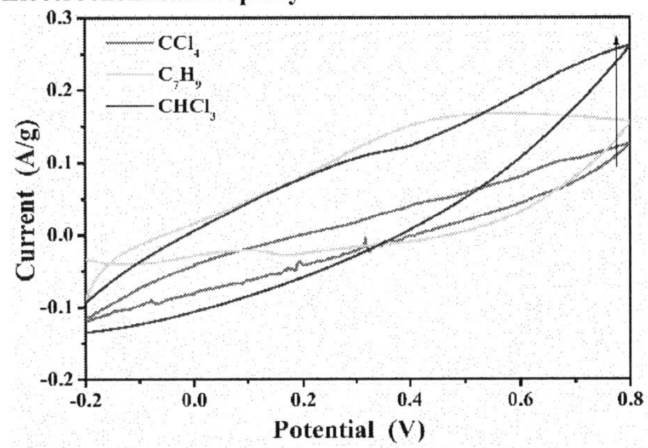

Fig. 4: Cyclic voltammograms of PANI nanofibers prepared with different organic solvents

Electrodes for electrochemical measurement were prepared by mixing electroactive materials with 10 wt.% acetylene black and 5 wt.% polytetrafluoroethylene emulsion to make homogeneous slurry in the presence of absolute ethyl alcohol. The cyclic voltammetry (CV) was carried out in three-electrode system with 0.5 M H_2SO_4 as electrolyte solution, and -0.2-0.8 V as the potential windows as well as the scan rate of 10 mV/s. As can be seen in Fig. 4, the voltammograms for all samples don't show a rectangular geometry, which indicates an ideal electronic double layer capacitance, demonstrating the pseudo-capacitance property of PANI nanofibers synthesized in different organic solvents. Additionally, the current density and closed area of CV curve of PANI nanofibers prepared at the aid of CCl_4 were less than the other two samples [4,5].

CONCLUSION

Polyaniline nanofibers with different diameters were synthesized via interfacial polymerization with different organic solvents. The lighter solvent benefits for the smooth nanofibers, whereas the denser one generates nanofibers with coarse surface. All the samples present similar electronic structure estimated from UV-Vis spectra. The organic solvent with high polarity can induce the formation of high crystalline structures, which harms for the capacitive property conversely.

ACKNOWLEDGMENT

This work has been financially supported by National Natural Science Foundation of China (21274019), Fundamental Research Funds for the Central Universities (2232013A3-02), and Program for Changjiang Scholars and Innovative Research Team in University (IRT1221)

REFERENCES

[1] Devillers, N., et al. *J. Power Sources*, 246, 2014: 596-608.
[2] Gordana, C. *Synth. Met.*, 177, 2013: 1-47.
[3] Zhang, D. *Adv. Polym. Technol.*, 32, 2013: 323-368.
[4] Chutia, P., et al. *Appl. Phys. A*, 115 (3), 2014: 943-951.
[5] Lei, W., et al. *J. Power Sources*, 266, 2014: 347-352.
[6] Tantawy, H., et al. *J. Phys. Chem. C*, 118 (2), 2014: 1294-1305.

Production of Ultra Strong Polymer Fibers from Extended Chains

Tom Wyatt, Xudong Fang, Donggang Yao
School of Materials Science & Engineering, Georgia Institute of Technology, Atlanta, GA, USA
yao@gatech.edu

INTRODUCTION

High performance polymer fibers can be produced using at least two approaches: rigid rod alignment and chain extension. The first approach is limited since it is largely dictated by the chemical and physical structure of the polymer. The latter represents a more versatile approach given the fact that most known polymers are linear polymers with flexible chains. However, creating high-strength polymer fibers by chain extension is not a simple process. Intra- and intermolecular interactions within and between polymer chains must be maneuvered for development of a capable process. Gel spinning [1] is specifically designed for achieving large extent of chain extension and has achieved notable commercial success. However, only a very few polymers have been successfully gel spun. Furthermore, the relatively complex processing route involved in production leads to costs 10 times more than conventional melt-spun fibers [2].

In the past decade, research has been conducted at Georgia Institute of Technology towards improved understanding of the gel spinning process and developing technical approaches for enhancing the process efficiency. With new knowledge applied to processing, the process economics has been significantly improved and more polymers have been successfully gel-spun into high-strength fibers. In particular, an environmentally benign process – twist-gel spinning process – was invented to enhance the productivity of gel spinning, and the new process was successfully used to make ultra-strong fibers with improved economics. The new process was also successfully applied to several other polymers that cannot be gel spun previously, including polyethylene oxide and polyoxymethylene, to produce ultra-strong polymer fibers.

GEL PREPARATION

It now becomes clearer that the so-called 'gel' material desired in gel spinning is not indeed a classical gel with elastic properties. The effects between gelation and phase separation must be judiciously controlled to achieve a gel-like material that is highly extendable. In order to produce a capable gel for gel spinning, control of the solution properties of the spin dope is crucial. It is desired to have a highly miscible solution prior to gel formation. Intramolecular interactions must be suppressed to allow polymer chains to swell to large extent in the spin dope. At the same time, intermolecular entanglements must be mediated to achieve a highly stretchable gel structure after gelation. The solution properties should be optimized so that gelation does not lead to significant regeneration of intramolecular bonds and molecules remain 'swollen' and lightly entangled. This is especially important to polar polymers. Suitable spinning dopes, in general, can be prepared by combination of solvents with different miscibility and polarity as well as ionic properties.

TWIST-GEL SPINNING

With proper control of spin dope composition and processing conditions, a phase-separated gel can be produced. In this case, the majority of solvent is phase-separated from the polymer while the desired swollen network still persists [3]. The solvent contained in such a gel can be removed effectively using mechanical twisting to obtain a nearly dried xerogel. The twisted gel-fiber can be directly hot-drawn, allowing conventional solvent extraction to proceed significantly faster after drawing [4].

SELECTED RESULTS

Polyethylene: A new twist-gel spinning process for ultrahigh molecular weight polyethylene (UHMWPE) fibers is demonstrated which significantly increases the extraction rate of non-volatile spin-solvent while simultaneously reducing the consumption of extraction solvent by a large amount. About 90% spin solvent can be removed simply by mechanical twisting. Applying twist to the gel-fiber enables it to be directly hot-drawn, allowing conventional solvent extraction to proceed significantly faster. While solvent extraction effectiveness is largely enhanced, the new process does not show reduced fiber properties. The tensile strength, Young's modulus, surface morphology, and fiber geometry are relatively unaffected when compared to fibers produced using the conventional gel-spinning process (refer to Table 1). The new twist-gel spinning process is expected to improve the processing efficiency of gel-spun high-strength fibers, promoting broad expansion of these high performance fibers into applications that were previously prohibitive due to extremely slow production.

Poly(ethylene Oxide): A new gel-spinning process for making high-strength poly(ethylene oxide) (PEO) fibers has been developed. The PEO gel-spinning process was enabled through an oligomer/polymer blend in place of conventional organic solvents, and the gelation and solvent-like properties were investigated. A 92/8 wt% poly(ethylene glycol)/PEO gel exhibited a melting temperature around 45 °C and was highly stretchable at room temperature. Some salient features of a gel-spun PEO fiber with a draw ratio of 60 are: tensile strength at break = 0.66 ±0.04 GPa, Young's modulus = 4.3 ±0.1 GPa, and a toughness corresponding to 117 MJ/m^3. These numbers are significantly higher than those previously reported. Wide-angle x-ray diffraction (Fig. 1) of the high-strength fibers showed good molecular orientation along the fiber direction. The results also demonstrate the potential of further improvement of mechanical properties.

Poly(oxymethylene): A spin dope for gel spinning of poly(oxymethylene) (POM) has been developed. In contrast to directly melt-spinning POM, the new method involves mixing POM with a variety of substances to swell POM molecules and form a gel network. This blend is subsequently extruded and cooled to form gel filaments which are further processed into high-modulus, high-strength fibers. Initial experiments using this method with an injection molding homopolymer POM resin demonstrated the production of POM fibers with modulus >25 GPa and strength >1.5 GPa (Fig. 2). Improvement of these tensile properties is anticipated if a higher-molecular-weight grade POM is used.

CONCLUSIONS AND OUTLOOK

A twist-gel spinning process has been developed for improving process efficiency and achieving direct drawing of gel fibers. For UHMWPE, approximately 90% of spin-solvent can be removed from gel-spun fibers by mechanical twisting without affecting the final fiber's tensile properties. New spin dopes have also been developed for other polymers and the applicability of gel spinning has been extended to other polymers including PEO and POM. Ongoing work is focused on developed effective gel spinning processes for more polar polymers including nylons, proteins and cellulose.

KEYWORDS

Gel spinning, solution spinning, extended chains, high molecular weight, high strength, polar polymers.

REFERENCES

1. Smith, P., Lemstra, P.J. "Ultra-high-strength polyethylene filaments by solution spinning-drawing." *J. Mat. Sci.*, 15, 1989: 505.
2. Mittal, V. *High Performance Polymers and Engineering Plastics*. Hoboken, NJ: John Wiley & Sons, 2011: 323.
3. Wyatt, T., Deng, Y., Yao, D. "Fast solvent removal by mechanical twisting for gel spinning of ultrastrong fibers." *Polym. Eng. Sci.* Online: May 11, 2014.
4. Wyatt, T., Gainey, T., Fang, X., Yao, D. "Direct drawing of gel fibers enabled by twist-gel spinning process." *Polym. Eng. Sci.*, 2014.
5. Wyatt, T., Chien, A.-T., Kumar, S., Yao, D. "Development of a gel spinning process for high-strength poly(ethylene oxide) fibers." *Polym. Eng. Sci.*, 54, 2014: 2839.

Table 1. Comparison between twist-gel spun fibers and top-notch commercial fibers.

Sample	Strength (GPa) / (g/denier)	Modulus (GPa)
Twist-gel spun fiber	4.68 / 53	181.09
Dyneema SK90 yarn	3.69 / 42	101.04
Spectra S2000 fiber	3.56 / 40	140.65

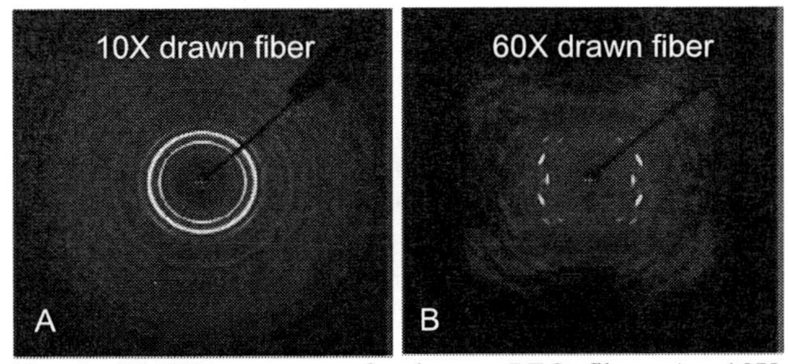

Fig. 1. WAXD patterns of gel-spun PEO fiber: A –10X drawn gel fiber; B–60X draw ratio fiber.

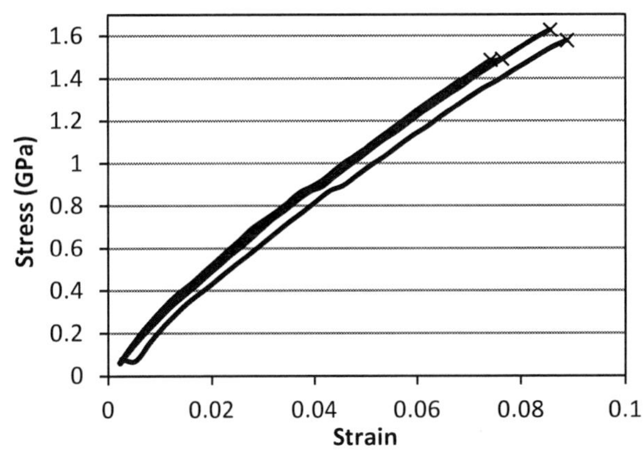

Fig. 2. Representative tensile curves for homopolymer POM fibers.

Morphology and Structure Development of Aromatic Polysulfonamide Fibers During Spinning Process

Yumei Zhang[1], Jinchao Yu[1], Xi Yang[1], Shenghui Chen[2], Xiaofeng Wang[2], Huaping Wang[1]
[1]State Key Laboratory for Modification of Chemical Fibers and Polymer Materials, Donghua University, Shanghai, China
[2]Shanghai Tanlon Fiber Co., Ltd, Shanghai, China
zhangym@dhu.edu.cn

OBJECTIVE

The morphology and structure development of the aromatic polysulfonamide fibers during wet-spinning process and heat-drawing process were studied to indicate the relationship between processing technology and structure, properties of the fibers.

INTRODUCTION

Due to their thermal stability, high glass transition, inflaming retarding and low dielectric constant, aromatic polysulfonamide (PSAs) have significant potential application in industrial and aerospace fields. Although the spinning of the some isomerically substituted PSAs had been carried out in a laboratory or pilot scale as early as 1970s[1], the industrialization of the aromatic copolysulfonamide (*co*-PSA) fibers was broken through by Shanghai Tanlon Fiber co. Ltd. till 2008 [2] with the commercial name of Tanlon®. Compared with the poly(4,4'-diphenylsulfonyl terephthalamide) fibers reported by Khudoshev et al [3], the tenacity of the *co*-PSA fibers Tanlon® increased by introducing *m*-phenylsulfonyl moieties into the relatively rigid backbone which improved the solubility and spinability of polymers while the thermal stability kept well [4]. However, the knowledge of the basic structure and properties of PSA is still limited. It is interesting to establish relationships between the physical properties and structure of PSA fibers, which is of great practical importance.

EXPERIMENTAL

Wet Spinning Process: The *co*-PSA was synthesized via polycondensation of 4, 4'-diaminodiphenylsulfone, 3, 3'-diaminodiphenylsulfone and terephthaloyl chloridein in DMAc at low temperature. The concentrated solution after polymerization was used as spinning dope directly. The *co*-PSA fibers were prepared by using the wet-spinning process shown in Figure 1.

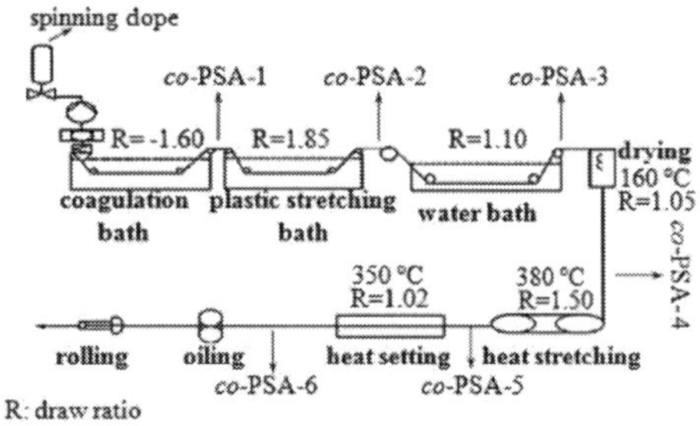

Fig. 1: Schematic of wet-spinning process of co-PSA fibers

Characterization: The cross-sections of a bundle of PSA fibers were cut by slicer and observed using a (Leica DM750P) microscope. S-S curves were obtained by XQ-1A mechanical tester.

Dynamic mechanical properties were performed on TA Q800 V7.5 instrument in the extension mode with a frequency of 1HZ and heating rate of 10°C /min.

Wide-angle X-ray diffraction (WAXD) was carried out at B14 beam line in shanghai Synchrotron Radiation Facility (SSRF) with a wavelength of 0.124nm.

SAXS profiles were obtained at Beamline (16B1) in Shanghai Synchrotron Radiation Facility (SSRF). The wavelength used was 0.124 nm.

RESULTS AND DISCUSSION

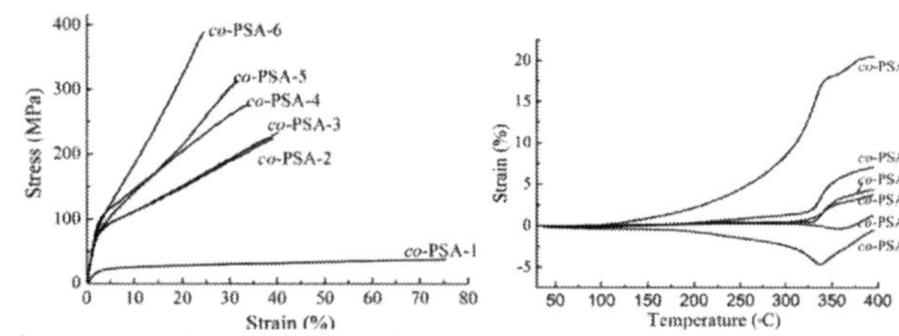

Fig. 2: Mechanical and thermo-mechanical properties of the *co*-PSA fibers on spinning line

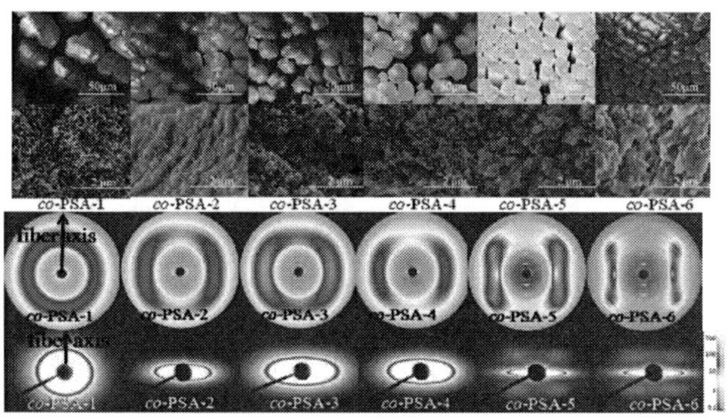

Fig. 3: Cross section, WAXS and SAXS patterns of the *co*-PSA fibers on spinning line

Morphology and Structure Development of co-PSA Fibers During Wet-spinning Process: The structure and performance changes of the *co*-PSA fibers that occurred during wet spinning process were studied. In the first coagulation, phase separation occurred within the *co*-PSA filaments and the molecules were arranged to form random network. Correspondingly, the *co*-PSA fibers had good extensibility but low strength. In the following plastic stretching process, the primary deformation process involved the molecular arrangement from disordered state (amorphous) to one or 2D ordering and pack along the fiber axis (mesophase). In the washing and drying process, the solvent and water was removed respectively. The structure of *co*-PSA fibers became much denser during these processes, but the amorphous orientation was almost unchanged. The *co*-PSA fibers before heat stretching were unstable. When the *co*-PSA fibers heat stretching at the temperature higher than T_g, the crystal structure formed within the fibers and became more perfect after heat setting, which contributed to the improvement of structural stability and mechanical properties.

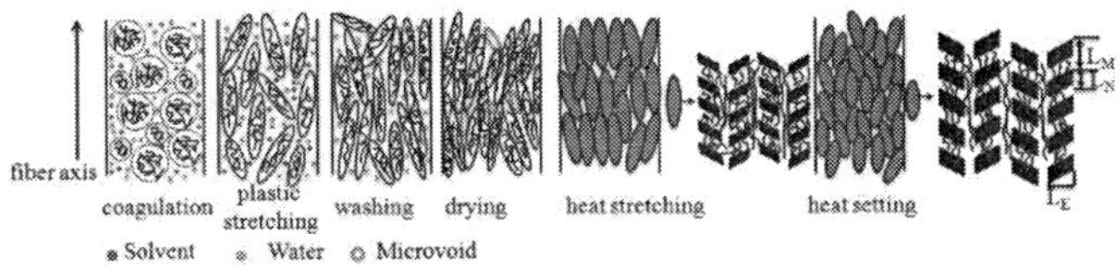

Fig. 4: Proposed structural models of the *co*-PSA fibers during wet-spinning process

Morphology and Structure Changes of the co-PSA Fibers Heat-Drawn at Different Temperatures: The washed *co*-PSA fiber was heat-drawn with the designed technology of multiple ratios at different temperature varying from 350 to 390℃. It was shown that the dependence of mechanical properties on temperature was not linear. It is found that the crystalline structure formed and the crystallinity, crystal orientation and crystallite size of the fibers increased with the heat-drawn temperature but the exorbitant temperature would lead to thermal oxidation and brittleness for the fibers.

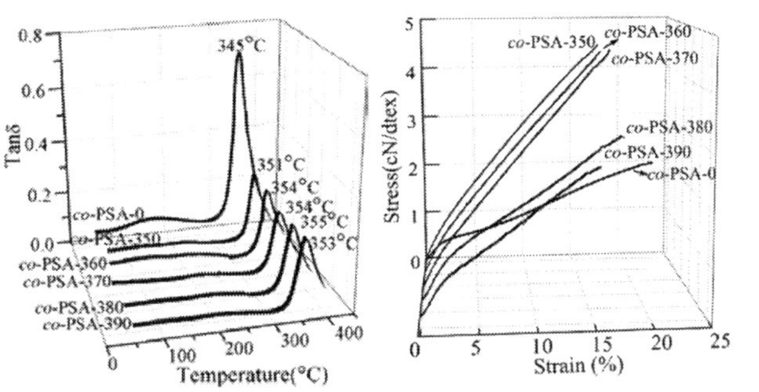

Fig. 5: DMA and S-S curves of *co*-PSA fibers heat-drawn at different temperatures

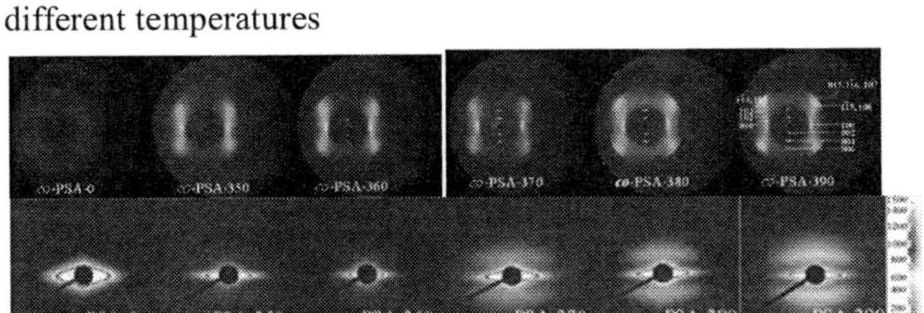

Fig. 6: WAXD and SAXS patterns of the *co*-PSA fibers heat-drawn at different temperatures

Morphology and Structure of the co-PSA Fibers Varied the Heat-Draw Ratios and Residence Time Change: By varying the heat-draw ratio of the wet-spun co-PSA fibers, the residence time of the fibers stayed in the heat tube was inevitably changed. It turned out that the tenacity of *co*-PSA fibers did not increase monotonously with the draw ratios. At low draw ratios, meaning small drawing stress, the long residence time was conductive to sufficient crystallization and high crystallinity, whereas the orientation in the amorphous region was relatively low. At high draw ratios, the crystallinity decreased because of the insufficient residence time while the amorphous orientation increased significantly under larger drawing stress.

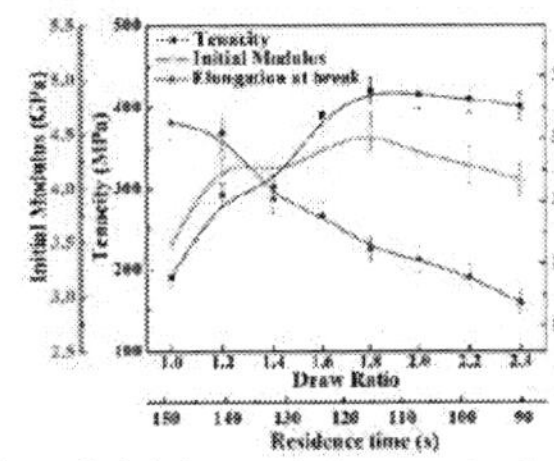

Fig. 7: Tenacity, initial modulus and elongation at break of co-PSA fibers heat-drawn at different ratios

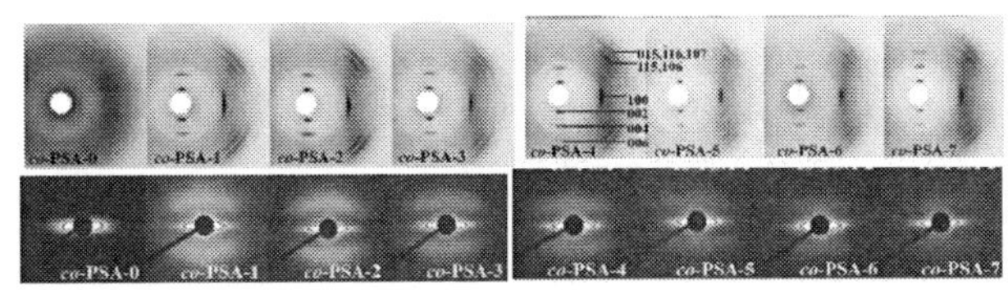

Fig. 8: WAXD and SAXS patterns of *co*-PSA fibers heat-drawn at different ratios

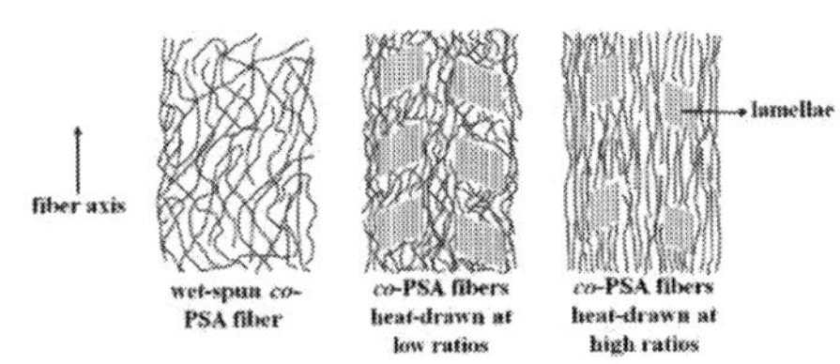

Fig. 9: Schematic structural models of *co*-PSA fibers heat-drawn at different ratios

CONCLUSIONS

During wet-spinning process, only amorphous orientation was happened before heat stretching and the crystal structure formed after the fibers were heat-drawn at the temperature higher than its Tg, which contributed to the improvement of structural stability and mechanical properties. The mechanical and thermal properties of the washed fiber improved by heat-drawing due to the formation of order crystalline structure but the exorbitant temperature would lead to thermal oxidation and brittleness for the fibers. It could be imaged that more perfect structure with larger crystalline size, mesophase and crystal phase, as well as the long period and fibril length, could only be formed with favorable draw ratio and residence time, which was corresponding to the maximum of strength of the heat-drawn fibers.

ACKNOWLEDGMENT

This work was financially supported National Natural Science Foundation of China (11079015 and 51273039).

REFERENCES

[1] Banduryan, S.I., Ivanova, N.A., Efimova, S.G., Sokira, A.N., Shchetinin, A.M., Frenkel, GG, et al. *Fibre Chemisty*, 9 (5), 1978: 484-485.

[2] Wang, J. *Huagong Keji Shichang*, 32 (2), 2009: 6-11.

[3] Kuznetsov, G.A., Nikiforov, N.I., Lebedev, V.P., Zhuravlev, V.G., Savinov, V.M. *Vysokomol. Soyed.*, 16, 1974: 2711-2716.

[4] Ding, X., Chen, Y., Chen, S., Li, J., Wang, X., Wang, H., et al. *J. Macromol. Sci. B*, 51, 2012: 1199-1207.

Study on the Silane Modification of PPTA Fibers

Junrong Yu, Na Lu, Yan Wang, Jing Zhu, Zuming Hu
State Key Laboratory for Modification of Chemical Fibers and Polymer Materials, College of Materials Science and Engineering, Donghua University, Shanghai, China
yjr@dhu.edu.cn

INTRODUCTION

Poly-p-phenylene terephthalamide (PPTA) fibers have been widely used in the national defense and military fields owning to their outstanding properties of high strength, high modulus, light weight and excellent chemical and thermal resistance. However, due to their smooth surface and chemically inertness, the bonding properties between the fibers and resin matrix are poor, which greatly limits fibers' application in the field of composite materials. Therefore, surface modification of PPTA fibers is necessary. Commercial PPTA fibers are dense and highly crystallized; it's difficult for the modifying agents to diffuse into fibers, so the fibers should be pretreated before surface modification. In this study, un-dried PPTA fibers obtained from fibers production line are used to be modified with silane agents.

APPROACH

Un-dried PPTA fiber bundles, which contained about 40~50% water, were rinsed in the solution of vinyl trimethoxysilane (VTMA) and benzoyl peroxide (BPO) under ultrasonication, then the treated fibers were dried under vacuum at 132°C for 2 min. The micropore structure in the un-dried PPTA fibers are observed by fluorescence microscope and laser scanning confocal microscopy after rhodamine marked. The changes in the surface functional groups, morphological structure, mechanical properties and creep behavior of the fibers before and after silane treatment were analyzed by using Fourier transform infrared spectroscopy (FTIR), scanning electron microscopy (SEM) and tensile tester. The interfacial shear strength (IFSS) between modified fibers and epoxy was tested by micro-debonding method.

RESULTS AND DISCUSSION

There existed about 40wt% water in the micropores of un-dried PPTA fibers. The micropore were open in un-dried fibers and became partially closed after air-dried. After treated with rhodamine solution, un-dried fibers showed obvious fluorescence emission signal, while the signal of air-dried fibers is very weak, as shown in Fig. 1. This indicated that modifier molecules could diffuse into un-dried PPTA fibers under ultrasonication.

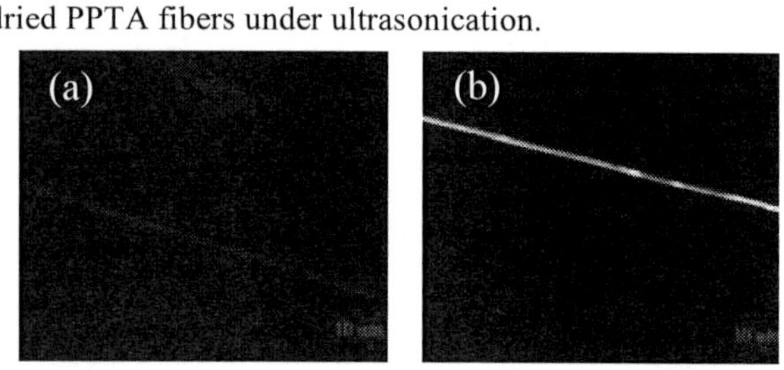

Fig. 1: Fluorescence microscope graph of rhodamine B treated dried PPTA fibers (a) and un-dried fibers (b)

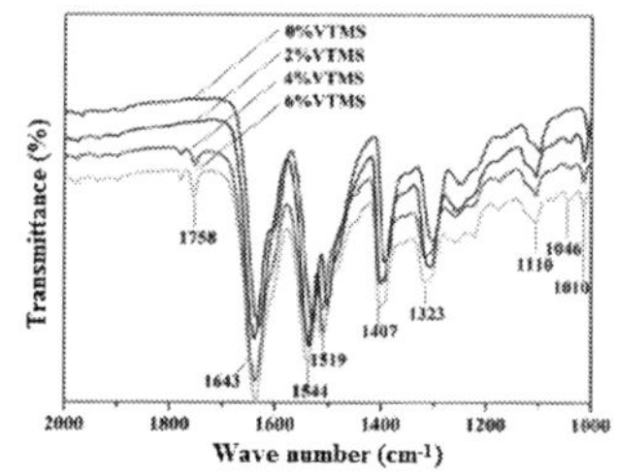

Fig. 2: FTIR spectra of different concentration of VTMS solution treated PPTA fibers

Fig. 2 was the FTIR spectra of untreated and VTMS treated PPTA fibers. The stretching vibration peak at 1758cm^{-1} and 1406cm^{-1} belonged to (O)C=O and Si-O-C, respectively, which indicated that new functional groups appeared on fibers' surface after silane modified. The fibers' surface also became rough after modified (shown in Fig. 3).

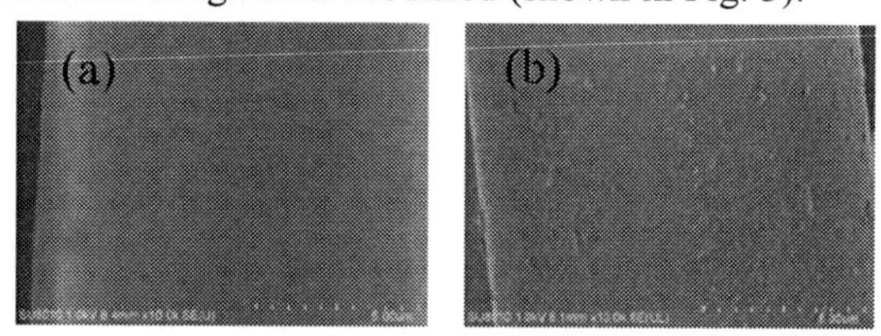

Fig. 3: Surface SEM photos of untreated (a) and 6wt% VTMS treated (b) PPTA fibers

So the fiber-epoxy interfacial bonding performance was improved obviously and the IFSS increased with the increasing of VTMS concentration, while the tensile strength slightly decreased after modified, as shown in Fig. 4(a). The creep resistance of silane modified fibers was also obviously improved because crosslinking structure has been formed between adjacent microfibers in PPTA fibers. The creep resistance of PPTA fibers was improved by more than 50% after 5 min treated with 6wt%VTMS solution, as shown in Fig. 4(b).

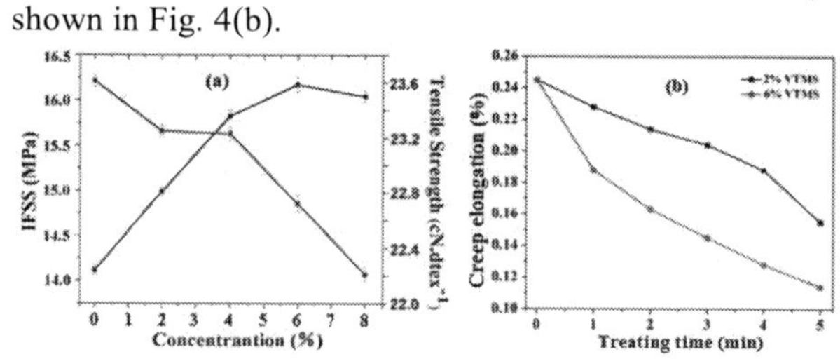

Fig. 4: IFSS and tensile strength (a), and creep elongation (b) of VTMS modified PPTA fibers

CONCLUSION

In this study, un-dried PPTA fibers was used to be treated with modifier agents solution, the fiber-epoxy interfacial bonding performance and creep resistance of modified fibers were obviously improved with only a slight damage on their tensile properties.

Functional Textiles and Clothing

Flow Asymmetricity in Fibrous Constructs

Jintu Fan[1], Dahua Shou[1], Jianlong Kou[2]
[1]Department of Fiber Science & Apparel Design, Cornell University, Ithaca, NY, USA
[2]Institute of Condensed Matter Physics, Zhejiang Normal University, Jinhua, China
jf456@cornell.edu

INTRODUCTION

For many different applications, it is desirable for the fibrous material to process differential fluid flow properties depending on the direction of the flow. For example, it is preferable to have apparel fabrics that can facilitate the flow of liquid sweat from the inner (i.e. next to skin) side to the external side, but prevent the flow of rain or hazardous liquid from the external side to the inner side. Water formed in fuel cell gas diffusion layer (GDL) should flow in a direction for removal so as to avoid "flooding", which is a major limiting factor of fuel cell performance. Medical wound dressings should allow the transfer of body fluids, but preventing the penetration of external bacteria.

Recently, we have carried out series of investigations on flow asymmetry in fibrous constructs [1-10]. In this paper, our study is further extended to analyze the effects of constructional parameters on the capillary-driven flow asymmetricity.

DEFINITION OF FLOW ASYMMETRICITY (F_A)

Consider a fibrous material with two sides (e.g. side A and B, see Fig. 1), we define the flow asymmetricity (F_A) as the ratio of the time required for a liquid to flow spontaneously from side B to side A (t_{BA}) over the time required for the same liquid to flow spontaneously in the opposite direction from side A to side B (t_{AB}). That is,

$$F_A = \frac{t_{BA}}{t_{AB}}$$

Therefore, the material is symmetric in terms of spontaneous flow if $F_A = 1$, preferential to spontaneous flow from side A to side B if $F_A > 1$, and preferential to spontaneous flow from side B to side A if $F_A < 1$.

FLOW ASYMMETRICITY IN FIBROUS CONSTRUCTS

Consider a porous fibrous material composed of two layers, each having different porosity and thickness as shown in Fig. 1a. The time that liquid water takes to flow spontaneously from one side to the other has been established [2]. Fig. 2 shows variations of F_A and t/t_0 against the thickness ratio h_B / h_A at $\varepsilon_B = \varepsilon_A = 0.5$ and the porosity ratio $\varepsilon_B / \varepsilon_A$ at $h_B = h_A$. Here, t_0 is the flow time required for the capillary flow from side A to side B in the control sample, which is a uniform single layer and has the same volume and thickness with the two-layer medium in Fig. 1a. As can be seen in Fig. 2a, the maximum value of F_A occurs when h_B / h_A is around 0.5, indicating the most asymmetric flow. This corresponding value of h_B / h_A is slightly different from those accounting for the minimum t_{AB} and the maximum t_{BA}. We found that the maximization of F_A is caused by the optimal interplay between capillary force and viscous resistance along the dynamic flow process. Fig. 2b shows that F_A decreases with the increase in $\varepsilon_B / \varepsilon_A$, whereas the minimum t_{AB} and t_{BA} are also observed.

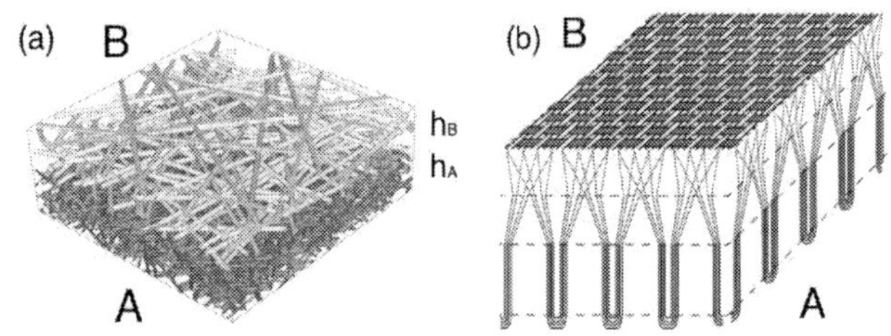

Fig. 1. (a) Two-layer fibrous media and (b) tree-like fabric

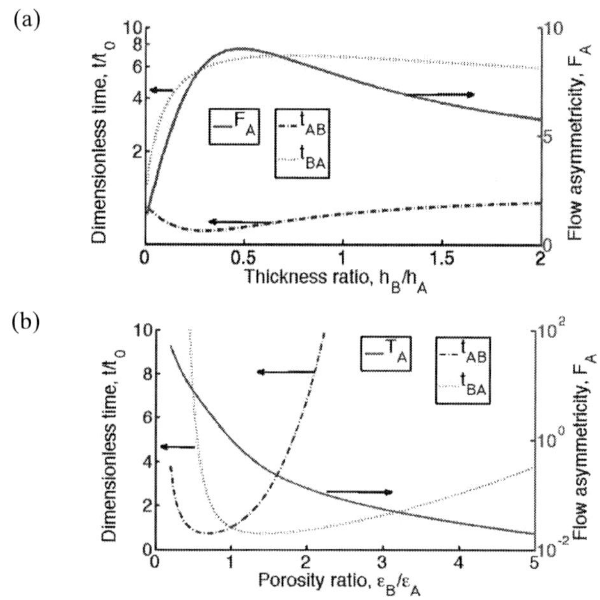

Fig. 2. Variations of F_A and t/t_0 against (a) the thickness ratio h_B / h_A and (b) the porosity ratio $\varepsilon_B / \varepsilon_A$

Next consider a multi-sectional structure consisting of the same porous fibrous material. The geometry of the structure has a great effect on the flow asymmetry [3]. We assume the tree-like fabric in Fig. 1b contains multi-level yarns with different fiber numbers or cross-sectional area. Fig. 3 demonstrates variations of F_A and t/t_0 against the length ratio l_B / l_A at $w_B = 0.5 w_A$ and the width ratio w_B / w_A at $l_B = l_A$ for a two-level tree-like fabric. The maximum F_A is found against l_B / l_A in Fig. 3a, and F_A decreases with w_B / w_A as shown in Fig. 3b. It is interesting to note that $l_B = l_A$ accounts for all the bound results of F_A, t_{AB}, and t_{BA} in Fig. 3a.

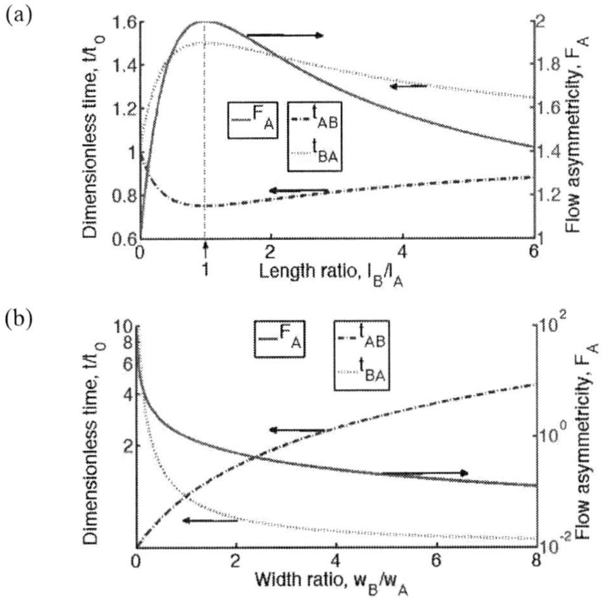

Fig. 3. Variations of F_A and t/t_0 against (a) the length ratio l_B / l_A and (b) the width ratio w_B / w_A

We also consider a porous material with tree-like branched flow network across its thickness, as shown in Fig. 1b. In Fig. 4, the values of F_A and t/t_0 in a three-level tree-like network vary against the length ratio l_B / l_A at a diameter ratio of 0.5 and the diameter ratio d_B / d_A a length ratio of 1. The maximum F_A is found against l_B / l_A in Fig. 4a, while F_A decreases with d_B / d_A in Fig. 4b.

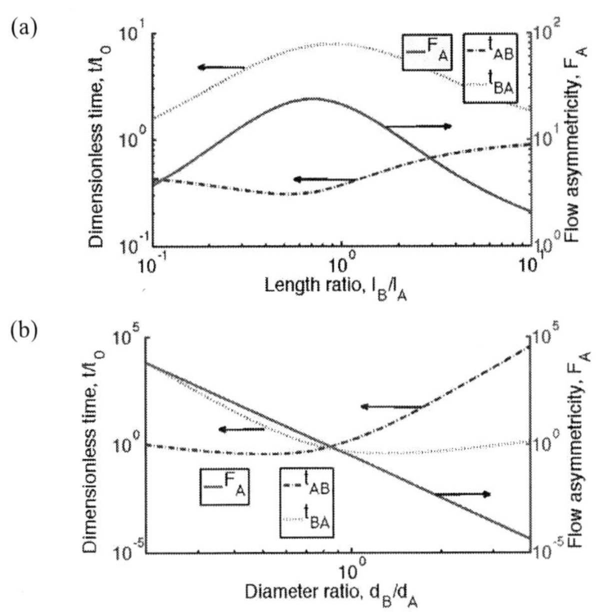

Fig. 4. Variations of F_A and t/t_0 against (a) the length ratio l_B / l_A and (b) the diameter ratio d_B / d_A

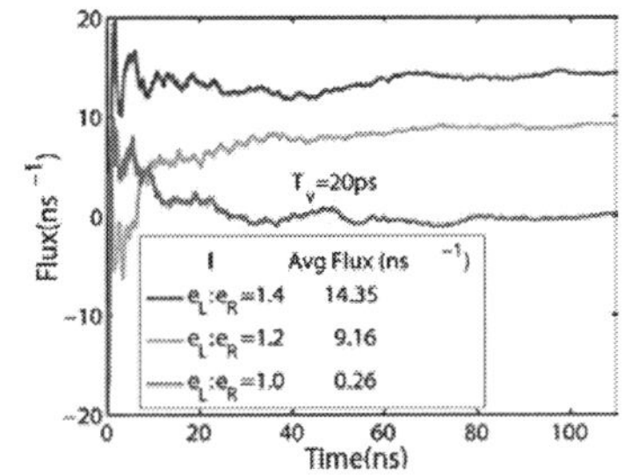

Fig. 5. Water net flux as a function of time for different degrees of the heterogeneity energy

Asymmetric flow also takes place in nanochannels due to surface tension gradients [7]. Fig. 5 shows the water flux rates in relation to the heterogeneity of surface energy (1.0, 1.2, and 1.4) with a vibrational period of 20 ps. The flux rate may also be dependent on the vibration frequency of the external charge [1].

CONCLUSIONS

Our studies have demonstrated the potential of engineering the heterogeneous structure and surface energy distribution to achieve asymmetric flow properties in fibrous porous materials for various applications. The present work is however limited to flow from an unlimited liquid reservoir; further work is being carried out on flow asymmetry for liquid droplets of confined sizes.

REFERENCES

1. Kou, , J.L., Lu, H..J., Wu, F.M., Fan, J.T., Yao, J. "Electricity Resonance-Induced Fast Transport of Water through Nanochannels." *Nano Lett.,* 14, 9, 2014: 4931-4936.
2. Shou, D.H., Ye, L., Fan, J.T., Fu, K.K. "Optimal Design of Porous Structures for the Fastest Liquid Absorption." *Langmuir*, 30, 1, 2014: 149-155.
3. Shou, D.H., Ye., L., Fan, J.T., Fu, K.K., Mei, M.F., Wang, H.J., Chen., Q. "Geometry-Induced Asymmetric Capillary Flow." *Langmuir*, 30, 19, 2014: 5448-5454.
4. Shou, D.H., Ye, L., Fan, J.T. "Treelike networks accelerating capillary flow." *Phys. Rev. E*, 89, 2014: 053007.
5. Shou, D.H., Ye, L., Fan, J.T. "The fastest capillary flow under gravity." *Appl. Phys. Lett.*, 104, 23, 2014: 231602.
6. Chen, Q., Fan, J.T., Sarkar, M.K. "Biomimetics of branching structure in warp knitted fabrics to improve water transport properties for comfort." *Tex. Res. J.*, 82, 11, 2012: 1131-1142.
7. Kou, J.L., Zhou, X.Y., Lu, H.J., Xu, Y.S., Wy, F.M, Fan, J.T. "A vibration-charge-induced unidirectional transport of water molecules in confined nanochannels." *Soft Matter*, 8, 2012: 12111-12115.
8. Chen, Q., Fan J.T., Sarkar, M.K., Bal, K. "Plant-based biomimetic branching structures in knitted fabrics for improved comfort-related properties." *Tex. Res. J.*, 81, 10, 2011: 1039-1048.
9. Bal, K., Fan, J.T., Sarkar, M.K., Ye, L. "Differential spontaneous capillary flow through heterogeneous porous media." *Int. J. Heat Mass Transf.*, 54, 2011: 3096-3099.
10. Sarkar, M., Fan, J., Szeto, Y.C., Tao, X.M. "Biomimetics of Plant Structure in Textile Fabrics for the Improvement of Water Transport Properties." *Tex. Res. J.*, 79, 7, 2009: 657-668.

Change the Fiber Inner Structure by Twisting Monofilament Under a Hot Condition

Jianjian Zhang[1,2], Yuanfeng Wang[1,2], Genyang Cao[1,2], Xin Liu[2], Weilin Xu[2], Zhigang Xia[1,2]
[1]College of Textile Science and Engineering, Wuhan Textile University, Wuhan, China; [2]Key Laboratory of Green Processing and Functional Textiles of New Textile Materials, Ministry of Education, Wuhan Textile University, Wuhan, China
896279961@qq.com; weilin-xu@hotmail.com; zhigang_xia1983@hotmail.com

STATEMENT OF PURPOSE

This study is aimed to get a α spiral phase inner macromolecular arrangement of fibers by twisting the mono-filament under a hot condition, which can provide fundamental references for change fiber inner structure to alter fiber properties.

INTRODUCTION

Fiber properties are determined by its inner structure. It is generally considered that the higher parallel and compact macromolecular arrangement in fiber lengthwise can result in more crystallization within fiber inner structure to better the fiber tenacity. To achieve this theory, hot drawing processing [1] is usually conducted to improve the orientation and flatness of macromolecule inside fibers and to increase fiber strength. Practical results [2] show that a single hot drawing processing does not only improve fiber strength, but also increase fiber rigidity and fiber modulus. Unfortunately, the increased fiber modulus and rigidity will deteriorate the fiber softness, leading to the corresponding fibers become unsuitable for producing fabric with the high softness and excellent drapability. Previous study [3-4] shows that when fiber macromolecules are spiral assembled to construct the fiber, the fiber modulus will be significantly reduced to improve fiber softness. However, the hot drawing of filaments fails to get the individual fiber rolling along its own axis; thus the conventional hot drawing processes cannot endow the fibers with a helical structure.
To settle aforementioned issues, this study exerts the fiber inner structure with α spiral phase macromolecular arrangements by twisting the mono-filament under a hot condition. The resultant filament tensile properties are also comparatively investigated.

APPROACH

Materials Polyester filament with 0.3mm diameter is provided by the Xieyongfu Industial Co. Ltd., Dongwan City, Guangdong Province.

Twisted filament preparation Self-constructed heating device is installed on an experimental ring frame shown in Figure 1. The monofilaments are inserted with 100 twists/meter, 200 twists/meter, 300 twists/meter, 400 twists/meter, and 500 twists/meter, respectively. The spindle speed is constant at 5115 rpm. During the twisting treatment, the monofilament keeps contacting on the hot surface of heating device. After the treatment, all the samples are conditioned for 6 hours in an oven with the temperature of 80 ℃

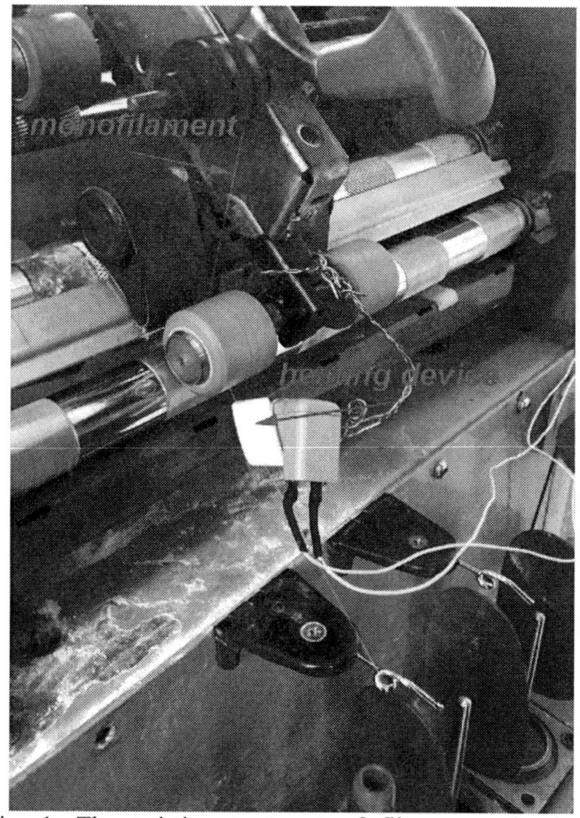

Fig. 1. The twisting treatment of filament under heat condition

Testing information The tensile property of filament fibers is tested on an INSTRON5500 type tester with the following input parameters: drawing gauge 20mm; drawing speed 50.00mm/min. All the testings are conducted under a standard atmospheric condition (20±2 °C and 65±2% RH). In specific, all the samples are conditioned more than 24 hours under a standard atmospheric condition (20±2 °C and 65±2% RH) before testing.

RESULTS AND DISCUSSION

The tensile parameters of monofilaments before and after twisting under a hot condition are listed in figure 1. It is obvious shown that breaking length of the monofilament increases slightly as the inserted twist ratio increases. This can be attributed to the increased fiber shrinkage of the twisting. However, the monofilament breaking force decrease significantly as the inserted twist ratio increases. This is because the higher twisted monofilament corresponds to the larger α spiral deformation of the inner macromolecular structure; the larger spiral deformation of the macromolecular arrangements will lead to more macromolecules shearing to decrease the breaking force.

Table I. Tensile parameters of monofilaments before and after the twisting treatment under a hot condition

Avg	max.disp (mm)	max.load (N)	Modulus (N/Tex)	energy.YLD (J)
100twist	35.88	51.91	13.54	1.117
200twist	36.07	48.14	12.85	0.943
300twist	37.60	48.21	12.39	0.941
400twist	37.47	45.52	11.92	0.795
500twist	37.21	42.35	11.80	0.821
0twist	37.30	50.29	13.22	1.099

In specialty, the monofilament initial modulus decreases significantly as the inserted twist ratio increase. This indicates that the higher twisting ratio corresponds to the larger α spiral deformation of the inner macromolecular structure; the larger spiral deformation of the macromolecular arrangements will better the softness of monofilament by decreasing the fiber initial modulus.

CONCLUSIONS

The result shows that breaking length and initial modulus of the monofilament increases slightly as the inserted twist ratio increases, while the monofilament breaking force decrease significantly as the inserted twist ratio increases. The results indicate that the higher twisting ratio corresponds to the larger α spiral deformation of the inner macromolecular structure; the larger spiral deformation of the macromolecular arrangements will lead to more macromolecules shearing to decrease the breaking force in one hand, and better the softness of monofilament by decreasing the fiber initial modulus in other hand.

ACKNOWLEDGMENT

We greatly acknowledge the support from the National Natural Science Foundation of China (Project No. 51403161) and Department of Science & Technology of Hubei Province (Project No.2014CFB755).

REFERENCES

[1] Buckley, C.P., Lew, C.Y. "Biaxial hot-drawing of poly(ethylene terephthalate): An experimental study spanning the processing range." *Polymer*, 52, 2011: 1803-1810.

[2] Vasanthan, N. "Effect of Heat Setting Temperatures on Tensile Mechanical Properties of Polyamide Fibers." *Textile Res. J.*, 74, 2004: 545-550.

[3] Tan, Qingbiao, He, Guo. "Stretching behaviors of entangled materials with spiral wire structure." *Materials & Design*, 46, 2013: 61-65.

[4] Nakamura, Y., Kurihara, C., Machida, T., Inada, R., Oota, A. "Effect of filament twist on the structure and properties of Ag-sheathed Bi2223 tape." *Physica C: Superconductivity*, 469, 2009: 1492-1495.

Flame-Resistant Protective Clothing in Industries

Zina Zhu, Shulong Li, Scott Margolin
Milliken & Company
Contact: zina.zhu@milliken.com; shulong.li@milliken.com; scott.margolin@milliken.com

OBJECTIVE

Flame-resistant (FR) and arc flash protective clothing has been widely used to protect workers against the flash fire and arc flash hazards in various industries such as oil&gas exploration, metallurgic, electric utility, power generation, chemical manufacturing, building and construction. It is important to recognize that fabric used to construct the garment is a vital factor in determining the amount of protection the garment will afford the wearer. This presentation will explore characterization of protection performance of fabric materials against industrial flame and electric hazards, to gauge real world performance. Some of the common national and international standard test methods and specifications, such as GB 8965.1, National Fire Protection Association (NFPA) 2112 and NFPA 70E will be compared. General trends for broader adoption and new development of protective clothing materials to save lives and reduce injuries will be presented as well.

INTRODUCTIONS

Every day, workers in maintenance and electrical, utilities, oil, gas, petrochemical, chemical, steel and many other industries work in environments that may expose them to hazards that could cause severe or fatal burns. Over the past decade, the FR clothing has drawn lots of attention due to relative regulations and an increased awareness of the importance of wearing FR clothing when a potential exposure to an electric arc flash, flash fire or molten metal splash hazard exists. FR clothing as personal protective equipment (PPE) have been widely used in developed countries as they are designed to protect and save lives.

FR PROTECTIVE CLOTHING

FR protective clothing is an engineered product to provide protection and save lives for workers exposed to flash fire/arc flash hazards in the working environment. Instead of not being ignitable, FR clothing suggests the garment will self-extinguish once the fire resource has been removed, which can offer secondary protection for the wearer to avoid severe burn. Fabric determines the protective performance the garment will conduct, at the same time, it is also a main factor in determining the comfort, durability and the overall value proposition of a FR clothing. Sometimes FR fabric is categorized via fiber's content, actually fiber is only one component of the fabric, which cannot directly predict the property of the fabric, not even the clothing. In other words, protection afforded by the garment depends almost entirely on the properties of FR fabric instead of fiber itself.

Diffusing fuel, air, and ignition source yield a rapidly moving flame front of about 84 kW (or 2 cal/cm^2) for relatively short durations, typically 3 seconds or less, is defined as flash fire by NFPA 2113 technically. The duration of heat levels is sufficient to ignite flammable clothing or cause 2nd degree burns to exposed skin. Via creating a chemical environment around the garment that dilutes the oxygen fuel source and limits combustion, FR clothing will self-extinguish and provide the protection.

Electric arc flash is a breakdown of the resistance of air resulting in an electric arc that can occur where there is sufficient voltage in an electrical system and a path to ground or lower voltage. Arc flash will release quite high energy, exposure of skin can result in severe thermal burns, and non-FR clothing can ignite and/or melt and contribute to burns. Arc flash protection clothing can create a barrier to limit burns to the wearer via reflection and absorption of arc energy release.

CHARACTERIZAITON OF PROTECTION PERFORMANCE

In order to understand protection performance of FR clothing, it is essential to understand what the standards are really measuring and how they are correlated to protection performance. It is also critical to keep in mind that compliance with standards is not enough, since there's such a wide disparity in performance among compliant fabrics. Specified common national and international standard test methods and specifications are discussed as below.

NFPA 2112 is US standard on flame-resistant garments for protection on industrial personnel against flash fire. The manikin test required by NFPA 2112 uses the ASTM (American Society of Tests and Methods) F1930 standard; ASTM F1930 features a full size manikin, wearing a standardized coverall, in a burn chamber with propane torches capable of fully engulfing it. NFPA 2112 standard requires a manikin test duration of 3 seconds precisely because it is viewed as the practical upper limit of a flash fire. Total body burn area (including 2nd and 3rd degree burn) cannot exceed 50% according to NFPA 2112, as body burn area directly affects the survival rate. The less the body burn area, the more protection the garment can provide. Manikin test results will directly indicate the protection performance of the protective clothing in the real hazardous environment. Thermal shrinkage is another vital factor which will affect the protection performance. When fabric shrinks under thermal exposure, the air gap between skin and clothing will be reduced, leading to additional burn, besides, due to the thermal shrinkage, skin may get exposed to the fire. Another risk is shrunk garment might not be easy to take off, which will stop the worker escaping from the hazard site quickly. Smaller thermal shrinkage can be correlated to better protection. GB 8965.1 is China national standard on flame-resistant protective clothing. Vertical burn test is required both in NFPA 2112 and GB 8965.1, which can tell if the fabric will self-extinguish or not after the flame is away, however, vertical burn test results only cannot predict the protection performance of the fabric.

NFPA 70E is US standard for electrical safety in the workplace. NFPA 70E, and ASTM F1959 standard on which it relies for arc ratings, are both excellent standards which save lives on a daily basis. The emphasis on de-energizing wherever possible, recognition of the catastrophic or fatal effect of arc flash on flammable clothing, and the requirement for flame resistant, arc rated clothing whose protection levels exceed the hazard are seminal changes in electrical safety. Arc rating is reported as either ATPV (arc thermal performance value) or Ebt (Energy to break open threshold), whichever is the lower value. ATPV essentially predicts the energy level below which a second degree burn is unlikely through the fabric. It is critical to choose the right arc rating protective clothing regarding to electric arc hazard energy level.

GENERAL TRENDS

With the development of industrialization, increasing awareness of safety mindset, promoted business ethics and improved relative laws and regulations, FR clothing has been an indispensable part in modern industries. More and more industries and companies have engaged in the application of FR protective clothing for providing personal protection. Due to the various working hazard, multi-hazard protection such as FR and arc flash protection, FR and molten metal protection have been urged accordingly for different industries. At the same time, comfort becomes more and more important during the development of FR protective clothing, since the protection can be fulfilled when workers are willing to wear them. According to the emphasis on environment protection and sustainability, protective clothing also need to be halogen or heavy metal like antimony free, which cannot harm environment or human's health.

Fully-Printed Flexible Temperature Sensors on Textile Substrates for Textronic Applications

Z. Stempień[1], E. Rybicki[2], K. Gniotek[1], J. Lesnikowski[1],
[1]Institute of Textile Architecture; [2]Department of Molecular Physics, Lodz University of Technology, Lodz, Poland
zbigniew.stempien@p.lodz.pl

INTRODUCTION

Body, underwear and ambient temperatures are one of the most commonly monitored parameters in smart and textronic clothing, especially in protective clothing dedicated to certain occupational groups like casters, steel workers, etc. Too high temperatures can lead to burns or hyperthermia whereas too low to hypothermia. In viewpoint of comfort, the sensors embedded to clothing should be flexible and thin and as small as possible. One way of obtaining such sensors is to use a digital ink-jet printing technique. It allows fabrication different sensors directly on the textile substrates.

We report on the fabrication of fully-printed temperature sensors, based on conductive polymers, for the integration into textronic clothing as a sensing device for monitoring ambient and/or underwear temperatures. For this purpose, temperature sensors were developed by deposition a conductive polymers which have a temperature dependent response, such as PANI or PEDOT, on textile substrates. Generally, the disadvantage of the temperature sensors based on conductive polymers is the dependence of polymer layer resistance on ambient temperature and humidity[1,2]. It causes that the practical applications of these types of temperature sensors are very limited. We propose to eliminate the influence of humidity on polymer layer resistance by covering the both sides of this layer by a thermal activated adhesive foils. The static and dynamic responses of temperature sensors were analyzed in the range of temperatures from 10°C to 70°C. The measured resistances of sensors versus temperature for PANI and PEDOT thermoactive polymer layers showed a non-linear distribution, lack of hysteresis and a very good repeatability over the tested temperature range.

DESIGN AND FABRICATION

Sensor Structure

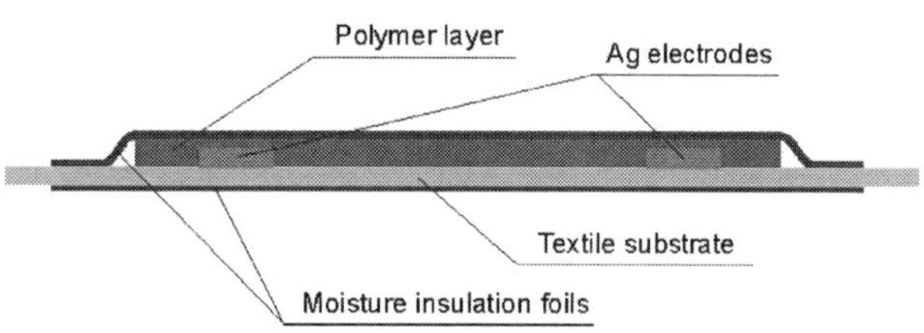

Figure 1. Printed temperature sensor cross section

In Figure 1, the proposed structure of fully printed temperature sensor on textile substrate is shown (Polish Patent Pending). In the presented sensor structure, the electrical conductivity response of thermoactive polymer layer, made of PANI or PEDOT, under the influence of temperature changes is analyzed. In this order, the resistance of the thermoactive polymer layer between two silver electrodes is measured. Both thermoactive polymer layer and silver electrodes are deposited directly on textile substrates by using digital printing techniques. It is well known that the resistance of conductive polymer layers depends on ambient humidity and temperature. It causes that their practice applications as an active element of humidity or temperature sensor are very limited. For this reason, in order to eliminate the influence of moisture on the resistance of conductive polymer layers, both sides of sensor are insulated by a thermal activated adhesive foils.

Fabrication Process

Basalt woven fabrics produced by the Basaltex a.s. Company (Czech Republic) were taken as a textile substrates. The fabric warp and weft densities, areal density and fabric thickness were 10 threads/cm, 350 g/m^2 and 0.47 mm respectively. This type of woven fabric was chosen because is usually used for making personal protective clothing for casters and steel workers. One of our main objectives of research conducted is to equip this occupational group with textronic system for monitoring temperature in an underwear area to protect workers against burns and hyperthermia.

For the fabrication of temperature sensors on textile substrates a prototype of the digital ink-jet printer was used. It contained: 1) nanodispensing hardware, represented by two one-nozzle valve-jet print-heads (ReaJet SK 1/080, Germany) and drive-electronics for generating electrical signals to the print-heads; 2) an ink pump system and two ink reservoirs; 3) a two-axis (x-y) motion system for print-head positioning with a step resolution of 35 μm, based on the step-motor linear drives.

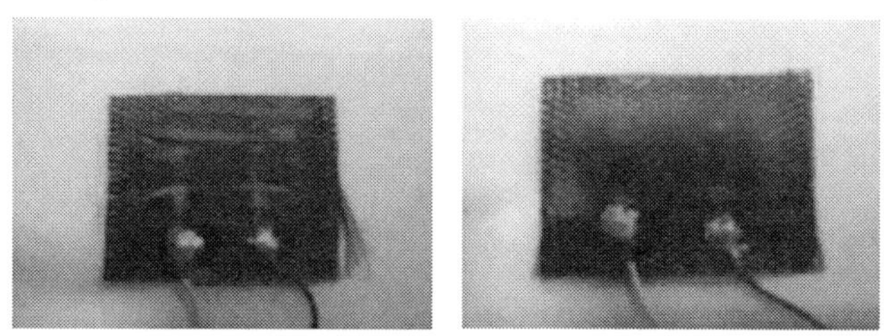

Figure 2. Examples of ink-jet printed temperature sensors on basalt fabrics with PANI (left) and PEDOT (right) layer

In the first step of fabrication, two silver line-shaped electrodes were printed on textile substrate by using the commercial UTDAg40 silver nanoink (UT DOTS Inc., USA). Next, the printed electrodes were sintered in an oven at 120°C for 10 minutes. The distance between electrodes was 10 mm. In the second step, the polymer PANI or PEDOT layer in size of 6 mm x 15 mm was printed on the top of silver electrodes. The PANI layers were deposited in-situ by the oxidative ink-jet printing of aniline hydrochloride at a concentration of 0.6 M/dm^3 with ammonium persulfate[3]. Finally, the printed temperature sensors were insulated from moisture by covering the polymer layers of thermal activated adhesive foils at two sides. In Figure 2, the examples of ink-jet printed temperature sensors on basalt fabrics with PANI and PEDOT layers are shown. The wires were connected to the silver electrodes by using a silver based glue.

TEST AND CHARACTERIZATION

The fabricated polymer temperature sensors were first calibrated and then characterized for reproducibility and hysteresis errors using an environmental test chamber Microcal T-100 (Eurotron, Italy) maintained at a series of temperatures which are measured with high accuracy temperature reference probe Tempmaster T100 (Labfacility, UK).

In the first step of calibration, the static characteristics for the sensors with PANI and PEDOT thermoactive polymer layers were determined in the range of temperature from 10°C to 70°C (Figure 3). When a specific test chamber temperature was set, the sensor resistance was measured when it has stopped changing. In order to analyze the temperature sensor hysteresis errors, the sensor resistance was measured when the test chamber temperature was gradually raised from 10°C to 70°C and then gradually lowered to 10°C. This measurement process was repeated two times for each sensor.

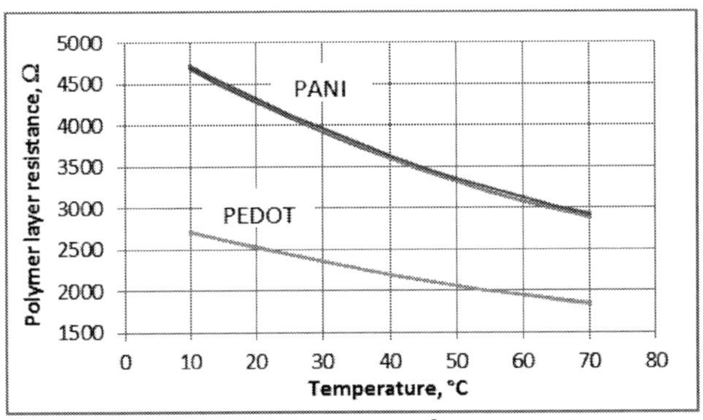

Figure 3. Resistance vs. temperature for temperature sensor with PANI or PEDOT layer

The measured resistance of the polymer temperature sensor was observed to vary non-linearly with temperature. The dependence could be very good represented by the second order polynomial relationship. For the both conductive polymers their resistances decreases when the temperature in the test chamber increases. The sensitivities of the temperature sensors with PANI and PEDOT layers are ca. 30 Ω/°C and ca. 15 Ω/°C respectively. As it is shown in Figure 3, no significant hysteresis errors can be observed for both temperature sensors. A small width of hysteresis loop is for the temperature sensor with PANI layer. For the temperature sensor with PEDOT layer, in practice, there is no hysteresis.

In the second part of characterization the dynamic properties of fabricated temperature sensor are analyzed. For this purpose, the tested temperature sensor was placed in the fitted foil bag and then placed in the glass beaker filled with silicon oil at a temperature of 22°C. Next, when the sensor resistance has stopped changing, the sensor was rapidly moved to the chamber of compact temperature calibration baths 7340 (Fluke, USA), filled with silicon oil at a temperature of 60°C. In Figure 4 the step responses of fabricated temperature sensors with PANI and PEDOT layers are shown when the measured temperature was rapidly changed from 22°C to 60°C and back.

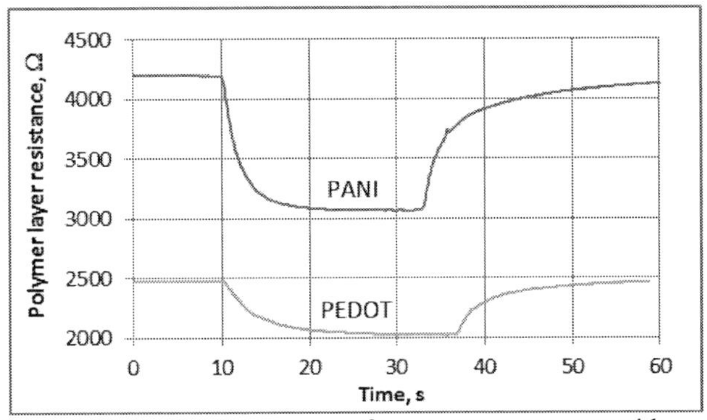

Figure 4. Step input response of temperature sensors with PANI or PEDOT layer

As it is shown in Figure 4, the temperature sensor with PANI layer is a little bit faster from the temperature sensor with PEDOT layer. The values of time constant for these sensors are ca. 3 s and ca. 5 s respectively.

CONCLUSIONS

The fully printed temperature sensors were developed by deposition the PANI or PEDOT conductive polymer on basalt woven fabrics. The output resistance of the fabricated temperature sensors does not depend on ambient humidity because of insulating the both sides of thermoactive polymer layer of thermal activated adhesive foils. The static characteristics of temperature sensors show a non-linear distribution, lack of hysteresis and a very good repeatability over the tested temperature range. It causes that presented sensors can be successfully applied in textronic clothing for monitoring an ambient, skin and underwear temperatures.

ACKNOWLEDGMENTS

The presented research was carried out within the scope of the third stage of the National Programme "Improvement of safety and working conditions" partly supported in 2014–2016 within the scope of state services by the Ministry of Labour and Social Policy. The Central Institute for Labour Protection National Research Institute is the Programme's main co-ordinator.

REFERENCES

[1] Daoud, W.A., Xin, J.H., Szeto, Y.S. *Sensors Actuators B*, 109, 2005: 329-33.

[2] Kinkeldei, T., Zysset, C., Cherenack, K.H., Troster, G. "A textile integrated sensor system for monitoring humidity and temperature." Transducers: 16th Int. Solid State Sensors, Actuators and Microsystems Conf., Beijing, 2011.

[3] Stempien, Z., Rybicki, T., Rybicki, E., Kozanecki, M., Szynkowska, M.I. *Synthetic Metals*, 202, 2015: 49-6.

Scalable and Controllable Fabrication of Conductive Fibers and Fabrics Through Facile Dip-Dying Approach: New Design and Potential Applications

Zong-Yi Qin, Qing-Qing Fan, Xiao-Lin Zhang, Xin Liu, Zhen-Jun Dou, Zhe Zhou, Long Chen, Mei-Fang Zhu
State Key Laboratory for Modification of Chemical Fibers and Polymer Materials, and College of Material Science and Engineering, Donghua University, Shanghai, China
phqin@dhu.edu.cn

ABSTRACT

With the progressing of information technology and growing awareness of healthcare, functional fibers and fabrics have attracted more and more attention. This paper introduces our recent works on the fabrication and potential application of conductive fibers and fabrics. This paper shows that the functionality can be created by extrinsic modifications to commercial fibers and fabrics by using a facile dip-dying approach, depending on the level of integration into the fibrous platform. These conductive composite fibers and fabrics appear to be more attractive in practical applications because of their good environmental stability, easy processing ability, low cost, wearable and tailorable properties.

INTRODUCTION

As the growing of electronic industry, conductive fibers and fabrics which combine the flexibility and weavability of traditional fibers and the sensing properties of functional materials, have been applied in many fields such as antictatic, electromagnetic shielding, smart clothing, sensors and super capacitor. Conductive fibers and fabrics are fabricated by integrating electronic functionality into fibrous structures through different methods such as polymerization, metal wire insertion and coating techniques. Coating techniques have become attractive in the field of conductive fibers and fabrics because fibrous materials with a large surface area possess several advantages over conventional materials, such as high flexibility and mechanical properties, making them good light-weight substrates to further deposit some functional materials for different applications. Moreover, the process of coating is relatively simple, cheap and materials obtained maintain their mechanical properties, along with electrical properties. Recently, our group has used a facile dip-dying as a noninvasive approach to fabricate electronically conductive and flexible polymer fibers by fixing conductive networks such as carbon nanotube (CNT), graphene and polyaniline (PANI) as a thin layer on the surface of commercial fibers and fabrics [1-4]. Herein, the fabrication, structures and potential applications for three kind of conductive polyurethane (PU)-based fibers and one kind of conductive poly(ethylene terephthalate) (PET) nonwoven fabrics would be introduced.

CNT/PU COMPOSITE FIBER

The conductive elastic CNT/PU fibers were prepared by immersing PU multifilaments into CNT dispersion in chloroform [1-2]. The anchoring of the CNT layer is achieved by partial embedding or penetrating of CNTs from dispersion into the swollen fiber surface. Thus a stable and high conductivity (up to 100 S/cm at 10 wt.% CNT loading) of the resulting CNT/PU fibers is realized while the mechanical properties of the PU fiber, especially the elongation at break of >1500 %, are not affected by increasing the thickness of the CNT layer as shown in Fig. 1.

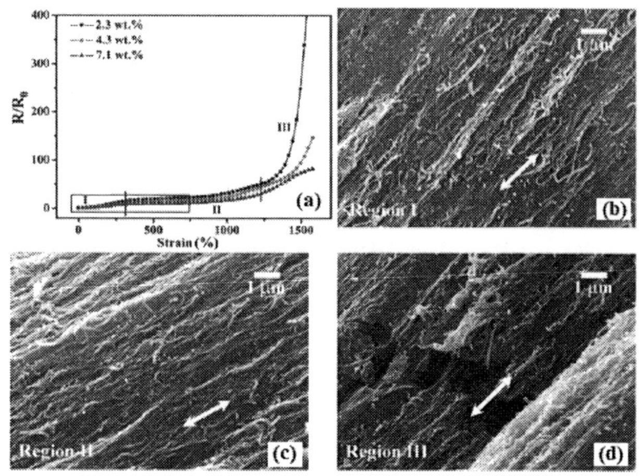

Fig. 1 Relative resistance of CNT/TPU fibers with various content of CNTs under stretching at a displacement rate of 5mm/min (a), and SEM images of stretched CNT/PU fibers with 4.3 wt.% CNT loading at low (150%) (b) , medium (500%) (c) , and high (1100%) strain (d) , respectively.

PANI/PU COMPOSITE FIBER

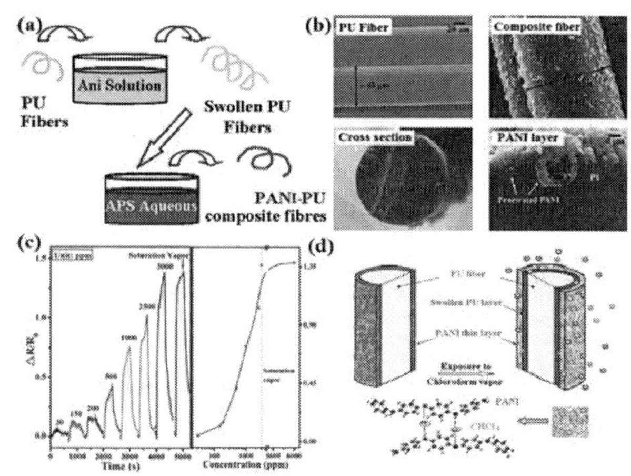

Fig. 2 Fabrication process (a), surface and cross section images (b) for PANI/PU composite fiber, and its sensing behaviors (c) and mechanism (d) for chloroform vapor detection.

PANI/PU composite fibers were prepared through placing PU fibers swollen by aniline monomers in 1 M HCL aqueous solution into 1 M aqueous solution of ammonium persulfate (APS) in an ice bath as shown in Fig. 2. The thickness of the PANI layer could be controlled by adjusting the content of aniline monomers. Some features for chloroform vapor detection such as high sensitivity (4×10^{-4} ppm^{-1}), fast response time (less than 30 s), and low detection limit (30 ppm) were achieved. The remarkable sensing performances of the composite fibers are not only related to the change in the conformation of the PANI chains, but also in the conducting pathway of the PANI layers due to the reversible solvent-swelling of the PU matrix [3].

PANI/CNT/PU COMPOSITE FIBER

Three phase PANI/CNT/PU composite fibers were prepared by in-situ chemical oxidative polymerization of PANI on CNT layers of CNT/PU fibers. The effective site-selective interaction between the π-bonds in the aromatic rings of the PANI and the graphitic structure of CNTs would strongly facilitate the charge-transfer reaction between the two components. As a result, the PANI/CNT/PU composite fibers shows improved sensing properties as shown in Fig. 3. For the strain sensing properties, reversible piezoresistivity is demonstrated and the strain gauge factor of the fiber under the strain deformation of < 200 % is about 20, which is much higher than those of PANI/PU and CNTs/PU fibers. Higher vapor sensitivity is also achieved for the PANI/CNT/PU fibers in various organic chemical vapors.

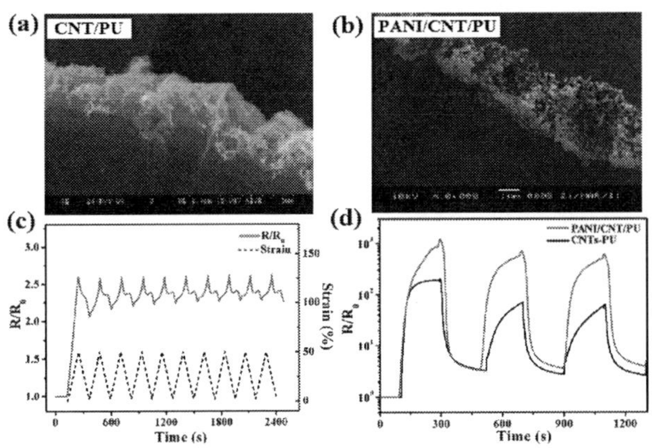

Fig. 3 SEM images of conductive layer for CNC/PU (a) and PANI/CNT/PU composite fiber, and the strain sensing (c) and gas response behaviors for chloroform vapor detection (d).

RAPHENE/PET NONWOVEN FABRIC

Graphene/PET composite fabrics were fabricated by using reduced graphene oxide (rGO) in aqueous solution as sheet dyestuff. To ensure the structure stability of composite fabrics, PU as thin adhesive layer was introduced [4]. These composite fabrics exhibited a rather low electrical percolation threshold due to the homogeneous coverage of graphene sheets on the surface of the PET nonwoven fabrics, and the highest conductivity was about 2.0×10^{-5} $S\cdot sq^{-1}$, ensuring feasibility to manufacture efficient heating elements. Importantly, the direct use of an aqueous dispersion of rGO in dyeing opens up new possibilities for the mass production of graphene-based innovative fabrics.

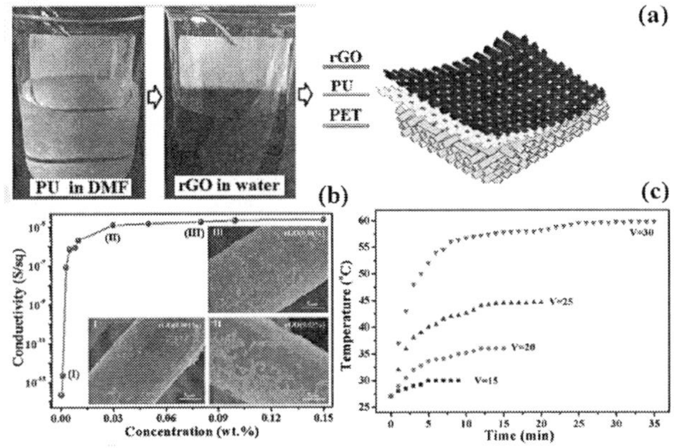

Fig. 4 Fabrication process of rGO-coated PET nonwoven fabrics (a), change in the surface conductivity and SEM observation of the composite fabrics with the increase in the weight fraction of rGO in the initial aqueous dispersion (b), and time dependence of temperature for the composite fabrics (c).

SUMMARY

The universal modification of commercial fibers and fabrics through the facile dip-dyeing approach by penetrating functional components into their surface layer can be extended to the designation of multi-functional polymer-based fibers by simply changing the solution solvent and the functional components. What is more, the direct use of rGO as a sheet dyestuff in an aqueous dispersion opens up new applications of nanotechnologies in fiber, fabric or textile production. For example, the interaction between graphene and metal oxide nanocrystals or conducting polymers provides hybrids with additional properties, offering rich opportunities for the composite fabrics to tune the materials' structure and properties. Thus, the dip-dyeing approach exhibited great potential for scalable and controllable fabrication of conductive fibers and fabrics, and would greatly extend their application as multi-functional flexible devices.

ACKNOWLEDGMENT

This work has been financially supported by National Natural Science Foundation of China (21274019), and Program for Changjiang Scholars and Innovative Research Team in University (IRT1221)

REFERENCE

[1] Fan, Q., et al. *Carbon*, 50, 2012: 4085-4092.

[2] Fan, Q., et al. *Sens. Act. B.*, 156, 2011: 57-61.

[3] Zhang, X., et al. *J. Mater. Chem. A.*, 1, 2013: 10327-10333.

[4] Liu, X., et al., *RSC Adv.*, 4, 2014: 23869-23875.

Synthesis of Triarylmethane Reversible Thermochromic Colorant Using Bromcresol Purple for Smart Fabrics

Wan Zhang, Chaoxia Wang, Yunjie Yin
College of Textiles and Clothing, Key Laboratory of Eco-Textile, Ministry of Education, Jiangnan University, Wuxi, China
wangchaoxia@sohu.com; zhangwan_wuxi@163.com

OBJECTIVE
A novel triarylmethane reversible thermochromic colorant was prepared using bromcresol purple to be cheaper to produce and obtain a brighter colour. The effects of monomer mass ratios, developer and solvent and synthesis conditions on thermochromics properties were investigated.

KEYWORD
thermochromism colorant; triarylmethane; reversibility

INTRODUCTION

1. The extents of reactions were evaluated on K/S values and the sharply chromatism ($\Delta E=((\Delta a^*)^2+(\Delta b^*)^2+(\Delta L^*)^2)^{1/2}$) would be presented by the better thermochromic colorant at thermochromics temperature.
2. The curves of DSC analysis demonstrated thermochromics temperatures can be regulated by mixing solvents.
3. The triarylmethane thermochromic colorant presented the same reflectance curve after 20 times reverse indicated the stable thermochromic property with the synthesized conditions: the mass ratio of bromcresol purple (BCP): stearic acid (SA): hexadecanol (HD) = 0.2:3:20 and reaction at 90°C for 24h with 1600r/min.

APPROACH

Bromcresol purple (BCP), acids and solvents were added into beaker and sealed, reaction for 3-36h at 50-90°C with magnetic stirrer. Colorants were tested after 24h reserved at low temperature and avoiding light [1], shown in Fig. 1.

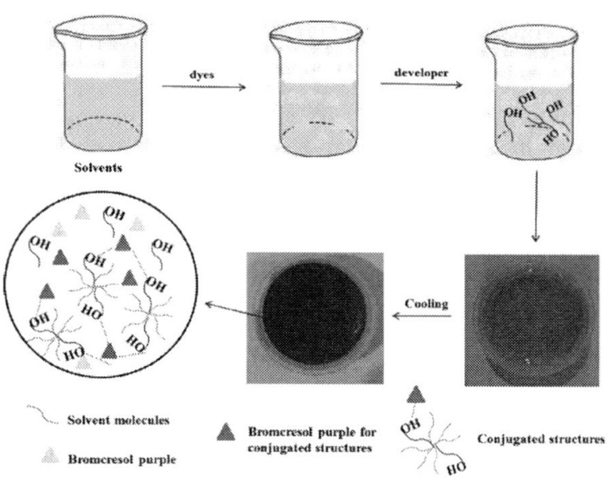

Fig. 1. Schematic diagram of preparation triarylmethane reversible thermochromic colorant

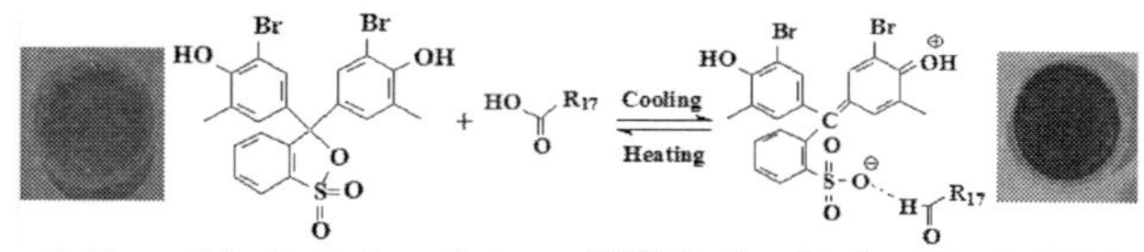

Fig. 2. Schematic illustration of thermochromic colorants synthesis by stearic acid and HD at 30 and 60°C.

RESULTS AND DISCUSSION

The influence of the solvents and the developers on thermochromism colorant

Chromatism temperatures were related with the melt points of solvents [2] and 1-10°C lower. The possibility of regulating chromatism temperature was realized via mixing solvents.

Colorants synthesis by bisphenol A showed chromatism in first 2s, materials synthetic by boric acid presented chromatism in 12s and 8s for materials synthesis by SA (heating in one water bath), shown in Fig. 3. Developer agents can influence the reacting time by the structures of developer agents as well as the cooperation with other materials.

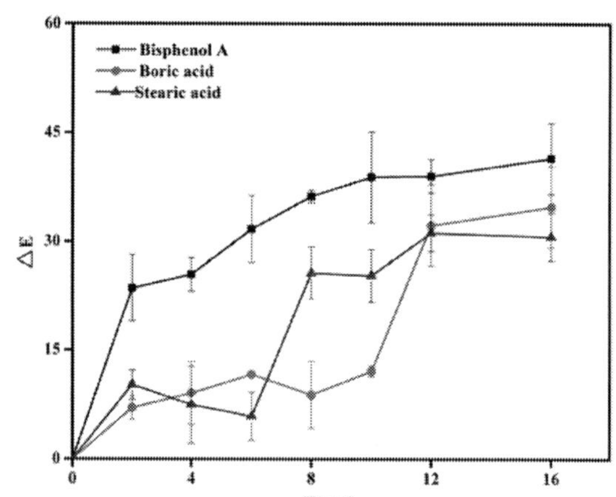

Fig. 3. The reacting time of colorant synthesized by bisphenol A, boric acid and SA

The influence of the mass ratios and the synthesis conditions on thermochromism colorant

K/S values increasing as bromocresol purple were added. However, when BCP was excessive, K/S values would decrease. Because the reactions can be hampered by aggregate BCP and resulted in scatting problems; K/S value increased with the adding of octadecanoic acid. Less HD can result in insoluble of octadecanoic acid leading to colorless but excessive HD can be remained leading to colorless too, shown in Fig. 4.

The synthesis conditions were necessary to discuss, as shorter synthesis time cannot scatter compounds well (BCP was only 0.8%) leading to incompletely reaction, but synthesis for a longer time can cause conjugated structures degradation. When temperature was low, octadecanoic acid was insoluble, when temperature was higher, octadecanoic acid would volatilize. Dispersion speed influenced the dispersion of reactants and the forming of conjugated structures. With high dispersion speed, reactants dispersed well but may result in conjugated structures degradation.

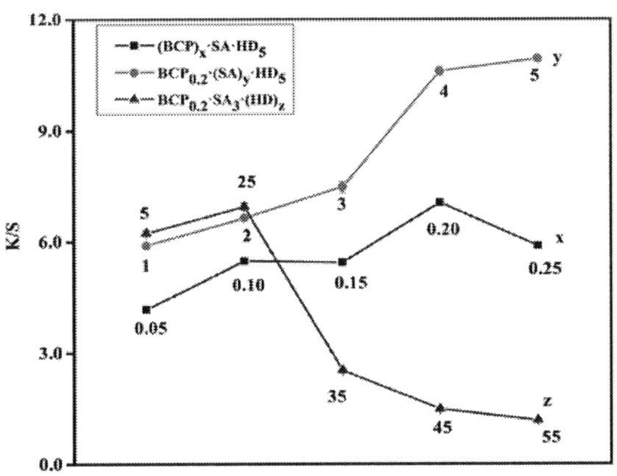

Fig. 4. The extents of reactions of colorants synthetic with different mass ratios

The thermodynamic behaviors of thermochromic colorants

Colours of colorant were almost unchanged after 10 times or 20 times heating-cooling, which indicated a good reversible property, shown in Fig. 5. And the maximum reflectance of thermochromic colorant was 650-700nm, which was indicated the red colour. ΔE reached 40.04 at chromatism temperature (35-40°C) of thermochromic colorants, which demonstrated a good thermochromic property.

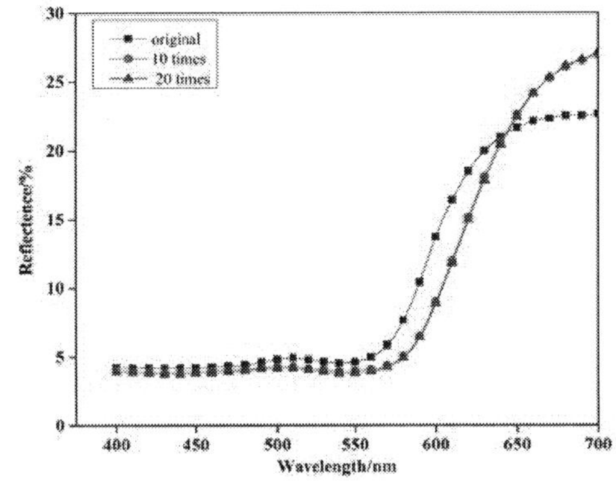

Fig. .5. Reflectance of thermochromic colorants

CONCLUSION

1. A cheaper triarylmethane reversible thermochromic colorant with a brighter colour was prepared by using bromcresol Purple.
2. The preparation conditions were concluded: the mass ratio of bromcresol purple (BCP): stearic acid (SA): hexadecanol (HD) = 0.2:3:20 and reaction at 90°C for 24h with 1600r/min.
3. Reacting time can be changed by using different developer agents and thermochromics temperature can be regulated by mixing solvents.
4. The triarylmethane thermochromic colorant presented stable thermochromic property after 20 times reverse, which provided a possible way using on smart fabrics.

ACKNOWLEDGMENT

The authors are grateful for the financial support of National Natural Science Foundation of China (21174055) and the International Joint Research Laboratory for Advanced Functional Textile Materials at Jiangnan University.

REFERENCES

1. Friškovec, M., Kulčar, R., Gunde, M.K. *Coloration Technology*, 129, 2013: 214-222.
2. Azizian, F., Field, A.J., Heron, B.M. *Dyes and Pigments*, 99, 2013: 432-439.

Simulation of Stoll's Empirical Burn Criterion Using a Mathematical Model for Evaluation of Thermal Protective Clothing

Li-Na Zhai[1,2], Jun Li [1,2,3]
[1]Fashion institute, Donghua University, Shanghai, China
[2]Protective Clothing Research Center, Donghua University, Shanghai, China
[3]Key Laboratory of Clothing Design &Technology, Donghua University, Shanghai, China
lijun@dhu.edu.cn; 03lena03@163.com

INTRODUCTION

In order to rank thermal protective performance of materials and clothing which are exposed to flash fire or radiation condition, prediction methods of skin burn injury are widely used in the related test methods. Stoll's empirical criterion is often used in the bench top tests while the mathematical model which is based on Henriques Integral equation are mainly used in the full-scaled manikin test.

Song[1] have compared the different burn prediction results using this two methods. However, little was known why the discrepancy was induced because it's hard to compare the Stoll's criterion and mathematical model in the same scale. In our study, we have tried to solve this problem by numerical simulation of the Stoll's curve so that these two methods can be compared directly.

APPROACH

In earlier test methods, temperature form of Stoll curve are suggested, but nowadays, in recent published standards, such as ASTM F2700-08[2], ASTM F1939-08[3], ASTM F2703-08[4], ASTM F2702-08[5], an energy form of Stoll curve are recommended. As the energy form is more flexible without considering the type of the used calorimeter, it is an idea target for the simulation.

The mathematical prediction model involves a bio-heat transfer model of skin and a burn injury model, many factors, such as partial differential equation and skin thermal properties, will influence the prediction results. In order to be less affected by other sophisticated factors, we chose the simple and effective model proposed by Torvi in 1992[6], which is also suggested by ISO 13506 standards[7] in our study.

The temperature field of skin is assumed as a transient one dimensional heat diffusion problem using the simplified Pennes bio-heat transfer equation. Parameters are as Table 1.

$$\rho C_p \frac{\partial T}{\partial t} = k\nabla^2 T + \omega_b \rho_b C_{p,b}(T_b - T) \qquad (1)$$

Initial and boundary conditions are as equation 2 to equation 5.

Table 1. Thermal physical properties of human skin

Property	Epidermis	Dermis	Subcutaneous Tisue
Thickness（μm）	80	2000	10000
Thermal conductivity(W/m·K	0.255	0.523	0.167
Density（kg/m^3）	1200	1200	1000
Specific heat	3598	3225	2760
Blood perfusion rate（$m_b^3/s/m^3$）	0	2000	0.00125

$$T(x,t=0)=T\_i\ (x,t=0),(0 \leqslant x \leqslant L) \qquad (2)$$

$$q(x=0,t)=q(t),(0 \leqslant t \leqslant t\_exp) \qquad (3)$$

$$q(x=0,t)=0,(t > t\_exp) \qquad (4)$$

$$T(x=L,t)=T\_c=37\,℃,(t \geqslant 0) \qquad (5)$$

Where t_{exp} is the exposure time and q_i represents heat flux incident on skin surface which is defined according to Stoll's 9 cases of experimental data[8]. The initial conditions of the skin ($T_i(x, t = 0)$) are represented by a linear temperature distribution from the epidermis surface (32.5°C) to the subcutaneous base (37°C).

In the simulation, grid size is 0.00001m and time step size is 0.01s. Thermal physical properties of skin are according to Table 1, which are obtained from Torvi[6]. Once the temperature filed of skin is solved, time to second degree burn will be determined when the integral parameter Ω reach 1.0 respectively at two depths (epidermis–dermis interface and dermis-subcutaneous interface) using equation 6. Parameters in the integral equation are according to ISO 13506[7].

$$\Omega = \int_0^t P e^{-\Delta E/RT} dt \qquad (6)$$

Where Ω is the burn injury parameter; P ,e , R is pre-exponential term, natural exponential, and the universal gas constant respectively; ΔE and T is the activation energy and absolute temperature at the basal layer or at any depth in the dermis for skin; t is the total time for which T is above 44℃

.

A similar process of developing the Stoll curve is used here. Firstly, skin is exposed to 9 simulated levels of heat exposure. Secondly, 2nd degree burn time is predicted using the mathematical model. Thirdly, accumulated energy of skin surface at the time when burn is induced could be calculate using both heat flux and predicted burn time as equation (7).

$$E_i=q_i\, t_i,\quad i=1,2,3,\ldots,9. \tag{7}$$

Where t_i is the second degree burn time predicted by mathematical model. E_i is the heat energy accumulated on the skin surface when the second degree burn is induced.

RESULTS AND DISCUSSION

The simulation results of 9 cases using CON boundary are shown as the square spot in Figure 1, together with a fitted curve. This curve is considered as a simulated criterion curve of Stoll's energy curve.

It is clear that in short time and high intensity conditions, when burn time is less than 10s, model predicts very well as the experimental data. However, when the time gets longer, i.e. in lower intensity conditions, the discrepancy between the two methods become larger. This is consistent with the fact that Torvi's model is designed for flash fire exposure, without more discussion in the relative lower intensity exposure. In fact, many other factors would influence temperature field and injury burn results. For instances, optical properties of skin will influence absorbed energy; the time dependent blood perfusion, water diffusion, and vaporization will influence heat transfer in skin; the selection of activation energy and pre-exponential term is also of significance.

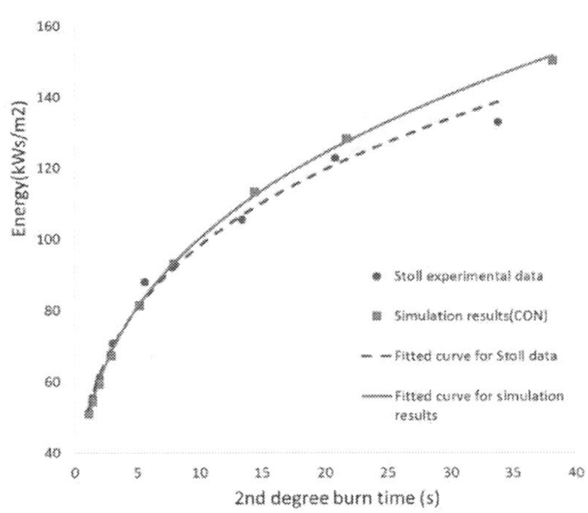

Figure 1. Simulated criterion curve and Stoll's experimental data[8]

CONCLUSION

We have successfully compared the empirical burn criterion and burn prediction model in the same scale by the simulation method. Torvi's model which is suggested by ISO standard is designed for flash fire exposure, without more discussion in the relative lower intensity exposure. Thus it does well in a higher intensity situation which seems week in lower intensities condition. In addition, there lies a discrepancy between these two prediction methods before material test begins which will contribute to the burn prediction results in practical use.

FUTURE WORK

New prediction models for the relatively longer duration of exposure at lower heat intensities should be developed for the increasing number of test methods of protective clothing which use lower heat intensities.

KEYWORDS

Thermal protective clothing; Skin burn injury; Stoll curve; Numerical simulation; bio-heat transfer

ACKNOWLEDGMENT

This work was financially supported by the Fundamental Research Funds for the Central Universities.

REFERENCES

[1] Song, G., Barker, R., Grimes, D., Thompson, D. "Comparison of Methods Used to Predict the Burn Injuries in Tests of Thermal Protective Fabrics." *Journal of ASTM International*, 2, 2, 2005: 12117.

[2] *Test Method for Unsteady-State Heat Transfer Evaluation of Flame Resistant Materials for Clothing with Continuous Heating.* ASTM F2700-08, 2013.

[3] *Test Method for Radiant Heat Resistance of Flame Resistant Clothing Materials with Continuous Heating.* ASTM F1939-08, 2008.

[4] *Test Method for Unsteady-State Heat Transfer Evaluation of Flame Resistant Materials for Clothing with Burn Injury Prediction.* ASTM F2703-08, 2013.

[5] *Test Method for Radiant Heat Performance of Flame Resistant Clothing Materials with Burn Injury Prediction.* ASTM F2702-08, 2008.

[6] Torvi, D.A. "A finite element model of heat transfer in skin subjected to a flash fire." MSc Thesis, University of Alberta, 1992.

[7] *Protective clothing against heat and flame-Test method for complete garments–Prediction of burn injury using an instrumented manikin.* ISO 13506-2008, 2008.

[8] Stoll, A.M., Chianta, M.A. "A method and rating system for evaluation of thermal protection." DTIC Document, 1968.

3D Clothes Shaping by Means of the Sewing Edges Configuration and Fabric Properties

Josefina Parodi[1], Samia Abes[1], Dominique C. Adolphe[1], Victor Kuzmichev[2], Laurence Schacher[1]
[1]Université de Haute-Alsace–Laboratoire de Physique et Mécanique Textiles–Mulhouse–France
[2]Ivanovo State Polytechnique University–Garment Department–Ivanovo–Russia
dominique.adolphe@uha.fr

INTRODUCTION

In garment industry, many assemblies deal with straight and curve sewing lines as it is shown in figure 1. In real case it could be the assembly between sleeve cup and arm hole, or in front of a dress for the bust darts, etc. This study is focused on this particular case of assembly. Through an experimental approach, 10 fabrics have been tested and quoted by using an index that has been defined especially for that purpose. Based on the obtained results correlation studies have been carried out in order to define the equation linking the mechanical properties of the fabric to the final aspect of the sewing line combination.

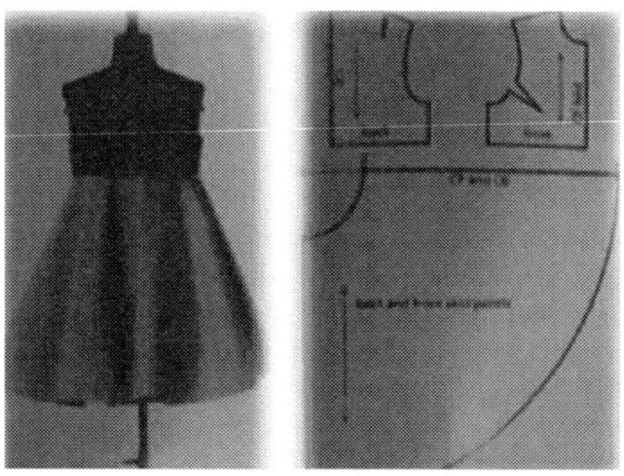

Figure 1: Example of straight-curve line assembly

MATERIALS AND METHODS

In order to carry out this study, 10 fabrics with the same weaving pattern (plain weave) and material (100% cotton) but with different characteristics have been chosen and are presented in Table I.

Table I: Chosen fabrics characteristics

Fabric	Mass per unit area (g/m²)	Number of warp yarns	Number of weft yarns
1	119	26	26
2	244	32	23
3	125	64	66
4	75	64	65
5	46	21	18
6	118	26	30
7	241	43	23.5
8	112	20	20
9	294	46	46
10	153	40	30

In order to introduce different curvature, different radiuses have been chosen using the scheme presented on Figure 2.

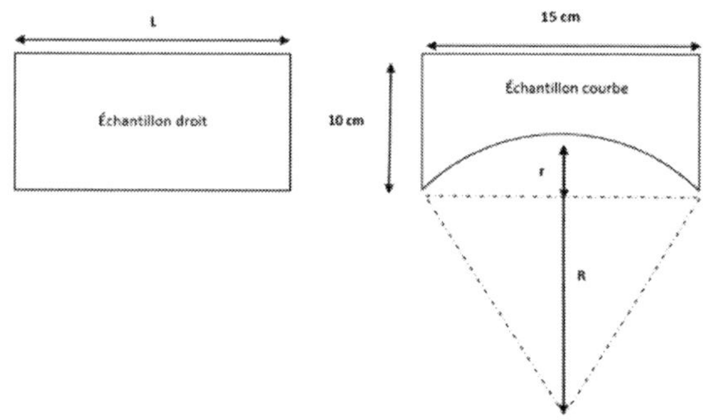

Figure 2: Shapes and sizes of samples

The different values of the radius r and R are displayed in Table II.

Table II: Sample sizes

Arrow "r" (cm)	Radius "R" cm	Length "L" cm
1	21.2	15.3
2	16.9	15.0
3	13.6	16.5
4	10.5	17.6
5	9.2	18.8

In this experiment, the straight and the curve lines have the same length and no fullness is added in the seams.

All the samples from the two pieces are sawn together thanks to 301 stitches (4 stiches/cm) with the same sewing thread (50/50 Cotton PET) and a seam allowance of 5 mm. This value of 5 mm has been defined after several tests as the most convenient for our purpose.

For every fabric, 15 assemblies have been done (3 per radius) and the final assemblies have been photographed and analyzed thanks to ImageJ software.

Figure 3 displays these steps.

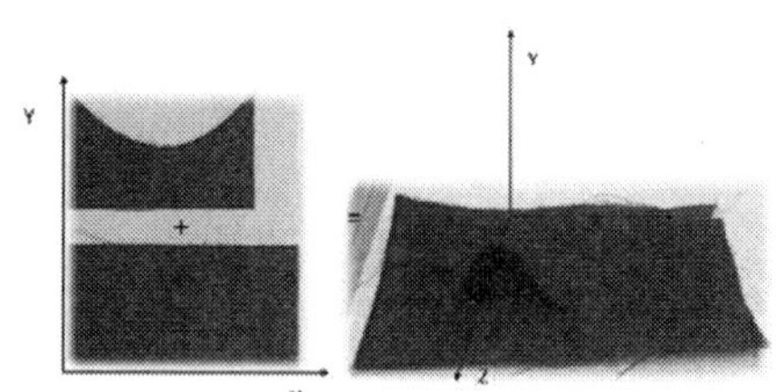

Figure 3: Assembly shape

Based on the obtained images the following parameters have been extracted: the numbers of pleat: "Nbre"; Height of pleat : "h"; Angle of each pleat: "α"; volume of pleat "V" as shown in figure 4. In the first step, the total volume of all the generated pleats is considered.

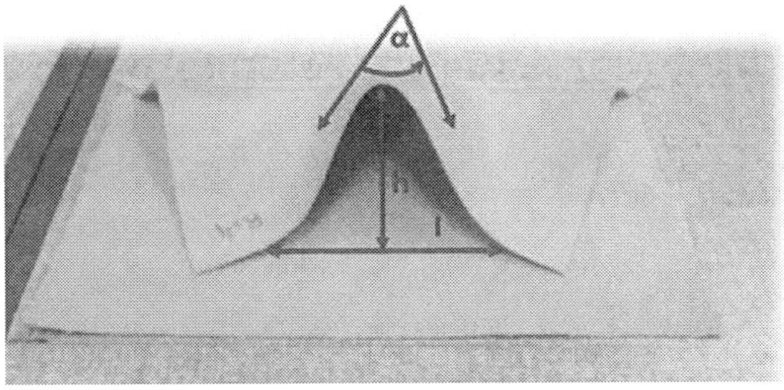

Figure 4: Parameters extracted from analyzed image

The 10 fabrics has been characterized thank to KES devices, the following parameters have been used for this first step of this study.

From the tensile tests: LT, WT (gf.cm/cm²), RT (%), EMT (%); from the shear tests: G (gf/cm/deg), 2HG (gf/cm) 2HG5(gf/cm), from the bending tests: B (gf.cm/cm²), 2HB (gf/cm/cm²); from the compression tests:LC, WC (gf.cm/cm²) RC(%), T_0 (mm) et T_M(mm).

In addition, the drape D is also evaluated.

RESULTS AND DISCUSSION

Five tests have been done for each fabric and for each property. The presented data are the average value of all these tests.

Statistical analyses through ANOVA study have shown that the following parameters of fabrics presented a good correlation with the seam's parameters defined in the introduction section. Considering the place in this abstract, only the results of the fabric 1 have been reported in Table III.

Table III: Correlation coefficient (CC) between mechanical parameters and experimental seam parameter.

	Nbre	Angle (°)	H (mm)	V (mm^3)
Mass per unit area (g/m²)	0.22	0.68	0.814	0.581
Drape F	0	0.494	0.482	0.675
LT	0.045	0.3	0.482	0.065
WT (gf.cm/cm²)	0.228	0.6465	0.737	0.41
RT (%)	-0.084	0.00	-0.424	-0.424
2HG (gf.cm)	0.	0.369		

Based on the all results obtained for the 10 fabrics, it has been highlighted that the ratio V/R is strongly linked to the final 3D shape behavior. This index (V/R) will be used in the further studies.

Table IV presents the obtained results.

Table IV: Correlation between mechanical properties, V/R index and mechanical properties (Correlation Coefficient)

	2HG	D	Mass	G	T_0	RT	WT	WC
V/R	0.767	0.70	0.683	0.66	0.55	0.501	0.467	0.38
2HG	1	0.866	0.71	0.93	0.35	-0.33	0.46	0
D		1	0.65	0.96	0.31	0	0.34	0.2
Mass			1	0.587	0.72	-0.36	0.792	0.49
G				1	0.2	-0.13	-0.32	0
T_0					1	-0.58	0.627	0.89
RT						1	-0.67	0.27
WT							1	0.30
WC								1

P = 95% - CP = 0.5494

Equations existing between V/R index and the mechanical properties issued of KES tests; drape "D" and mass per unit area "M" are presented in Table V.

Table V: Proposed equations

Factors	CC	Fisher criteria	Equation
D	0.70	7.57	-0.804 + 5.508 (D)
M	0.68	6.93	1.658 + 0.009 (M)
2HG	0.76	11.43	2.05 + 0.364 (2HG)
2HG-T_0	0.82	7.34	1.05 + 0.319 (2HG) + 2.097 (T_0)
2HG-T_0-WT	0.82	4.24	1.097 + 0.328 (2HG) + 2.343 (T_0) -0.15 (WT)
2HG-T_0-RT	0.83	4.46	2.218 + 0.308 (2HG) + 1.6 (T_0) -0.2 (RT)
2HG-WC-WT	0.827	4.317	1.153 + 0.35 (2HG) + 4.147 (WC) + 0.009 (WT)

Based on these equations, validations have been performed thanks to 3 additional fabrics (in the same range of mass per unit area). Good agreement between the results of the theoretical models and experimental data has been found.

CONCLUSION

Thanks to this experimental approach, few models have been proposed to predict the volume generated by a straight –curve line assembly in relation the fabric properties. For fabrics in a range of 75 g/m² to 300g/m² these models have been validated by using additional samples.

REFERENCES

1. Kawabata, S., Postle, R., Niwa, M. "Objective specification of fabric quality, mechanical properties and performance." Japan–Australian Joint Symposium, Kyoto, The Textile Machinery Society of Japan. 1982.

2. Guowen, Song. *Improving Comfort in Clothing.* United Kingdom: Woodshead Publishing Series in Textiles, 2011: 459.

3. Das, Apurba, Alagiruamy, R. *Science in Clothing Comfort.* India: Woodshead Publishing, 2010.

4. Aldrich, W. "Fabric, Form, and Flat Pattern Cutting." *Blackwell Science*, 1996: 205.

Applications in Medical, Transportation, and Energy

Functional Polymer Nanofibers for Water Purification

Guishan Hong, Xuefen Wang
State Key Lab for Modification of Chemical Fibers and Polymer Material,
Donghua University, Shanghai, China
wangxf@dhu.edu.cn

OBJECTIVE

Construction of hierarchically structural, functional, nanofiber-based materials and development of their applications acting as the affinity membrane for adsorption to remove the contaminants, such as hazardous dyes, metal ions, and water purification.

INTRODUCTION

Nowadays, the industrial polluted effluent containing heavy metal ions is a worldwide environmental threat, which will contaminate the surface water, groundwater, or sea water. These heavy metal ions tend to accumulate into human body via food chains and cause serious toxic effects. Adsorption is generally regarded as the most effective and economical technique for the removal of heavy metals from various matrices[1-3]. Here, our work will focused on the construction of hierarchically structural functional nanofiber-based materials and their applications acting as the affinity membrane for adsorption to remove the contaminants such as hazardous dyes, metal ions.

APPROACH

Here, the 20 wt% homogeneous polymer blend solution for electrospinning was obtained by mixing polymer powders of PAN and PVP with PAN/PVP weight ratio of 1:2.5 and dissolving in DMF at 50°C for 24 h. Highly porous PAN nanofibrous membranes were fabricated by in-situ pore-forming method via wet-electrospinning technique (shown in Figure 1)[4]. The typical parameters for wet-electrospinning experiments were as follows: the applied electric voltage was 30 kV, the distance between the spinneret and collector was 15 cm, and the solution feed rate was 0.5 ml/h. The leaching bath was hot water with the temperature at 85°C for PVP removal during the electrospinning process. The porous PAN nanofibrous membrane was aminated with DETA at 90°C for 2 hours. For adsorption of the heavy metal ions, a series of static adsorption experiments were performed.

The morphology of nanofibrous membranes were examined by scanning electron microscopy (SEM) (JSM-5600LV, Hitachi, Japan). The concentrations of metal ions in aqueous solution were investigated by inductively coupled plasma-atomic emission spectroscopy (ICP-AES, Prodigy, USA).

RESULTS AND DISCUSSION

The setup of wet-electrospinning was shown in Figure1. Highly porous PAN nanofiber membranes were fabricated via one-step wet-electrospinning technique from PAN/PVP blend solution with hot water acting as the leaching bath. The typical SEM images of porous PAN nanofibers before and after amination were shown in Figure 2. As can be seen, the numerous nanopores and nano-protuberances were observed both on surface and inner of the fibers for as prepared PAN nanofibers, and the pores were shrunk into very tiny and the porous structure was substituted with a homogeneous nano-protuberant structure on the surface of APAN nanofibers. This special micro/nano structure will ensure APAN nanofiber membrane a large specific surface area.

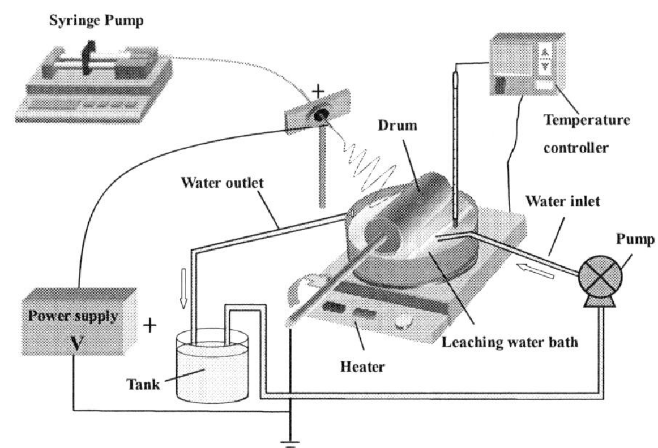

Figure 1. Schematic diagram of the wet-electrospinning setup for the in-situ pore-formation[4].

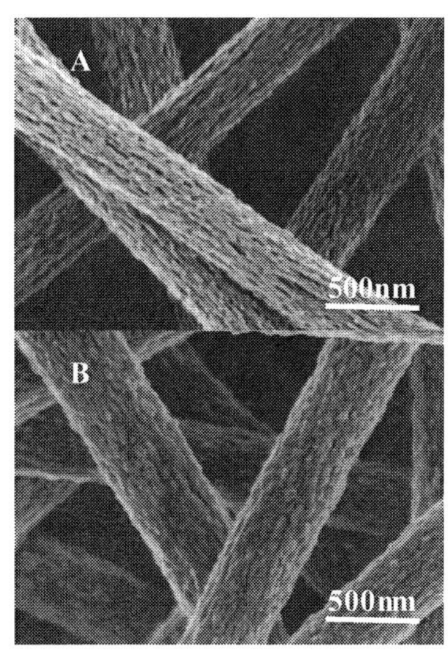

Figure 2. Typical FESEM images of porous PAN nanofibrous membrane before (A) and after (B) amination.

Also the conversion degree of nitrile group on these porous nanofibers was high up to 88.0% to provide sufficient amine and imine groups for APAN nanofiber mats. Obviously, the unique morphology combined with abundant amine and imine groups will endow APAN nanofibrous membrane extra high extraction capability for the removal of metal ions from wastewater. In this work, hexagonal crystals were observed on the surface of APAN nanofibers with micro/nano structure after the nanofibrous membrane was dipped into the $Pb(NO_3)_2$ aqueous solution for very short time (Figure 3). As compared to the conventional aminated PAN fibers reported by other groups, the maximum extraction capacity for Pb(II) of the present work is 1520.0 mg/g (7.34 mmol/g), which is 5 times more than that reported in the literatures[5-7].

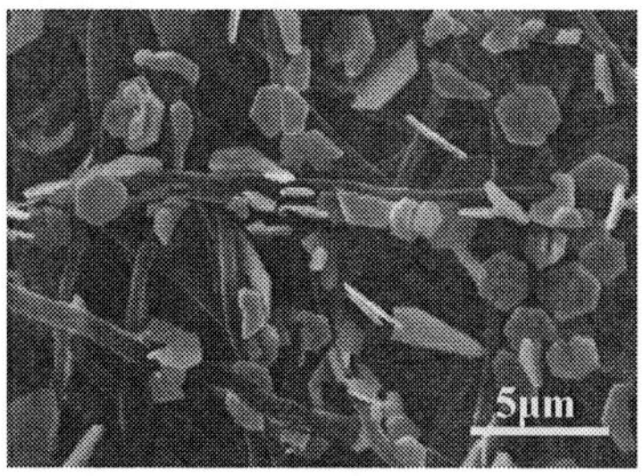

Figure 3. Typical SEM image of Pb^{2+} extracted onto micro/nano structured APAN nanofibrous membranes for very short time (5 mins.).

CONCLUSION

Highly porous polyacrylonitrile (PAN) nanofibrous membranes were successfully fabricated by wet-electrospinning technique using hot water bath as the collector from PAN and poly(vinyl pyrrolidone) (PVP) blended solution, followed by the amination with diethylene triamine (DETA). The aminated PAN (APAN) nanofibrous mats showed unique micro/nano structures and possessed extra high extraction capability for the removal of lead ions from wastewater (maximum uptake capacity of Pb^{2+} was 1520 mg/g, which is the highest uptake capacity reported). Interestingly, the hexagonal crystals of basic lead (II) carbonate ($Pb_3(CO_3)_2(OH)_2$) were grown on micro/nano structured APAN nanofibers when APAN membrane was immersed in Pb (II) ions aqueous solution. The results provided new insights for the removal of metal ions by metal crystal growth from wastewater with high recovery.

KEYWORDS

Wet-electrospinning, hierarchical structure, metal ions, extraction.

REFERENCES

[1] Hong, G., Wang, M., Li, X., Shen, L., Wang, X., Zhu, M., Hsiao, B.S. *RSC Advances*, 5, 2015: 21178-21188.

[2] Min, M., Shen, L., Hong, G., Zhu, M., Zhang, Y., Wang, X., Chen, Y., Hsiao, B.S. *Chem. Eng. J.*, 197, 2012: 88-100.

[3] Hong, G., Shen, L., Wang, M., Yang, Y., Wang, X., Zhu, M., Hsiao, B.S. *Chem. Eng. J.*, 244, 2014: 307-316.

[4] Hong, G., Wang, M., Wang, C., Yu, X., Wang, X. *J. Hazardous Mater.*, 295, 2015: 161-169.

[5] Kampalanonwat, P., Supaphol, P. *ACS Appl. Mater. Interf.*, 2, 2010: 3619-3627.

[6] Deng, S., Bai, R., Chen, J.P. *Langmuir*, 19, 2003: 5058-5064.

[7] Liu, C., Bai, R., San Ly, Q. *Water Res.*, 42, 2008: 1511-1522.

Melt-Spun Nanofibrous Filtration Media for Smog Control for Personal Respiratory Protection and Environmental Air Quality

Tao Huang, Jay N. Croft-Jr, Zachary R. Dilworth
DuPont Central Research and Development, Experimental Station, Wilmington, DE, USA
tao.huang@dupont.com

ABSTRACT

DuPont melt-spun nanofibrous electret filtration media consists of the stand-alone electret nanofibrous web comprising a single source randomly intermingled fiber network with self-composite of nanofibers and microfibers with the average diameter of total fibers less than 1000 nm. As improved air filtration media for personal respiratory protection and environmental air quality, it yields the better breathability with lower resistance due to the much lower packing density of fibers, the better protection with the improved filtration efficiency due to the more effective electrostatic charge to surfaces of individual fibers.

INTRODUCTION

Recent years, many places in Asia were blanketed with thick yellow fog. Smoggy weather has become a lingering problem in China and the country faces a tough battle in its fight against serious air pollution. The China's Haze or Smog is air pollution in which mixed up with dust, smoke and fine suspended particles. The large amount of pollutant emissions, including coal-fired pollution, motor vehicle emissions, industrial pollution and dust, is the key reason. Wearing the respirator and face mask to protect ourselves defensively is not enough; we should pay our attention offensively to decrease the PM2.5 by advanced air filtration media to against the sources of industrial pollution.

The increased surface to volume ratio afforded by nanofibers and their enhanced mechanical properties has had significant influences on a broad range of applications. In particular in filter performance, which is based on producing the highest flow rate while trapping and retaining the finest particles without blocking the filter, nanofibers have improved interception and inertial impaction efficiencies and result in slip flow at the fiber surface, resulting in better performance at a given pressure drop. As a result of this, nanofibers are currently incorporated into commercial filters in air, liquid and automotive applications in both industrial and consumer markets. Filtration media is typically defined by its rating or the particle size where it exhibits its max filtration efficiency. DuPont has been developing a nozzle-less melt-based nanofiber (MSNF) spinning and nanofibrous web technology. Advanced air filtration media have been made by this technology.

RESULTS AND DISCUSSION

The MSNF filtration media comprises a single layer stand-alone polypropylene nanofibrous web comprising a single source randomly intermingled fiber network wherein the fibers of the intermingled fiber network have a number average fiber diameter of less than about 1000 nm and a median fiber diameter less than about 500 nm.

Figure 1. a. The nanofibrous web comprising with self-composite of nanofibers and microfibers with the average diameter of total fibers less than 1000 nm.; b. the 3D image of a nanofibrous web with very low packing density of fibers (porosity > 95%).

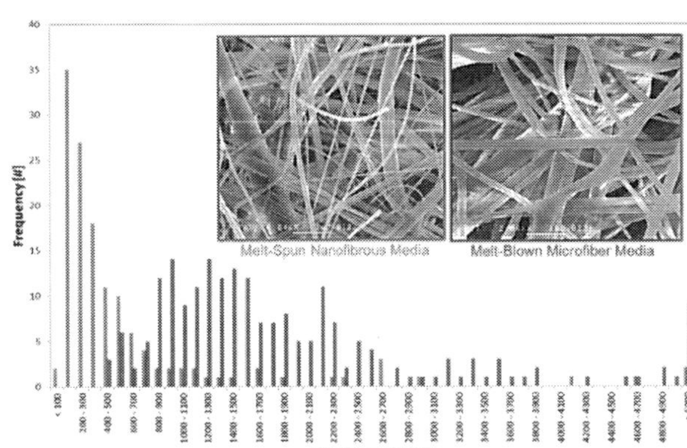

Figure 2. Comparison of MSNF and Melt-blown media: SEM images and fiber diameter distributions.

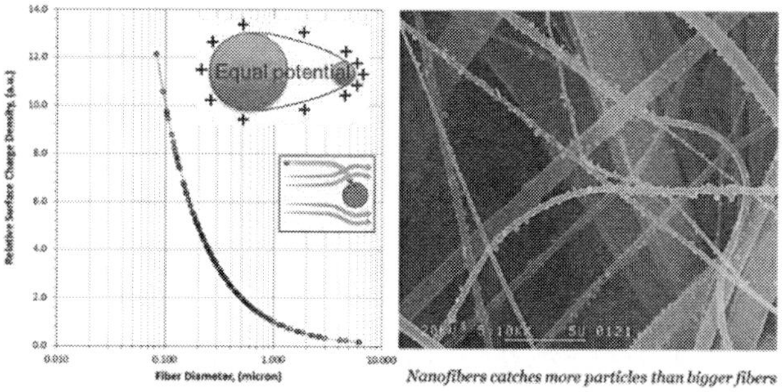

Figure 3. The plot of the relative surface charge density as function of fiber diameter of the web and a SEM image of web after filtration test showing the particles captured on fiber surface.

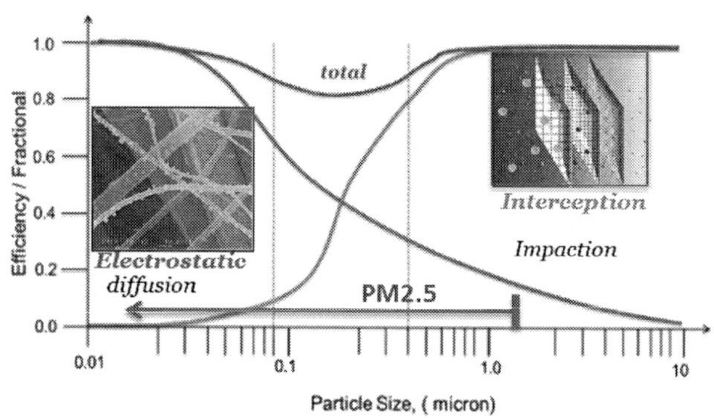

Figure 4. The filtration efficiency as function of particle size. The electrostatic charge contributes to capture the small particles (<0.1 micron). The mechanical interception contributes to capture the big particles (>1.0 micron). The MSNF nanofibrous electret media can effectively stop PM2.5.

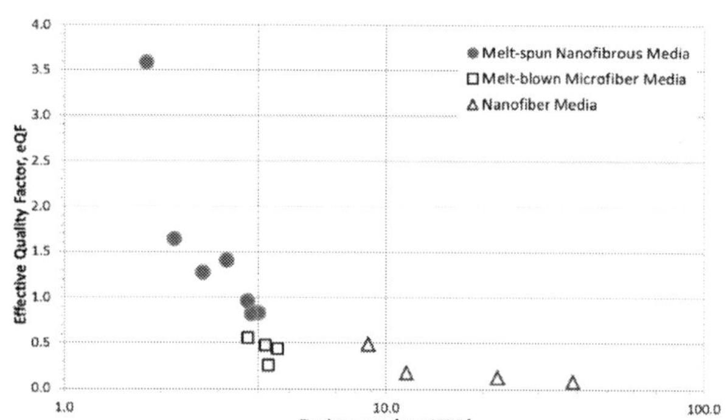

Figure 5. The filtration performance of effective quality factor vs. resistance comparing melt-spun nanofiber media with incumbent melt-blown microfiber media and electroblown nanofiber media.

Figure 1-a. shows the nanofibrous web comprising with self-composite of nanofibers and microfibers with the average diameter of total fibers less than 1000 nm. The web fibers comprise: (i) at least about 70% nanofibers with a number average fiber diameter from about 330 nm to about 450 nm; (ii) at most about 25% microfibers with an average fiber diameter from about 1.0 μm to about 3.0 μm; and (iii) at most about 5% coarse fibers of a number average diameter greater than about 3.0 μm. Figure 1-b shows the 3D image of a nanofibrous web with very low packing density of fibers (porosity > 95%).

During the spinning process, the electrostatics has been applied to individual fibers by an in-situ charging method before fibers laydown into web. Figure 3 shows the plot of the relative surface charge density as function of fiber diameter of the web and a SEM image of web after filtration test. The SEM shows that the nanofibers capture more particles than the microfibers due to the greater charging density on the nanofiber surface. Figure 4 shows the filtration efficiency as function of particle size. The electrostatic charge contributes to capture the small particles (<0.1micron). The mechanical interception contributes to capture the big particles (>1.0 micron). For electrospun nanoweb, reducing pore size by nanofibers of the media gives increased filtration efficiency in terms of interception at the cost of loss of permeability. However, the MSNF nanofibrous electret media shows the lower resistance due to the nanofibrous web has the porosity > 95%. Therefore, the MSNF nanofibrous electret media can effectively stop PM2.5 due to the surface charge on individual fibers and very low packing density of fibers. Figure 5 shows that the MSNF media has better performance than both incumbent and melt-blown microfiber media and electroblown nanofiber media.

CONCLUSIONS

In this study, we have demonstrated that the performance nanofiber media produced using nozzle-less melt-spinning technology as a media for air filtration displays improved filter performance than current commercially available incumbent media. As improved air filtration media for personal respiratory protection and environmental air quality, it yields the better breathability with lower resistance due to the much lower packing density of fibers, the better protection with the improved filtration efficiency due to the more effective electrostatic charge to surfaces of individual fibers.

REFERENCES

[1] Huang, Tao. "Advanced Nanofiber Technology and NanoWeb as Energy and Environmental Materials." *International Symposium for Energy and Environmental Materials,* QLD, Australia, Feb. 18-22, 2013.

[2] Huang, Tao. "Centrifugal Melt-Spun Nanofibrous Web." *Fiber Society Conference, Clemson*, SC, USA, Oct. 23-25, 2013.

Nanofibrous Scaffolds from Polycaprolactone as Wound Dressings

Kateřina Pilařová[1], Věra Jenčová[1], Jana Horáková, Silvie Krchová[2], Petr Mikeš[1], David Lukáš[1], Ladislav Dzan[2]
[1]Technical University of Liberec; [2]Regional Hospital of Liberec
katerina.pilarova1@tul.cz

ABSTRACT

In our study we compared our nanofibrous scaffold made by electrospinning from ε-polycaprolactone with commercially available wound dressing XeDerma® *in vitro*. Our results show that the PCL scaffold support cell adhesion and proliferation better than the reference material XeDerma®. According to our results we decided to try the healing capability of our scaffold on two patients with badly healing wounds. Our preliminary clinical trials show that the PCL electrospun scaffold is suitable as wound dressing, both patients healed completely without any complications and side effects.

INTRODUCTION

It is well known that nanofibrous scaffolds are capable of cell adhesion and proliferation mainly because they mimic extracellular matrix. But these scaffolds have also many other advantages. They have very small pores, so they work as antibacterial barrier. But still they keep high porosity, which enables the gas exchanges and access of oxygen [1]. We are able to produce very thin materials, so the dressing can easily adhere to the wound scum. Another advantage is that these dressing can be combined with other therapeutic approaches, with local antibiotics or disinfections. Moreover, polycaprolactone is FDA approved polyester and its advantage is that it is biocompatible and biodegradable [2].

EXPERIMENTAL APPROACH

Scaffold Fabrication

Materials: For fabrication of our scaffolds we used ε-polycaprolactone (Sigma Aldrich, Mw 45.000). We prepared 16% solution in a mixture of chloroform, ethanol and acetic acid in a ratio of 8:1:1. The scaffolds were electrospun on Nanospider^{TM} (1WS500U, Elmarco). The PCL wound dressings were tested for residual solvents with negative results (data not shown).

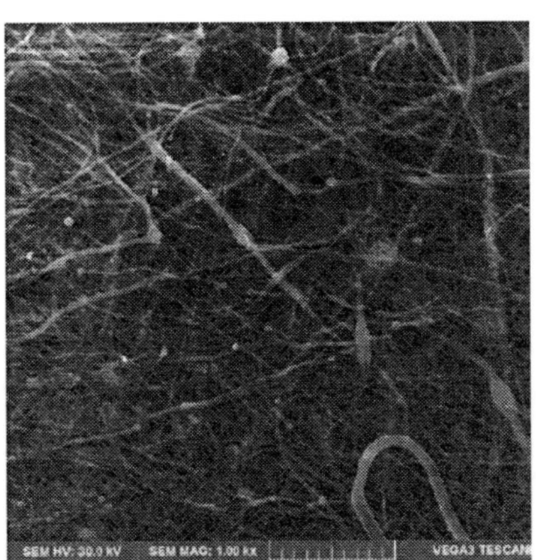

Fig. 1. SEM of electrospun PCL, scalebar: 50 µm.

In Vitro Experiments

In the *in vitro* experiments we compared our PCL eletrospun scaffold with commercially available wound dressing XeDerma® (Medicem), an acellular porcine dermis. We tested both scaffolds for cell adhesion and proliferation with MTT assay, fluorescence and electron microscopy. For all experiments we used 3T3 mice fibroblasts. We tested both scaffolds at day 1, 3, 7 and 14 after cell seeding.

In Vitro Cell Culture: 3T3 mice fibroblasts (3T3 Swiss Albino, ATCC) were cultivated in Dubelcco's modified Eagle's medium (DMEM, Lonza) with 10% (v/v) of fetal bovine serum (Lonza) and 1% (v/v) of penicillin / streptomycin / amphotericin B (Lonza). Cells were maintained at 37°C, 5% of CO_2. The medium was changed three times a week.

Samples: From our scaffolds we cut samples 0,5 cm in diameter. The samples were sterilized for 30 minutes in 70% ethanol and washed three times with sterile phosphate buffered saline (PBS, pH 7,4). After wash, the scaffolds were seeded with cells in the concentration of $1*10^4$ per well in 96-well plates.

Case Reports

According to our *in vitro* experiments we decided to use our PCL electrospun scaffold as temporary wound dressing. In collaboration with our colleagues from Regional Hospital of Liberec we tested our PCL wound dressings on two patients with bad healing wounds. The PCL wound dressings were sterilized by ethylene oxide prior to use and they were used in combination with Hyiodine gel. Both patients were controlled once to twice per week and the PCL wound dressing was changed if necessary.

RESULTS AND DISCUSSION

Regarding to our results from MTT assay (Fig. 2) as well as from fluorescence (Fig. 3) and electron microscopy, our PCL electrospun scaffold shows better results than XeDerma® considering both cell adhesion and proliferation. Cells adhere to XeDerma very unwillingly and after the cell adhesion the cells do not proliferate. Moreover, between individual samples of XeDerma® are big differences in cell adhesion (data not shown). On our PCL electrospun scaffold, cells were proliferating within all testing days.

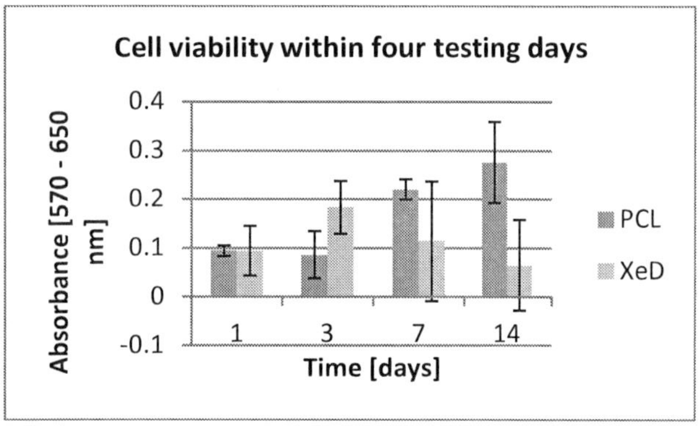

Fig. 2. MTT assay: cell viability within four testing days on electrospun PCL (PCL) and commercially available XeDerma (XeD).

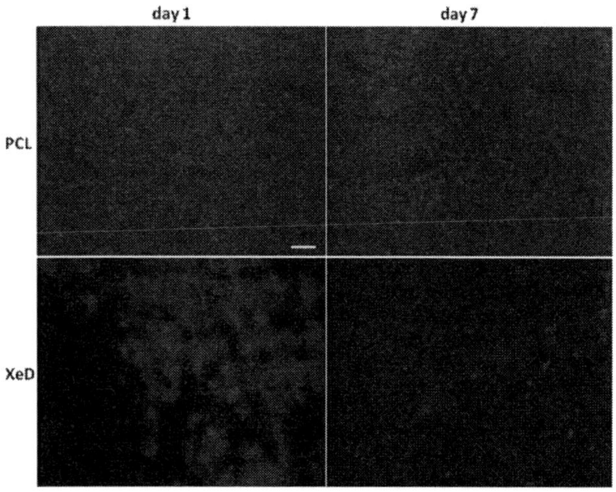

Fig. 3. Fluorescent microscopy analysis of PCL electrospun wound dressing (PCL) and XeDerma® (XeD) on day 1 and 7 after cell seeding, scale bar: 100 μm.

Regarding our case reports, both patients had long-term poorly healing wounds. Patient no. 1 (P1) had non-healing bazalioma, patient no. 2 (P2) had a varicose ulcer. They were treated with our PCL wound dressing in combination with Hyiodine gel. Patient no. 1 healed comletely in 5 weeks and the PCL wound dressing was changed every four days. Patient no. 2 healed completely in 20 weeks and the PCL wound dressing was changed twice per week (Fig. 4).

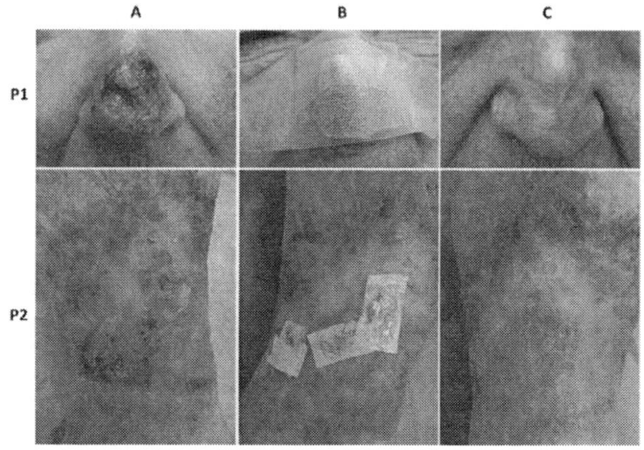

Fig. 4. Patient one (P1) and two (P2) before healing (A), with PCL dressing (B) and after completely healed injury (C).

CONCLUSIONS

1. We prepared PCL electrospun scaffold, which posses many properties supporting the cell adhesion and proliferation
2. Our PCL electrospun scaffold shows better results *in vitro* than commercially available XeDerma®.
2. The PCL wound dressing was used for the case study in two patients with bad healing wounds.
3. Both patients were healed completely without any side effects and complications and with very small scarring.

FUTURE WORK

The mechanisms of healing properties of our PCL electrospun wound dressing are still poorly understood. A big influence on the healing time will also have the health conditions of the patients. That is why much more complex clinical study is needed. And even though our electrospun PCL scaffold has unique properties, which are themselves advantageous for cell adhesion and proliferation, there is a lot of space for different kinds of modifications and functionalization.

KEYWORDS

wound dressings, nanomaterials, polycaprolactone

ACKNOWLEDGMENT

This research was supported by the project "Nanofibre materials for tissue engineering" reg. no. CZ.1.05/3.1.00/14.0308, which is co-financed by the European Social Fund and the state budget of the Czech Republic. The authors would like to acknowledge also to the doctors from Regional Hospital of Liberec, Ladislav Dzan and Silvie Krchová.

REFERENCES

[1] Zahedy, P., Rezaeian, I., Ranaei-Siadat, S. O., Jafari, S. H., Supaphol, P. *A Review on Wound Dressings with an Emphasis on Electrospun Nanofibrous Polymeric Bandages.* 2009.

[2] Woodruff, M.A., Hutmacher, D.W. *The Return of Forgotten Polymer—Polycaprolactone in the 21st Century.* 2010.

Development of Poly (L-lactide-co-caprolactone) Multichannel Nerve Conduit with Aligned Electrospun Nanofibers for Schwann Cell Proliferation

Bhutto M. Aqeel[1], Jianguang Zhang[1], Binbin Sun[1], Hany El-Hamshary[2,3], Salem S. Al-Deyab[2], Xiumei Mo[1,2]

[1]Biomaterials and Tissue Engineering Lab, College of Chemistry, Chemical Engineering and Biotechnology, Donghua University, Shanghai, China
[2]Department of Chemistry, College of Science, King Saud University, Riyadh, Kingdom of Saudi Arabia
[3]Department of Chemistry, Faculty of Science, Tanta University, Tanta, Egypt

aqeel_bhutto@hotmail.com

ABSTRACT
To biomimic the structure of the autograft nerve conduit is one of the most promising strategies for nerve regeneration. In present work, we reported a novel fabrication method by using special collector for the 3D multichannel nerve conduit, which composed of longitudinally oriented channels with aligned electrospun nanofibers. The conduit contained a high number of channels (vary from 7 to 19) and each channel showed a good morphology, which separated from each other. We fabricated nerve conduits with the varying length ranging from 4cm to 9cm and total diameter ranging from 2.20 ± 0.04mm to 3.951±0.196mm while the channel diameter ranging from 0.35±0.005mm to 0.78±0.02mm. It was clearly shown that the average porosity of nerve conduits has reached nearly 89%. In this study, we optimized the parameters to control the structural stability, including the size and the number of the channel in the nerve conduit. We also studied the cell biocompatibility of multichannel nerve conduit in vitro. The result demonstrated that Schwann cells have the tendency to grow along the direction of nanofibers and high cells growth were observed in high number of channels compared to low number of channels conduit. These results elaborated the potential use of this biocompatible multichannel nerve conduit for nerve tissue engineering.

KEYWORDS
Multichannel nerve conduit; P(LLA-CL); aligned electrospun nanofibers; Schwann cells; electrospinning.

Development of Biocompatible Hydrogels for Biomedical Applications

Bhuvanesh Gupta[1], Sadiya Anjum[1], Mythili Tummalapalli[1], Anupama Sharma[2], Surya Bhan[2]

[1]Bioengineering Laboratory, Department of Textile Technology, Indian Institute of Technology
New Delhi, India
[2]Department of Biochemistry, North Eastern Hill University, Shillong, India
bgupta@textile.iitd.ernet.in

ABSTRACT
Polymeric materials, especially hydrogels and hydrocolloids have been a boon to human healthcare due to their very innovative and beneficial life support features. Maintaining a moist environment around the wound area enhances wound healing and reduces trauma. Hydrogel dressings, which are transparent semi-occlusive dressings retain moisture and warmth, have been shown to reduce the incidence of wound infection and are a more appropriate treatment than gauze dressings. Infection by pathogenic microorganisms is a major concern in human healthcare industry, particularly in post-surgical wound care. It is therefore imperative that infection resistant materials be developed. However, the development of such materials requires a close collaboration between medicine, microbiology and materials chemistry.

The development of polymers with bioactive coatings is an important area of research focused on solving the problem of infection in sutures and in wound care systems. A bioactive component may be incorporated within the hydrogel matrix to make it infection resistant. In one instance, nanosilver nanohydrogels (nSnH) of poly(methacrylic acid) were synthesized and stabilized using gamma irradiation by water in oil nanoemulsion polymerization. The nanohydrogels thus synthesized exhibited excellent antimicrobial activity against both Gram negative and Gram positive bacteria. Because of the inert nature of most polymers, they must undergo surface functionalization prior to attachment of a bioactive compound. Plasma functionalization is an effective method by which desired features can be achieved on a material surface. Desired chemistry can be generated on the polymer surface by appropriate selection of the gaseous medium and has been very effective in developing sutures and wound dressings. Several bioactive agents, such as chitosan, chlorhexidine, aloe vera and curcumin have been used for this application. Herbal materials such as aloe vera and curcumin have been observed to be very interesting components of a bioactive chemistry. These herbal drugs were incorporated into natural hydrogels like pectin and carboxymethyl cellulose. The wound dressings thus fabricated from the herbal drug incorporated hydrogels have been evaluated for their efficiency in wound healing using Swiss albino mice. Excellent healing by hydrogel dressing has been observed with complete control over infection at the wound site.

CONCLUSION
Biocompatible hydrogels were developed using both natural and synthetic hydrogels for biomedical applications. nSnH designed using nanosilver entrapped within nanohydrogel matrix exhibited very high antimicrobial activity and therefore could be used for wound dressing applications. Likewise, herbal bioactive components, aloe vera and curcumin were incorporated within a hydrogel matrix to fabricate wound care systems. Such dressings led to effective infection control and rapid wound healing when tested on animal models. Therefore, our studies have demonstrated the potential of bioactive agent incorporated hydrogels in various aspects of human healthcare.

REFERENCES
[1] Tummalapalli, M., Deopura, B.L., Alam, M.S., Gupta, B. *Mater. Sci. Eng. C*, 50, 2015: 31.
[2] Gupta, B., Anjum, S., Gautam, D., Kapil, A. *Radiat. Phys. Chem.*, 54, 2013: 213.

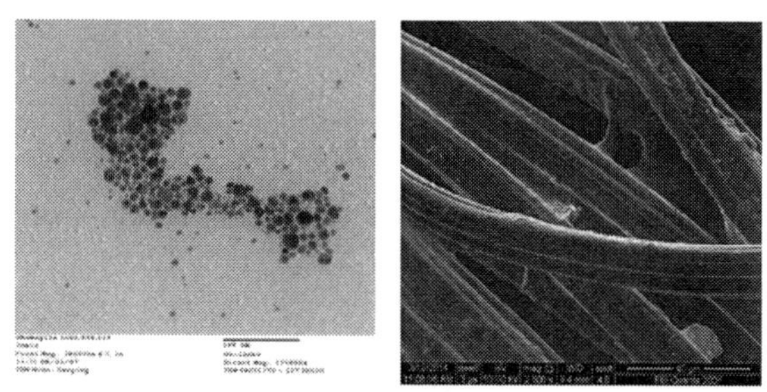

(a) (b)

Figure 1. (a) TEM image of nSnH particles and (b) SEM image of nSnH coated cotton fabric

Electrospun Porous Carbon Nanofiber@MoS_2 Core/Sheath Fiber Membranes as Highly Flexible and Binder-Free Anode for Lithium-Ion Batteries

Yue-E Miao, Yunpeng Huang, Longsheng Zhang, Wei Fan, Feili Lai, Tianxi Liu
State Key Laboratory of Molecular Engineering of Polymers, Department of Macromolecular Science, Fudan University, Shanghai, China
txliu@fudan.edu.cn

ABSTRACT

With severe stress from energy crisis and global environmental concerns, lithium-ion batteries (LIBs) have attracted extensive attention as power sources for electric vehicles due to their high energy density, low gravimetric density, long cycle life, and flexible design.[1] As a typical two-dimensional transition-metal sulfide, MoS_2 shows analogous layered structure of graphene in which molybdenum atoms are sandwiched between two layers of sulfur atoms via weak van der Waals interactions, thus giving rise to wide interlayer spaces that allow facile insertion and extraction of Li^+ ions during charge/discharge processes to achieve high specific capacity for LIBs.[2-5] However, MoS_2 nanostructures suffer from severe restacking and condensation during charge/discharge processes due to the high surface energy, interlayer van der Waals attraction and poor conductivity, which results in drastically reduced accessibility of electrolyte and fast capacity fading in practical applications.[6,7] Therefore, self-standing membranes of porous carbon nanofiber (PCNF)@MoS_2 core/sheath fibers have been facilely obtained in this work through the combination of electrospinning and solvothermal reaction (Figure 1). Thin MoS_2 nanosheets are uniformly distributed on the surface of the PCNF core with porous channels to form a three-dimensional network which is favorable for rapid charge-transfer reactions and fast lithiation/delithiation of Li^+ ions. Thus, synergistically improved electrochemical performance is achieved with high specific capacity of 954 mAh g^{-1} upon the initial discharge and high rate capability of 475 mAh g^{-1} even at a high current density of 1 A g^{-1}. Moreover, the highly conductive PCNF core can effectively prevent the restacking and volume change of MoS_2 nanosheets during charge/discharge processes which results in a good cycling stability with almost 100% reversible retention after 50 cycles, making the PCNF@MoS_2 composite membrane a promising binder-free anode for high-energy lithium-ion batteries.

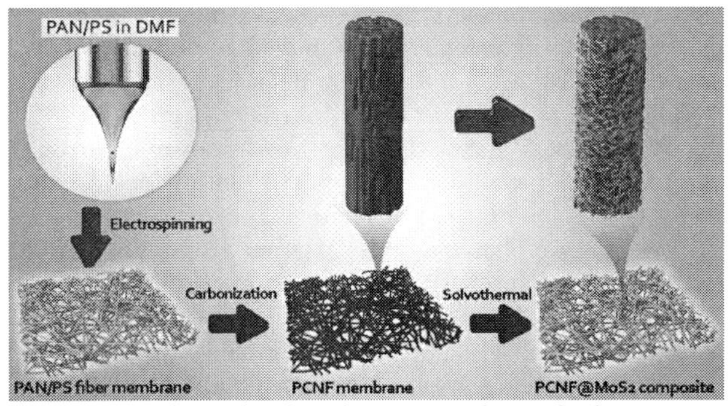

Figure 1. Schematic for the preparation of electrospun PCNF@MoS_2 core/sheath fiber membranes.

KEYWORDS

Porous carbon nanofiber@MoS_2; core/sheath structure; electrospinning; flexible; lithium-ion battery

ACKNOWLEDGMENT

The authors are grateful for the financial support from the National Natural Science Foundation of China (51125011, 51373037, 51433001).

REFERENCES

1. Chhowalla, M., Shin, H.S., Eda, G., Li, L.J., Loh, J.P., Zhang, H. *Nat. Chem.*, 5, 2013: 263-275.
2. Zhou, S.F., Xu, Y., Zhao, W.W., Zhang, J., Zhang, B. *Angew. Chem. Int. Ed.*, 52, 2013: 8602-8606.
3. Ding, Ding, Zhang, D.Y., Chen, J.S., Lou, X.W. *Nanoscale*, 4, 2012: 95-98.
4. Feng, C.Q., Ma, J., Li, H., Zeng, R., Guo, Z.P., Liu, H.K. *Mater. Res. Bull.*, 44, 2009: 1811-1815.
5. Li, H., Li., W., Ma., L., Chen, W.X., Wang, J.M. *J. Alloy. Compd.*, 471, 2009: 442-447.
6. Xiao, J., Choi, D., Cosimbescu, L., Koech, P., Liu, J., Lemmon, J.P. *Chem. Mater.*, 22, 2010: 4522-4524.
7. Du, G.D., Guo, Z.P., Wang, S.Q., Zeng, R., Chen, Z.X., Liu, H.K. *Chem. Commun.*, 46, 2010: 1106-1108.

Hierarchical $ZnCo_2O_4$@$NiCo_2O_4$ Core-Sheath Nanowires: Bifunctionality Toward High-Performance Supercapacitors and Oxygen Reduction Reaction

Yunpeng Huang, Yue-E Miao, Hengyi Lu, Tianxi Liu
State Key Laboratory of Molecular Engineering of Polymers, Department of Macromolecular Science,
Fudan University, Shanghai, China
txliu@fudan.edu.cn

ABSTRACT

Increasing energy demands and worsening environmental issues have stimulated intense research on alternative energy storage and conversion systems including supercapacitors and fuel cells. Mixed transition-metal oxides (MTMOs), typically designated as $A_xB_{3-x}O_4$ (A, B = Co, Ni, Zn, Mn, Fe, etc.) with stoichiometric or non-stoichiometric compositions, have recently captured worldwide interests due to their unlimited potentials in energy storage and conversion applications in terms of remarkable electrochemical properties, low cost and environmental friendliness.[1-3] Among various MTMOs, spinel nickel cobaltite ($NiCo_2O_4$) has been suggested as a promising alternative candidate for pseudocapacitors and ORR catalyst because of its better electrical conductivity and richer redox chemical valences over its two corresponding single-component oxides.[4, 5] Despite all the merits delivered by MTMOs, the relative poor electrical conductivity, low surface area and large volume change during charge-discharge processes are still great challenges limiting their capacitive behavior and ORR catalytic activity.

Here, we proposed a rationally designed hierarchical structure of $ZnCo_2O_4$@$NiCo_2O_4$ core-sheath nanowires synthesized through facile electrospinning combined with a simple coprecipitation method. The obtained core-sheath nanostructures consisting of mesoporous $ZnCo_2O_4$ nanowires as the core and uniformly distributed ultrathin $NiCo_2O_4$ nanosheets as the sheath, exhibit excellent electrochemical activity as bi-functional materials for supercapacitor electrodes and oxygen reduction reaction (ORR) catalysts. Compared with the single component of either $ZnCo_2O_4$ nanowires or $NiCo_2O_4$ nanosheets, the hierarchical $ZnCo_2O_4$@$NiCo_2O_4$ core-sheath nanowires demonstrate higher specific capacitance of 1476 F g^{-1} (1 A g^{-1}) and better rate capability of 942 F g^{-1} (20 A g^{-1}) while maintaining 98.9% capacity after 2000 cycles at 10 A g^{-1}. Meanwhile, the $ZnCo_2O_4$@$NiCo_2O_4$ core-sheath nanowires reveal comparable catalytic activity but superior stability and methanol tolerance over Pt/C as ORR catalyst. The impressive performance may originate from the unique hierarchical core-sheath structures that greatly facilitate enhanced reactivity, faster ion and electron transfer.

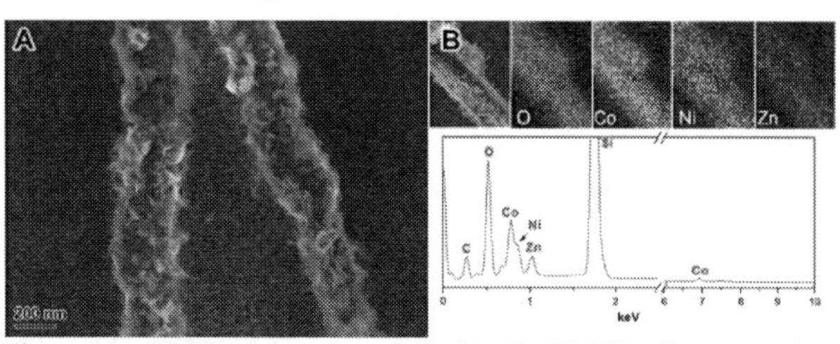

Figure 1. FESEM image of $ZnCo_2O_4$@$NiCo_2O_4$ nanowires and corresponding elemental mapping.

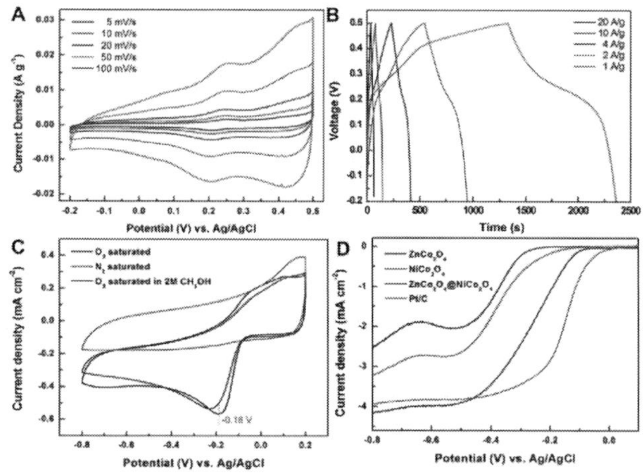

Figure 2. Electrochemical characterizations of the supercapacitor: (A) CV curves of $ZnCo_2O_4$@$NiCo_2O_4$ nanowires at various scan rates ranging from 5 to 100 mV s^{-1}; (B) CP curves of $ZnCo_2O_4$@$NiCo_2O_4$ nanowires at various current densities. And electrochemical characterizations of ORR: (C) CV curves for $ZnCo_2O_4$@$NiCo_2O_4$ nanowires; (D) RDE voltammograms of $NiCo_2O_4$ nanosheets, $ZnCo_2O_4$, $ZnCo_2O_4$@$NiCo_2O_4$ nanowires and Pt/C electrodes in O_2 saturated 0.1 M KOH with a scan rate of 5 mV s^{-1} at 1600 rpm.

KEYWORDS

Electrospinning; Hierarchical structure; $ZnCo_2O_4$@$NiCo_2O_4$; Supercapacitors; Oxygen reduction reaction

ACKNOWLEDGMENT

The authors are grateful for the financial support from the National Natural Science Foundation of China (51433001, 51373037, 51125011).

REFERENCES

[1] Yuan, C.Z.,. Wu, B., Xie, Y., Lou, X.W. *Angew. Chem., Int. Ed.*, 53, 2014: 1488-1504.
[2] Liu, B., Tan, D.S., Wang, X.F., Chen, D., Shen, G.Z. *Small*, 9, 2013: 1998-2004.
[3] Kim, T.W., Woo, M.A., Regis, M., Choi, K. *J. Phys. Chem. Lett.*, 5, 2014: 2370-2374.
[4] Wang, J.P., Wang, S.L., Huang, Z.C., Yu, Y.M. *J. Mater. Chem. A*, 2, 2014: 17595-17601.
[5] Xiao, Y., Hu, C.G., Qu, L.T., Hu, C.W., Cao, M.H. *Chem.–Eur. J.*, 19, 2013: 14271-14278.

Controllable Preparation of Multidimensional Hybrid Materials of Nickel-Cobalt Layered, Double Hydroxide Nanorods/Nanosheets on Electrospun Carbon Nanofibers for High-Performance Supercapacitors

Feili Lai, Yunpeng Huang, Yue-E Miao, Tianxi Liu
State Key Laboratory of Molecular Engineering of Polymers, Department of Macromolecular Science, Fudan University, Shanghai, China
txliu@fudan.edu.cn

ABSTRACT

Nowadays, high-performance energy storage technologies are widely explored in order to remit the increasingly tensive energy situation.[1] Supercapacitors, also called as electrochemical capacitors, have attracted extensive attention because of their fast charge-discharge process, long lifespan and high power density.[2] Nickel-cobalt layered double hydroxides (Ni-Co LDHs) are one kind of the most promising candidates among binary metal hydroxides due to their low cost, high redox activity and electrical conductivity.[3] However, the low electrochemical stability can hardly be avoided for pure Ni-Co LDH due to the conglomeration of nanoparticles during the electrochemical processes, which severely limits their wider applications.

Recently, hybrid nanomaterials with hierarchical structures have been considered as one kind of the most promising electrode materials for high-performance supercapacitors with high capacity and long cycle lifetime.[4] In this work, multi-dimensional hybrid materials of nickel-cobalt layered double hydroxide (Ni-Co LDH) nanorods/nanosheets on carbon nanofibers (CNFs) were prepared by electrospinning technique combined with one-step solution co-deposition method as shown in Figure 1. Carbon nanofiber membranes were obtained by electrospinning of polyacrylonitrile (PAN) followed by pre-oxidation and carbonization. The successful growth of Ni-Co LDH with different morphologies on CNF membrane by using two kinds of auxiliary agents reveals the simplicity and universality of this method. The uniform and immense growth of Ni-Co LDH on CNFs significantly improves its dispersion and distribution. Meanwhile the hierarchical structure of carbon nanofiber@nickel-cobalt layered double hydroxide nanorods/nanosheets (CNF@Ni-Co LDH NR/NS) hybrid membranes provide not only more active sites for electrochemical reaction but also more efficient pathways for electron transport. Galvanostatic charge-discharge measurements reveal high specific capacitances of 1378.2 F g-1 and 1195.4 F g-1 (based on Ni-Co LDH mass) at 1 A g-1 for CNF@Ni-Co LDH NR and CNF@Ni-Co LDH NS hybrid membranes, respectively. Moreover, cycling stabilities for both hybrid membranes are significantly enhanced compared with those of Ni-Co LDH NR and NS powders. This facile method provides a new strategy for designs and applications of binary transition metal oxides/hydroxides deposited on various substrates for next-generation energy storage devices.

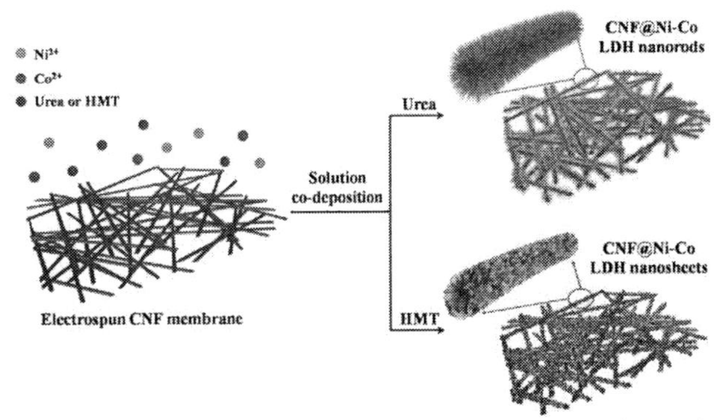

Figure 1. Schematic illustration of the preparation of CNF@Ni-Co LDH NR and CNF@Ni-Co LDH NS hybrid membranes.

KEYWORDS

Nickel-cobalt layered double hydroxides; hierarchical nanostructure; electrospinning; carbon nanofiber membrane; supercapacitors

ACKNOWLEDGMENT

The authors are grateful for the financial support from the National Natural Science Foundation of China (51125011, 51373037, 51433001).

REFERENCES

(1) Tarascon, J.M., Armand, M. "Issues and Challenges Facing Rechargeable Lithium Batteries." *Nature*, 414, 2001: 359-367.

(2) Arico, A.S., Bruce, P., Scrosati, B., Tarascon, J.M., Schalkwijk, W. Van. "Nanostructured Materials for Advanced Energy Conversion and Storage Devices." *Nat. Mater.*, 4, 2005: 366-377.

(3) Zhang, G. Q., Lou, X.W. "Controlled Growth of NiCo2O4 Nanorods and Ultrathin Nanosheets on Carbon Nanofibers for High-performance Supercapacitors." *Sci. Rep.*, 3, 2013: 1470.

(4) Guo, Q.H., Zhou, X.P., Li, X.Y., Chen, S.L., Seema, A., Greiner, A., Hou, H.Q. "Supercapacitors Based on Hybrid Carbon Nanofibers Containing Multiwalled Carbon Nanotubes." *J. Mater. Chem.*, 19, 2009: 2810-2816.

Functional and Smart Fibers

Fascinating Electroactive Hydrogels-Based Bacterial Cellulose

Zhijun Shi[1,2], Sixiang Li[1,2], Guang Yang[1, 2]
[1]Department of Biomedical Engineering, College of Life Science and Technology, Huazhong University of Science and Technology, Wuhan, China
[2]National Engineering Research Center for Nano-Medicine, Huazhong University of Science and Technology, Wuhan, China
yang_sunny@yahoo.com

KEYWORDS: Biphasic, hydrogel, bacterial cellulose, biocompatibility, electroactive

ABSTRACT

Bacterial cellulose (BC) is secreted by microorganism. Its biocompatibility, mechanical strength, chemical and morphologic controllability make it a natural choice for adoption in biomedical fields, including use as biomaterial for wound dressing, artificial blood vessels, vascular grafts, scaffolds for tissue engineering and controlled-release drug carriers. BC combine with carbon nanotubes or conductive polymers (CP), which has electractive property, can be used as flexible supercapacitor [1], electrodes.

Biphasic electroactive hydrogels are we design and synthesis of bacterial cellulose (BC) hydrogel with two distinct phases [2,3]. One of phases is a layer of electroactive hydrogels combined with BC and CP. The electroconductivity provides the composites with capabilities for voltage and current response; and the BC hydrogel layer provides good biocompatibility, biodegradability, bioadhesion and mass transport properties. Such a system might allow selective biological functions such as molecular recognition and specific catalysis and also for probing the detailed genetic and molecular mechanisms of life. This biphasic Janus of bacterial cellulose hydrogel, which intergrates electroactivity with biocompability, have potential to be used to build a biology–device interface to produce implantable biosensors, electrostimulated drug release devices, and implantable devices for personalized and regenerative medicine.

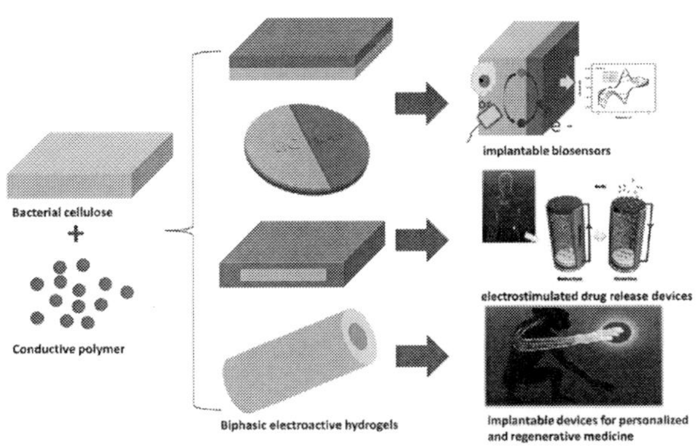

Figure 1. Biphasic electroactive hydrogels

REFERENCES

1. Li, S., Huang, D., Zhang, B., Xu, X., Wang, M., Yang, G., Shen, Y. "Flexible Supercapacitors Based on Bacterial Cellulose Paper Electrodes." *Adv. Energy Mater.* , 4, 2014: 1301655-61.

2. Shi, Z., Phillips, G.O., Yang, G. "Nanocellulose Electroconductive Composites." *Nanoscale*, 5, 2013: 3194-201.

3. Shi, Z., Li, Y., Chen, X., Han, H., Yang, G. "Double Network Bacterial Cellulose Hydrogel to Build a Biology–device Interface." *Nanoscale*, 6, 2014: 970-7.

ACKNOWLEDGMENT

This work was supported by the National Natural Science Foundation of PR China (21074041, 31270150).

Shape Memory Effect of Biopolymer Fibers Responsive to Humidity

Jinlian Hu, Xueliang Xiao, Xiaoting Gui, Yong Zhu
Institute of Textiles and Clothing, the Hong Kong Polytechnic University, Hung Hom, Hong Kong
tchujl@polyu.edu.hk; xueliang.xiao@polyu.edu.hk

INTRODUCTION AND DISCOVERY

In addition to the amazingly higher elasticity and toughness than any other existing artificial materials, biopolymer fibers, such as spider silk, wool and collagen fiber, have fascinated scientists by an intriguing feature called super-contraction[1-3]. Super-contraction is a huge shrinkage of "native" fibers when immersed in water or subjected to a critical humidity[2]. Although there exists a bulk of works[4-6] being intended to reveal the nature of this wonder since its discovery, a number of issues still cannot be explained thoroughly and succinctly by any theory. Here, we unravel the mystery of super-contraction of biopolymer fibers by examining existing data in literature including our recent work[7-9] in water switchable shape memory polymers (SMPs). Our conclusion is that super-contraction is clearly the water triggered shape recovery to its original shape (ground state) from a temporary shape fixed due to the stress-residual produced from biopolymer's growing, spinning or applied fixation. For example, a morning's washing to the curly hairs due to a whole night's temporary fixation can backwards the hair's original straight state. This means that biopolymer fibers are typically a group of natural SMPs sensitive to water, leading to its adaptability to many needs in nature. Natural SMPs usually have the capability of memorizing their permanent shapes where fixity, recovery stress, and shape recovery are the key parameters[10]. These key parameters of SMPs are resulted from the combination of their molecular architecture and corresponding programming, normally following a cycle of deformation → shape fixation → shape recovery. Various temporary shapes of SMPs can be programmed and subsequently recover their original permanent shapes upon exposure to an external stimulus. Traditionally external stimulus were reported that can be heat, electricity and magnetic field[11].

Subsequently, a structural model[12,13] of biopolymer fiber responsive to humidity is established, which integrates a number of phenomena such as tunable mechanics, the origin, functions and ground state of super-contraction into one concept, that is, shape memory. This discovery has also resulted in our successful synthesis of a new SMP using β-sheet crystals existing in spider silks as net-points. Furthermore, it can further lay foundation for man-making and bio-mimicking biopolymer fibers using chemical approaches rather than genetic ones[14].

RESULTS AND DISCUSSION

Here, we attempted two cases of biopolymer fibers, spider silk and wool fibers, for investigating their shape memory effect (SME) induced by water, which represent β-sheet and α-helical protein fibers, respectively. Their pre-stretched dry fibers were immersed in different humidity conditions, and their shape recovery performances can be found in Fig. 1a and b, where the equivalence of shape recovery of water switchable SMPs to the super-contraction of spider silk and wool yarn is shown.

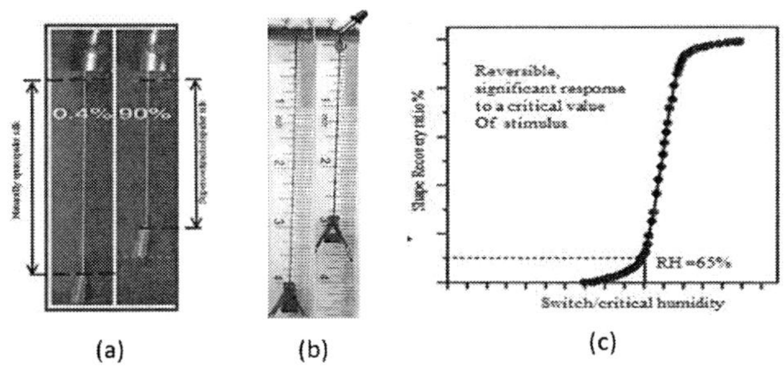

Fig. 1 Water-induced super-contraction of (a) spider silk and (b) wool yarn, (c) silk SME sensitivity to humidity

Fig. 1a and c show a spider silk super-contracts to its "ground state" when the relative humidity (RH) of the air changes from dry (0.4%) to wet (90%). It is found that 65% of humidity is a critical value deciding silk's shape recovery performance. Fig. 1b shows the contraction of wool yarn reaches 25% when it is immersed in water. Apart from the slippage of fibers, the image discloses a super-contraction ability of wool fiber when encountered with water.

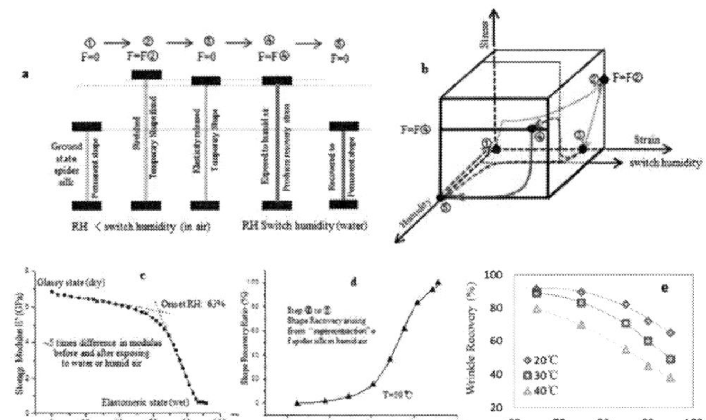

Fig. 2 a, SME Programming of a biopolymer fiber; b, General Hydro-mechanical SME Relationships, ① original shape and stretching starts, recovery and drying ends; ② stretching ends and fixation starts; ③ temporary shape; ④ wetting produces recovery stress; ⑤ shape recovered in wetting; c and d, DMA characteristics of a spider silk and shape recovery of silk in humid air from[15]; e, wrinkle recovery of wool fabric along relative humidity[15].

Fig. 2a is the programming cycle for water switchable SMEs of a biopolymer fiber and we can generalize the humidity-stress-strain relationship in Fig. 2b, while Fig. 2d gives specific shape recovery ratio from step ② to ⑤[20,21], Fig.2e shows the shape recovery of wool fabric decreased with the increase of RH, indicating the switch opened in higher RH value. Fig. 2c demonstrates the water sensitivity of a spider silk in the sharp change of storage modulus *E'* and a rubbery plateau achieved above the critical humidity[15]. *E'* can be switched easily between two states through two simple operations, namely drying and wetting by immersion in water. Thus a spider silk has dual identity: plastic when dried and elastomeric when wetted. The net-points in spider silk are verified as β-sheet crystals while collective hydrogen bonding (HB) among protein chains in amorphous regions is the switch. The HB leads to the fixation for a temporary shape in dry state while entropic elasticity in amorphous regions is responsible for shape recovery to original shape when wetted. Besides the net-points and switch are confirmed as crystal and switch, the discovered net-points and switches of other biopolymer fibers are listed in the Table I, where we conclude that the net-points are normally strong chemical bonds, crystals, cross-links, while the switches are normally weak attractive bonds such as HBs between macromolecules.

Table I Structural constitution of natural biopolymer fibers

	Net-points	Switches from hydrogen bonding in
Spider silk[15]	β-sheet	α-helix (Gly-Gly-X)
Elast-fiber[16]	Cross-links by lysine	Val-pro-Gly-val-Gly
Wool[17]	Disulfide or crystal	Gly-X
Collagen[18]	Cross-links by lysine	Gly-Pro-Hyp
Regenerated cocoon sil [6]	β-sheet (Alan-Gly) rich	(ser-gly-phe-gly-pro-tyr-cal-ala-asn-gly-gly-try-ser-arg-arg-glu-gly-)

On the basis of our concluded net-points and switches in each biopolymer fiber, a structural model for describing SME programming is shown in Fig. 3.

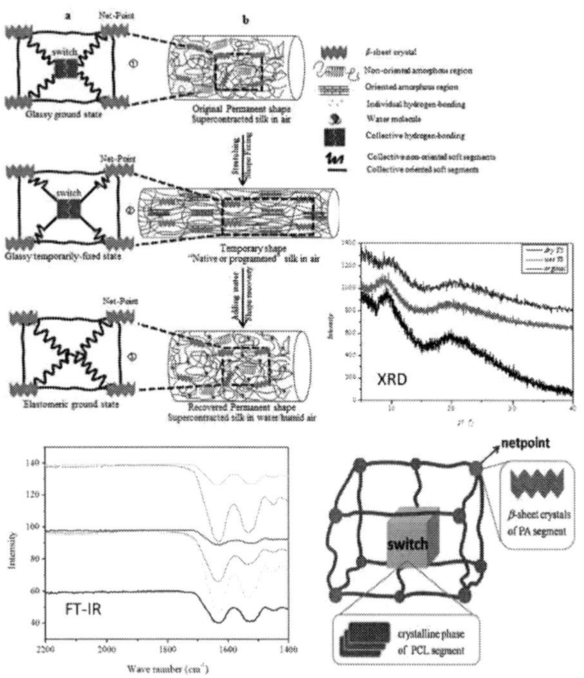

Fig. 3 Structural model of shape memory spider silk, ①, Switch is on with non-oriented soft segments; ②, switch is on with oriented soft segments; ③, switch is off with non-oriented soft segments; X-ray diffraction and FT-Infrared for characterizing net- points (crystal and disulfide bonds) and switch (hydrogen bond); bio-inspired synthetic SMP based on spider silk structure.

Fig. 3 shows a structural model for biopolymer fiber (silk) SME responsive to water. As a whole, the combination of a switch, that is, the collective HB in amorphous regions and elastomeric network with β-sheet crystals as net-points, is their microstructural prerequisite. At step ①, the switch is on and the silk is glassy due to the existence of the collective HBs. The protein chains in amorphous regions at step ② are extended to a certain level by stretching, thus the highly ordered structure makes the switch even tighter. Then immersion into water at step ③ can turn off the switch by disrupting HBs in the amorphous regions and leads to the recovery from the temporary shape by elastic entropy of soft segment molecular chains, thus the super-contraction. Under this wet condition, the silk becomes an elastomer free from the restriction of switch. XRD and FTIR characterizations prove the on and off states of net-points and switch during SME programming. As a reference furthermore, our design of a new SMP inspired by spider silks supports the model in terms of the effectiveness of β-sheet crystals as net-points. A new class of SMPs containing poly(ε-caprolactone) segment as switches and β-sheet crystals as net-points have been found to exhibit excellent shape fixation and recovery behaviors[21]. From this model, it is reasonable to anticipate that anything elastic in water and with HBs forming groups should have SMEs. This is particularly meaningful for protein biopolymers such as elastin and collagen in human body and animals with evident SME discovery.

ACKNOWLEDGMENT

This study is supported by the Hong Kong Government RGC research projects (PolyU 5161/11E and PolyU5182/09E) and the Niche Area Fund of the Hong Kong Polytechnic University (J-BB6M) and by the projects of JC-201104210132A and 2012 BAI17B06 from China. The authors would like to express their appreciations to the current and the past team members of the Shape Memory Textile Centre of PolyU, who contributed to this work as a whole in an accumulated approach in the past 15 years.

REFERENCES

[1] Agnarsson, I., et al. *J. Exp. Biol.*, 212 (13), 2009: 1989-93.
[2] Bell, F.I., et al. *Nature*, 416 (6876), 2002: 37-37.
[3] Vollrath, F., et al. *Soft Matter*, 7 (20), 2011: 9595-9600.
[4] Heim, M., et al. *Angew. Chem. Int. Ed.*, 48 (20), 2009: 3584-96.
[5] Liu, Y., et al. *Nature Materials*, 4 (12), 2005: 901-905.
[6] Plaza, G.R., et al. *J. Polym. Sci. Part B*, 44 (6), 2006: 994-999.
[7] Chen, S.J., et al. *Polymer*, 50 (19), 2009: 4424-4428
[8] Luo, H., et al. *Macromol. Chem. Phys.*, 212 (18), 2011: 1981-86.
[9] Zhu, Y., et al. *Soft Matter*, 8 (8), 2012: 2509-2517.
[10] Ji, F.L., et al. *High Perform. Polym.*, 23 (4), 2011: 314-325.
[11] Xie, T. *Polymer*, 52 (22), 2011: 985-5000
[12] Cranford, S.W., et al. *Nature*, 482 (7383), 2012: 72-76.
[13] Hu, J.L., et al. *Prog. Polym. Sci.*, 37 (12), 2012: 1720-63.
[14] Foo, C. et al. *Adv. Drug Deliv. Rev.*, 54 (8), 2002: 1131-43.
[15] Guan, J. et al. *Biomacromolecules*, 12 (11), 2011: 4030-35.
[16] Elices, M., et al. *Mater. Sci. Eng. C.*, 31 (6), 2011: 1184-88.
[17] Vollrath, F., et al. *Soft Matter*, 7 (20), 2011: 9595-9600.
[18] Work, R.W. et al. *Text. Res. J.*, 52 (5), 1982: 349-356.
[19] Vollrath, F. et al. *Soft Matter*, 2 (5), 2006: 377-385.
[20] Vollrath, F. et al. *Nature*, 340 (6231), 1989: 305-307.
[21] Blackledge, T.A., et al. *J. Exp. Biol.*, 212 (13), 2009: 1980-88.
[22] Huang, H.H., et al. *Macromol. Biosci.*, 13 (2), 2012: 161-166.

Nanocellulose Hierarchical Structures and Functional Materials

You-Lo Hsieh
University of California, Davis, CA 95616, USA
ylhsieh@ucdavis.edu

Nanocellulose is the highly crystalline (65%-95%) cellulose nanocrystals or nanofibers that can be isolated from a broad range of cellulose sources including tunicate, algae, bacteria and wood. Nanocellulose has been reported to be 2 to 20 nm wide with aspect ratios varying from 40 (~200 nm long and 5 nm wide) to around 66 (~1 μm long and 15 nm wide) for tunicin whiskers[1]. The bending strength and modulus of these nanocellulose have been estimated by Raman spectroscopy to be ~10 GPa and ~150 GPa,[2] respectively. This impressive bending strength is about one-seventh of 63 GPa for the carbon nanotubes and much higher than high strength fibers including Kevlar.

Due part to lacking free rotation along the β-(1→4)-glycosidic linked D-anhydroglucopyranose and the extensive inter-molecular and intra-molecular hydrogen bonds from the equatorial primary C6 and secondary C2 and C3 hydroxyls, native cellulose are rigid, highly crystalline, strong and, in particularly, challenging to dissolve and process into other forms. To take full advantage of unique cellulose chemistry and crystalline structure, direct construction of functional materials from nanocellulose is a superior approach to fulfill their greatest potential. The unique chemical and physical structures of cellulose make it the most significant renewable resources for a great variety of advanced materials beyond fiber, paper and biofuel.

This paper presents approaches by which to isolate pure cellulose from biomass, to generate crystalline nanocellulose and to create new nano-materials.

We have developed streamlined processes to isolate cellulose from under-utilized agricultural and processing byproducts. Crystalline nanocellulose can then be derived by chemical, physical and a combination chemical-physical methods. Pure cellulose has been isolated from rice straw at 36% yield. Cellulose nanocrystals (CNCs) and nanofibrils (CNFs) can be derived from pure rice straw cellulose *via* sulfuric acid hydrolysis, mechanical blending and TEMPO-mediated oxidation to 16.9%, 12% and 19.7% yields, respectively. Sulfuric acid hydrolysis produced highly crystalline (up to 90.7% CrI) rod-shaped (3.96-6.74 nm wide, 116.6-166 nm long) CNCs with similarly negative surface charges (-44 to -38 mV) and sulfate contents but decreasing yields and dimensions with increasing hydrolysis time. Mechanical defibrillated CNFs were 82.5% crystalline and bimodally distributed in sizes (2.7 nm wide and 100-200 nm long; 8.5 nm wide and micrometers long). TEMPO mediated oxidation liberated the most uniform, finest (1.7 nm) and micrometer long, but least crystalline (64.4% CrI) CNFs (Figure 1).[4,5]

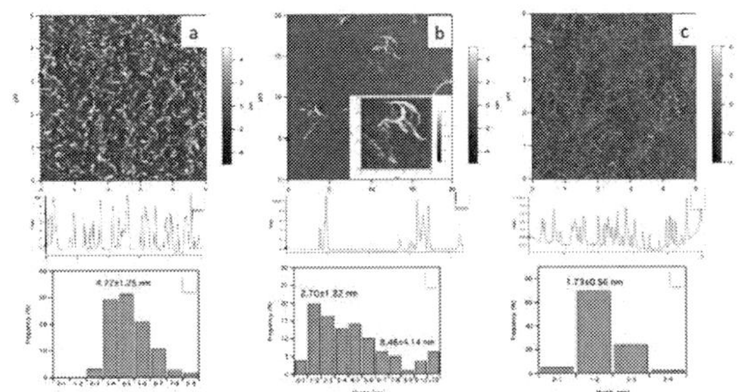

Figure 1. AFM images of nanocellulose: (a) CNCs; (b) CNFB; (c) CNFT.

Combined TEMPO mediated oxidation (5 mmol/g NaClO/cellulose) and mechanical blending (30 min) produced highly uniform (2.09 nm wide, 1.52 nm thick, up to 1 μm long) cellulose nanofibrils (CNFs) at an impressive 96.8% yield and contained 1.29 mmol surface carboxyls per g of cellulose or 0.21 COOH/anhydroglucose (AG), representing 70.9% surface C6 primary hydroxyl to carboxyl conversion.[6]

All nanocellulose have cellulose Iβ crystalline structure and are more crystalline than its source. Rapid freezing of aqueous nanocellulose suspensions in liquid nitrogen and freeze-drying induced self-assembly into nanometer to sub-micron wide fibers (Figure 2).

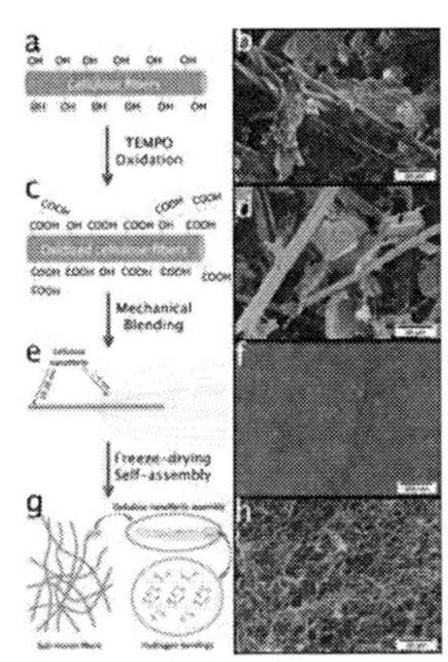

Figure 2. Defibrillation of cellulose and self-assembly of cellulose nanofibrils: (a, c, e, g) (b) SEM of pure cellulose; (d) SEM of oxidized CNF1.5; (f) TEM of CNF1.5; (h) SEM of assembled CNF1.5.

Slow freezing, on the other hand, induce ice-crystal templating to generate hydrogels and aerogels (Figure 3)[7]. The self-assembled fiber morphologies showed strong dependence on nanocellulose morphologies and extents of surface functionalization.

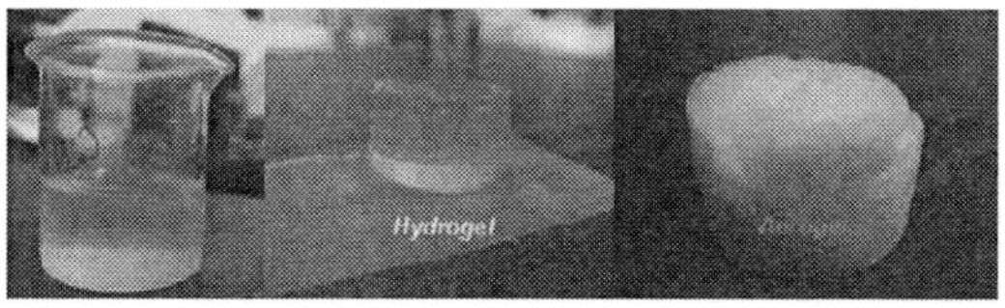

Figure 3. Freeze-thaw induced hydrogels and amphiphilic aerogels.

Ultra-lightweight (1.7 mg/cm^3) and highly porous (99.9% porosity) aerogels are facilely fabricated by freezing induced self-assembling of ultrathin CNFs. The as prepared aerogels are amphiphilic and absorb over 200 to nearly 300 mL/g of aqueous as well as non-polar hydrocarbon liquids. Silanization can convert the hydrophilicity toward hydrophobicity, making aerogel superior oleophilic super-absorbents for hydrocarbons, excellent for selective oil removal and recovery (Figure 4)[8].

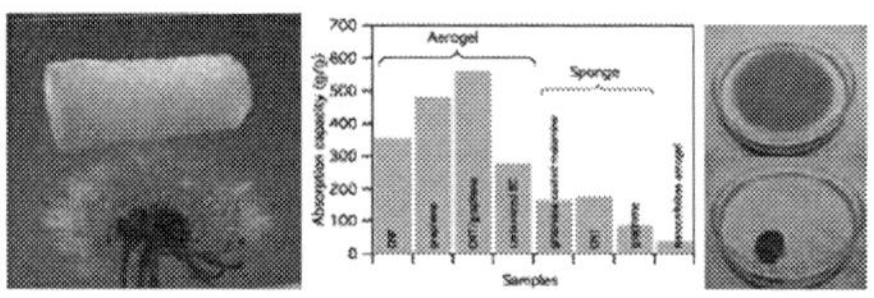

Figure 4. Ultra-light (1.7 mg/cc) and ultra-porous (>99.995%) aerogels capable of absorbing 350 g/g chloroform and selective removal of hydrocarbons from water.

The surface charges on CNFs may be tuned to allow assembling into other controlled hierarchical structures. CNFs have also been demonstrated to serve as dual capping and shape-regulating agent for the synthesis of silver nanoparticles as well as nanoprisms. CNFs significantly improved the stability and dispersibility of Ag nanoparticles in physiological medium, validating the antimicrobial potential of CNF-AgNPs, and provided an insight on the physio-chemical interactions between bacteria and CNF-AgNPs in a physiological growth medium[9]. Ag nanoprism bound CNFs have shown excellent surface enhanced Raman scattering of Rhodamine 6G with analytical enhancement factor of 5x10^3 to potential applications as senors.

Further stream-lined processes to isolate and derive nanocelluolose from other food and beverage processing wastes will be presented so are other advanced carbon materials such as activated carbon fibers for supercapacitors.

ACKNOWLEDGMENT

Funding from USDA NIFA and California Rice Research Board is greatly appreciated. Research was conducted by P. Lu, F. Jiang and S. Hu.

REFERENCES

1. Samir, M.A.S.A., Alloin, F., Dufresne, A. "Review of recent research into cellulosic whiskers, their properties and their application in nanocomposite field." *Biomacromolecules*, 6 (2), 2005: 612-626.
2. Sturcova, A., Davies, G.R., Eichhorn, S.J. "Elastic modulus and stress-transfer properties of tunicate cellulose whiskers." *Biomacromolecules*, 6 (2), 2005: 1055-1061.
3. Lu, P., Hsieh, Y.-L. Preparation and properties of cellulose nanocrystals." *Carbohydrate Polymers*, 82 (2), 2010: 329-336.
4. Lu, P., Hsieh, Y.-L. "Preparation and characterization of cellulose nanocrystals from rice straw." *Carbohydrate Polymers*, 87, 2012: 564-573.
5. Jiang, F., Hsieh, Y.-L. "Chemically and mechanically isolated nanocellulose and their self-assembled structures." *Carbohydrate Polymers*, 95 (1), 2013: 32-40.
6. Jiang, F., Han, S., Hsieh, Y-L. "Controlled defibrillation of rice straw cellulose and self-assembly of cellulose nanofibrils into highly crystalline fibrous materials." *RSC Advances*, 3 (30), 2013: 12366-12375.
7. Jiang, F., Hsieh, Y.-L. "Super water absorbing and shape memory nanocellulose aerogels from TEMPO-oxidized cellulose nanofibrils via cyclic freezing-thawing." *Journal of Materials Chemistry A*, 2, 2013: 350-359.
8. Jiang, F., Hsieh, Y.-L. "Amphiphilic superabsorbent cellulose nanofibril aerogels." *Journal of Materials Chemistry A*, 2, 2014: 6337-6342.
9. Wang, M.S., Jiang, F., Hsieh, Y.-L., Nitin, N. "Cellulose nanofibrils improve dispersibility and stability of silver nanoparticles and induce production of bacterial extracellular polysaccharides." *Journal of Materials Chemistry B*, 2 (37), 2014: 6226-6235.
10. Jiang, F., Hsieh, Y.-L. "Synthesis of cellulose nanofibril bound silver nanoprism for surface enhanced Raman scattering." *Biomacromolecules*, 15 (10), 2014: 3608-3616.

Molecular Orientation in Melt-Spun P3HB Fibers

Rudolf Hufenus, Felix A. Reifler, María P. Fernández-Ronco
Empa, Swiss Federal Laboratories for Materials Science and Technology, Switzerland
rudolf.hufenus@empa.ch

OBJECTIVE

The aim of this work was to develop an upscalable melt-spinning method for high-strength P3HB fibers by studying the effects of additives and filament drawing on fiber performance. The additives were analyzed either in combination with P3HB powder extracted from bacteria or using pellets extruded from native polymer powder. In order to address the poor melt strength and the particular crystallization behavior of P3HB, the godet set-up of our draw-off unit was modified. Finally, the physical and structural properties of the produced filaments were characterized and compared to those of conventional fibers.

INTRODUCTION

The use of inherently biodegradable polymers is an opportunity to reduce the amount of plastic waste that cannot be assimilated by microorganisms. Poly(3-hydroxybutyrate) (P3HB) is a sustainable and biocompatible[1] thermoplastic polyester produced by bacteria as intracellular carbon and energy storage compound[2]. The bacteria build the polymer chains perfectly linear and isotactic, which guarantees superior properties[3]. Due to its biological background, P3HB is truly biodegradable in aerobic (e.g. in soil or compost) and anaerobic (e.g. in drainage pits or sea mud) conditions without forming toxic by-products[4]. In addition, P3HB is water insoluble, hydrophobic, and comparatively resistant to hydrolytic degradation[5]. Due to its exceptional properties, as well as reasonable production costs via the relatively simple biosynthesis process[6], P3HB is a promising substitute for conventional petrochemical plastics[7], and it qualifies for the use in numerous textile and medical applications.

Being a thermoplastic polymer, P3HB is suitable for a continuous production of filaments via melt-spinning[8], which is probably the most efficient process to produce fibers[9]. However, its rapid thermal degradation at temperatures just above the melting temperature, and the brittleness of native P3HB pose technical and scientific challenges[10]. As a result, melt-spinning of P3HB into fibers at large scale is not feasible to date. The issue has been previously addressed in the scientific literature, but just at the laboratory scale[7].

The nucleation step in the crystallization process is generally associated with the presence of inhomogeneities or foreign particles in the melt[11]. In the case of P3HB, nucleating agents may be helpful to increase the amount of crystal nuclei, and therefore to decrease the dimension of unwanted spherulites[12]. Common nucleating agents are non-soluble inorganic fillers like metal oxides, metal salts, silicates or boron nitride with particle sizes of about 3 μm[7]. For the present study, boron nitride (BN) was chosen since it is one of the best nucleating agents reported for P3HB[12]. The inherent brittleness of native P3HB can be overcome by adding plasticizers[13]. To maintain biodegradability, non-toxic citrate plasticizers derived from naturally occurring citric acid are favored for this purpose[13]. Thus, tri-n-butyl citrate (TBC) was chosen as plasticizer in the present study.

EXPERIMENTAL APPROACH

Biomer P209 (with nucleating agent and plasticizers) and Biomer P309 (with nucleating agent, without plasticizers), both pelletized P3HB compounds, were provided by Biomer (Krailling, Germany). For reference purposes, Biomer also provided the raw polymer (T19) extracted from bacteria, which was used to produce the compounds P209 and P309. Plasticizer tri-n-butyl citrate (TBC) with the trade name Citrofol BI was provided by Jungbunzlauer (Ladenburg, Germany). Nucleating agent boron nitride (BN) was provided by Biomer.

Fiber melt-spinning was carried out on Empa's custom-made pilot melt-spinning plant[14]. The analytical tools applied included differential scanning calorimetry (DSC), gel permeation chromatography (GPC), scanning transmission electron microscopy (STEM), and wide angle X-ray diffraction (WAXD).

RESULTS AND DISCUSSION

Factors that impede spinnability of P3HB are manifold. The first challenge is the unsatisfactory consistency of the raw polymer regarding purity, thermal behavior, viscosity, molecular weight, and crystallization rate. As this is an intrinsic handicap of a product originating from natural resources, we propose the use of additives to improve melt-spinnability of P3HB. DSC and extrusion trials showed that nucleating agents control and stabilize crystallization. From our findings it can be concluded that, for a given polymer formulation containing BN, the onset of crystallization is largely independent of the temperatures the polymer experiences during extrusion. As a result, the onset is insensitive to small changes in the temperature of the polymer melt, which effectively stabilizes the melt-spinning process.

Another major drawback that limits the processability of P3HB is thermal degradation of the polymer near the melting point (170-200 °C). It is known that its thermal degradation mainly occurs via an ester decomposition mechanism involving a six-membered ring transition state, which is a non-radical mechanism and a random chain scission process[15]. GPC measurements showed that the extrusion steps required for fiber melt-spinning strongly contributed to the decay of molecular weight and, as a consequence, lowered the achievable tensile strength of the fibers. DSC analysis, on the other hand, revealed that the presence of the plasticizer TBC reduces the melting point by almost 15 °C, which signifies a gain in the processing and melt-spinning performance, particularly in view of the general proximity of melting and degradation temperatures with P3HB.

In addition to the incorporation of additives, we successfully modified the draw-off unit of the extrusion line to address the poor melt strength and the particular crystallization behavior of P3HB (Figure 1). By introducing an additional godet that enabled an oriented crystallization before complete solidification, the extrudate could withstand the drawing forces induced by melt-spinning, and filaments dominated by longitudinally oriented lamellae rather than spherulitic structures could be achieved. Secondary crystallization, which would result in brittleness and poor mechanical performance of as-spun fibers, could be prevented in short and long term. The filaments produced exhibit long term stability with tensile strengths up to 215 MPa at 20 % tensile strain, and thus comply with requirements of textile manufacturing and application.

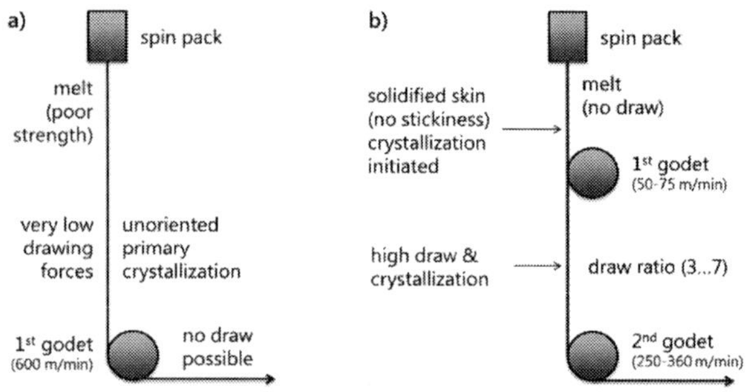

Figure 1: Melt-spinning of P3HB in the pilot melt-spinning plant with draw-off unit in (a) standard and (b) modified configuration.

In the region from 8.2° to 9.4° 2θ in the equatorial 2Theta scan of the WAXD patterns, we observed a series of local maxima for the fibers melt-spun and drawn in this work. From this observation, and in contrast to the previous literature where diffraction signals in this region are commonly described as one reflection and assigned to the so-called β-form of P3HB, we postulate a highly ordered amorphous phase, which is locally trapped between the aligned lamellae of the crystalline α-phase. The local maxima in the 2Theta scan would correspond to preferred distances between polymer chains that are oriented nearly parallel to the fiber direction, but irregularly arranged along this direction.

CONCLUSION

Bacterial poly-3-hydroxybutyrate (P3HB) is a thermoplastic polyester with perfectly isotactic structure; it has the potential for fiber melt-spinning and thus is qualified for textile applications. Aiming at developing an upscalable melt-spinning method to produce high-strength fibers, we evaluated the effect of the additives tri-n-butyl citrate (TBC) as plasticizer, and boron nitride (BN) as nucleating agent, on the melt-spinning performance of P3HB. In addition, the draw-off unit has been modified in order to enable a complete primary crystallization of the fibers during drawing. This allows obtaining P3HB fibers exhibiting promising tensile strengths up to 215 MPa. It is proposed that these fibers contain domains of highly oriented molecules between crystallites. The wide-angle X-ray diffraction (WAXD) data suggest that besides the oriented crystalline α-phase, a highly oriented amorphous phase is as well present.

KEYWORDS

Melt-spinning; biopolyester; biopolymer; processing; wide-angle X-ray diffraction

ACKNOWLEDGMENT

Part of this study was kindly supported by the Zurichian Textile Research Foundation. The authors thank Urs Hänggi from Biomer (Krailling, Germany) for providing materials, Benno Wüst for operating the spinning plant, Pierluigi Barbadoro for physical fiber characterization, Liliane Diener and Patrick Rupper for STEM support, Karl Kehl for GPC measurements, Boguslawa Gradzik, Seraina Bokanyi and Youna Maussion for DSC analysis, Beatrice Gutt for GC measurements, Thomas Weber, Christian Baerlocher and Walter Steurer of the Laboratory of Crystallography (Department of Materials, ETH Zürich, Switzerland) for providing access to their four-circle diffractometer and for generous support.

REFERENCES

[1] Leja, K., Lewandowicz, G. *Polish Journal of Environmental Studies*, 19, 2010: 255-266.
[2] Sudesh, K., Abe, H., Doi, Y. *Progress in Polymer Science*, 25, 2000: 1503-1555.
[3] Pan, P., Inoue, Y. *Progress in Polymer Science*, 34, 2009: 605-640.
[4] Bonartseva, G.A, Myshkina, V.L., Nikolaeva, D.A., Kevbrina, M.V., Kallistova, A.Y., Gerasin, V.A., Iordanskii, A.L., Nozhevnikova, A.N., *Applied Biochemistry and Biotechnology–Part A Enzyme Engineering and Biotechnology*, 109, 2003: 285-301.
[5] Zinn, M., Hany, R. *Advanced Engineering Materials*, 7, 2005: 408-411.
[6] Choi, J.L., Lee, S.Y. *Bioprocess Engineering*, 17, 3, 1997: 35-342.
[7] Laycock, B., Halley, P., Pratt, S., Werker, A., Lant, P. *Progress in Polymer Science*, 38, 2013: 536-583.
[8] Hinüber, C., Häussler, L., Vogel, R., Brünig, H., Heinrich, G., Werner, C. *Express Polymer Letters*, 5, 2011: 643-652.
[9] Ziabicki, A. *Fundamentals of Fibre Formation*. London: John Wiley & Sons, Ltd., 1976.
[10] Barham, P.J., Keller, A. *Journal of polymer science: Part A-2, Polymer Physics*, 24, 1986: 69-77.
[11] Walczak, Z.K. *Processes of Fiber Formation*. Amsterdam: Elsevier Science Ltd., 2002.
[12] Kai, W., He, Y., Inoue, Y. *Polymer International*, 54, 2005: 780-789.
[13] Erceg, M., Kovacic, T., Klaric, I. *Polymer Degradation and Stability*, 90, 2005: 313-318.
[14] Hufenus, R., Reifler, F.A., Maniura-Weber, K., Spierings, A., Zinn, M. *Macromolecular Materials and Engineering*, 297, 2012: 75-84.
[15] Grassie, N., Murray, E.J., Holmes, P.A. *Polymer Degradation and Stability*, 6, 1984: 127-134.

Changes in Microclimate Temperature by Temperature-Adaptable Materials Under Transient Conditions

Guira Park, Eunae Kim
Clothing and Textiles, Yonsei University, Seoul, Korea
eakim@yonsei.ac.kr

STATEMENT OF PURPOSE

In our lab, we have developed an instrument, Human-Clothing-Environment Simulator (US Patent 7680638 B2, Korea patent 0483996) and have shown the usefulness of it in previous studies. In this study, the surface temperature or air gap change by the temperature adaptable fibers or fabrics and their effects on the microclimate temperature and humidity is investigated.

INTRODUCTION

Heat releasing materials, phase change materials and two way shape memory alloy (TWSMA) spring embedded fabrics are all recognized as temperature adaptable materials. Their mechanisms and temperature range to adapt the environmental temperature are all different. But there is no standard method to test the temperature adaptability yet. Recently, heat releasing property measurement was standardized by BS ISO16533:2014. This method is quite useful to determine the exothermic and endothermic reactions of the material by the humidity change. In real life, however, the environment temperature and humidity change simultaneously. Therefore results obtained from the standardized method show the material property and do not show how much they will affect to the microclimate in the clothing system at a certain environmental condition and how the subject will perceive the difference. As there are various temperature adaptable materials, and they are always controversial we need to develop a standardized method to quantify the temperature adaptability

EXPERIMENTS

Specimens

Heat releasing fibers vs wool fibers; 2, 5 and 10g of fibers are prepared as fiber webs and covered with net fabrics.

Table I. Specifications for the heat releasing fiber webs

specimen	Fiber type	Weight(g)	Pouch thickness(mm)
HE2	Hydrophilic polyacrylonitrile	2	2
HE5		5	5
HE10		10	8
WL2	Wool	2	6
WL5		5	16
WL10		10	23

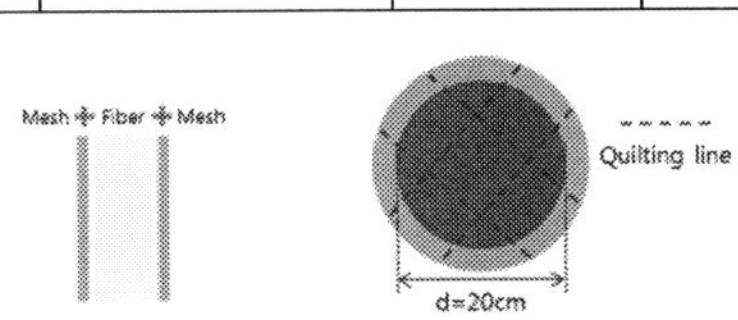

Figure 1. Specimen assembly for the fiber webs.

TWSMA springs embedded fabrics; low temperature adaptable7 springs were attached between two layers of the 20cm diameter fabrics. As control, two layers of same fabrics were used without springs.

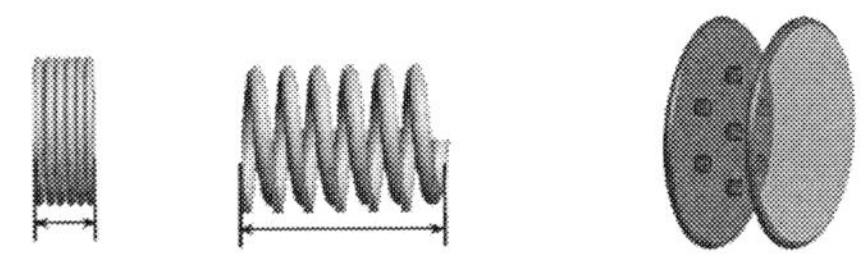

Above 24.5℃ Below 24.5℃ Attachments

Figure 2. Low temperature adaptable springs with two way shape memory effect and their attachment onto fabric layers.

Instruments

Human-Clothing-Environment Simulator was used to measure the changes in the microclimate temperature and humidity in the air layer or the surface temperature at transient conditions from high temperature to low temperature or vice versa. The guarded sweating hot plate is used as skin part. The temperature of the hot plate can be controlled independently and the required input heat flux to maintain this accurately measured with time. Sweat is simulated by supplying water from six nozzles connected to a micro pump (ISM937, ISMATEC, Switzerland). On the surface of the fabrics and/or between fabric layers, temperature and humidity sensors can be located. Needle sensor (IT-18, Physitemp Instrument, USA) was attached on the surface of the testing materials.

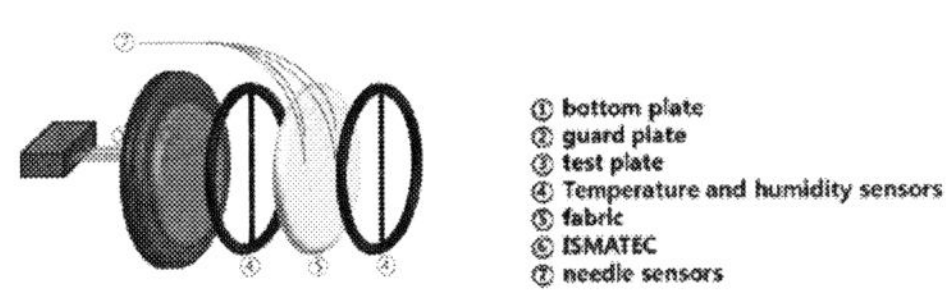

Figure 3. Schematic diagram of the vertical sweating skin model and the needle sensor locations.

Test protocol

For each test, different test protocol was used ; (a) was for isothermal condition, (b) was for practical use condition and (c) was for TWSMA spring embedded fabrics.

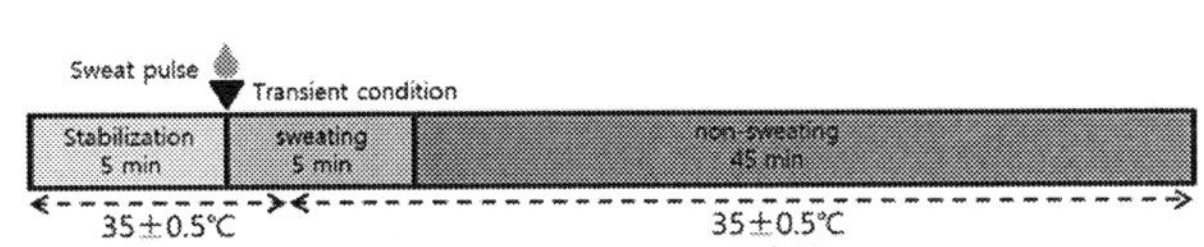

(a)

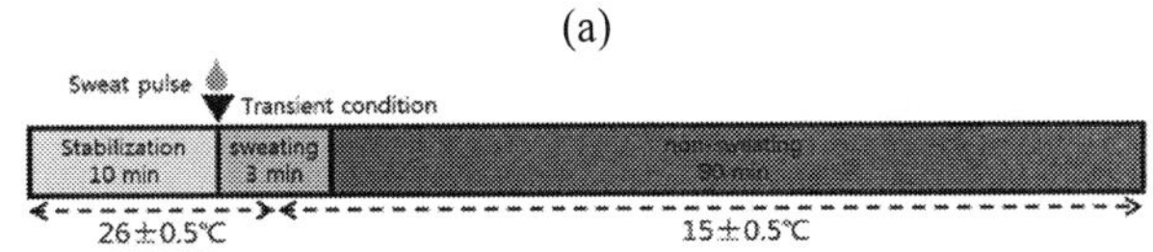

(b)

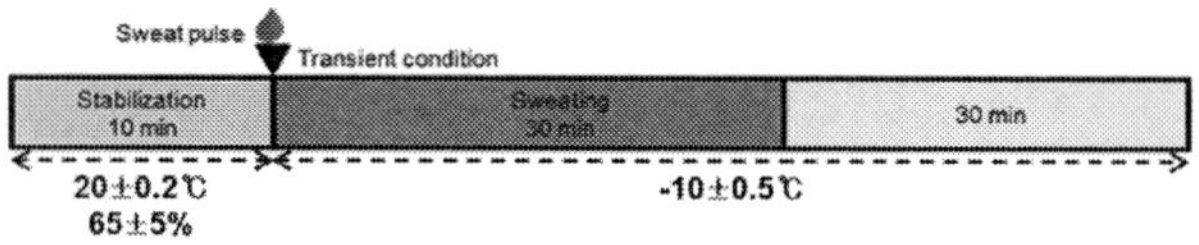

(c)

Figure 4. Test protocol for each test.

RESULTS AND DISCUSSION

By heat releasing materials, the surface temperature increased abruptly as high as 16 degrees as is shown in Figure 5. The temperature increase for heat releasing material is higher than the wool fibers. As the weight increased, the temperature increased. The same increase of temperature was observed by method ISO 16533 (Figure 6).

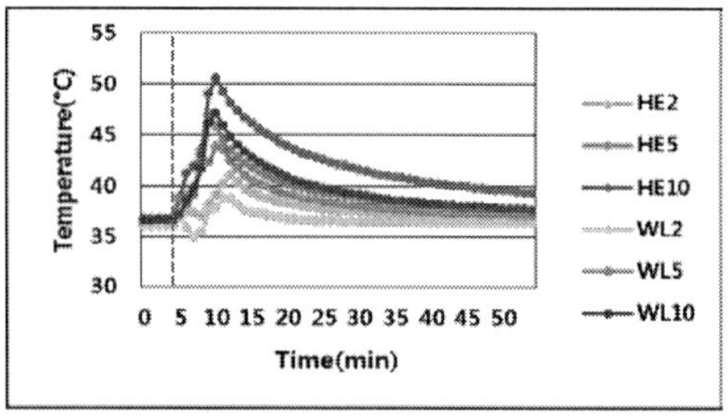

Figure 5. Surface temperature changes of wool and heat releasing fiber webs by humidity changes.

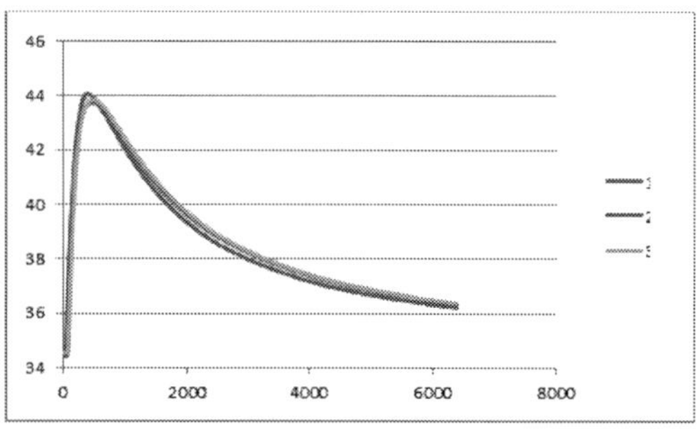

Figure 6. Temperature increase by heat releasing material tested by ISO16533:2014 (3 repeats).

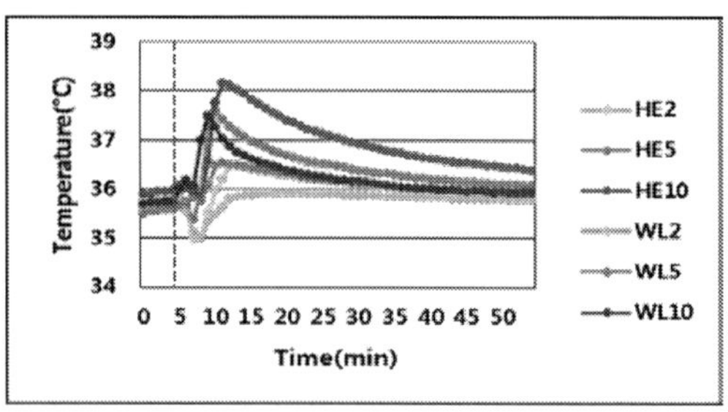

Figure 7. Microclimate changes of wool and heat releasing fiber webs by humidity changes.

Figure 7 shows the increase in the temperature in the microclimate. The maximum increase was only 2.5 °C when 20g heat releasing material was used. In other words, 13-14 °C of energy was lost to the environment. As these results were obtained at the isothermal condition and changed only the humidity, we tested at the practical use condition where the temperature was lowered from 26 to 15 °C and the results are shown in Figure 8. Temeprature increase was almost the same but as the environmental temperature was lower, the microclimate temperature was lower. Figure 9 is a result obtained from the TWSMA spring embedded fabrics. As the air layer is formed by the expansion of the springs, microclimate temperature was less decreased than the control fabric.

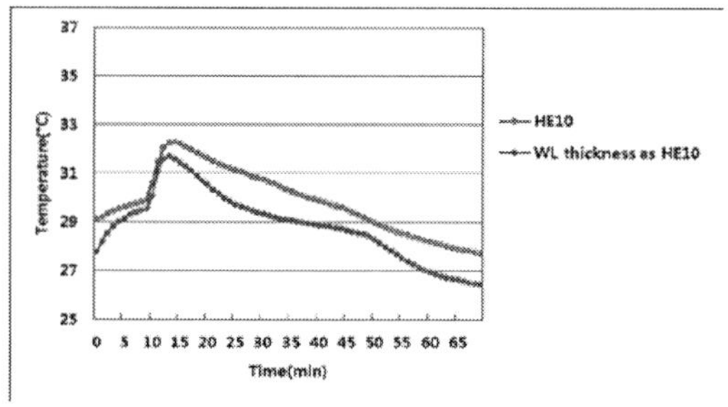

Figure 8. Microclimate changes of wool and heat releasing fiber webs at transient conditions.

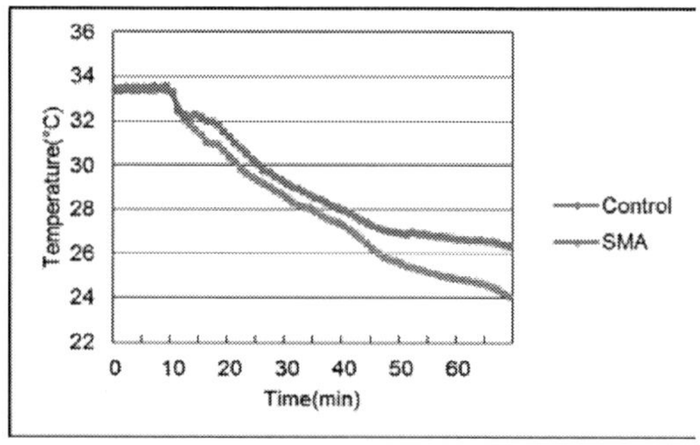

Figure 9. Difference in temperature decrease after exposure to the low environment using TWSMA spring embedded fabrics.

CONCLUSION/FUTURE WORK

As bulk of material was used to verify the effect, we need further study using commercially available specimens. Wear test is also required to understand how the temperature difference is perceived by the human subjects.

KEYWORDS

Heat releasing material, TWSMA springs, Human-Clothing-Environment Simulator, temperature adaptability

ACKNOWLEDGMENT

This work was supported by the National Research Foundation of Korea (NRF) grant funded by the Korea government (MEST) (No. NRF-2012R1A2A2A03047265).

REFERENCES

[1] HCE Simulator; Korea Patent 0483996; U.S. Patent 7680638B2.

[2] Kim, Eun Ae, Yoo, Shinjung, Kim, Jeongjin. "Development of a Human-Clothing-Environment Simulator for Dynamic Heat and Moisture Transfer Properties of Fabrics." *Fibers and Polymers*, 4, 4, 2003: 215-2.

Effect of Hot Drawing on the Mechanical Properties, Moisture Absorption, and Dyeability of Novel Nylon 5,6 Fibers

Yassir A. Eltahir[1,2], Haroon A. M. Saeed[1,2], Yumin Xia[1], He Yong[1], Wang Yimin[1]
[1]Collage of Materials Science and Engineering, State Key Laboratory for Modification of Chemical Fibers and Polymer Materials, Donghua University, Songjiang, Shanghai, China, [2]Faculty of Textiles, University of Gezira, Wad-Medani, Sudan
ymw@dhu.edu.cn; yassir_eltahir@yahoo.com

THE PURPOSE OF THE STUDY

The purpose of this study is to prepare nylon 5,6 and nylon 6 fibers by melt spinning process and to investigate the effect of drawing conditions on its mechanical properties, moisture absorption and dyeing behavior.

INTRODUCTION

Since the invention of nylon 66 by Carothers in 1934[1], many kinds of nylons have been developed, and most of them have been commercialized in the past 80 years. Polyamides have found their way in various applications in many industries, such as automotive, electrical& electronics, general industries, and packaging and (film, fiber, and sheet) extrusion. This is merely a result of their excellent performance balance as related to temperature resistance, mechanical properties, chemical resistance, and processability[2]. The effect of melt spinning and drawing variables on the crystalline orientation and mechanical properties of synthetic fibers were devised by Carothers and Coworkers in 1932[3]. Since this work, many studies have been carried out to investigate the effect of drawing of polyamide fibers on the structure and mechanical properties[4, 5]. However, up to now, only a few studies have been reported on the nylon 5,6. Upon to our knowledge the preparation of nylon 5,6 fibers and their mechanical properties have not been reported so far.

EXPERIMENTAL APPROACH

Materials

Nylon 5,6 pellets with an intrinsic viscosity of 0.68 dL/g measured in dichloroacetic acid at 25 °C, and nylon 6 pellets (Model: Bright R.V.2.4-2.8) were kindly supplied by Guangdong Xinhui Meida Nylon Co. Ltd, China.

Methods

Nylon 5,6 and nylon 6 fibers were prepared by melt-spinning technique at take-up speed of 1000 m/min. and subsequently drawn with various draw ratios with the same drawing temperature (70/130°C). The effects of draw ratios on the mechanical properties, moisture absorption and dyeability were investigated by means of tensile test measurements, moisture absorption test and dyeablity tests in terms of Exhaustion% and color strength measurements.

RESULTS AND DISCUSSION

Mechanical Properties

The tenacity, elongation at break and tensile modulus of the nylon 5,6 and nylon 6 fibers prepared at different draw ratio (DR) at the same drawing temperature (70/130°C) are shown in Figure 1 and 2, respectively. As shown in Figure 1 and 2, the tenacity and tensile modulus of the nylon 5,6 and nylon 6 drawn fibers increased continuously with the drawing process, while the elongation at break decreased at the same drawing temperature (70/130°C).

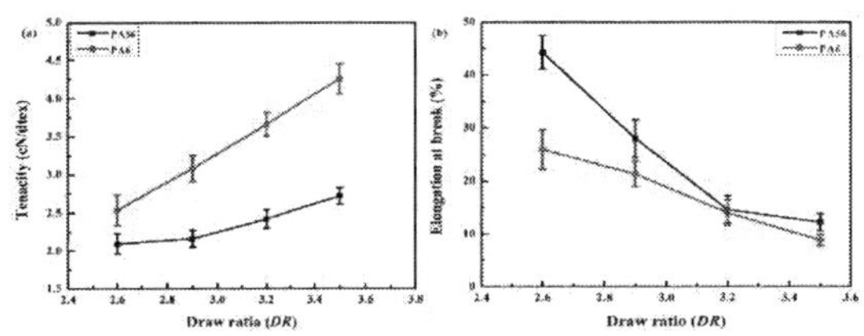

Figure 1: Effect of draw ratios on the nylon 5,6 and nylon 6 fibers mechanical properties: (a) tenacity and (b) elongation at break.

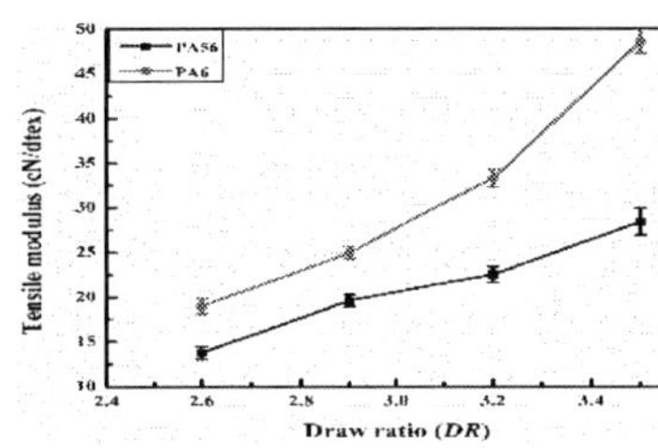

Figure 2: Effect of draw ratios on the nylon 5,6 and nylon 6 fibers tensile modulus.

These results clearly indicate that the improvement of molecular orientation along the drawing direction, the development of more ordered crystalline structure and the increase in crystallinity, have a positive influence on the mechanical properties of nylon 5,6 and nylon 6 fibers when subjected to the drawing process[6]. It is worthy noted that the nylon 6 fibers show high tenacity and tensile modulus than nylon 5,6 fibers at different draw ratios. This can be ascribed to the regularity of molecules conformation and crystallinity of nylon 6.

Moisture Regain Properties

Table I shows the standard moisture regain evaluation results for the nylon 5,6 and nylon 6 fibers drawn at different draw ratios. Nylon 5,6 fibers showed a moisture regain of 4.48, 3.87, 3.43, 3.38, and 3.07% for undrawn, 2.6, 2.9, 3.2, and 3.5, respectively. From Table 1, it can be seen that both nylon 5,6 fibers and nylon 6 fibers show significant decrease in moisture regain with increase of draw ratios. This results indicates that the increase in crystallinity and molecular orientation decrease the moisture absorption of nylon 5,6 and nylon 6 fibers, because in semi-crystalline fibers, moisture is not easy to access the densely packed crystalline regions, while it is absorbed into the amorphous regions[7]. On the other hand, nylon 5,6 fibers exhibit much higher moisture regain than nylon 6 fibers at different draw ratios. This can be attributed to the amide group content[8] in the polymer backbone, which yield higher moisture regain in nylon 5,6 fibers than nylon 6 fibers.

Table I Moisture regain property of nylon 5,6 and nylon 6 fibers at different draw ratios

	Moisture regain (%)	
Draw ratio (DR)	PA56	PA6
Undrawn	4.48	3.26
2.6	3.87	3.12
2.9	3.43	3.10
3.2	3.38	3.03
3.5	3.07	2.99

Dyeability of Nylon 5,6

In order to obtain fundamental knowledge of the dyeability of nylon 5,6, the dyeing process was carried out using an acid dyes, which are the most widely used dyes for nylons. Figure 3(a) shows the dyebath exhaustion% of nylon 5,6 and nylon 6 knitted fabrics prepared at different draw ratios. It can be seen that the dyebath exhaustion% decreased continuously with increased the draw ratios for both nylon 5,6 and nylon 6. Therefore, the decrease of dye exhaustion can be attributed to the increased in crystallinity and molecular orientation with the draw ratios. On the other hand, nylon 5,6 shows high dyebath exhaustion% than that of nylon 6, which can be attributed to the amide group content in nylon 5,6.

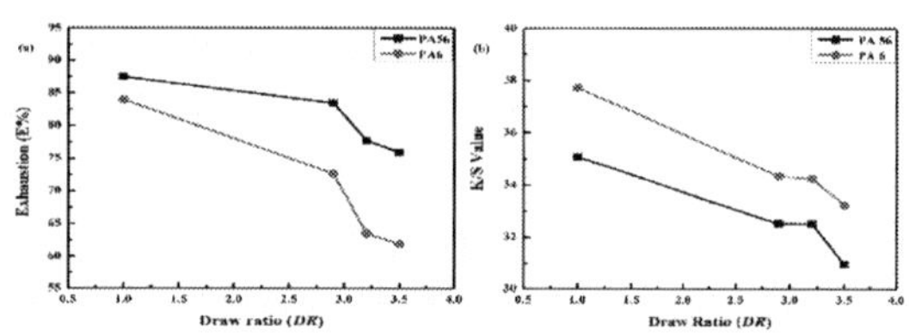

Figure 3: Exhaustion% and Color strength (K/S) values for nylon 5,6 and nylon 6 knitted fabrics at different draw ratios

Figure 3(b) shows the color strength (K/S)λ obtained for nylon 5,6 and nylon 6 knitted fabrics dyed with an acid red 88 dye. The color strength ($K/S_{ma}x$) values were determined at λ_{max} equal to 520nm. From Figure 3(b), a significant decrease has been observed in the color strength (K/S) values for both nylon 5,6 and nylon 6 fibers with the draw ratios. The decrease in color strength is high at the beginning and the decrease of tendency starts to full by increasing the draw ratio. The decrease in color strength (K/S) values with the draw ratios can be attributed to the structural and morphological changes in terms of changes in crystallinity, crystal sizes and molecular orientation. On the other hand, nylon 6 shows high color strength (K/S) values than those of nylon 5,6 fibers at different draw ratios.

CONCLUSIONS

Nylon 5,6 and nylon 6 fibers were successfully prepared by the melt spinning process. The tenacity, Young's modulus was found to be significantly increased, while the elongation at break decreased as the draw ratio increased. Nylon 5,6 fibers exhibits much higher moisture regain than nylon 6 fibers at different draw ratios. The dyeing behavior of nylon 5,6 in terms of dyebath exhaustion, color strength, imply that the nylon 5,6 behaved in a manner that was to be expected of its conventional nylon 6.

KEYWORDS: Nylon 5,6, Tenacity, Moisture regain, Color Strength.

REFERENCES

[1] Carothers, W.H., U.S. Patent 1938: 2,130,948.

[2] Kohan, M.I. *Nylon Plastics Handbook.* Vol. 378. Munich, Germany: Hanser Publishers, 1995.

[3] Carothers, W.H., Hill J.W. "Studies of polymerization and ring formation. XV. Artifical fibers from synthetic linear condensation superpolymers." *J. Am. Chem. Soc.*, 54, 1932: 1579-1587.

[4] Yeh, J.-T., Wang, C.-K., Liu, Z.-W., Li, P., Tsou, C.-H., Lai, Y.-C., Tsai, F.-C. "Drawing and ultimate tensile properties of nylon 6/nylon 6 clay composite fibers." *Polym. Eng. Sci.*, 52, 2012: 1348-1355.

[5] Rahbar, R.S., Mojtahedi, M.R.M.. "Influence of Hot Multistage Drawing on Structure and Mechanical Properties of Nylon 6 Multifilament Yarn." *J. Eng. Fibers Fabr.*, 6, 2011: 7-15.

[6] Yeh, J.T., Wang, C.K., Liu, Z.W., Li, P., Tsou, C.H., Lai, Y.C., Tsai, F.C. "Drawing and ultimate tensile properties of nylon 6/nylon 6 clay composite fibers." *Polym. Eng. Sci.*, 52, 2012: 1348-1355.

[7] Murthy, N.S., Stamm, M., Sibilia, J.P., Krimm, S. "Structural changes accompanying hydration in nylon 6." *Macromolecules*, 22, 1989: 1261-1267.

[8] Zhu, L., Teng, W., Xu, H., Liu, Y., Jiang, Q., Wang, C., Qiu, Y. "Effect of absorbed moisture on the atmospheric plasma etching of polyamide fibers." *Surface and Coatings Technology*, 202, 2008: 1966-1974.

Thermal-Stored Energy and Performance of Protective Clothing Upon Hot Water Exposure Using Instrumented Spray Manikin

Farzan Gholamreza[1], Guowen Song[2], Yehu Lu[3]
[1]Department of Human Ecology, University of Alberta, Edmonton, Canada
[2]Department of Apparel, Events, ands Hospitality Management, Iowa State University, Ames, Iowa, US
[3]Laboratory for Clothing Physiology and Ergonomics, Soochow University, Suzhou, China
gwsong@iastate.edu

INTRODUCTION

Industrial workers might be exposed to thermal hazards and sustain burn injuries in their occupational environments. Thermal protective clothing is designed to provide protection from thermal hazardous environments and slows down the heat transfer to the workers' skin. However, while working in thermal exposures, the clothing is heated and stores thermal energy. Depending on the thermal intensity and the fabric systems' structure, a large amount of thermal stored energy during exposure can be discharged to the skin after the termination of the thermal exposure and can contribute to burn injuries (Song et al.).

Studies on stored energy have focused on the thermal performance of fabric systems against convective-radiant heat exposure and hot liquid splash on bench-scale tests. There are limited numbers of research on full scale hot water spray using a manikin in order to investigate the effect of clothing design features and fabric properties on protective performance of the coveralls (Ackerman et al.; Lu, Song and Li). Nonetheless, no systematic study has been carried out to investigate the effect of body geometry on thermal stored energy developed in the garment during exposure and its discharge to the skin in cooling phase of the garment. In addition, in the existing full scale manikin studies, the average of total absorbed energy (TAE) and the percentage of second and third degree burn have been used as parameters to evaluate the thermal performance of the garment. These parameters do not address the contribution of the thermal stored energy to thermal performance of the garment.

The aim of this study is to investigate the contribution of stored energy effect in full scale garment on burn injuries. As such, new predictive parameters such as the average of the weighted second degree burn which occurred during cooling period of the garment, the stored energy contribution in total second degree burn injury (SEC_{2nd}), average total energy discharged to the manikin during cooling period of the garment (TED), and the stored energy contribution in total energy absorbed throughout the test (SEC_E) are introduced. In addition, the effects of the fabric properties and garment design on the proposed predictive parameters are analyzed.

EXPERIMENTAL APPROACH

Materials

The garments selected for the study were the thermal protective garments typically worn by industrial workers. Garment G-5 is a double layer coverall with G-4 as the outer layer. The details of the coveralls are presented in Table I.

Table I. Specification of the thermal protective garments

Garment Code	Fiber content	Size	Weight (g/m^2)	Air permeability (cm^3/cm^2s)	Average air gap (mm)
G-1	100% Nomex	42	169	25.8	25.1
G-2	88% Cotton and 12% Nylon	42	305	18.3	27.1
G-3	88% Cotton & 12% Nylon (polymer finishing)	42	412	0	27.8
G-4	100% cotton	42	360	2.97	33.7
G-5	100% cotton/ Quilted lining Arcxel	42	730	2.16	38.5
G-6	88% Cotton and 12% Nylon	42	305	5.62	28

Methods

The performance of the garments upon hot water splash was measured by the instrumented spray male manikin developed by The University of Alberta Protective Clothing and Equipment Research Facility (Ackerman et al.). The manikin with size of 40R was equipped with 110 skin simulant sensor uniformly distributed over its surface (Dale et al.). Four groups of cylinder spray jet with a set of three-automatically-controlled-valve-bottom-up-nozzle assembly in each group were employed to spray the manikin. Hot water at 85°C and were ejected from the 12 nozzles with the pressure of 250 kPa in order to simulate hot water splash in industrial occupational environment. The data acquisition system was recording the calorimeter output and the programmed software was employed to obtain the skin burn distribution over the body, the transmitted and the discharge energy during and after exposure. The bio-heat transfer skin model in conjunction with Henriques Burn Integral was employed to predict second degree burn time.

A Vitus 3-D whole body scanner (Human Solutions GmbH, Germany) was used for the determination of air gap between the clothing and the manikin surface. A motionless male manikin was used, scanned nude and dressed with each thermal protective coveralls.

RESULTS AND DISCUSSION

The percentage of the total second degree burn throughout the test and average total absorbed energy throughout the test (TAE) are shown in Table II. Also, the average of the weighted second degree burn which occurred during cooling period of the coverall, the stored energy contribution in total second degree burn injury (SEC_{2nd}), average total energy discharged to the manikin during cooling period of the garment (TED), and the stored energy contribution in total heat absorbed throughout the test (SEC_E) are introduced as parameters to evaluate the role of the thermal stored energy on protection and the total thermal absorbed energy in full scale manikin test (Table II).

Table II. Thermal performance of the tested garments

Garment Code	2nd Degree Burn throughout the test (%) (SD)	2nd Degree Burn during cooling Period (%) (SD)	SEC_{2nd} (%) (SD)	TAE (kJ/m^2) (SD)	TED (kJ/m^2) (SD)	SEC_E (%) (SD)
G-1	48.07 (4.51)	4.17 (1.30)	8.57 (2.07)	173.3 (6.89)	67.36 (1.56)	38.9 (0.83)
G-2	47.64 (1.56)	6.03 (2.79)	9.58 (8.16)	153.02 (7.69)	64.47 (3.10)	42.2 (1.82)
G-3	6.35 (1.92)	2.25 (1.01)	34.88 (7.65)	49.77 (2.61)	28.52 (0.62)	57.4 (2.22)
G-4	16.75 (5.50)	3.94 (1.50)	24.14 (1.16)	91.7 (4.61)	56.53 (2.79)	61.7 (3.29)
G-5	0.00	0.00	---	18.67 (0.38)	12.75 (0.19)	68.3 (0.95)
G-6	37.83 (2.42)	9.78 (0.78)	25.84 (0.46)	133.40 (3.97)	68.66 (3.15)	51.4 (1.14)

Using StatCrunch software, the results from stepwise multiple linear regression and correlation between the dependant (SEC_{2nd}, TED and SEC_E) and independent variables (weight, air permeability and air gap size) are shown in Tables III and IV. The P-values and the coefficient of determination (R^2) obtained for air permeability show that this property is a significant factor and is an important physical property in SEC_{2nd} and SEC_E. According to the P-values in Table III, weight also affects the average total energy discharged to the manikin during cooling period of the garments.

Table III. Stepwise multiple regression for physical properties of the garments upon hot water spray

Dependent Variable	Independent Variable	P-value	Tstat	R^2
SEC_{2nd}	Air Per	0.0142	-5.1497	0.90
TED	Weight	0.0146	-4.1251	0.81
SEC_E	Air Gap	0.0165	5.7774	0.80
	Air Per	0.0105	-5.7446	0.98

Table IV. The correlation between the fabric properties and SEC_{2nd}, SEC_E and TED

	SEC_{2nd}	SEC_E	TED
Weight	0.83	0.86	-0.90
Air Gap	0.43	0.89	-0.69
Air Permeability	-0.95	-0.89	0.63

The tested thermal protective garments provided a distinct thermal response upon pressurized hot water spray in terms of second degree burn, TDE, SEC_E and SEC_{2nd}. The stored energy contribution in total heat absorbed throughout the test shows that there is a significant amount of thermal energy within the garment (38.9% to 68.3%). The maximum amount of SEC is 68.3% which is observed in G-5. Although there was no burn injury predicted in this garment, the garment system has the capability to store significant amount of thermal energy within its structure.

Air gap size is positively correlated with SEC_E ($P<0.0165$). This correlation shows that as the air gap size in permeable garment increases, the stored energy contribution in TAE increases significantly. As the air gap size increases, there is more insulation generated and hence more thermal energy can be stored in fabric. Therefore, mass and heat convection develops and causes more thermal energy to be discharged to the sensors during cooling period and may lower thermal performance of the garment. As such loose fitted permeable garments are not recommended upon hot water spray.

Fabric weight is one of the contributing factors in the amount thermal energy stored in the garment upon hot water spray. Garments G-5 which is the heaviest garment stores the highest amount of stored energy among the tested coveralls. Also, garments G-6, G-3 and G-4 have relatively higher values of SEC_E than the other garments which show the influence of garment weight on the accumulation of the thermal stored energy. In the heavy weight garment such as G-3 and G-4, the stored energy contribution in second degree burn (SEC_{2nd}) is almost 50%. This indicates that stored energy plays a significant role in the total second degree burn injuries in these garments.

The fabric air permeability also plays an important role in the prediction of second degree burn injuries occurred due to the discharge of the stored energy. Garments G-2 and G-6 have almost the same weight and air gap size but their air perm abilities are 18.3 and 5.62 respectively. The SEC_E in garment G-6 with lower air permeability is higher than garment G-2. This shows that the stored thermal energy is more in the structure of garment G-6. This may be due to the fact that the air permeable garment with lower air permeability (G-6) is capable of blocking more mass through the fabric's porous structure than the garment with relatively higher air permeability (G-2). Therefore, more mass is trapped and accumulated in garment G-6 fabric system during exposure. As a result, garment G-6 discharges more thermal energy in cooling phase and causes more second degree burn injuries (TDE= 68.66 kJ/m^2). The SEC_{2nd}'s values of G-2 and G-6 are 9.6% and 25.8% respectively which confirm this phenomenon.

CONCLUSIONS

The findings demonstrate that the thermal stored energy contributes significantly in the total absorbed energy by the sensors during the exposure and cooling period of the garment which may lower the thermal performance of the garment. Also, it is confirmed that the SEC_E, TED and SEC_{2nd} can be used as critical parameters to evaluate the thermal performance of the fabric system in full scale tests. The results also indicate that air permeability and weight of the fabrics are the key factors that need to be engineered in the protective clothing system upon hot water spray. The understanding of the thermal stored energy in the fabric system and the associated discharge after exposure will lead to the development of more advanced textile materials.

REFERENCES

Project Update: "Protection from Steam and Hot Water Hazards." *Protective Clothing System for Safety '11*, 2011.

Dale, J. D., et al. "Instrumented Mannequin Evaluation of Thermal Protective Clothing." *Performance of Protective Clothing : Fourth Volume, ASTM STP.* West Conshohocken, PA: American Society for Testing and Materials, 1133, 1992: 717-33.

Lu, Yehu, Song, Guowen, Li, Jun. "Analysing Performance of Protective Clothing Upon Hot Liquid Exposure Using Instrumented Spray Manikin." *Annals of Occupational Hygiene*, 57.6, 2013: 793-804.

Song, Guowen, et al. "Thermal Protective Performance of Protective Clothing Used for Low Radiant Heat Protection." *Textile Research Journal*, 81.3, 2011: 311-23.

Centrifugally-Spun Energy-Storage Nanofibers

Xiangwu Zhang
Fiber and Polymer Science Program, Department of Textile Engineering, Chemistry and Science,
North Carolina State University, Raleigh, NC, USA
xiangwu_zhang@ncsu.edu

ABSTRACT

Centrifugal spinning is a simple and versatile technique for producing nanofibers from various materials. One major application of centrifugal spinning is to produce advanced functional nanofibers for energy storage. Among the various energy storage technologies, supercapacitors have been considered as effective solution to the increasing need for high-power energy storage. Novel centrifugally-spun nanofibers with functional properties can dramatically alter surface reaction rates and charge transport throughout the supercapacitors, causing significant improvement in energy storage efficiency. Here, we present our work on the development of advanced functional nanofibers and the integration of these materials into advanced supercapacitors to achieve high system performance.

INTRODUCTION

Ever-growing concerns over environmental issues and fossil fuel depletion continue to invoke researchers' interest in developing alternative energy storage systems. Recently, supercapacitors have gained extensive attention because of their extremely rapid charge/discharge capability, considerably high power density, and satisfactorily long cycle life. Electric double-layer capacitors (EDLCs) are one important type of supercapacitors, where the storage of energy is realized through the reversible adsorption/desorption of electrolyte ions onto the conductive carbon electrodes that have large specific surface area. Among the diverse carbonaceous materials for EDLC electrodes, carbon nanofibers have been extensively studied because their one-dimensional architecture not only offers shortened pathways for electron transportation, but also contributes to high ion-accessible surface sites and facilitates the electrolyte penetration from perpendicular to longitudinal directions with respect to the fibers[1,2]. Consequently, carbon nanofibers are extraordinarily desirable for EDLCs with high capacitance and high rate capability *simultaneously*. Moreover, carbon nanofibers are flexible and they self-assemble to form free-standing mats that can be directly tailored to binder-free electrodes without the addition of other conductive agents and polymeric binding agents, reducing the mass of electrochemically inactive components and sustaining the high specific capacitance with sufficiently high conductivity.

So far, carbon nanofibers for EDLC electrodes have been mostly fabricated through the electrospinning of precursor polymer fibers and subsequent heat-treatment. Electrospinning technique is versatile in tuning fiber structure and morphology, however, it cannot meet the increasing demand of mass production of carbon nanofibers due to its low production rate, safety concern and high cost[3-5]. Recently, centrifugal spinning has been investigated as an alternative approach to fabricate nanofibers in a large-scale, low-cost fashion[5]. In this work, centrifugal spinning was employed in synthesizing polyacrylonitrile (PAN)/polymethylmethacrylate (PMMA) precursor fibers with different weight ratios, and different thermal treatment conditions were used to carbonize the precursor fibers, during which PAN was converted to carbon while PMMA was decomposed to generate small open pores in the resultant carbon nanofiber matrix. Results demonstrated that porous carbon nanofibers (PCNFs) prepared by centrifugal spinning of PAN/PMMA binary precursors could be promising EDLC electrode material and this method would also be of potential interest for researchers who are searching for alternative approach in fabricating PCNF-based materials for other applications.

APPROACH

Figure 1 shows the schematic of the centrifugal spinning device and the preparation of PAN/PMMA fibers and PCNFs. The major components of the centrifugal spinning device include a perforated Teflon spinneret residing in the center of the spinning platform, a motor beneath the spinneret to provide high-speed rotation, a speed controller connected to the motor to tune the rotational speed, and a series of fiber collectors installed radially in the outer perimeter of the spinning platform. To prepare PAN/PMMA precursor fibers, PAN/PMMA solutions were loaded into the spinneret and the rotation was initiated. As the rotational speed increased to a critical value, where the centrifugal force overcame the solution surface tension, liquid jets were formed at the nozzles, stretched in the flight of the circular trajectory accompanied by solvent evaporation, and deposited on the collectors in the form of fiber mat. This simple centrifugal spinning process is low cost and is two orders of magnitude faster than conventional electrospinning[5].

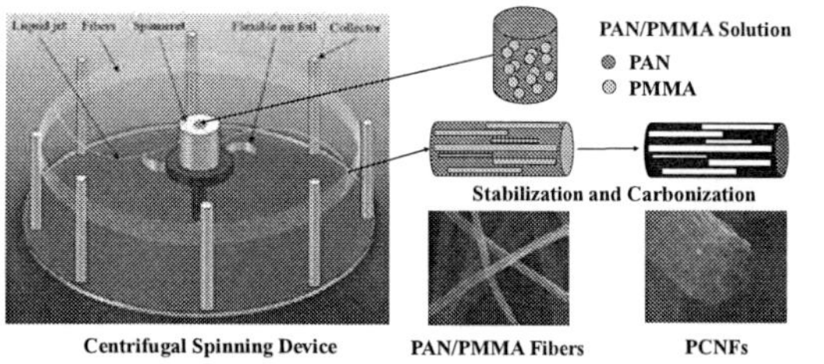

Figure 1. Schematic of the centrifugal spinning device and the preparation of PAN/PMMA precursors and PCNFs.

RESULTS AND DISCUSSION

Figure 2 shows the plane-viewed SEM, cross-sectional SEM and TEM images of PCNFs fabricated from precursor fibers with PAN/PMMA weight ratio of 7/3 at carbonization temperature of 900 °C. The low-magnification, plane-viewed SEM image shows long fibers without much detailed surface or internal characteristics (Figure 2a), while the cross-sectional SEM image reveals the development of large quantities of surface roughness and internal pores (Figure 2b). The TEM image, as complimentary evidence in addition to the cross-sectional SEM image, further demonstrates that a multi-channel porous architecture is successfully created for PCNFs (Figure 2c).

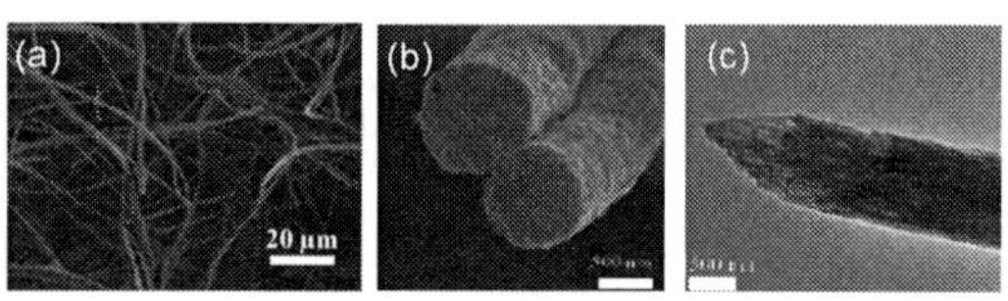

Figure 2. Plane-viewed SEM (a), cross-sectional SEM (b) and TEM (c) images of PCNFs prepared from precursor fibers with PAN/PMMA weight ratio of 7/3 at carbonization temperature of 900 °C.

To evaluate the electrochemical performance of the PCNFs, symmetrical cells were assembled using 6 M KOH aqueous electrolyte and glass fiber separator. The cells were charged and discharged in a voltage range of 0 - 1 V at current densities from 0.1 A g^{-1} to 2 A g^{-1}. The rate capabilities of the PCNFs prepared from precursor fibers with different PAN/PMMA weight ratios and different carbonization temperatures are shown in Figure 3. The specific capacitance of the PCNFs decreases with increase in current density. For PCNFs prepared from precursor fibers with PAN/PMMA weight ratio of 7/3 at 700 °C, 800 °C and 900 °C, the capacitance retention percentages at 2 A g^{-1} with respect to those at 0.1 A g^{-1} are 0.1%, 69% and 71%, respectively. For PCNFs synthesized from precursor fibers with PAN/PMMA weight ratios of 9/1, 7/3 and 5/5 at 900 °C, the capacitance retention percentages at 2 A g^{-1} compared to those at 0.1 A g^{-1} are 69%, 71% and 74%, respectively. These results indicate that PCNFs have excellent electrochemical performance.

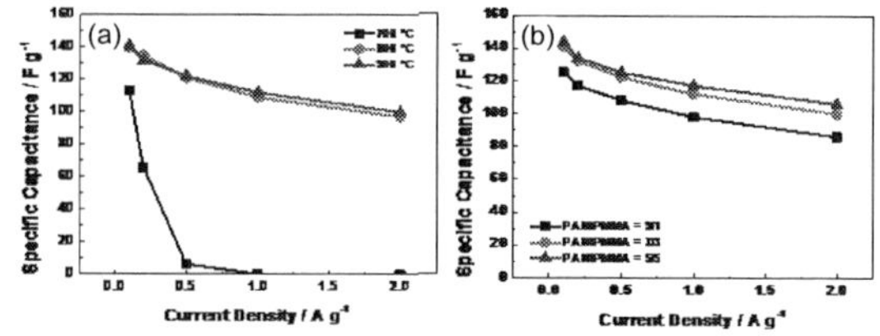

Figure 3. Rate capabilities of (a) PCNFs prepared from precursor fibers with PAN/PMMA weight ratio of 7/3 at different carbonization temperatures and (b) PCNFs prepared from precursor fibers with various PAN/PMMA weight ratios at carbonization temperature of 900 C.

CONCLUSIONS

PCNFs were prepared by centrifugal spinning of PAN/PMMA precursor solutions followed by two-step heat treatment and were used as binder-free electrode materials for EDLCs. The results demonstrate that the PCNFs offer promising potential for EDLC electrode applications and centrifugal spinning provides a simple, cost-effective strategy to produce PCNFs.

ACKNOWLEDGMENT

This research was funded through the U.S. National Science Foundation under Award Number CMMI-1231287.

REFERENCES

[1] Bonso, J.S., Kalaw, G.D., Ferraris, J.P. *J. Mater. Chem. A2*, 2014: 418-424.

[2] Wang, K., Wang, Y., Wang, Y., Hosono, E., Zhou, H. *J. Phys. Chem. C*, 113, 2009: 1093-1097.

[3] Badrossamay, M.R., McIlwee, H.A., J.A. Goss, Parker, K.K. *Nano Lett.*, 10, 2010: 2257-2261.

[4] McEachin, Z., Lozano, K. *J. Appl. Polym. Sci.*, 126, 2012: 473-479.

[5] Lu, Y., Li, Y., Zhang, S., Xu, G., Fu, K., Lee, H., Zhang, X. *Eur. Polym. J.*, 49, 2012: 3834-3845.

Nanofibrous Materials

High-Performance Fibers and Composites

Electrospun Nanoyarn Scaffold for Tissue Engineering

Xiumei Mo[1], Jinglei Wu[1], Yuan Xu[2], Qiang Zhou[2]
[1]College of Chemistry, Chemical Engineering & Biotechnology, Donghua University, Shanghai, China
[2]Department of Orthopedics, Southwest Hospital, Third Military Medical University, Chongqing, China
xmm@dhu.edu.cn

INTRODUCTION

Recent approaches in tissue engineering using electrospun scaffolds showed the superiority of electrospinning in preparation of scaffolds. Yet some limitations appeared due to the tightly layer-by-layer process of electrospinning. For example, electrospun nanofiber scaffolds allow little cellular infiltration into their inside. This study fabricated a porous nanyarn scaffold, which allows cell infiltration and growth in the scaffold three dimensionally; this nanoyarn scaffold has been successfully used for tendon tissue engineering.

EXPERIMENTAL

Collagen and P(LLA-CL) were dissolved in 1,1,1,3,3,3-hexafluoro-2-propanol to obtain blended solution at the concentration of 8%(wt/v) with a collagen/P(LLA-CL) weight ratio of 10:90. A custom-made dynamic water electrospining instrument was used to fabricate the nanoyan scaffold. That is, Electrospun fibers were deposited in water vortex which twinst the nanofiber into nanoyarn, the scafffold were obtained by collacte the naoyarns.

Tendon cell of 10^4 were seeded on the electrospun nanofibers and nanoyarns separately and culture in high glucose DMEM medium. Histological analysis was performed at 4, 7, and 14 days. Longitudinal sections were stained with hematoxylin and eosin (H&E) to evaluate cell infiltration in scaffold.

Tendon stem cells (TDSCs) were seeded on the nanoyarn scaffold after culture for 3 days, it was then implanted onto the back of nude mouse, at weeks 2, 4, and 8 to observe the tendon regeneration.

RESULTS AND DISCUSSION

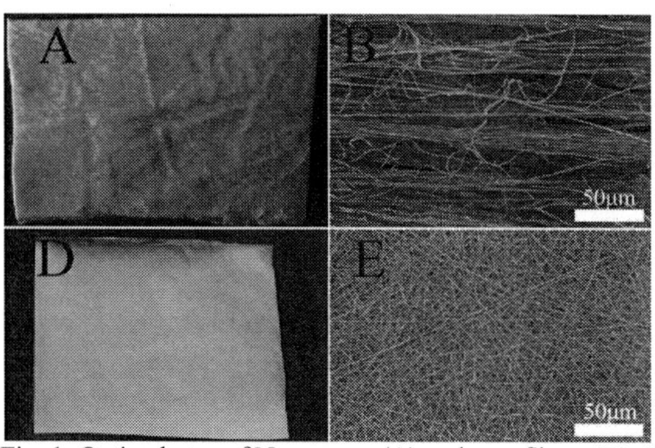

Fig. 1. Optic photos of Nanoyarns (A) and nanofibers (D), SEM of Nanoyarns (B) and nanofibers (E).

From Fig. 1 it can be seen nanofibers and nanoyans have different morphology, nanoyarn scaffold have bigger pore and higher porosity comparing with nanofiber scaffold. The high porosity of the nanoyarn benefited the cell growth in the nanoyarn scaffold. As show in Fig. 2, with the cell culture time increased from 4 days to 14 days tendon cells infiltrated from one side of the nanyarn scaffold to the other side (Fig. 2 A, B, C), but the cells only grow on one side of nanofiber and did not infiltrate inside of nanofiber (Fig. 2 D, E, F) because of tightly package of nanofibers.

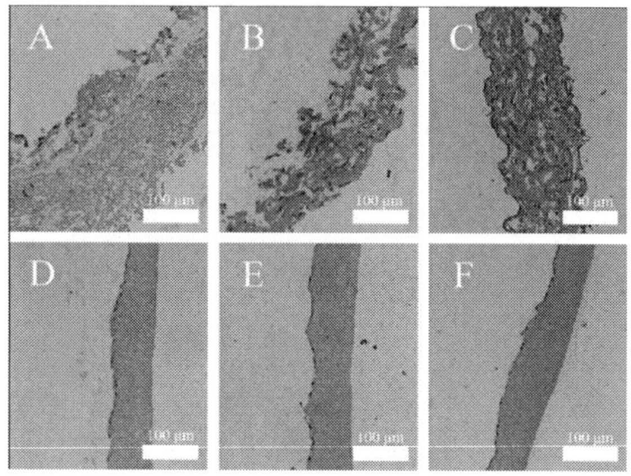

Fig. 2. HE histology of tendon cell cultured on nanoyarns for 4 (A), 7 (B) and 14 days (C); on nanofibers for 4 (D), 7 (E) and 14 days (F).

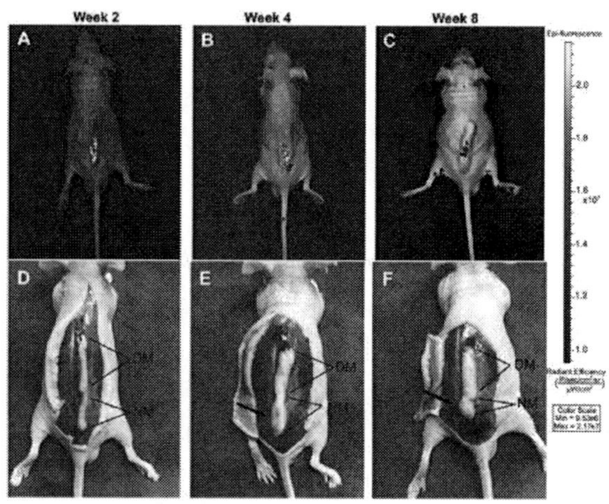

Fig. 3. Photographs showed gross observations and in vivo fluorescence imaging of the neo-tendon formation by TDSCs-P(LLA-CL)/Col constructs in nude mice. (A-C) in vivo fluorescence imaging at weeks 2, 4, and 8 after implantation; (D-F) gross observations at weeks 2, 4, and 8 after implantation.

From Fig. 3 it can be seen that neo-tendon formed after implant nanoyarn seeded with TDSCs in nude mouse for 2 weeks and it become thinker with implantation time increasing.

CONCLUSION

Nanoyarn scaffold has big pores and high porosity for cell growth and neo-tendon regenerated from nanoyarn scaffold with TDSCs in nude mouse.

Air-Impedance Electrospinning Technique for Cartilage Regeneration

Gretchen E. Schreyack, Allison E. Fetz, Gary L. Bowlin

Department of Biomedical Engineering, The University of Memphis and Joint University of Memphis-UTHSC-Memphis Biomedical Engineering Program
Memphis, Tennessee, USA
glbowlin@memphis.edu

SUMMARY
Here we present a system for air-impedance electrospinning to generate templates with optimized pore sizes that allow for directed 3D Articular Cartilage (AC, a.k.a. hyaline cartilage) engineering.

INTRODUCTION
AC is a smooth tissue that covers the ends of bones, reducing friction and distributing the load at an articulating joint. Restoring or replacing damaged or diseased AC is critical to relieve pain and reestablish normal tissue function. To achieve this, a custom system was developed for air-impedance electrospinning. Air-impedance electrospinning is a new fabrication method that has been recently published with non-optimized parameters [1]. The premise of the technique is that air-impedance electrospinning generates areas of low and high density fiber deposition. This regulated and enhanced pore size has been shown to be conducive for rapid cellular infiltration to depth unachieved to this point. Airflow scaffolds are structurally stable templates (non-cotton ball-type structure), which is especially crucial for scaffold preservation, and ultimately, scaffold efficacy.

EXPERIMENTAL METHODS
The system consists of a mounted funnel with a concatenate insert designed to hold changeable, porous mandrels at the mouth of the funnel and a nipple for tube attachment at the stem of the funnel for pressurized airflow through the system (Figure 1). Preliminary studies were conducted by dissolving Bovine gelatin in 1, 1, 1, 3, 3, 3-hexafluoro-2-propanol (HFP) at a concentration of 140 mg/mL. Solutions were electrospun using the following parameters: 5 mL/hr, +22 kV, and 5 inch air gap distance. Mandrel 1 and Mandrel 2 were porous stainless steel meshes with pore sizes of 0.5 mm and 1 mm, respectively. For mandrel 1, the system was translated with flexible tubing attached, but no airflow was administered. During use of mandrel 2, the funnel was translated, and airflow was administered through the flexible tubing at 7 SCFM. In addition, gelatin was also electrospun onto a solid, non-porous mandrel for comparison. Scanning electron microscopy (SEM) images loaded in FibraQuant™ software were used for fiber diameter and porosity analysis. Light microscopy was used to illustrate the differences in fiber collection density on the mandrels (Figure 2).

RESULTS AND DISCUSSION
Both mandrels 1 and 2 yielded smaller average fiber diameters, 3.3 µm and 2.7 µm, respectively, than the solid non-porous mandrel with fiber diameter of 4.3 µm. Scaffold porosity varies as a function of the mandrel porosity. Figure 2 shows the qualitative differences in fiber deposition through light microscopy (mandrel 1(A) and mandrel 2 (B,C,D)). Mandrel 1, with pore sizes of 0.5 mm and no airflow, and mandrel 2, with pore sizes of 1 mm and an airflow rate of 7 SCFM, resulted in scaffold porosity with an average pore size of 19 µm and 23 µm, respectively. These results indicate that mandrels with larger pore size in conjunction with the airflow decreased fiber deposition in the pore areas, in turn resulting in larger pore diameters.

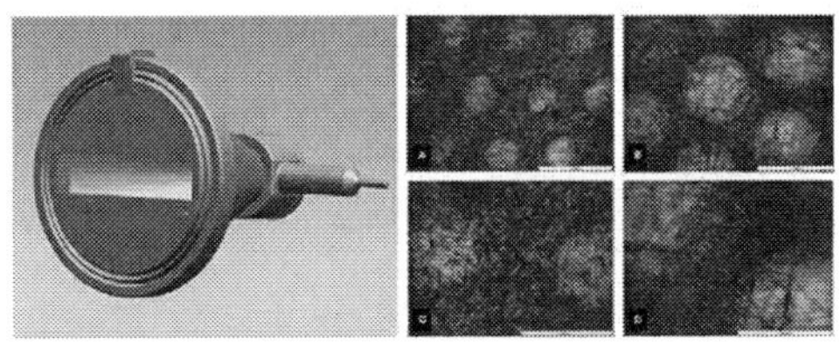

Figure 1 (left) and Figure 2 (right)

CONCLUSION
Enhanced scaffold porosity has the potential to be advantageous for use in many tissue engineering applications. Use of other natural and synthetic polymers and polymer blends for fabrication will be explored as well as adjusting the fiber alignment. Fiber diameter and porosity analysis as well as mechanical testing will be compared for scaffolds of various thicknesses. Mechanical testing methods will be optimized so that scaffold strength can be determined and to ensure that the deliberate formation of pores in scaffolds does not cause catastrophic differences in scaffold integrity. Acellular implants for *in situ* regeneration or for cell seeding and *in vitro* tissue development intended to mimic the middle zone of AC using chondrocytes remain the focus of this project.

REFERENCES
1. McClure, M.J., Sell, S.A., Simpson, D.G., Bowlin, G.L. "The Use of Air-flow Impedance to Control Fiber Deposition Patterns during Electrospinning." *Biomaterials*, 33, 2012: 771-779.

Polydopamine-Derived Porous Carbon Fiber Embedded with CoO Nanoparticles for Efficient Oxygen Reduction Reaction

Jiajie Yan, Hengyi Lu, Yunpeng Huang, Tianxi Liu
State Key Laboratory of Molecular Engineering of Polymers, Department of Macromolecular Science,
Fudan University, Shanghai, China
txliu@fudan.edu.cn

ABSTRACT

With the increasing demand for clean and sustainable energy, a great deal of efforts have been devoted to the energy conversion and storage technologies, such as supercapacitors, lithium ion batteries, fuel cells and water splitting. Among them, electrocatalyst for cathodic oxygen reduction reaction (ORR) is undoubtedly the key of the fuel cell. Platinum-based materials have usually been recognized as the most active catalysts for ORR, but their high cost, sensitivity to methanol and CO and weak durability largely limit their large-scale application. To address these problems, non-precious metal/metal composites, e.g., Co or Fe-based catalysts, and metal-free materials, e.g., carbon based materials (graphene, carbon nanotubes and carbon fibers), have been investigated as alternatives for platinum-based catalysts. However, the activity and stability of these materials for ORR catalysts are still hard to meet the requirements for practical applications in fuel cell.

In this work, highly mesoporous fibers in the manner of freestanding fibrous mats are successfully fabricated by single-spinneret electrospinning of polystyrene (PS) combined with a mussel-inspired biomimetic polydopamine (PDA)-coating process. The PDA coating offers functional groups (–OH, –NH_2), exhibiting an extraordinary versatile active nature. Then the designed mesoporous composites composed of nitrogen-doped carbon fiber (NCF) and cobalt oxide (CoO) is produced through a simple solution deposition method followed by subsequent annealing. The N-doped carbon fiber embedded with uniformly distributed CoO nanoparticles (NCF-CoO) possesses a large surface area, exhibiting excellent electrochemical activity as ORR catalysts. Compared with NCF, the NCF–CoO composites shows a more positive onset potential (-0.165 V vs Ag/AgCl) and a higher current density. The superior stability and methanol tolerance over Pt/C endow NCF–CoO as promising ORR electrocatalyst in practical applications.

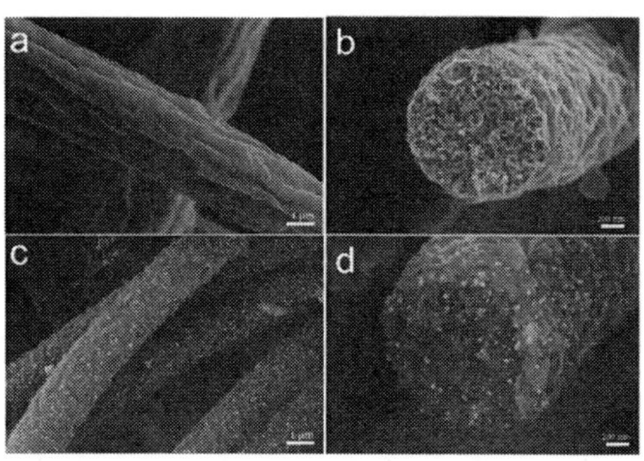

Fig. 1. FESEM and the cross-sectional images of the porous NCF (a, c) and NCF-CoO composites (b, d).

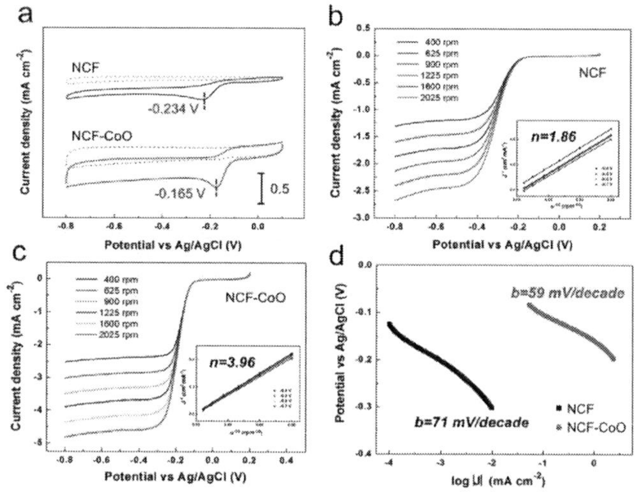

Fig. 2. (a) CV curves of the NCF and NCF–CoO in O_2-saturated (solid line) and N_2-saturated (dashed line) 0.1 M KOH solution. Rotating-disk voltammograms of the (b) NCF and (c) NCF–CoO in O_2-saturated 0.1 M KOH solution with a sweep rate of 5 mV s^{-1} at various rotating speeds. The insets show corresponding Koutecky–Levich plots at different potentials. (d) Tafel plots of the NCF and NCF–CoO at a rotating speed of 1600 rpm in O_2-staturated 0.1 M KOH solution.

KEYWORDS

Electrospinning; Dopamine, Porous Fiber, CoO Nanoparticles; Oxygen Reduction Reaction

ACKNOWLEDGMENT

The authors are grateful for the financial support from the National Natural Science Foundation of China (51433001, 51373037, 51125011).

REFERENCES

[1] Mao, S., Wen, Z., Huang, T., Hou, Y., Chen, J. *Energy & Environmental Science*, 7, 2014: 609-616.
[2] Guo, S., Zhang, S., Sun, S. *Angewandte Chemie, International Edition*, 52, 2013: 8526-8544.
[3] Zhang, C., Antonietti, M., Fellinger, T.P. *Advanced Functional Materials*, 24, 2014: 7655-7665.
[4] Kong, J., Zho, C., Wei, Y., Phua, S.L., Dong, Y., Lu, X. *Journal of Materials Chemistry A.*, 2, 2014: 15191-15199.

Spherical Nanocellulose Formates as Green Fillers to Improve Crystallization and Mechanical Property of Poly(lactic acid)

Chen-feng Yan[1], Hou-yong Yu[1], Ju-ming Yao[1,2]

[1]The Key Laboratory of Advanced Textile Materials and Manufacturing Technology of Ministry of Education, College of Materials and Textiles, Zhejiang Sci-Tech University, Hangzhou , China

[2]National Engineering Lab of Textile Fiber Materials & Processing Technology, Zhejiang Sci-Tech University, Hangzhou, China

phdyu@zstu.edu.cn; yaoj@zstu.edu.cn

ABSTRACT

The green nanocomposites of poly(lactic acid) (PLA) with various spherical nanocellulose formates (SCNFs) contents were prepared by solution casting method. It is found that the SCNFs act as an effective nucleation agent for crystallization of PLA. A study of the isothermal crystallization kinetics further illustrated that overall crystallization rate of PLA in the nanocomposites was faster than that of neat PLA. Furthermore, the tensile strength and Young' modulus increased by 130% and 116% for the nanocomposites with 10% SCNFs (mass fraction) compared to the neat PLA, respectively.

KEYWORDS: nanocomposites; poly(lactic acid); spherical nanocellulose formates; crystallization behavior; mechanical property

INTRODUCTION

PLA is becoming increasingly popular as a biodegradable engineering plastic owing to its easy processability compared to other biopolymers. Today, PLA has been widely used in packaging, medical but also in automotive applications where also other properties than strength and stiffness are important. Nonetheless, the slow crystallization rate and poor mechanical properties of PLA restrict its specific applications [1]. Therefore, considerable efforts have been made to improve its properties in order to compete with commodity polymers. Strong research efforts have been focused on the use of cellulose nanocrystals (CNCs) as reinforcements for new and inexpensive biodegradable materials, because cellulose nanocrystals possess numerous advantages such as high aspect ratio, good mechanical properties and fully degradable and renewable character [2-4]. As expected, the incorporation of CNCs could efficiently improve the PLA properties [5-6]. However, due to the hydrophilic nature of the CNCs, they are difficultly dispersed into relatively hydrophobic PLA matrix, and thus they often aggregate easily, leading to low reinforcement efficiency of CNCs in the nanocomposites due to poor interfacial compatibility between two components. Therefore, the improvement for dispersion of CNCs in PLA is a key problem to develop the PLA/CNC nanocomposites as food packaging materials.

In this study, SCNFs were partially surface esterified by the reaction between hydroxyl groups and mixtures of formic and hydrochloric acid (HCOOH/HCl) to improve their dispersion in PLA. X-ray diffraction was used to study the structure of the SCNFs and nanocomposites. The important industrial problem of slow crystallization of PLA was addressed by the use of SCNFs as biobased nucleation agents and the isothermal crystallization behavior of binary systems was studied.

RESULTS AND DISCUSSION

XRD analysis of neat PLA and PLA/SCNFs nanocomposites was carried out in order to study the crystalline structure (Fig. 1). Neat PLA was mainly comprised of the amorphous polymer structure, and the crystalline characteristic was ambiguous. SCNFs display the typical diffraction peaks of cellulose II crystalline form (20.2°, 21.8°, and 12.2°). The SCNFs diffraction signals are masked by PLA crystalline diffraction at low SCNFs content (< 10%). In particular, when enough SCNFs was introduced, as 10%, the presence of SCNFs made the crystalline character assigned to the cellulose nanocrystals in the nanocomposite system more clear and definite. At last, it is found that the crystal structure of the PLA has not been changed by the addition of SCNFs.

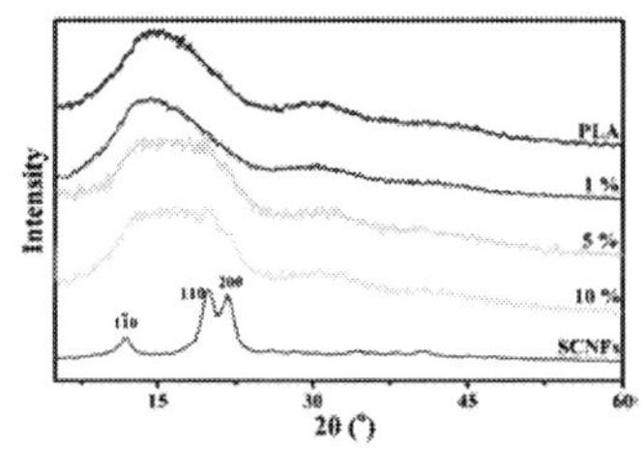

Fig. 1. X-ray diffraction (XRD) patterns of neat PLA, SCNFs and the nanocomposites with various SCNFs contents.

In order to fully understand the effect of the addition of esterified SCNFs on the bulk crystallization rate of PLA, isothermal crystallization kinetics of the PLA/SCNFs nanocomposites were studied. The relative crystallinity (X_t) as a function of crystallization time (t) is calculated as follows

$$X_t = \left(\int_0^t (dH/dt)dt\right) / \left(\int_0^\infty (dH/dt)dt\right) \qquad (1)$$

Fig. 2(a) gives the evolution of X_t with t for neat PLA and the nanocomposites with various SCNFs contents at their crystallization temperatures. As can be seen from Fig. 2(a), the crystallization finished within 50.72 min for neat PLA while within 50.11, 28.55 and 26.63 min for 1, 5 and 10%, respectively. It is clear that the addition of SCNFs could enhance the isothermal crystallization of PLA, moreover, the SCNFs content with 10% (mass fraction) was more efficient in accelerating the crystallization process of PLA than the other nanocomposites.

To describe the isothermal crystallization kinetics, the classical Avrami equation was employed:

$$\log[-\ln(1-X_t)] = \log k + n \log t \qquad (2)$$

where k is the overall kinetic constant depending on the geometry of the growing crystalline phase, and n is the Avrami exponent correlating with the nucleation mechanism and crystal growth dimension. The values of log[-ln(1-X_t)] were plotted versus log t in Fig. 2(b), and the Avrami exponent n and the crystallization rate constant k were calculated from the slope and intercept of the linear fit. The Avarmi equation rarely describes the whole crystallization process and is usually valid at the primary crystallization stage [7].Thus, the initial linear portions of the plots were analyzed and the results are summarized in Table I. The k values of the PLA/SCNFs samples were higher compared to those of the PLA and PLA/CNC samples. The Avrami exponent, n, relates to the nature of nucleation and the dimensionality of the growing crystals. The n values of PLA were around 2.8–3.3, suggesting a three-dimensional crystallization growth and homogeneous nucleation mechanism [8]. The obtained n values were similar to those reported for pure PLA. The n values of PLA/SCNC bionanocomposites were around 3.0–4.9, indicating a crystallization mode of heterogeneous nucleation [7].

Half-time of crystallization ($t_{1/2}$) is defined as the time when X_t is equal to 0.5 and is calculated as follows [7]:

$$t_{1/2} = \left(\frac{\ln 2}{k}\right)^{1/n} \quad (3)$$

Table I. Avrami kinetic parameters for the isothermal crystallization of the PLA and PLA/SCNFs nanocomposites.

Sample	T_c (°C)	n	k (min^{-n})	$t_{1/2}$ (min)
PLA	110	2.88	3.24×10^{-5}	31.9
	115	2.80	2.13×10^{-5}	40.1
	120	/	/	/
	125	/	/	/
1 %	110	4.01	1.06×10^{-6}	28.2
	115	3.37	4.59×10^{-6}	34.1
	120	3.06	9.68×10^{-6}	38.6
	125	2.82	1.40×10^{-5}	46.2
5 %	110	4.31	3.36×10^{-6}	17.1
	115	3.42	1.78×10^{-5}	22.0
	120	3.37	5.34×10^{-6}	32.9
	125	3.14	5.67×10^{-6}	41.7
10 %	110	4.52	2.12×10^{-6}	16.6
	115	4.12	2.95×10^{-6}	20.1
	120	3.79	4.07×10^{-6}	24.0
	125	3.48	2.99×10^{-6}	34.8

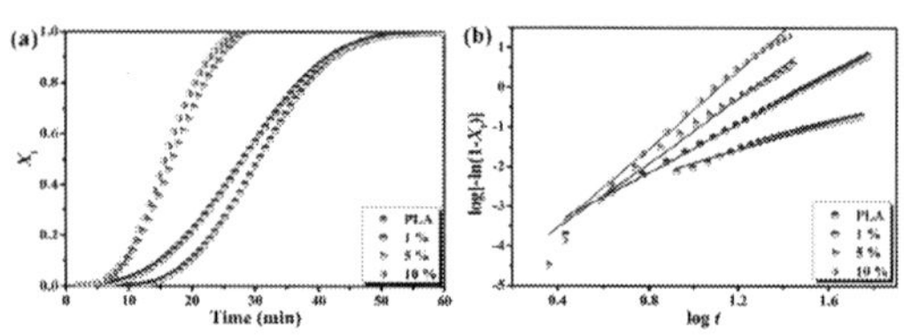

Fig. 2. Relative crystallinity plots (a) and Avrami plots (b) versus crystallization time for neat PLA and its nanocomposites at 110 °C.

The results are also listed in Table I. Generally, the $t_{1/2}$ value increased with increasing of T_c. The $t_{1/2}$ values of PLA/SCNFs biocomposites were lower than those of pure PLA, indicating rapid bulk crystallization rate by the addition of well-dispersed surface esterified cellulose nanocrystals in the PLA matrix.

In general, food packaging materials should have enough strength to be self-supporting and resist handling damage [9]. Thus PLA and the nanocomposite films were examined by tensile testing. Fig. 3 gives the tensile strength, elongation to break and Young's modulus for neat PLA and the nanocomposites. It is clear that a strong reinforcement effect in the mechanical properties of the nanocomposite films was achieved by adding SCNFs. When an increase of SCNFs contents, both the tensile strength and Young's modulus of the nanocomposites were increased gradually, which was ascribed to an efficient dispersion of the SCNFs and efficient stress transfer at the interface, resulting from a strong adhesion between the reinforcing phase and the matrix. The addition of 10 wt% SCNFs promoted 130% and 118% increments in the tensile strength and Young's modulus, respectively. These improvements could be attributed to the increased crystallinity and refined crystalline structure of the nanocomposite during the deformation.

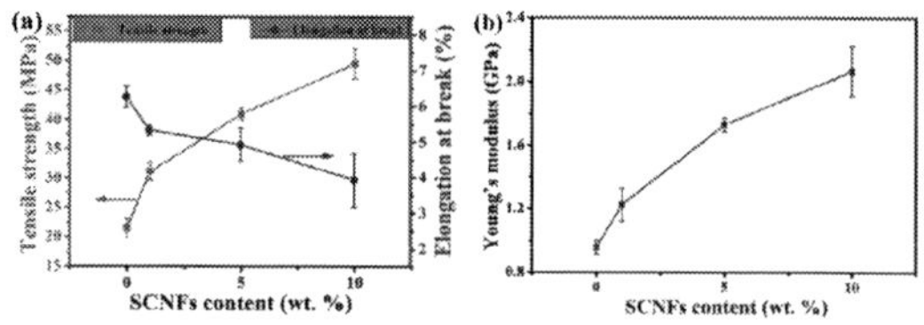

Fig. 3. Tensile strength and elongation at break (a) and Young's modulus (b) as a function of SCNFs contents for neat PLA and the nanocomposites.

REFERENCES

1. Saeidlou, S., Huneault, M A, Li, H, et al. "Poly (lactic acid) crystallization." *J. Prog. Polym. Sci.*, 37, 12, 2012: 1657-1677.
2. Fortunati, E, Armentano, I, Zhou, Q., et al. "Multifunctional bionanocomposite films of poly(lactic acid) cellulose nanocrystals and silver nanoparticles." *J. Carbohydr. Polym.*, 87: 2012: 1596-605.
3. Kargarzadeh, H., Sheltami, R.M, Ahmad, I., et al. "Cellulose nanocrystal: A promising toughening agent for unsaturated polyester nanocomposite." *J. Polymer*, 56, 2015: 346-357.
4. Bettaieb, F., Khiari, R., Dufresne, A., et al. Mechanical and thermal properties of Posidonia oceanica cellulose nanocrystal reinforced polymer." J. *Carbohydr. Polym.*, 123, 2015: 99-104.
5. Bitinis, N., Verdejo, R, Bras, J., et al. Poly(lactic acid)/natural rubber/cellulose nanocrystal bionanocomposites. Part I. Processing and morphology." *J. Carbohydr. Polym.*, 96, 2013: 611-620.
6. Raquez, J.M., Murena, Y, Goffiin, A.L., et al. Surface-modification of cellulose nanowhiskers and their use as nanoreinforcements into polylactide: A sustainably-integrated approach." *J. Compos. Sci. Tech.*, 72: 2012: 544-549.
7. Pei, A, Zhou, Q., Berglund, AL. "Functionalized cellulose nanocrystals as biobased nucleation agents in poly(L-lactide) (PLLA)–Crystallization and mechanical property effects." 70, 2010: 815-21.
8. Lin, Y, Zhang, K.Y., Dong, Z.M., Dong, L.S., Li, Y.S. "Study of hydrogen-bonded blend of polylactide with biodegradable hyperbranched poly(ester amide)." *J. Macromolecules*, 40, 17, 2007: 6257-67.
9. Pascual, A.M., Díez-Vicente, A.L. "ZnO-reinforced poly(3-hydroxybutyrate-co-3-hydroxyvalerate) bionanocomposites with antimicrobial function for food packaging." *ACS Appl. Mater. Interfaces*, 6, 12, 2014: 9822-9834.

Construction Nanomaterials for Modification of General Thermal Plastic Polymer Resin and Fiber

Bin Sun, Yuyuan Shi, Zhe Zhou, Meifang Zhu
State Key Lab for Modification of Chemical Fibers and Polymer Materials, College of Material Science and Engineering, Donghua University, Shanghai 201620, China
dhu.edu.cn; sunbin@dhu.edu.cn

ABSTRACT
Organic–inorganic hybrid materials have a promising-potential of application since they could combine the good properties of organic and inorganic materials. The nano-material's dispersibilty in and miscibility with polymer matrix are the two most important factors, which influence the structure and properties of nanocomposite. In our research, two typical nanomaterials, nano-Ag particles with high organic grafting ratio (Oc-OS-Nano-Ag), which can be dissolved in alkane[1], and the nanosheet-like titanium organic compound (NS-TOC)[2], which has similar component with poly elthylene terephthalate (PET), and respectively.were constructed and composited with genernal thermal plastic polymer, PET and polyproplene (PP)[4], The structure and properties of the resulted nanocomposites have been studied in detail.

The Oc-OS-Nano-Ag with narrow scale (average size 6 nm) and about 18wt% of surfactants capped on its surface was soluble in typical nonpolar solvent, for example cyclohexane. The structure and properties of polyester/Oc-OS-Nano-Ag composite resin prepared by melt blending had been examined. The results showed that Oc-OS-Nano-Ag had fine and homogeneous dispersion and size less than 20nm in polyester matrix (polybutylene terephthalate (PBT) and polyethylene terephthalate (PET)) as shown by TEM images. The DSC and TGA results showed that there was no obvious influence on thermal degradation and melting properties, and a weak crystallization acceleration due to its heterogenous nucleating action. The polyester nanocomposites with lower content (≤500ppm) of Ag loadings had good spinnability. What's more, the resulted polyester fibers had higher antibacterial rate (more than 96%) against Staphylococcus aureus and Escherichia coli under lower content of Oc-OS-Nano-Ag (100ppm).
The NS-TOC shows excellent mono-dispersibilty and its PET base nanocomposite shows higher viscosity than pristine PET as well as temperature increasing under lower shearing rate, but strong shearing thin tendency under higher shearing rate. These intersting rehology properties indicate that NS-TOC has strongger interaction with PET chain due to their similar component, and strong orientation tendency during melt-processing is also verified by 2D-WAXD pattern.

ACKNOWLEDGMENT
This research is financially supported by the Shanghai Natural Science Foundation (13ZR1400700), the Program for Changjiang Scholars and Innovative Research Team in University (T2011079, IRT1221),

REFERENCES
1. Shi, Yu-yuan, Sun, Bin, Zhou, Zhe, Wu, Yongtao, Zhu, Meifang. "Size-controlled and large-scale synthesis of organic-soluble Ag nanocrystals in water and their formation mechanism." *Progress in Natural Science: Materials International*, 21, 2011: 447-454.
2. Xin, Guoying, Sun, Bin, Wang, Wei, Zhu, Liping, Wu, Wenhua, Zhu, Meifang, Chen, Yanmo, Adler, Hans-JÜrgen P. *Macromolecular Symposium*, 254, 2007: 173-179.
3. Sun, Bin, Hua, Beibei, Ji, Xiaohuan, Shi, Yuyuan, Zhou, Zhe, Wang, Qingmiao, Zhu, Meifang. "Preparation of silver nanoparticles with hydrophobic surface and their polyester based nanocomposite fibres with excellent antibacterial properties." *Materials Research Innovations*, 18 (S4), 2014: 869-874.
4. Shi, Yu-yuan, Sun, Bin, Zhou, Zhe, Wangatiam, Lodrick M., Chen, Long, Zhu, Meifang. "Polypropylene Nanocomposites Based on Synthetic Organic-Soluble Ag Nanocrystals with Prominent β-nucleating Effect: Quiescent Crystallization and Melting Behavior." *Journal of Macromolecular Science, Part B*, 51 (12), 2012: 2505-2518.

Silk Fibroin Fibers and Scaffolds from Aqueous Solutions

Zhaobo Li, Yaopeng Zhang
State Key Laboratory for Modification of Chemical Fibers and Polymer Materials, College of Materials Science and Engineering, Donghua University, Shanghai, China
zyp@dhu.edu.cn

ABSTRACT

Tissue engineering aims at developing functional substitutes for damaged or diseased tissues and organs. Electrospinning has emerged to be a simple and effective technique to fabricate tissue engineering scaffolds, which were widely used in tissue engineering.

Silk fibroin (SF) from *Bombyx mori* has been investigated widely by electrospinning to fabricate tissue engineering scaffolds. The mechanical properties of most electrospun SF scaffolds are still poor, although people have tried many ways to reinforce the scaffolds, such as using organic solvents, adding some reinforcing agents, post-treated with different methods, or using the cylinder to collect alighed SF fibers instead of the random fibers. Bladder acellular matrix graft (BAMG), a tough natural material, was already used in urethral reconstruction, penile reconstruction, myocardial repair, esophageal repair and fascial tissue reconstruction. In this study, we used BAMG to toughen the electrospun SF scaffolds with random or aligned fibers. Considering the essential role of vascularization in the success of tissue regenerating, we fabricate SF scaffolds loaded VEGF (Vascular endothelial growth factor), which is essential for tissue vasculogenesis and angiogenesis, by blend and coaxial electrospinning.

Compared with the conventional SF mats, the SF/BAMG composite scaffolds were significantly reinforced. The morphology, microstructure and biocompatibility of the prepared scaffolds were characterized by Scanning Electron Microscopy (SEM), Transmission Electron Microscopy (TEM), Fourier Transform Infrared Spectroscopy (FTIR), Wide angle X-ray diffraction (WAXD), Capillary Flow Porometer, Instron 5969 material testing machine and Laser Scanning Confocal Microscope (LSCM). An Elisa kit was adopted to investigate the release behaviors of VEGF loaded in the scaffolds.

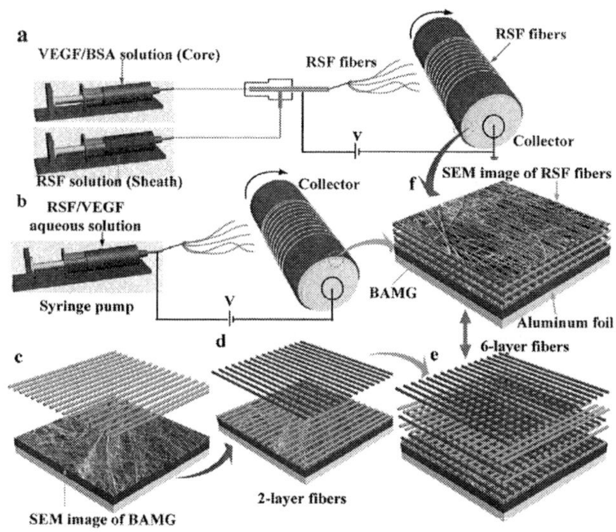

Fig. 1. Schematics of the (a) coaxial and (b) blend electrospunning, (c~f) the structures of SF/BAMG composite scaffolds in multiple layers.

.

In order to find the best parameters of coaxial electrospinning, the concentration of the SF aqueous solutions and the flow rates of the core and sheath spinning dopes were studied. Results show that when the concentration of the SF aqueous solutions was 39 wt% and flow rates were 0.3 and 1.2 mL/h for core and sheath electrospinning dopes, we can gain SF fibers with stable core-sheath structure. The voltage applied was 20 kV and the distance from the spinneret to collector was 10 cm. The electrospinning surroundings were 5 ± 5 °C and 50 ± 5 % relative humidity. For aligned fibers collected, the rotating speed of the cylinder was 2000 r/min while other conditions were kept constant for SF mats.

The best conditions of water vapor annealing were 37 °C, 90 % relative humidity, and post-treated time of 36 h. The breaking energies of the composite scaffolds ranged from 458 to 970 $J \cdot kg^{-1}$. The breaking strength and elongation at break of the blend and coaxially electrospun SF/BAMG composite scaffolds with random fibers reached 14.4 MPa and 12.0 %, 12.9 MPa and 10.7 % in dry state, respectively. In wet state, the breaking strength and elongation at break of the blend and coaxially composite scaffolds were 3.9 MPa and 53.7 %, 3.3 MPa and 48.2 %, respectively. These were much better than the bare SF scaffolds electrospun in the same conditions. The suture retention strength of the composite scaffolds was reinforced ranging from 4.1 N to 4.3 N in dry state and 2.0 N to 2.3 N in wet state, respectively. These satisfied the requirements for suturing in tissue engineering (2 N).

For the composite scaffolds with aligned SF fibers in multiple layers collected by the cylinder (Fig. 1), the breaking strength and breaking elongation of the blend and coaxially electrospun composite scaffolds annealed in water vapor were 22.7 MPa and 18.8 %, 21.0 MPa and 13.2 % in dry state, respectively. In wet state, the breaking strength and breaking elongation of the blend and coaxially electrospun composite scaffolds were 5.9 MPa and 87.2 %, 5.5 MPa and 84.5 %, respectively. These were higher than those of the composite scaffolds with random fibers. In addition, the suture retention strength of the blend composite scaffolds was 4.3 N and 1.9 N in dry and wet states, respectively. For the coaxially electrospun composite scaffolds, it is 4.1 N and 1.8 N in dry and wet states, respectively.

The cumulative release profiles of VEGF from different scaffolds indicate that VEGF was successfully loaded and released for more than 2 weeks. Comparing the two types of post-treatments, we found that water vapor annealing was a much better method than immersion in ethanol to keep VEGF alive and reinforce the SF/BAMG composite scaffolds. The proliferation of PIECs (porcine iliac endothelial cells) cultured on the scaffolds proved that both the random and aligned-fiber scaffolds had a good biocompatibility. MTT assay indicates that VEGF-loaded scaffolds could promote PIECs to adhere, spread and proliferate better than their counterparts without VEGF. In addition, the well-aligned fibers are suitable to guide the cell growth with desired anisotropy. Furthermore, PIECs could not only grow on the surface of the scaffolds, but also infiltrate into the mats (Fig. 2).

In a word, these SF/BAMG electrospun composite scaffolds were successful to deliver and release VEGF in a sustained manner, and could be considered as attractive candidates for tissue engineering applications.

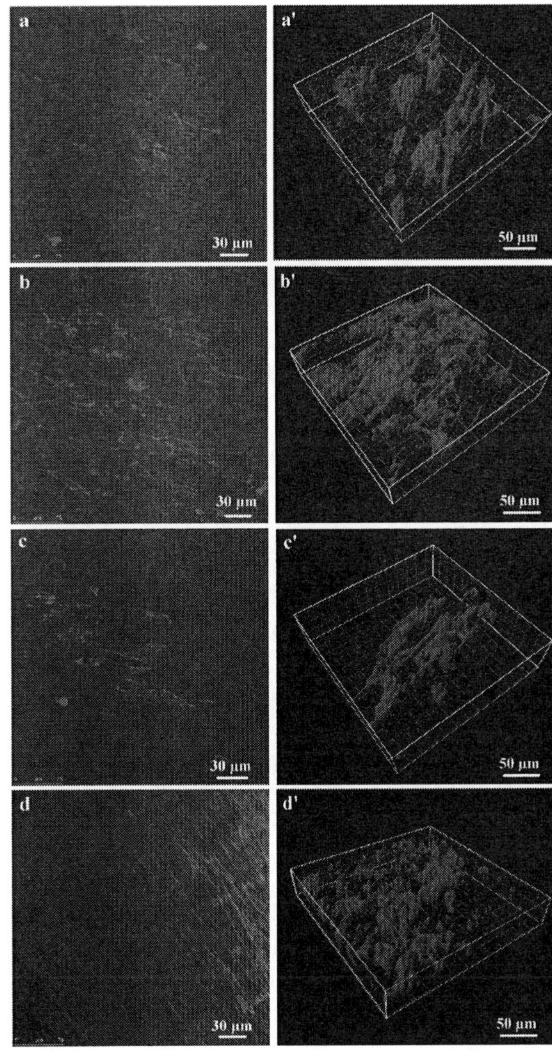

Fig. 2. 2D and 3D LSCM images of PIECs cultured for 3 days on the composite scaffolds of SF and BAMG with aligned SF fibers in multiple layers: (a, a') Blend electrospun composite scaffolds without VEGF, (b, b') Blend electrospun composite scaffolds with VEGF, (c, c') Coaxially electrospun composite scaffolds without VEGF, (d, d') Coaxially electrospun composite scaffolds with VEGF, respectively.

ACKNOWLEDGMENT

This work is supported by the National Natural Science Foundation of China (21274018, 81170641), DHU Distinguished Young Professor Program (A201302), and the Fundamental Research Funds for the Central Universities (2232013A3-11).

REFERENCES

1) Li, Z., Liu, Q., Wang, H., Song, L., Shao, H., Xie, M., Xu, Y. Xu, Zhang, Y. *ACS Biomat. Sci. Eng.*, 2015, http://dx.doi.org/10.1021/ab5001436.

2) Li, Z., Song, L., Huang, X., Wang, H., Shao, H., Xie, M., Yu, Y., Zhang, Y. *RSC Adv.*, 5, 22, 2015: 16748-16758.

Green Graphene-Silk Nanocomposite Materials

Xiao Hu[1,2], Fang Wang[1,3], S.S. Jyothirmayee Aravind[1,4], Hao Wu[5], Joseph Forys[1], Venkat Venkataraman[5], Kandalam Ramanujachary[4]

[1]Department of Physics and Astronomy, Rowan University, Glassboro, NJ, USA; [2]Department of Biomedical and Translational Sciences, Rowan University, Glassboro, NJ, USA; [3]Center of Analysis and Testing, Nanjing Normal University, Nanjing, China; [4]Department of Chemistry and Biochemistry, Rowan University, Glassboro, NJ, USA; [5]Department of Cell Biology, Rowan University School of Osteopathic Medicine, Stratford, NJ, USA
hu@rowan.edu

KEYWORDS
Silk, Graphene, DSC, FTIR, TGA, SEM, Cell Biocompatibility

INTRODUCTION
Two dimensional graphene materials with only an atom thickness have paved novel ways for the development of new functional materials. There has been a great interest in developing graphene-polymer composites to completely exploit superior properties of graphene. In this study, green graphene materials prepared by photo-reduction of graphite oxide were first time blended with silk fibroin to improve the thermal and mechanical properties of silk biomaterials. Powder X-ray Diffraction (XRD) analysis confirmed the complete exfoliation of graphite oxide to graphene in presence of focused pulses of solar radiation. By varying the concentration of grapheme (0.1 wt% to 10 wt%) in silk, different free standing graphene-silk films were prepared and were systematically characterized by Fourier Transform Infrared Spectroscopy (FTIR), Scanning Electron Microscope (SEM) and nanoindentation measurements. The homogeneous mixing of graphene in silk as well as the thermal stability of the composite films was demonstrated by Thermal Gravimetric (TG) and temperature-modulated Differential Scanning Calorimetry (TMDSC) analysis. Fibroblast cells studies proved that the graphene-silk blend materials can also support and facilitate cell adhesion, growth and proliferation. This optimum test of graphene in silk polymer provide a useful model to understand how to control the hydrophobic-hydrophobic and polar-polar interactions between high surface area-amphiphilic inorganic nanomaterials and the protein polymer chains, which has many emerging applications in biomaterial science and engineering.

EXPERIMENTAL APPROACH
Silk-graphene composite soultion–Bombyx mori silkworm fibers were first boiled in a $NaHCO_3$ solution and then dissolved in a LiBr solution. After dialyzing against distilled water, a pure silk fibroin aqueous solution was obtained. Graphene from photo-reduction of graphite oxide powders was then slowly mixed with the 1 wt% silk aqueous solution homogeneously. The final blended solutions were based on a mass ratio of graphene to silk (0.1~10 wt%). Films were obtained by casting solutions on polydimethylsiloxane (PDMS) substrates with 30 μm thickness.

RESULTS
The standard DSC and temperature modulated DSC (TMDSC) was used to understand the molecular interactions between silk proteins and graphene sheets in the formed SF/Graphene systems with different mixing ratios.

The TGA results demonstrated that the silk and graphene components were well blended in the film systems, and did not follow their original individual thermal stability profiles.

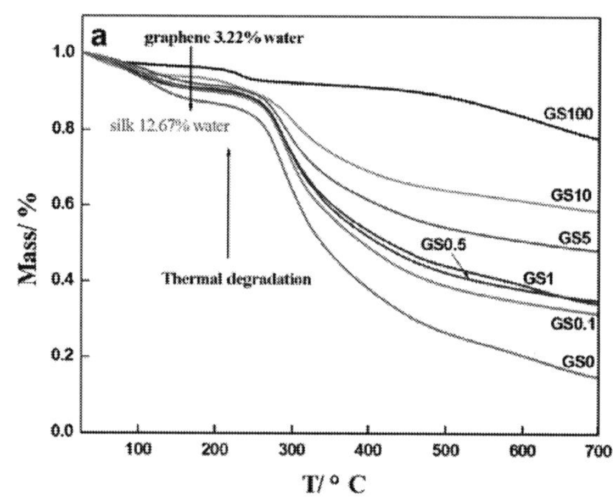

Figure 1a. Mass percentage change of the grapheme-silk films (GS100, GS10, GS5, GS1, GS0.5, GS0.1, GS0) measured by TGA during heating from room temperature to 800 °C at 10 °C/min. It shows that a small amount of graphene nano-materials can significantly improve the thermal stability of composite samples.

The IR results indicated that the conformations of silk structures are generally stable, with small amount of Silk II structure appeared after mixing (1625 cm^{-1}, 1540 cm^{-1} and 1267 cm^{-1}). Some structure contents, such as the Tyr side chains (1514 cm^{-1}) were also enhanced after silk mixing with graphene. Besides, a deconvolution method based on curve fitting FSD spectra in the Amide I region was performed to quantify the percentage of the secondary structures in the SF/graphene samples. SEM showed that the morphologies of composite films are uniform in macroscopic scale, but with interesting micro-meter patterns.

After 30mins methanol-annealing treatment, soluble silk-graphene films can be physically crosslinked by the beta-sheet crystals (peak is around 1620 cm^{-1}), and were used for fibroblast cell biocompatibility tests.

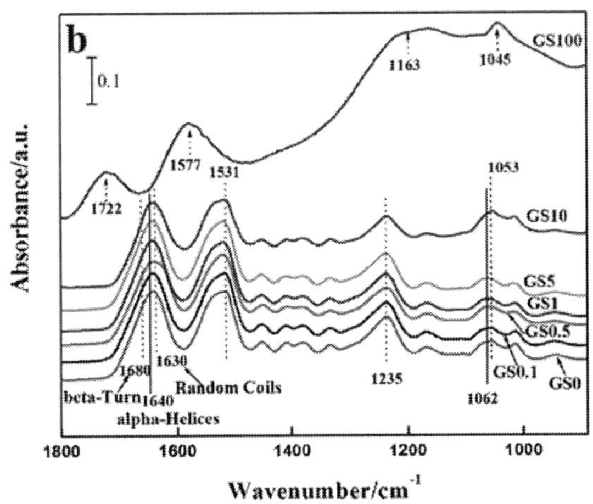

Figure 1b. FTIR absorbance spectra of graphene-silk films (GS10, GS5, GS1, GS0.5, GS0.1) and controls (GS100, and GS0) for the wavenumber region 1800-850 cm^{-1}, with the regions assigned as random coils, alpha-helix, and beta-turns. FTIR study reveals the structures of different silk-graphene films.

CONCLUSION

Graphene prepared by graphite oxide reduction in presence of focused pulses of sunlight has been embedded in silk matrix by a simple strategy and the improved thermal/mechanical properties of silk fibroin with the addition of graphene are demonstrated. The synergistic enhancement from graphene and silk, facilitated by hydrophobic-hydrophobic interaction between graphitic areas and hydrophobic silk blocks along with hydrophilic-hydrophilic interactions between oxidized areas of 2D graphene and hydrophilic silk backbone is achieved with very low concentration of graphene in silk matrix. Detailed thermal, spectroscopic and morphological characterizations and cell studies were carried out to further understand the properties of the films. With their excellent properties, easy and fast fabrication and inherent biocompatibility, the flexible graphene-silk fibroin composite films can be valuable for potential applications in many emerging biomedical fields.

ACKNOWLEDGMENT

This study was supported by Rowan University Faculty Research Seed Grants and NSF-MRI Program (DMR-1338014).

REFERENCES

[1] Hu, X., Cebe, P., Weiss, A., et al. *Materials Today*, 15, 2012: 208-215.

[2] Jyothirmayee Aravind, S.S., Eswaraiah, V., Ramaprabhu, S. *J. Mater. Chem.*, 21, 2011: 17094-17097.

[3] Hu, X., Kaplan, D.L. "Silk Biomaterials." *Comprehensive Biomaterials*, 2, 2.212, Elsevier, 2011.

[4] Hu, X., Wang, X., Rnjak, J., Weiss, A., Kaplan, D. *Biomaterials*, 31, 2010: 8121-8131.

[5] Hu, X., Park, S., Gil, E., et al. *Biomaterials*, 32, 2011: 8979-8989.

[6] Hu, X., Kaplan, D., Cebe, P. *Macromolecules*, 39, 2006: 6161-6170.

[7] Hu, X., Kaplan, D., Cebe, P. *Macromolecules*, 41, 2008: 3939-3948.

[8] Hu, X., Lu, Q., Kaplan, D., Cebe, P. *Macromolecules*, 42, 2009: 2079-2087.

[9] Sutherland, T., Young, J., Weisman, S., Hayashi, C., Merritt D. *Annual Review of Entomology*, 55, 2010: 171-188.

[10] Wray, L., Hu, X., Gallego, J., et al. *J. Biomed. Mater. Res. B: Appl Biomater.*, 99, 2011: 89-101.

In situ Self-Assembly Nanocomposites from Eletrospun Poly(lactic acid)/Lipid and Bacterial Cellulose Nanofibers

Chunhui Xiang[1], Nuria Acevedo[2]
[1]Department of Apparel, Events and Hospitality Management, Iowa State University, USA
[2]Department of Food Science and Human Nutrition, Iowa State University, USA
chxiang@iastate.edu

OBJECTIVE

The purpose of the study is to develop novel bacteria cellulose (BC) nanocomposites with optimal mechanical and environmental performance. To develop these cellulose-based nanocomposites is necessary to increase the hydrophobicity of the nanofibers as well as to improve the interface between matrix and fibers. In this study, pure poly(lactic acid) (PLA) and PLA/lipid nanofibers were initially produced by electrospinning and used as mechanical support for the BC growth. The lipid used in this research is stearic acid. Lipid was dissolved into PLA/chloroform/acetone solution before electrospinning. The presence of stearic acid may favor cohesion and water permeability.

INTRODUCTION

The growing concern of environmental problems and exhausting natural resources caused by non-biodegradable petrochemical-based materials has led to a growing interest in developing biodegradable constituents. Biopolymers obtained from natural resources are a promising alternative due to their environmentally friendly nature (Faruk, et al. 2012; Tharanathan 2003). Bacterial cellulose (BC) is a promising, sustainable, and biodegradable nanofibrous material which has high thermal stability, excellent environmental biodegradability and strong biocompatibility (Qiu and Netravali, 2014).

APPROACH

PLA nanofibers were electrospun from 8 wt.% PLA dissolved in chloroform/acetone (3:1, v/v). PLA/Lipid nanofibers were electrospun from 8 wt.% PLA in chloroform/acetone (3:1, v/v) containing 5% lipid (PLA w/w). The lipid used in this study was stearic acid. During electrospinning, the polymer solution was introduced into 10 mL glass syringe (VWR Scientific, West Chester, PA). The syringe was attached to a metal needle (ID=0.60 mm). Electrospinning was conducted at 15 kV by a high voltage supply (Gamma High Voltage Research Inc., FL) and at 10 µL/min feed rate driven by a programmable syringe micropump (Harvard Apparatus, MA). A rotating aluminum drum (diameter = 10cm) covered with aluminum foil was used to collect nanofiber mats at a 15 cm distance away from the needle tip. Each sample was collected for 2 hours. The electropsun PLA and PLA/lipid nanofiber mats were produced and incorporated to the surface of the bacteria and yeast culture media (fermented tea) to act as a base/foundation for the BC nanofibers growth. The growing period was three weeks.

RESULTS AND DISCUSSION

Structural properties

Surface morphology of the nanocomposites were examined using Scanning Electron Microscopy (SEM) and Quanta-250 Field Emission Scanning Electron Microscopy (FESEM). All samples were mounted on 1-inch stubs and coated with a 5nm layer of Gold/Palladium to increase conductivity and prevent any charging during the imaging process. Fig. 1 shows the morphology of PLA nanofibers (Fig. 1a), PLA/lipid (Fig. 1b) and bacterial cellulose (BC) (Fig. 2c-d). As expected, both PLA nanofibers and PLA/lipid nanofibers showed a porous structure attributed to the chloroform/acetone solvent system (Xiang, et al., 2007). The BC sample shows a highly fibrous network-like structure consisting of ultrafine nanofibrils (Fig. 1c). The typical lateral dimension of the BC nanofibrils is 30 – 40 nm. These results are in good agreement with the work previously reported by Feng, et al. (2012). A dense network of BC fibers was observed, as described by Bielecki et.al. (2005). The cellulose mat is comprised of multiple layers of fiber networks of about 1µm thick.

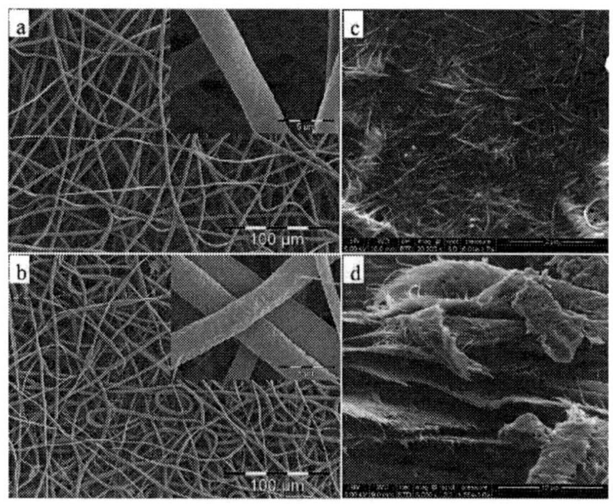

Fig 1. SEM images of nanofibers. a. PLA from 8 wt. % PLA in chloroform/acetone (3:1, v/v) solution; b. PLA/Lipid from 8 wt. % PLA in chloroform/acetone (3:1, v/v) containing 5% lipid (PLA w/w); c. BC sample surface; d. cross-section of the BC mat showing multiple layers.

The morphology of the BC/PLA and BC/PLA/lipid nanocomposites is shown in Fig 2. Fig 2a depicts the cross-section of BC/PLA nanocomposite. The cellulose nanofibers grew in the space among the electrospun PLA nanofibers (Fig. 2b). The entanglement of the PLA and BC nanofibers favored the structural utility of the nanocomposite. Fig. 2c shows the cross-section structure of the BC/PLA/lipid nanocomposite. A similar entanglement of the PLA/lipid electrospun nanofibers and BC nanofibers was observed (Fig. 2d).

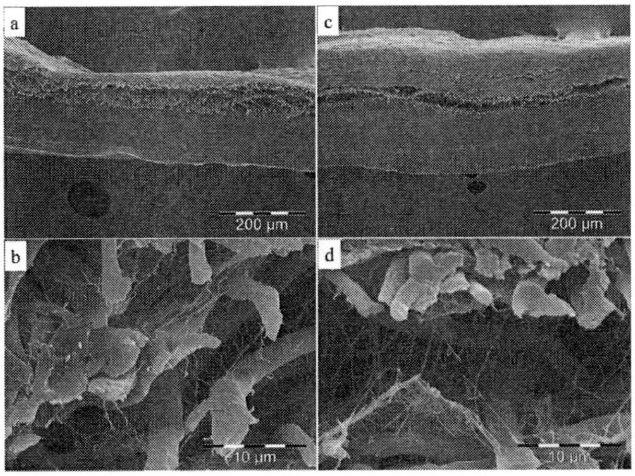

Fig. 2. SEM images of the cross-section of BC/PLA and BC/PLA/lipid nanocomposites. a-b. BC/PLA nanocomposite; c-d. BC/PLA/lipids nanocomposite.

Mechanical properties

Tensile properties of the BC/PLA and BC/PLA/lipid nanocomposites were determined using an Instron universal testing machine (model 4502) following the guidelines of ASTM D638-10. Tensile test were based on the test conditions using a 10k N load cell and a strain rate of 10mm/min. All samples were tested for at least five replicate measurements to obtain reliable data. The mechanical properties of BC, BC/PLA, and BC/PLA/lipid nanocomposites grown for 3 weeks are summarized below (Table I). The Young's modulus of BC, BP/PLA, and BC/PLA/lipid nanocomposites are 576, 302 and 734MPa respectively. The addition of the electrospun PLA nanofiber mat to the surface of the culture medium at the beginning of the BC growing process decreased the stiffness of the nanocomposites, while the incorporation of PLA/lipid increased the stiffness of the nanocomposites. These results are consistent with the elongation results of the nanocomposites. The elongation of BC/PLA nanocomposites increased compared to pure BC mat;while the incorporation of lipid to the electrospun PLA nanofiber decreased the elongation of the nanocomposites. This is likely due to the lipid making the structure rigid. (Qui & Netravali, 2014). Among BC, BC/PLA, and BC/PLA/lipid nanocomposites, the tensile strength did not change significantly.

Samples	Young's Modulus (Mpa)	Tensile strenth (Mpa)	Elongation at break (%)
BC	576	20.4	12.4
	(151.3)	(7.1)	(0.3)
BC/PLA	302	19.5	17.5
	(60.7)	(6)	(2)
BC/PLA/lipid	734	19.2	7
	(138.3)	(6.9)	(1.2)

Table I. Mechanical properties of BC, BC/PLA, and BC/PLA/lipid nanocomposites. Values in the parenthesis are % coefficient of variation values.

Water vapor transmission rate (WVTR, g/m^2 h)

WVTR was determined gravimetrically using a modification of ASTM E 96-95 (1995b). The procedure involves using a Petri dish filled with distilled water and covered and sealed with a film. The mass of water lost from the dish was monitored as a function of time when placed in a convection oven at 37.0 °C. WVTR was calculated using: WVTR = mass of water lost/time/ area = flux/area with units of g/hm^2.

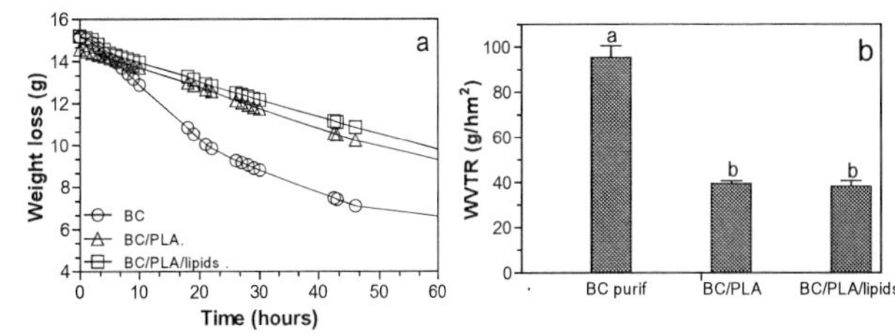

Fig. 3. Mass of water loss as a function of time (a) and WVTR values (g/hm^2) values of the produced nanocomposites (b)

Figure 3 shows the growth of BC using electrospun PLA fibers as support impacted water transmission properties. For instance, the presence of PLA fibers in the BC film induced a significant decrease in WVTR; the calculated values were 2.5 times lower for the BC/PLA nanocomposites. On the other hand, the addition of 5% lipid did not significantly affect the WVT properties. It is possible that lipids were negatively affected during purification of the mats after their production.

CONCLUSION

The results showed that cellulose nanofibers grew in the space among the electrospun PLA and PLA/lipid nanofibers. The entanglement of the PLA and PLA/lipid and BC nanofibers favored the structural utility of the nanocomposite. PLA and PLA/lipid nanofiber mats in general improved the hydrophobicity and mechanical properties of the nanocomposites.

REFERENCES

[1] Bielecki, S., Krystynowicz, A., Turkiewicz, M., Kalinowska, H. "Bacterial Cellulose." In *Biopolymers Online*. John Wiley and Sons, Inc., 2005.

[2] Feng, Y., Zhang X., Shen Y., Yoshino K., Feng W. *Carbohydrate Polym.*, 87, 2012: 644-649.

[3] Xiang C., Frey M., Taylor A., Rebovich M. *J Applied Polym. Sci.*, 106, 2007: 2363-2370.

[4] Faruk, O., Bledzki, A.K., Fink, H.P., Sain, M. Biocomposites reinforced with natural fibers: 2000–2010." *Prog. Polym. Sci.*, 37, 2012: 1552-1596.

[5] Tharanatha, R.N. "Biodegradable films and composite coatings: past, present and future." *Trends Food Sci. Technol.*, 14, 2003: 71-78.

[5] Qiu, K., Netravali, A.N. "A review of fabrication and applications of bacterial cellulose based nanocomposites." *Polymer Review*, 54, 2012: 598-626.

Poly(ether ester) Alternative Multiblock Copolymers: Synthesis and High Resilience Properties

Jun Wu, Weichun Huang, Jianying Chen, Yingfeng Tu
College of Chemistry, Chemical Engineering and Materials Science, Soochow University, Suzhou, China
tuyingfeng@suda.edu.cn

KEYWORDS
Poly(ether ester), Block Copolymer, Resilience

STATEMENT OF PURPOSE

Synthesize poly(ether ester) alternative multiblock copolymers with better structural control and test their properties as high resilience fiber.

INTRODUCTION

Poly(ether ester) copolymers were discovered when scientists searching for high resilience fibers. These copolymers are failed in the application as high resilience fibers but have shown great elastomers properties, and now are a class of important industrial thermal plastic elastomers with polyethers as soft segments and crystalline polyesters as hard segments. Traditional synthetic method by mixing dihydroxide-terminated polyether with small molecular diol and dimethyl terephthalate *via* condensation polymerization produces random block copolymers with unclear structure. Recently, we have discovered a new method to synthesize poly(ether ester) alternative multiblock copolymers using a condensation polymerization after ring opening polymerization method (CPROP), by simply mixing cyclic oligo(aromatic ester)s (COAEs) with dihydroxide-terminated polyether macroinitiators in a one pot process.[1-2] This new polymerization method have the advantage of better structural controlling, simple polymerization technique, easy manufacturing, no post-polymerization process needed etc. and can produce different types of poly(ether ester) alternative multiblock copolymers, including poly(trimethylene terephthalate)-*block*-poly(tetramethylene oxide) $(PTT\text{-}b\text{-}PTMO\text{-}b\text{-}PTT)_x$, poly(ethylene terephthalate)-*block*-poly(tetramethylene oxide) $(PET\text{-}b\text{-}PTMO\text{-}b\text{-}PET)_x$, and poly(butylene terephthalate)-*block*-poly(tetramethylene oxide) $(PBT\text{-}b\text{-}PTMO\text{-}b\text{-}PBT)_x$ alternative multiblock copolymers, respectively. The structure of the alternative multiblock copolymers can be determined and confirmed by the quantitative ^1H NMR characterization technique and two dimensional (2D) ^1H-^1H COSY and ^{13}C-^1H HSQC techniques, and also by the viscometry and gel permeation chromatography (GPC) measurements. We would like to study the properties of these poly(ether ester) alternative multiblock copolymers, and how the structure (block segment's length, number and structure etc.) affects these properties. Also, the properties from the fibers made by these alternative multiblock copolymers are interesting.

APPROACH

The synthetic approach for the poly(ether ester)s using the synthesis of $(PET\text{-}b\text{-}PTMO\text{-}b\text{-}PET)_x$ as example is listed in Scheme 1.

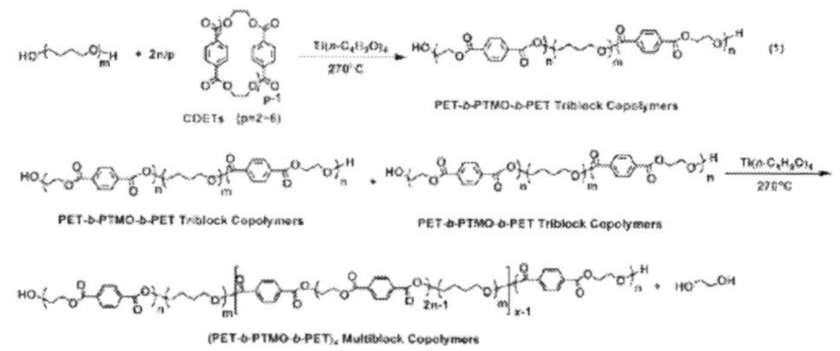

Scheme 1. The synthetic approach of $(PET\text{-}b\text{-}PTMO\text{-}b\text{-}PET)_x$ alternative multiblock copolymers.

RESULTS AND DISCUSSION

Three series of poly(ether ester) alternative multiblock copolymers, (PET-b-PTMO-b-PET)x, (PTT-b-PTMO-b-PTT)x, and (PBT-b-PTMO-b-PBT)x, were synthesized by corresponding COAEs with PTMO diol via the CPROP method. The quantitative 1H NMR technique was applied to investigate the multiblock copolymers' structure by the end group and functional groups estimation. The results show that the CPROP method can form multiblock copolymers with molecular weights up to 105 g/mol. The obtained multiblock copolymers show good thermal stability with 5% weight loss temperature ~340 °C. The segmental number's influence on the block copolymer properties is roughly studied. The mechanical property measurements reveal that the block copolymers with larger segment number having the better mechanical properties. Among them, the (PTT-b-PTMO-b-PTT)x alternative multiblock copolymers show the best resilience properties (elongation can approach 1700%), better than the industrial used thermoplastic elastomers. Fibers were prepared from these copolymers, and they show excellent shape recovery properties.

CONCLUSION

Three series of poly(ether ester) alternative multiblock copolymers are synthesized, and they show good resilience properties.

ACKNOWLEDGMENT

The financial support from the National "973" Project (No. 2011CB606004) and the National Natural Science Foundation of China (Grant No. 21274099, 21474067) is gratefully acknowledged.

REFERENCES

(1) Xu, Q., Chen, J., Huang, W., Qu, T., Li, X., Li, Y., Yang, X., Tu, Y. *Macromolecules*, 46, 2013: 7274-7281.
(2) Huang, W., Wan, Y., Chen, J., Xu, Q., Li, X., Yang, X., Li, Y., Tu, Y. *Polymer Chemistry*, 5, 2014: 945-954.

ZnS/Bacterial Cellulose/Epoxy Resin (ZnS/BC/E56) Nanocomposites with Good Transparency and Flexibility

Fangyi Guan, Shiyan Chen, Jingjing Yao, Weili Zheng, Huaping Wang
State Key Laboratory for Modification of Chemical Fibers and Polymer Materials, Key Laboratory of High-Performance Fibers and Products, Ministry of Education College of Materials Science and Engineering, Donghua University, Shanghai, China
chensy@dhu.edu.cn

Multifunctional composite material is growing rapidly driven by advancement of optoelectronic information technology. Continuous "roll-to-roll" processing using flexible substrates is taking place of the conventional "batch" one on glass substrates[1]. Also, it is relatively easier and less expensive, which is suitable for the future technologies.

Bacterial cellulose (BC) has an ultrafine nanosized three-dimensional (3D) fibrous network comprised of the ribbon-shaped nanofibers with diameters of 30-100 nm[2,3]. The light-scattering between interfaces could be negligible owing to the nanosize effect of BC[4]. Its toughness, Young's modulus, tensile strength are also very high. BC could also maintain the flexibility and ductility of the polymer matrix[4]. In our previous report[5], BC could obviously reduce coefficient of planar thermal expansion (CTE) of bisphenol A epoxy resin (E56). And its transparency was good and insensitive to temperature in certain range. Therefore, the BC nanofibers reinforced polymer composites are expected as potential alternatives in the application of flexible electronic materials.

ZnS are known as semiconductors with a direct wide band-gap energy, high refractive ndex, high transparency in visible spectrum. In this work, firstly, ZnS/BC composite membranes were prepared. They were successively exchanged with ethanol, cold-pressed, impregnated with E56, and then cured under hot-compressed at 140–150°Cfor 3h.

The structures and properties of BC reinforced nanocomposites were characterized by the element analysis (EA), UV-Vis spectroscopy, Spectroscopic ellipsometer (Ellip-SR-II), tensile mode, WDW 3020 Universal Testing Machine , and transmission electron microscopy (TEM).

Figure 1 shows the morphology of ZnS/BC nanocomposites using TEM by ultra-thin sections. From the images, we can see that ZnS was round nanoparticles. With the precursor Zn^{2+} concentration between 0.1 and 1 wt%, their sizes were small (<50 nm) and the distribution was homogeneous, while few nanoparticles were observed (Figure d, e) and the distribution became heterogeneous with the concentration increasing above 2.5 wt%. This observation indicated Zn^{2+} concentrations had an great effect on the size of ZnS nanoparticles. Increase of the precursor Zn^{2+} concentration led to larger nanoparticles, which were difficult to enter the 3D network of BC, thus attaching to the BC surface.

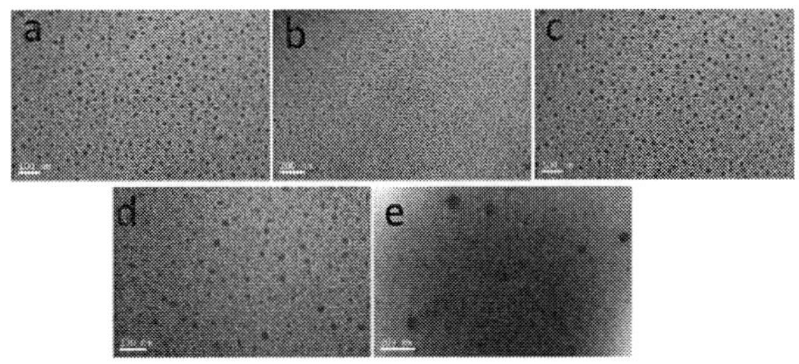

Figure 1. The TEM images of ZnS/BC/E56 prepared in different Zn^{2+} concentrations (a) 0.1 wt%, (b) 0.5 wt%, (c) 1 wt%, (d) 2.5 wt%, (e) 5 wt%.

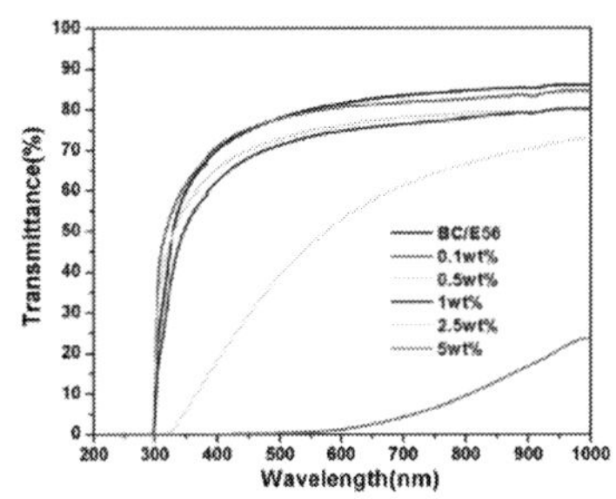

Figure 2. Transmittance spectra of ZnS/BC/E56 membranes.

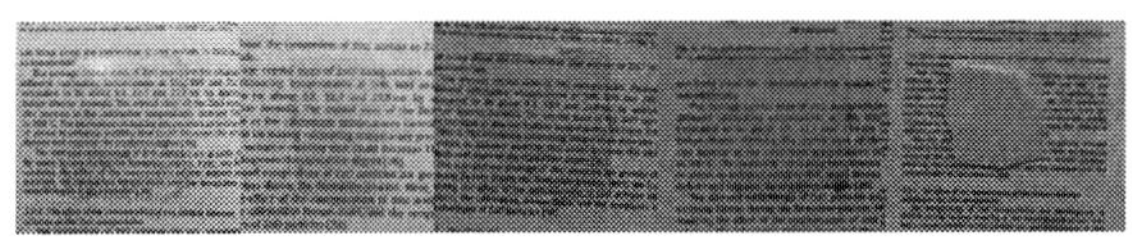

(a) (b) (c) (d) (e)

Figure 3. Optical transparency of ZnS/BC/E56 membranes with the precursor of Zn^{2+} concentrations (a) 0.1 wt%, (b) 0.5 wt%, (c) 1 wt%, (d) 2.5 wt%, (e) 5 wt%.

We investigated the transmittance of ZnS/BC/E56 membranes from 200 to 1000 nm with the different precursor Zn^{2+} concentration of 0.1, 0.5, 1, 2.5, 5 wt %. As shown in Figure 2, with the precursor Zn^{2+} concentration less than 1 wt%, ZnS/BC/E56 membranes exhibited outstanding transmittance over 70% in the visible light. When the concentration increased to 2.5 wt %, the light transmittance decreased seriously to 50% at 600 nm. When the concentration was 5wt %, the membranes finally became opaque completely.

Figure 3 shows the photo images of ZnS/BC/E56 with Zn^{2+} concentration from 0.1 to 5 wt %. It can be observed that the membranes presented a good transparence when Zn^{2+} concentration was lower than 1 wt %.

The transmittance of ZnS/BC/E56 membranes at 600 nm under different temperature was shown in figure 4. It was observed that the ZnS/BC/E56 membranes maintained a good transmittance in a range from 10 to 100 °C.

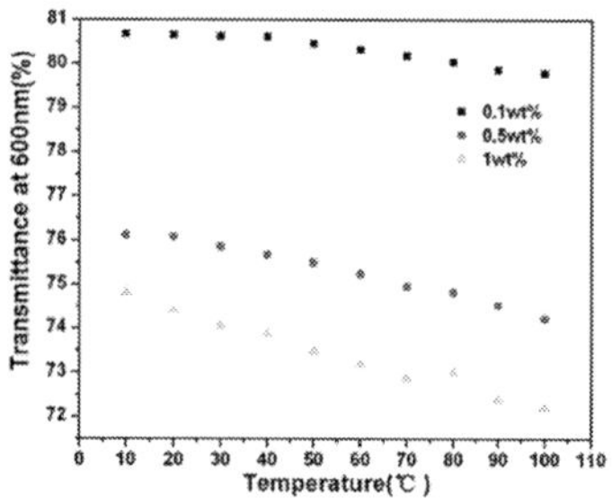

Figure 4. Transmittance of ZnS/BC/E56 membranes at 600 nm under different temperature.

Table 1 showed the effect of ZnS on the RI and the thermo-optical instability of the composites. The BC content of the membranes was 24.8% and the thickness was same. The RI increased from 1.5535 to 1.5785 with the increase of ZnS content to 8.95%. By comparison with BC/E56, the thermo-optical coeffecient of the composites obviously increased from –361 ppm/K (BC/E56) to –310 ppm/K.

Table 2. The refractive index and thermo-optical coefficient for ZnS/BC/E56 membranes.

Zn^{2+} concentration (wt%)	Membrane Thickness(μm)	RI (600nm)	dn/dT (ppm/K)
0	189	1.5525	–361
0.1	191	1.5555	–354
0.5	193	1.5602	–342
1	198	1.5663	–336
2.5	195	1.5722	–318
5	202	1.5785	–310

The tensile stress–strain behaviors of BC, E56, BC/E56 and ZnS/BC/E56 at different ZnS concentrations were depicted in Figure 5, and the mechanical data of the samples calculated according to the stress–strain curves were listed in Table 2. It can be seen that the Young's modulus and tensile strength of ZnS/BC/E56 nanocomposites were lower than that of the BC/E56. With the increase of Zn^{2+} concentrations, the mechanical properties of ZnS/BC decreased.

Due to BC, BC/E56 exhibited excellent flexibility as shown in Figure 5. Incorporated with ZnS, the membranes exhibited excellent flexibility and the strain didn't decrease obviously especially for the nanocomposites with precursor Zn^{2+} concentration of < 1wt% compared with that of BC/E56. When the concentration of Zn^{2+} precursor was 1 wt%, the elastic modulus of membrane were the largest.

Table 2. Tensile strength, elongation, and Yong's modulus of membranes determined from strain-stress curves.

Sample	Stress (MPa)	Strain (%)	Young's modulus (MPa)
BC	34.6	5.4	635.2
E56	20.6	17.6	117.5
BC/E56	28.3	15.2	349.6
0.1wt%	25.9	13.2	315.5
0.5wt%	22.8	11.9	288.1
1wt%	24.6	9.2	337.4
2.5wt%	17.5	8.8	217.6
5wt%	11.8	8.03	147.3

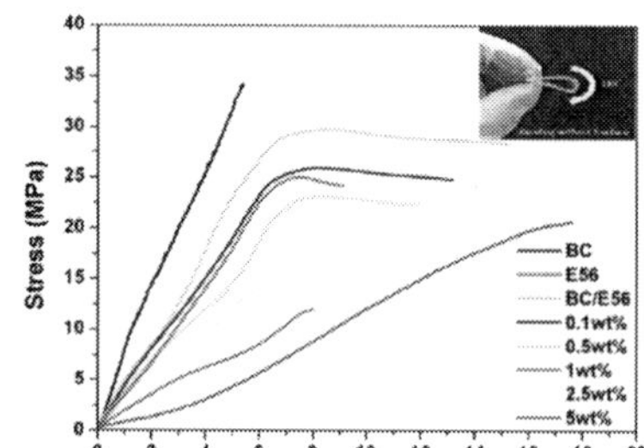

Figure 5. Tensile stress–strain behaviors of the pure BC membrane and ZnS/BC/E56 membranes at different concentrations of Zn^{2+} precursor. Inset shows the representative flexibility of BC/E56 composites with BC content of only 2.6%.

In conclusion, the composite ZnS/BC/E56 had good transparency and flexibility when the precursor Zn^{2+} concentration was less than 1wt%. Its excellent thermo-optic stability and mechanical properties endow the nanocomposites potential application in the flexible optoelectronic materials.

ACKNOWLEDGMENT

This work was financially supported by Program of Introducing Talents of Discipline to Universities (B07024), Shanghai Leading Academic Discipline Project (B603), the National Natural Science Foundation of China (51273043), Project of the Action on Scientists and Engineers to Serve Enterprises (2009GJE20016).

REFERENCES

1. MacDonald, W.A. *J. Mater. Chem.* 14, 2004: 4-10.
2. Li, X., Chen, S., Hu, W., Shi, S., Shen, W., Zhang, X., Wang, H. *Carbohyd Polymer*, 76, 2011: 509-512.
3. Zheng, Y., Yang, J., Zheng, W., Wang, X., Xiang., C., Tang, L., Zhang, W., Chen, S., Wang, H. *Mat. Sci. Eng. C-Mater.* 33, 2013: 2407-2413.
4. Nogi, M., Yano, H. *Adv. Mater.* 20, 2008: 1849-1852.
5. Zheng, W., Zhao, S., Zheng, Y., Chen, S., Wang, H. *Mater. Sci. Forum,* 789, 2014: 262.

Functional Textiles and Clothing

Functional Finishing and Dyeing

Research on the Nanogenerators Based on Electrospun Polymeric Nanofiber Mats

Hao Yu, Bin Yu, Tao Huang
College of Materials Science and Engineering, Donghua University, Shanghai, China
yuhao@dhu.edu.cn

ABSTRACT

In this work, environmental mechanical energy, including human motion, was harvested and transferred to electronic energy by nanogenerators with electrospun polymeric mats. Considering the high electrostatic field during the process, electrospinning was ideally suited for producing piezoelectric polyvinylidene fluoride (PVDF) nanofibers with β-form crystal phase. The inorganic salts and multiwalled carbon nanotubes (MWCNTS) were doped into the polymer nanofibers mats to further enhance the output of nanogenerators, respectively. High-alignment was proved to improve the output voltage and power of bend-mode piezoelectric nanogenerator (PENG).

As an electronegative polymer, PVDF nanofiber mats were coupled with different electronegativity polymer materials to set up contact-mode triboelectric nanogenerators (TENG). The TENG assembled with poly(hydroxybutyrate-co-valerate)(PHBV) for positive electrode material was selected with its exciting output voltage, which could reach more than 110 V of output voltage. The mat thickness, morphology and friction angle of TENG were studied to optimize the electrospun mats-based TENG. Enhancement of output voltage was found when decreasing mat thickness. Meanwhile uniform nanofibers with smaller diameter and higher friction angle could improve the output voltage of TENG. Human walking-driven wearable all-fiber triboelectric nanogenerator containing electrospun PVDF piezoelectric nanofibers was prepared, and it could lighten 214 LEDs simultaneously when harvesting human-walking energy. Doped with graphene oxide, the PVDF/GO electrospun mats made the output voltage and power of TENG enhanced. It is prospective that the PENG and TENG with electrospun mats can contribute to the wearable energy supplement in the future.

Characterizing Thermal Protective Fabrics of Firefighters' Clothing in Hot Surface Contact

Sumit Mandal[1], Guowen Song[2]
[1]Department of Human Ecology, University of Alberta, Canada
[2]Department of AESHM, Iowa State University, USA
gwsong@iastate.edu

INTRODUCTION

Firefighters are often exposed to hot surfaces of various non-combustible solid substances (e.g., iron or steal furniture/windows) during firefighting, rescuing fire victims and/or property in a structural fire hazard [1-3]. The nature of this exposure is a direct physical contact between hot surfaces of solid substances and fabrics of firefighters' clothing. At the area of contact, heat is imposed from hot surfaces onto fabrics, and fabrics physical properties changed as a result of compression. Therefore, heat stored from solid substances gradually transfer through fabrics towards firefighters' bodies and causes burn injuries. Considering this exposure, Mandal, et al. (2013) comprehended thermal protective performance of fabrics (used in firefighters' clothing) under laboratory simulated hot surface contact [2]. They found that thickness of a fabric has a significant impact on the conductive heat transfer through the fabric; in turn, fabric thickness strongly affects the thermal protective performance. By employing the fabric's thickness and thermal resistance, Mandal and Song (2014) further developed MLR (Multiple Linear Regression) and ANN (Artificial Neural Network) models to predict the thermal protective performance [3]. However, these studies [2, 3] investigated only a few factors affecting the thermal protective performance. Additionally, these studies did not elaborate about heat transfer mechanisms through thermal protective fabrics under hot surface contact. Thus, previous studies are limited and fragmented in terms of full understanding of the characteristics of thermal protective fabrics under hot surface contact.

OBJECTIVE

This study aims to characterize thermal protective fabrics of firefighters' clothing under hot surface contact. To fulfill this aim, thermal protective performance of various fabrics is examined and factors affecting the thermal protective performance are statistically explored. Furthermore, heat transfer mechanisms associated with hot surface contact are analytically explained for better understanding of the thermal protective performance.

APPROACH

Different types of thermal protective fabrics (that are commercially available and commonly used in firefighters' clothing) were selected based on their constructional features. By assembling these fabrics, single- and multi-layered (double- and triple-layered) fabric systems were prepared, and properties (air permeability, weight, thickness, and thermal resistance) of these assembled fabric systems were evaluated. Three specimens (10×15cm) of each selected fabric systems were prepared and conditioned in a standard atmosphere (temperature = 21°C and relative humidity = 65%) for 24 hours. To characterize fabric systems under hot surface contact, the conditioned specimens were tested using the modified ASTM (American Society for Testing Materials) F 1060 standard (Figure 1) and thermal protective performances of these specimens were measured in terms of time required to generate second-degree burn on wearer's body. Next, the properties and thermal protective performance values of all the fabric systems were statistically normalized and t-tests were carried out. The associations between fabric systems' properties and thermal protective performance were inferred based on the sign (+ or −) of the T-stat values obtained from the t-tests. P-values obtained from the t-tests for each fabric property were also analyzed. For the P-value < 0.05, the property was identified as the key factor for thermal protective performance. The coefficients of determination (R^2) of the relationship between the fabric properties and thermal protective performance were calculated. A R^2 value with proximity to 1 for a particular relationship was inferred to be a strong association between the respective fabric property and thermal protective performance. Furthermore, inference tests [hypothesis test (P-value) and 95% confidence interval (upper and lower limit)] were carried out to understand the differences in thermal protective performances between two types of fabric systems. Finally, heat transfer mechanisms during hot surface contact with thermal protective fabrics were analytically explained based on the test configuration shown in Figure 1. Here, heat flow and temperature distribution during the testing were identified and few basic assumptions regarding the heat transfer were stated.

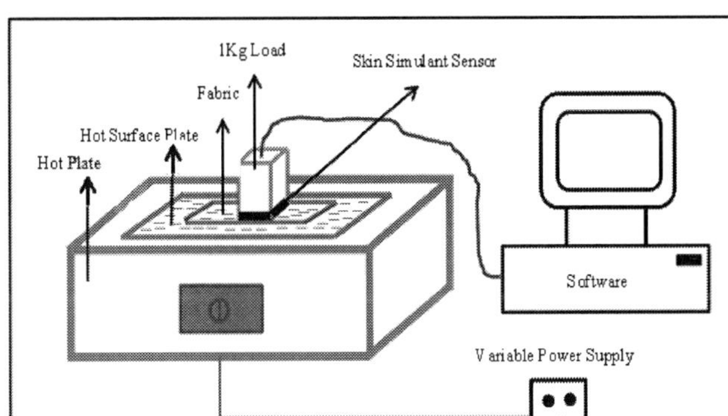

Figure 1. Hot surface contact test (modified ASTM F 1060)

RESULTS AND DISCUSSION

In this study, triple-layered fabric systems demonstrated higher thermal protective performance than single- and double-layered fabric systems. The air layers and dead air trapped by triple-layered fabric systems contributed to the higher performance. However, triple- and double-layered fabric systems with moisture barriers in the outer layers can easily burn on contact with the hot surface. Eventually, the integrity of the fabric systems gets damaged and their thermal protective performances reduce. Additionally, multi-layered fabric systems with thinner thermal liner demonstrated much lower performance than fabric systems with thicker thermal liners. Through inference testing, it can also be confirmed that performance of non-thermal liner based fabric systems significantly differs from performance of thermal liner based fabric systems (P-value < 0.05). And, the differences between the performances of non-thermal and thermal liner based fabric systems are always negative. Thus, it can be deduced that performance of thermal liner based fabric systems is generally high. This is because thermal liners are nonwoven fabrics and can trap a larger amount of dead air in their structures. This dead air trapped results in high thermal protective performance of fabric systems. Furthermore, results (T-stat and P-value) of the t-test between normalized values of fabric systems' properties and thermal protective performance show that T-stat value between air permeability and thermal protective performance is negative. It means that air permeability has a negative relationship with thermal protective performance. This is because a highly air-permeable fabric can blow away the heat through its constituent pores. This heat transmitted can generate quick burns on wearers' bodies and lower the thermal protective performance of the fabric. In this regard, the R^2 value between air permeability and the performance is not close to 1 (R^2 = 0.19), which indicates a weak negative relationship between air permeability and thermal protective performance. It is also found that the P-value of air permeability is 0.22, which is much greater than 0.05; consequently, air permeability is an insignificant property to influence the thermal protective performance. Moreover, the T-stat values of weight, thickness, and thermal resistance are positive. It means that a positive relationship exists between these properties and thermal protective performance. This is because fabric systems get compressed between hot surfaces and wearers' bodies; in this compressed situation, a fabric with high weight and thickness can trap more dead air within its structure than a fabric with low weight and thickness. As the dead air acts as thermal insulator, the thermal resistance of the weighty and thicker fabric becomes high; eventually, the thermal protective performance of such fabric increases. In this context, it has been confirmed that R^2 values between these properties and thermal protective performance are all close to 1 (> 0.7); hence, a strong positive relationship exists between these properties and thermal protective performance. Also, the P-value of thickness (0.001) is the minimum among all these properties. Thus, thickness is the most significant property to affect performance of fabrics under the exposure of hot surface contact.

Based on the hot surface contact test shown in Figure 1, heat transfer occurs from the uniform high temperature hot surface plate towards low temperature skin simulant sensor (or human body) via fabric. In this heat transfer process, the hot surface plate and the sensor act as solid bodies. On the other hand, the fabric acts as a porous media because its structure is a combination of solid fiber and gaseous air phases. As all the objects (hot surface plate, fabric, and sensor) are in contact with each other in the test configuration (Figure 1), so conduction is the prime mode of heat transfer. Analytically, this conductive heat transfer takes place mainly through four regions (R_1 to R_4), (i) R_1 = between the hot surface plate and fabric (q′); (ii) R_2 = within the fabric (q′′); (iii) R_3 = between the fabric and sensor (q′′′); and (iv) R_4 = within the sensor (q′′′′). Here, it can be assumed that the conductive heat transfer is one-dimensional i.e., in the X-direction along the objects' thickness (plate thickness = X_P, fabric thickness = X_F, and sensor thickness = X_S), and the thermal properties i.e., thermal conductivity (for plate = k_P, fabric = k_F, and sensor = k_S), density (for plate = ρ_P, fabric = ρ_F, and sensor = ρ_S), specific heat (for plate = C_{PP}, fabric = C_{PF}, and sensor = C_{PS}) of each object do not change with temperature. Additionally, it can be hypothesized that all the objects are in thermal equilibrium conditions (i.e., temperature within the objects are spatially and temporarily uniform at any time). Here, a notable point is that thickness of a fabric may change upon contact with hot surface due to compression. This situation can accelerate the heat transfer through the fabric towards firefighters' bodies. If the surface roughness of the hot surface and/or fabric is high, this causes high thermal contact resistance and can reduce the heat transfer.

CONCLUSION

It can be concluded that constructional features and properties of fabrics affect the thermal protective performance. Generally, a multi-layered fabric system consisting of a thick thermal liner shows high thermal protective performance. Moreover, the fabric system having a moisture barrier in the outer layer can easily ignite, which can decrease the thermal protective performance. Additionally, a fabric system with low air permeability, high weight, high thickness, and high thermal resistance can possess a commendable thermal protective performance for firefighters' clothing. Moreover, heat transfer through fabrics is dependent upon the thermal properties of fabrics, changes in fabric thickness due to compression, and/or, surface roughness of the hot surface/fabrics.

FUTURE WORK

This study demonstrated the key factors affecting thermal protective performance of fabrics under hot surface contact. A further study on changes in thermo-physical properties of a multi-layered fabric system could enhance the understanding on characteristics of thermal protective fabrics under hot surface contact. The findings from this study could also help textile/material engineers to develop fabrics that can provide better occupational health and safety for firefighters.

REFERENCES

1. Veghte, J.H. "Effect of moisture on the burn potential in fire fighters gloves." *Fire Technology*, 23, 1986: 313-322.
2. Mandal, S., et al. "Characterization of textile fabrics under various thermal exposures." *Textile Research Journal*, 83, 2013: 005-1019.
3. Mandal, S., Song, G. "An empirical analysis of thermal protective performance of fabrics used in protective clothing." *Annals of Occupational Hygiene*, 58, 2014: 1065-1077.

Fabrication of ZnO&SiO_2 with Eliminated Photocatalytic Activity for Preventing Wool Photoyellowing

Mingwen Zhang[1], Bin Tang[1,2], Lu Sun[1,2], Xungai Wang[1,2]
[1]Institute for Frontier Materials, Deakin University, Geelong, Victoria, Australia
[2]School of Textile Science and Engineering, Wuhan Textile University, Wuhan, China
lu.sun@deakin.edu.au.

INTRODUCTION

Wool fiber is well known as a superior natural textile material due to its resilience and softness. However the photodegradation caused by sunlight, particularly from UV rays, makes the fiber weak and yellow [1-4], which is a critical problem for the wool industry. The most widely used protective method is application of UV absorbers to wool [5-7]. Compared with the organic UV absorbers, inorganic UV absorbers, such as ZnO and TiO_2, are generally non-toxic, more stable and effective. Development of nano-science and nano-technology provides new approaches to a better application of these inorganic UV absorbers [8-12]. However, wool is so highly sensitive and delicate that the process of photoyellowing can be greatly accelerated by reactive radicals generated by the photosensitized inorganic UV absorbers due to their strong photocatalysis [13, 14].

In this study, mono-dispersed core-shell structured ZnO&SiO_2 nanoparticles were fabricated using a two-step process and their photocatalytic properties were measured. Firstly, mono-dispersed ZnO nanoparticles were prepared using a hydrothermal process. The conditions for generating ZnO nanocrystallites with different dimensions were studied. The UV absorption and visible transparency spectra of aqueous dispersions of the nanoparticles were investigated together with their photocatalytic activity. Secondly samples of the ZnO nanoparticles were coated with SiO_2 shells using a sol-gel procedure. Relationships between photocatalytic activity, core-particle size and silica shell thickness as well as the UV absorption and visible transparency of aqueous dispersions of the nanoparticles were investigated. The effect of the ZnO&SiO_2 nanoparticles on wool yellowing was evaluated.

APPROACH

ZnO nanoparticles were prepared using a hydrothermal method. And then SiO_2 shells were coated onto the surfaces of synthesized particles by stöber method [16]. Wool was treated with ZnO&SiO_2 nanoparticles by dip coating method.

RESULTS AND DISCUSSION

Figure 1 shows TEM images of the ZnO core-particles before and after the silica coating process.

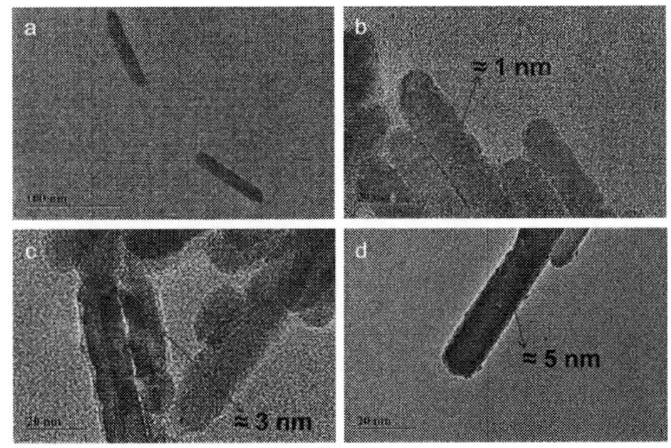

Figure 1. TEM images of (a) ZnO core-particles before silica coating, and SiO_2 coated nanoparticles prepared with different aging times of (b) 4 h (Z&S1), (c) 5 h (Z&S2), and (d) 6 h (Z&S3).

It can be seen that, the average length/width of the ZnO nanoparticles is 90/15 nm. And the thickness of outmost silica layer is from 1 nm to 3 nm and 5 nm respectively, which can be controlled by the sol-gel reaction time.

The UV/Vis absorption spectra and photocatalytic activity measurements of the silica coated of samples derived from the ZnO, are shown in Figure 2.

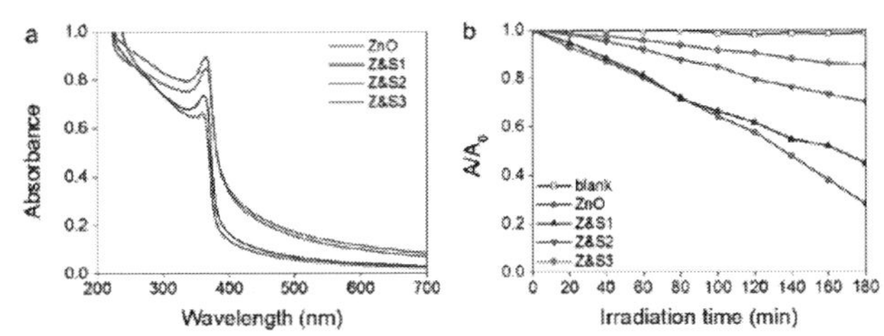

Figure 2. (a) UV-Vis absorption spectra of ZnO, Z&S1, Z&S2 and Z&S3 NP nanoparticles, and (b) photodegradation profiles of Rhodamine B in the presence of sample ZnO (●), as well as Z&S1 (▲), Z&S2 (▼) and Z&S3 (♦), and a blank control (□).

Compared with the uncoated ZnO nanoparticles, the intensity of the peak absorption of Z&S1 (362 nm) increased by about 11%, while that of the Z&S2 (364 nm) increased by 36%, while the growth of the intensity of the absorption peak of Z&S3 (at 365 nm) fell by 29%. However the photocatalysis effects from ZnO core-particles were gradually suppressed with thickening of the SiO_2 shell.

After 180 mins irradiation, the catalytic system showed reducing degradation levels of 45%, 70% and 86% respectively as the thickness of the silica shells increased from 1 nm (Z&S1) to 3 nm (Z&S2) and 5 nm (Z&S3).

The protection of Z&S3 nanoparticles on wool fabric whiteness is represented in Figure 3, which indicates the extent of discoloration of the fabric after 168 h simulated solar irradiation.

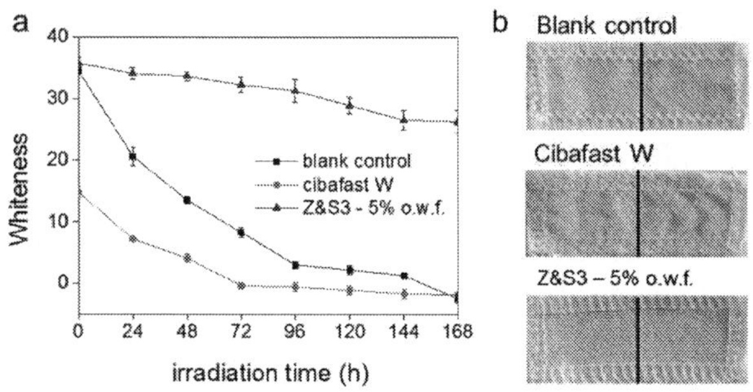

Figure 3. (a) Whiteness change and (b) images of fabrics with no treatment, cibafast W and 5% o.w.f. Z&S3 coating before (left) and after (right) 168 h simulated solar irradiation.

It can be seen that, the treatment of Z&S3 nanoparticles indeed retarded the yellowing process of fabrics than untreated and commercial Cibafast W treated fabric. After 168 h simulated solar irradiation, the whiteness of untreated fabric is decreased greatly from 34.5±0.6 to -2.5±0.6. An apparent protective result was observed with the treatment with 5% o.w.f. of Z&S3 nanoparticles, with which the Whiteness Index decreased from 35.7±1.0 to 26.2±1.8.

CONCLUSIONS

Core-shell structured $ZnO\&SiO_2$ nanoparticles have been prepared successfully. These structured nanoparticles had high UV absorbing ability and low photocatalytic activity, as well as good transparence. These nanoparticles have excellent applications in UV protection of organic substrates such as wool fabrics.

ACKNOWLEDGMENT

This research was financially supported by a Ph.D. Scholarship from Deakin University for the first author. It was also partially supported by the National Natural Science Foundation of China (NSFC 51302197, NSFC 51273153).

REFERENCES

[1] Launer, H.F. *Textile Research Journal*, 35, 1965: 395.
[2] Launer, H.F. *Textile Research Journal*, 35, 1965: 813.
[3] Millington, K.R. *Coloration Technology*, 122, 2006: 169.
[4] Millington, K.R. *Coloration Technology*, 122, 2006: 301.
[5] Evans, N.A., Waters, P.J. *Textile Research Journal*, 51, 1981: 432.
[6] Evans, N.A., Waters, P.J., Wilshire, J.F.K. *Textile Research Journal*, 56, 1986: 203.
[7] Carr, C.M., Leaver, I.H. *Journal of Applied Polymer Science*, 33, 1987: 2087.
[8] Sun, L., Rippon, J.A., Cookson, P.G., Wang, X., King, K., Koulaeva, O., Beltrame, R. *International Journal of Technology Transfer and Commercialisation*, 7, 2008: 224.
[9] Becheri, A., Dürr, M., Lo Nostro, P., Baglioni, P. *Journal of Nanoparticle Research*, 10, 2008: 679.
[10] Mosimann, W., Benisek, L., Burdeska, K., Leaver, I.H, Myers, P.C, Reinert, G., Wilshire, J.F.K. "A New Commercial UV Absorber for the Protection of Wool and Wool Dyeings." *Proceedings 8th International Wool Textile Research Conference*, 1990.
[11] Montazer, M., Pakdel, E., Moghadam, M. *Fibers and Polymers*, 11, 2010: 967.
[12] Chen, Xianqiong; Liu, Yuyang, Shi, Haifeng, Wang, Xiaowen, Qi, Kaihong, Zhou, Xiang; Xin, J.H. *Textile Research Journal*, 80, 2010: 2214.
[13] Sun, L., Rippon, J.A., Cookson, P.G., Koulaeva, O., Wang, X. *Chemical Engineering Journal*, 147, 2009: 391.
[14] Zhang, H., Millington, K.R., Wang, X. *Polymer Degradation and Stability*, 94, 2009: 278.
[15] Li, X., Leng, Y., Li, J.G., Xiu, Z., Huo, D., Sun, X., Ishigaki, T. *Current Nanoscience*, 6, 2010: 110.
[16] Stöber, W., Fink, A., Bohn, E. *Journal of Colloid and Interface Science*, 26, 1968: 62.

Dynamical Wettablity Switch via Photoinduced F/TiO_2 for Cellulose Substrates

Yunjie Yin, Yuanhui Chen, Xuan Zhang, Chaoxia Wang
Key Laboratory of Eco-Textile, Ministry of Education, Jiangnan University, Wuxi, China
wangchaoxia@sohu.com; yinyunjie@jiangnan.edu.cn.

OBJECTIVE

The ability to deposit surfaces with switchable wettability assumes an actively pursued research area in materials science due to its implications in disparate technological fields. The reversible wetting behavior has been extensively researched for their potential applications in self-cleaning surfaces, intelligent microfluidic devices, lab-on-a-chip systems, and so on. During the numerous switchable wettability pathways, a more powerful strategy relies on the modifiable capability of some transition-metal oxides, namely TiO_2, to increase their hydrophilicity under band-gap photoexcitation, and to revert back to the initial conditions upon storage in dark. To prepare a functional surface with UV-switchable wettability on cotton fabric by hybrid coating, the hybrid sol containing TiO_2 and silane coupling agent, such as Triethoxyoctylsilane, Dodecyl triethoxy silane, Hexadecyltrimethoxysilane and 1H, 1H, 2H, 2H-perfluorooctyltrimethoxysilane via sol-gel technology. The maximum contact angles of cotton fabric coated with functional TiO_2 hybrid sol were more than 150°, respectively. With irradiation under UV light, the decay of the contact angle from superhydrophobicity to superhydrophilicity on the irradiated fabric treated with functionial TiO_2 hybrid sol was rapid, and in the black environment, the superhydrophilic fabric was recover to superhydrophobic state.

KEYWORDS

TiO_2, Dynamical wettablity switch, UV irradiation

INTRODUCTION

The ability to deposit surfaces with switchable wettability assumes an actively pursued research area in materials science due to its implications in disparate technological fields[1-3]. There are two approaches to obtain switchable wettability: by microstructure or surface free energy transitions with the external stimulations. The reversible wetting behavior has been extensively researched for their potential applications in self-cleaning surfaces, intelligent microfluidic devices, lab-on-a-chip systems, and so on[4,5]. For the UV-switchable wettability, most substrates are glass, silicon wafer or metal sheet, which are even and thermoduric. The literature about UV-switchable wettability on fiber substrates, such as cotton, is very few. In this paper, a photoconductive F/TiO_2 hybrid sol was synthetized and coated on cotton fiber substrate. The superhydrophobic-superhydrophilic switchable property was represented by the contact angle under the response condition of UV irritation or dark store.

MATERIALS

The poplin 100% cotton fabric weighting 141.0 g/m^2 is produced by Jiangsu Hongdou Industrial Co., Ltd (China). The tetrabutyl titanate (TBT), ethanol and HCl are supplied by Sinopharm Chemical Reagent Co., Ltd (China). All the chemicals are analytical reagent grade. The 1H,1H,2H,2H-perfluorooctyltrimethoxysilane is offered by Harbin Xeogia fluorine-silicon chemical Co. (China), and it is technical grade.

PREPARATION OF F/TIO_2 HYBRID SOL

The F/TiO_2 hybrid sol was prepared by adding diluted TBT 23 g (TBT: ethanol 1:3, weight ratio) into the mixture (71.0 g) containing HCl (1 mol/L) and H_2O ($HCl \cdot H_2O$ = 2:8, weight ratio) in a flask at 500 rpm with a magnetic stirring apparatus at room temperature. Six grams 1H,1H,2H,2H-perfluorooctyltrimethoxysilane solution (50 wt% in ethanol) was added into the flask. The mixture was stirred at 30 °C for 8 hours with a magnetic stirring apparatus and then aged for 120 hours. Before used on fabric, the F/TiO_2 hybrid sol was neutralized by $NH_3 \cdot H_2O$ (1 mol/L) to the pH value about 6.0.

FABRIC MODIFICATION

The cotton fabric were pre-cleaned in deionized water and then dried at 50 °C for 2 h before other treatment. The fabric was immersed in the F/TiO_2 hybrid sol at room temperature for 30 minutes, and then dried at 60 °C for 20 minutes and lastly baked at 150 °C for 3minutes.

IRRADIATION

To induce hydrophilicity, the fabrics coated with F/TiO_2 hybrid sol were irradiated with the UV light (light intensity 21.0 mW/cm^2, distance 40.0 cm, predominantly wavelength 253.7 nm). The contact angle was measured during the irradiating process. When the contact angle was decreased to 0°, the fabrics coated with F/TiO_2 hybrid sol and irradiated with the UV light were placed in dark box under ambient environment (with relative humidity of 30-40%), and the contact angle was also measured.

RESULTS AND DISCUSSION

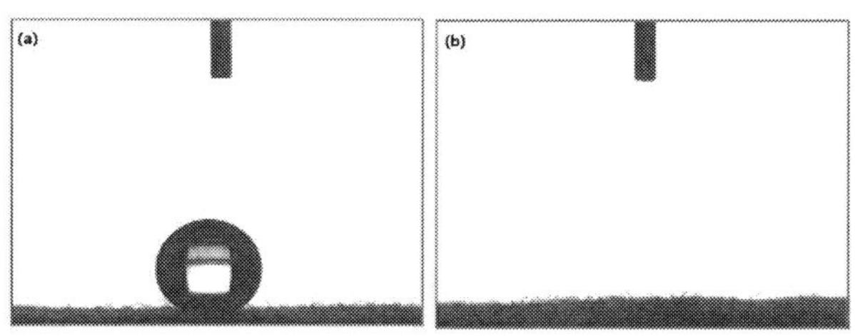

Fig. 1. Contact angle changes: (a) coated cotton fabric; (b) coated cotton fabric irradiated for 24 hours; (c) coated and irradiated cotton fabric in dark for 60 hours.

The contact angle of cotton fabric coated with F/TiO_2 hybrid sol was 161.8° (Fig. 1). Under UV light for 24 h, the contact angle of the cotton fabric coated with F/TiO_2 hybrid sol decreased to 0° after 24 hours and the coated fabric was superhydrophilic. The superhydrophilic cotton sample was placed in dark, and the contact angles were increased slowly. The contact angle of the coated cotton sample was increased to 94.7° after stored in dark for 36 hours. Continuing to store in dark, the contact angles were further risen. While stored in dark for 60 hours, the contact angle of the coated cotton sample was increased to 133.6°, and it did not increase sequentially when kept on in dark again.

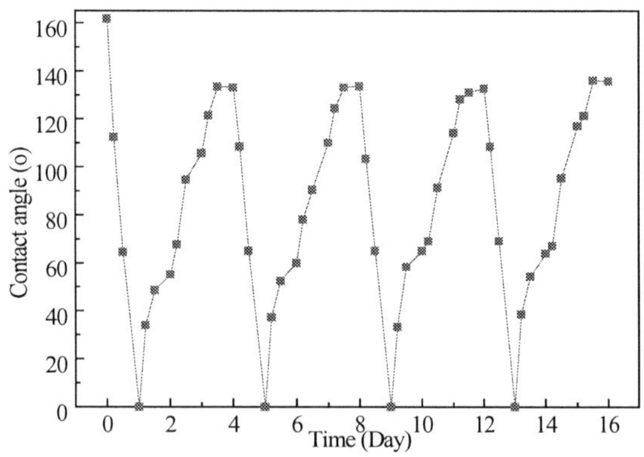

Fig. 2. Contact angle curves of the coated fabric in UV light and dark

From Fig. 2, the process was repeated for several cycles. The cycle time of the cotton was 84 hours. The cycle might be affected by the fabric construction and fiber morphology.

Permeability of H_2O

To confirm the wettability and content of the air pore, the water diffusion on absorbent paper was tested.

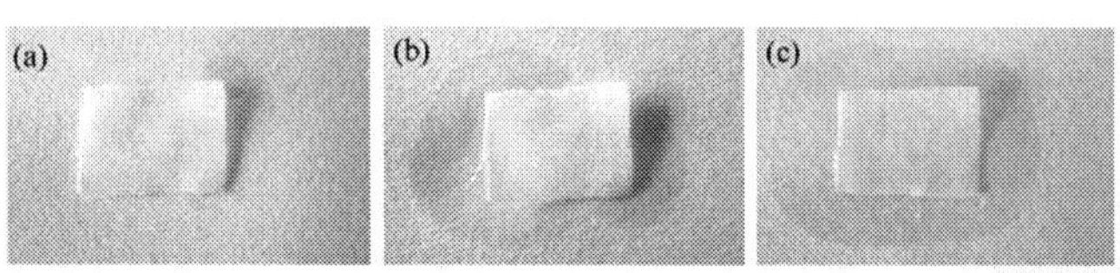

Fig. 3 Water diffusion on paper from different fabrics: (a) coated cotton fabric; (b) coated cotton fabric irradiated for 14 h; (c) coated cotton fabric irradiated for 24 hours.

From Fig. 3, there were nearly no water diffusion on the absorbent paper while the cotton fabric coated with F/TiO_2 hybrid sol were placed on the paper. It indicated that there was nearly no water in the fabric. The fabrics were irradiated 14 hours for cotton, and the water diffusion was obvious around the fabric. It revealed that there was some water in the fabrics. The coated cotton fabric was irradiated 24 hours, the water stain area on the absorbent paper was increased, which indicated the water was more in the fabric. The permeability of the H_2O was another support to the wettability. Without UV irradiation, the permeable capacities of cotton coated with F/TiO_2 hybrid sol were 0 mL, and water was wholly resisted by the fabric. With UV irradiation for 12 hours, the permeable capacity of cotton fabric was 8.1 mL and continuing to treat in UV light for another 6 hours, the permeable volume increased to 46.0 mL. The permeable volume reached to 79.2 mL when the irradiated time was more than 24 hours. The changes of the permeable capacities were related to the wettability of the F/TiO_2 hybrid coating within irradiation in UV light and revert back in dark.

CONCLUSION

F/TiO_2 hybrid sol was synthesized via sol-gel technology and coated on cotton fabric to present UV-switchable wettability. The contact angle of cotton fabric coated with F/TiO_2 hybrid sol was 141.8°, and it was decreased to 0° by UV irradiation for 24 hours. The contact angle of the coated cotton sample was increased to 133.6° after stored in dark for 72 hours, revealing that the contact angle was nearly completely reverted. The fiber morphology had a significant effect on the hydrophobicity and switchable wettability. The F/TiO_2 showed different valence state in UV irradiation and dark storage. The floatability of powder and fabric, water diffusivity and permeable capacity were significantly affected by UV irradiation or storage in dark.

ACKNOWLEDGMENT

The authors are grateful for the financial support of National Natural Science Foundation of China (51403083), Fundamental Research Funds for the Central Universities (JUSRP11446) and the International Joint Research Laboratory for Advanced Functional Textile Materials at Jiangnan University

REFERENCES

[1] Zhang, J., Lu, X., Huang, W., Han, Y. "Reversible Superhydrophobicity to Superhydrophilicity Transition by Extending and Unloading an Elastic Polyamide Film." *Macromolecular Rapid Communications*, 26, 6, 2005: 477-480.

[2] Guan, K.S., Lu, B.J., Yin, Y.S. "Enhanced effect and mechanism of SiO2 addition in super-hydrophilic property of TiO2 films." *Surface & Coatings Technology*, 173, 2-3, 2003: 219-223.

[3] Wu, J.F., Fernando, S., Jagodzinski, K., Weerasinghe, D., Chen, Z. "Effect of hyperbranched acrylates on UV-curable soy-based biorenewable coatings." *Polymer International*, 60, 4, 2011: 571-577.

[4] Jiang, M.J., Dang, Z.M., Yao, S.H., Bai, J.B. "Effects of surface modification of carbon nanotubes on the microstructure and electrical properties of carbon nanotubes/rubber nanocomposites." *Chemical Physics Letters*, 457, 4-6, 2008: 352-356.

[5] Chang, C.J., Kuo, E.H. "Roughness-enhanced thermal-responsive surfaces by surface-initiated polymerization of polymer on ordered ZnO pore-array films." *Thin Solid Films*, 519, 5, 2010: 1755-1760.

Methodologies for High-Performance Electrospun Fibers

Ton Peijs[1], Jian Yao[1], Cees W.M. Bastiaansen[1,2]
[1]Queen Mary University of London, School of Engineering and Materials Science, London, UK
[2]Eindhoven University of Technology, Chemistry & Chemical Engineering., Eindhoven, The Netherlands
t.peijs@qmul.ac.uk

ABSTRACT
Electrospinning is a rapidly growing polymer processing technology as it provides a viable and simple method to create ultra-fine continuous fibers. This paper presents a review of the mechanical properties of electrospun fibers and particularly focuses on methodologies to generate high strength and high modulus nanofibers.

INTRODUCTION
Over the last two decades, electrospinning, has been receiving renewed interest due to its huge potential of producing polymer nanofibers. Compared to corresponding textile fibers made from the same polymers, the mechanical properties of electrospun nanofibers are often found to be poor. Electrospun nanofibers typically display tensile strengths below 300 MPa and Young's moduli below 3 GPa [1], which can be mainly ascribed to the low degree of orientation and chain extension of the polymer chains along the fiber axis. The objective of this paper is to report on mechanical properties of electrospun nanofibers using different processing concepts to obtain high performance synthetic fibers, focusing on flexible chain versus rigid chain polymers.

BASIC CONCEPTS
In 1932, Staudinger postulated the basic requirements for producing a high strength and high modulus synthetic fiber. In his structural model for such a fiber, all molecular chains should be fully extended and perfectly aligned along the fiber axis. Simultaneously, few chain end defects should exist as the tensile strength is determined by secondary bonds rather than primary bonds. The fully extended and oriented polymer chains will provide ultimate stiffness to the fiber while the few chain end defects as in the case of high molecular weight polymers provides high tenacity, the combination of both making the ideal fiber (Figure 1).

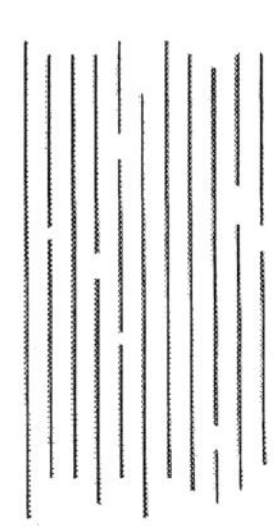

Figure 1. Ideal polymer chains model for producing high strength and high modulus fibers.

It has long been recognized that only extended polymer chains will provide high stiffness. In 1960, Terloar estimated the longitudinal modulus of a single extended polyethylene chain to be 182 GPa in comparison to the Young's modulus of isotropic polyethylene being less than 2 GPa. More detailed calculations showed that the Young's modulus of a single extended polyethylene chain can reach even higher values. Elastic modulus of crystalline regions in the direction parallel to the chain axis have since also been measured experimentally using X-ray diffraction (Table 1), providing great insight into the potential of certain polymers to generate high performance fibers.

Table 1. Theoretical crystal modulus of polymers derived from X-ray diffraction studies [1].

Polymer	Theoretical modulus (GPa)
Polyethylene (PE)	235
Poly(vinyl alcohol) (PVA)	250
Polyamide-6 (PA 6)	157
Poly(ethylene terephthalate) (PET)	108
Polypropylene (PP)	40
Polyacrylonitrile (PAN)	35–55
Thermotropic polyester (Vectran)	126
Poly(*p*-phenylene terehthalamide) (PPTA)	156
Poly(phenylene benzobisoxazole) (PBO)	478

All the above studies indicated that high modulus and high strength in fibers mainly relies on high polymer chain orientation and extension, combined with sufficient chain length. Inspired by this, polymer scientists have attempted to improve the mechanical properties of fibers by pursuing two routes to reach Staudinger's ideal model, i.e., flexible chain polymers and rigid chain polymers.

FLEXIBLE CHAIN POLYMERS
Although it has been shown by various researchers that electrospinning can induce some level of chain orientation in fibers based on flexible chain polymers, these levels are often rather low. In conventional spinning, the solid-state drawing step of the as-spun fiber can be regarded as the single most important process step needed to create such high performance flexible chain fibers. Since the relaxation times of flexible chains are typically very short, the elongational flow induced orientation in as-spun fibers rapidly disappears due to relaxation before solidification. High performance fibers based on flexible chain polymers are therefore typically post-drawn in the solid-state below the melting temperature as this will prevent chain relaxation after orientation and chain extension. As a post-drawing step in commercial electrospinning processes is technologically challenging, the mechanical properties of electrospun fibers solely rely on spinning induced chain orientation which is relatively ineffective compared to solid-state drawing processes. Nevertheless, a certain degree of orientation of

polymer chains has been observed in electrospun fibers by various authors when their diameter is decreased. Arinstein et al. [1], for example, reported an increase in Young's modulus in electrospun nylon 6,6 nanofibers with decreasing fiber diameter (Figure 2). Here the authors ascribed the increase in mechanical properties to orientation of amorphous regions when their sizes are comparable or smaller than the nanofiber diameter. Most studies pursuing property improvements through diameter dependent chain confinement approaches show similar results, i.e., a 3-5 fold increase in Young's modulus of thin electrospun fibers compared to that of bulk polymer. These improvements are however rather modest when compared to property improvements after solid-state drawing, i.e. typically 10-100 times the bulk modulus.

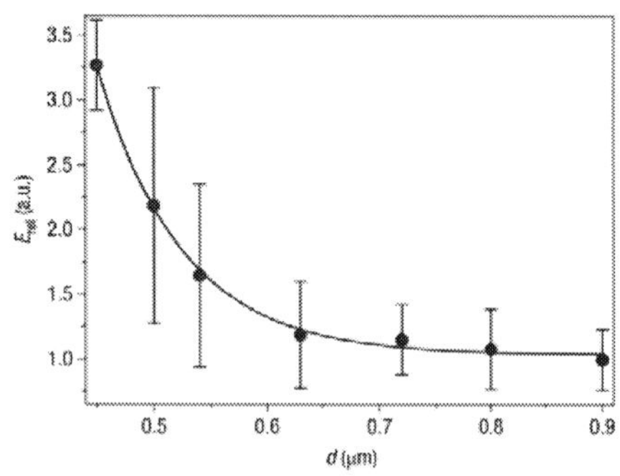

Figure 2. Relative Young's modulus (E_{rel} = E/E_{bulk}) of electrospun PA 6,6 nanofibers as a function of their diameters [1].

RIGID CHAIN POLYMERS

While chain extension in the case of flexible chain polymers can effectively be only achieved by solid-state drawing, the extension of polymer chains can be readily build in by polymer scientists when using rigid rod polymers. Rigid rod polymers have a significant advantage over flexible chain polymers as these systems can be oriented during the spinning process without the need of a post-drawing process to induce chain extension. Electrospinning of poly(p-phenylene terephthalamide) (PPTA) solutions an anisotropic phase has been attempted [2]. *p*-Aramid fibers with diameters between 275 nm and 15 μm were obtained from anisotropic phase solutions, independent of solution concentration. Mechanical properties of the electrospun p-aramid single fibers were evaluated and it was found that Young's modulus and tensile strength increased dramatically with decreasing fiber diameter, suggesting a higher degree of chain orientation in thinner fibers. A maximum Young's modulus and tensile strength of respective ~60 GPa and ~1.1 GPa were found and a geometrical model based on a reduction for the size-dependent Young's modulus for rigid rod polymers was developed to predict the properties of fibers down to the nanoscale [2].

FROM FLEXIBLE TO RIGID CHAIN

As an alternative third method, flexible polyamic acid chain polymers were spun into precursor fibers which were transferred into rigid rod polyimide fibers by heat treatment. Here, a flexible chain polymer is used to create slightly oriented precursor fibers, with chemistry (imidization) being used to transfer the flexible chains into highly aligned and extended semi-rigid chains. Co-polyimide nanofibers were produced by electrospinning followed by imidization from precursor polyamic acid (PAA) nanofibres [3]. The aligned co-polyimide nanofiber mats possessed a modulus, strength and strain-at-break of respective 10 GPa, 1 GPa, and 14%. These co-polyimide nanofibers exhibited similar toughness levels as Kevlar fibers. A novel and efficient way to evaluate mechanical properties of aligned nanofiber bundles (~30 nanofibers in a bundle) by virtue of a micro-tensile tester is also reported. Nanofiber bundles exhibited a Young's modulus of 40 GPa and tensile strength of 1.6 GPa. Further evaluation based on composite data [4] and Weibull statistics resulted in a predicted Young's modulus and tensile strength of 60 GPa and 1.9 GPa for single co-polyimide fibers, which are among the highest ever reported for electropun fibres. An overview of mechanical properties of various electrospun flexible chain polymers, rigid chain polymers and polyimides, in relation to conventional high performance fibers such as UHMW-PE and aramids is shown below in Figure 3.

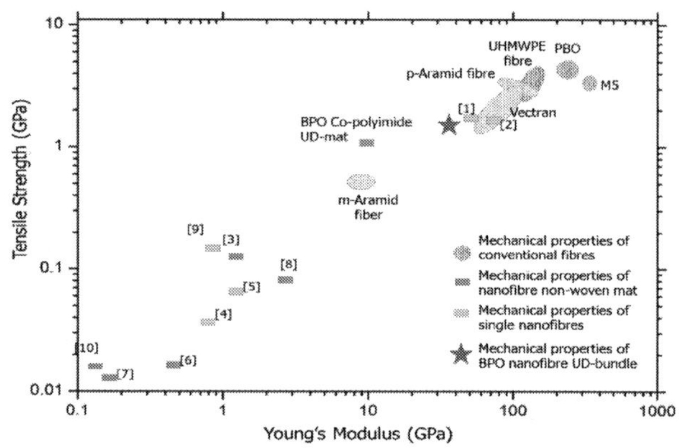

Figure 3. Tensile strength versus Young's modulus of different electrospun fibers and conventional fibers [1].

CONCLUSIONS

As the introduction of a post-drawing step in commercial electrospinning processes may prove technologically challenging, the use of rigid-rod polymers or semi-rigid polyimides as an alternative to flexible chain polymers may be more promising as here chains extension is either already build in or can be induced chemically through a post-heating step.

REFERENCES

1. Yao, J,. Bastiaansen, C.W.M., Peijs, T. "High strength and high modulus electrospun nanofibres." *Fibers*, 2, 2014: 158-186.
2. Yao, J., Jin, J., Pugno, N.M., Bastiaansen, C.W.M., Peijs, T. "Electrospinning of *p*-aramid fibers." Submitted.
3. Yao, J., Pugno, N.M., Bastiaansen, C.W.M., Peijs, T. "Electrospinning of co-polyimide nanofibres." Submitted.
4. Yao, J., Pugno, N.M., Bastiaansen, C.W.M., Peijs, T. "Composites based on co-polyimide nanofibres." Submitted.

Role of Relative Humidity and Ca^{2+} Concentration on Molecular Assembly of Silk Fibroin

Qingfa Peng, Huili Shao, Xuechao Hu, Yaopeng Zhang
State Key Laboratory for Modification of Chemical Fibers and Polymer Materials,
College of Materials Science and Engineering, Donghua University, Shanghai, China
zyp@dhu.edu.cn; pengqingfa08@126.com

ABSTRACT

Silk fibroin heavy chain is made up of hydrophilic and hydrophobic segments. During the dry-spinning process of silk fibroin aqueous solution, the interactions between the hydrophilic segments and water diminished gradually while those among the hydrophobic segments were improved for the silk fibroin assembly. However, the effect of water removal and Ca^{2+} concentration was not well understood. In this paper, we constructed a dry-spinning apparatus mimicking the spinning duct of *Bombyx mori* to investigate how relative humidity and Ca^{2+} concentration affect RSF assembly. The results of FTIR, WAXD and tensile test suggest that low relative humidity and higher Ca^{2+} concentration promote RSF assembly.

INTRODUCTION

Recent advances in our understanding of native spinning process have shown that properties of silk fiber are controlled by a combination of changes in chemistry and extrusion conditions.[1] The feedstock of silkworm is synthesized at the posterior gland, stored in gel state several weeks *in vivo*, and then changed into a dry silk fiber at ambient temperature and pressure in air. During this process, both chemical stimuli (pH and metal ions) and physical effect (shear stress, extensional flow and drawing force) [2, 3] have been realized to play an important role in the molecular assembly, phase transition and formation of tough fiber. Silk fiber is composed of ordered and amorphous fractions, which give rise to outstanding strength and flexibility. Vollrath[4] attributed this to highly controlled nanoscale folding structure and the interaction of silk proteins with water. Researchers also reported that there are huge differences between natural feedstock and regenerated silk fibroin (RSF) either in molecular weight or in rheology.[4] Native dope, sensitive to pH and shear stress, can easily be induced to crystalize with spontaneous removal of water. However, RSF has distinct rheological feature, which shows that RSF molecule is not so sensitive as in native dope.[5]

In this paper, we constructed a new dry-spinning apparatus (Figure 1) according to natural silk gland and duct. We mimicked native silk spinning process both in chemical changes and spinning condition. In addition to these processes, post-drawing process was also adopted to reinforce the as-spun silk fiber.

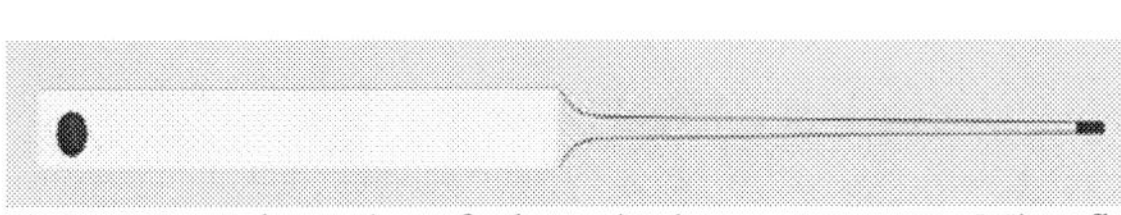

Figure 1. Schematic of dry-spinning apparatus. Microfluidic channel is composed of three segments, including dope store segment (yellow), elongation segment (light) and shear segment (blue).

METHODS

Bombyx mori cocoons were firstly degummed twice using Na_2CO_3 solution. Then degummed fibers were dissolved in 9.0 mol/L LiBr aqueous solution at 40 °C for 2 hours and dialyzed in deionized water for 3 days. When RSF aqueous solution was concentrated to 20 (w/w) %, $CaCl_2$ aqueous solution of 3 mol/L was added into RSF solution to obtain 1.0, 1.6, and 2.4 mmol/g calcium ion in RSF solution. Mechanical properties of fibers were tested using an Instron 5565, with a load cell of 2.5N, at 24±1℃ and (50±5) % RH. Infrared spectra were recorded with a Nicolet iN10 MX (Thermo Fisher, USA) with a resolution of 4 cm^{-1} at 25 ℃ and 50 % RH. Wide-Angle X-ray Scattering (WAXD) patterns were obtained at the BL15U1 beamline at the Shanghai Synchrotron Radiation Facility (SSRF). The wavelength (λ) of the X-ray is 0.07746 nm.

RESULTS AND DISCUSSION

Role of water

It is widely known that water plays an important role in the protein phase transition from sol-gel to silk fiber. Researchers[6] have studied the protein-water interaction through thermal analysis of the secondary structure changes during the removal of water and isothermal crystallization. Evidently this is not sufficient to understand the role of water in controlling the protein structure *in vivo*. Here, we studied the properties and structure of fibers spun at different humidity and RSF concentration. Table I lists the mechanical properties and structure parameters of the resultant fibers. Results show that the diameter of the fiber is sensitive to the surrounding atmosphere condition and RSF concentration.

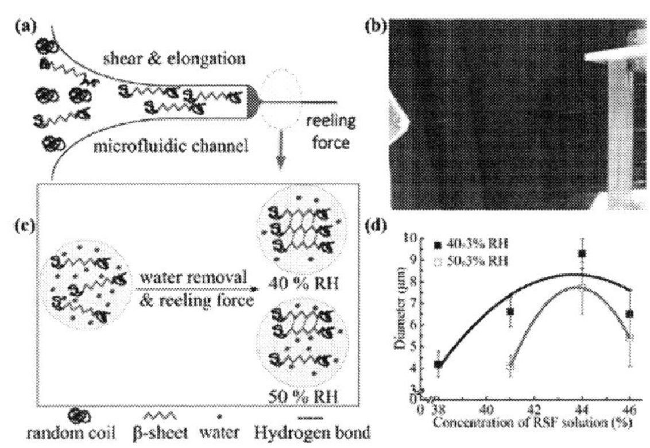

Figure 2. (a) Schematic diagram of dry spinning. Spinning dope flows through the microfluidic channel, in which RSF is subjected to shear and elongation, and then it is reeled onto the roller as shown in (b). (c) Schematic diagram of the effect of RH on the diameter of RSF fiber. At the moment of RSF fiber was reeled into air, high concentration RSF transformed from sol-gel to solid fiber, accompanied with the removal of water. The faster removal of water molecules at 40 % RH will facilitate the formation of

hydrogen bonds between neighboring RSF molecules. And the water molecules that cannot remove promptly at 50% RH will play a role of lubricant in the formation of solid fiber. (d) Diameter of post-treated RSF fiber spun at different RH and RSF concentration.

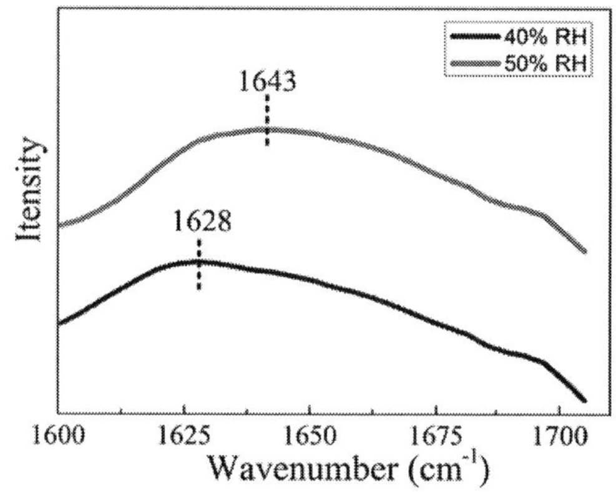

Figure 3. FTIR spectra of post-treated RSF fibers.

When relative humidity was changed from 50 % RH to 40 % RH, diameter of post-treated RSF fiber significantly increased. FTIR peaks shift from 1643 cm-1 to 1628 cm-1, respectively, which corresponds to random coil and β-sheet (Figure 3). Moreover, WAXD data (Table I) shows that the crystallinity of fiber was also increased. Thus we may assume that at low humidity water removes faster than it does at high humidity, and the removal of water benefits the interaction between protein amide groups and facilitates the assembly of adjacent protein molecular to increase fiber diameter.

Role of calcium ion

There is controversy to the role of calcium ion *in vivo* and *in vitro*. Some researchers considered that calcium ions had no significant effect to the formation of silk fibroin aggregates. While other researchers[7] insisted that calcium ions were used to form a stable protein network and promoted gelation of RSF. In this paper, we studied the effect of calcium ion concentration on the properties of RSF fiber with fixed RSF concentration (43 (w/w) %) at 40 % RH. Results in Figure 4 indicate that the fibers were of high strength and good toughness. Fiber diameter increased while birefringence index decreased with the increase of calcium ion concentration (Table II). Calcium ions may facilitate the protein assembly through ionic interaction and decrease the interaction between water and protein molecules.

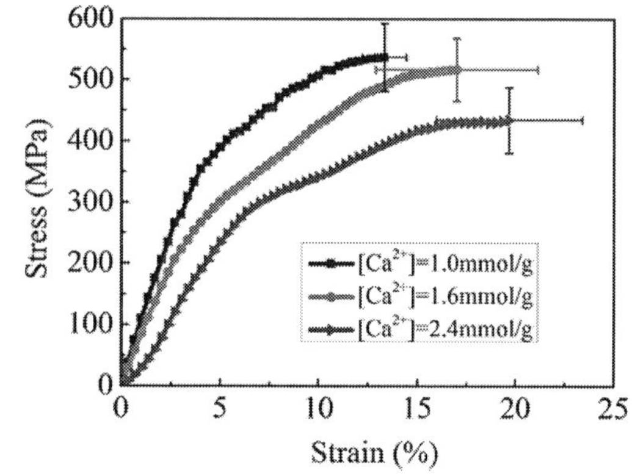

Figure 4. Stress-strain curves of post-treated RSF fibers with different calcium contents.

CONCLUSION

Low relative humidity and higher Ca^{2+} concentration was considered to promote RSF assembly. Shearing and elongation flow together with removal of water promote protein molecules to orient along the flow direction and finally induce phase transition of protein solution to solid fiber.

REFERENCES

1.Omenetto, F.G., Kaplan, D.L. "New Opportunities for an Ancient Material." *Science*, 329, 5991, 2010: 528-531.
2. Jin, H.J., Kaplan, D.L. "Mechanism of silk processing in insects and spiders." *Nature*, 424, 6952, 2003: 1057-1061.
3. Greving, I., et al. "Shear-Induced Self-Assembly of Native Silk Proteins into Fibrils Studied by Atomic Force Microscopy." *Biomacromolecules*, 13, 3, 2012: 676-682.
4. Porter, D., Vollrath, F. "Silk as a Biomimetic Ideal for Structural Polymers." *Advanced Materials*, 21, 4, 2009: 487-492.
5. Holland, C., et al. "Comparing the rheology of native spider and silkworm spinning dope." *Nature Materials*, 5, 11, 2006: 870-874.
6. Hu, X., Kaplan, D., Cebe, P. "Dynamic protein-water relationships during beta-sheet formation." *Macromolecules*, 41, 11, 2008: 3939-3948.
7. Li, Z., et al. "Effect of Metallic Ions on Silk Formation in the Mulberry Silkworm, Bombyxmori." *Journal of Physical Chemistry B*, 109, 35, 2005: 16937-16945.

Table I. Structure and properties of RSF fibers spun at different humidity.

Sample	Diameter [μm]	Strain [%]	Stress [GPa]	Crystallinity [%]	Crystalline orientation factor
C44R40, $[Ca^{2+}]$=1.6 mmol/g	9.0±1.3	20.8±5.5	0.53±0.04	52.7	0.854
C44R50, $[Ca^{2+}]$=1.6 mmol/g	7.5±0.9	17.2±4.3	0.48±0.05	47.8	0.848

Table II. Diameter and birefringence index of RSF fibers spun from the dope with different calcium concentrations.

Sample code	Diameter [μm]	Birefringence index (Δn)
C44R40, $[Ca^{2+}]$=1.0 mmol/g	6.0±0.7	0.0480±0.0039
C44R40, $[Ca^{2+}]$=1.6 mmol/g	7.8±1.2	0.0482±0.0031
C44R40, $[Ca^{2+}]$=2.4 mmol/g	9.8±0.6	0.0404±0.0034

Applications in Medical, Transportation, and Energy

Natural Fiber Spinning, Weaving, and Knitting

Hydrogen-Bonding Polymer Complexation for Low-Dimensional Materials

Shuguang Yang, Sunjia Xing, Chao Su, Xuejian Zhang
State Key Laboratory for Modification of Chemical Fibers and Polymer Materials,
College of Material Science and Engineering, Donghua University, Shanghi, China
shgyang@dhu.edu.cn

ABSTRACT

Polymer complexation is a type of polymer aggregation process. Two initially soluble polymers are bound together through cooperative interactions, such as electrostatic force and hydrogen bonding, along the polymer chains to form a polymer complex which generally precipitates out from the solution. The recently developed layer-by-layer assembly technique can realize the multi-time polymer complexation at interface with a controllable way to prepare functional thin films, coating and microcapsules. Meanwhile, the simple sing-time polymer interfacial complexation has been utilized to construct the special structures, such as vesicle and raspberry-like particles. With special processing, the hydrogen bonding complexation can also apply to produce fibers. This presentation will introduce both single-time and multi-time hydrogen bonding polymer complexation to prepare low-dimensional materials such as micellar particle, fibers, and thin film.

KEYWORDS

Polymer complexation, film, fiber and capsule, hydrogen bond

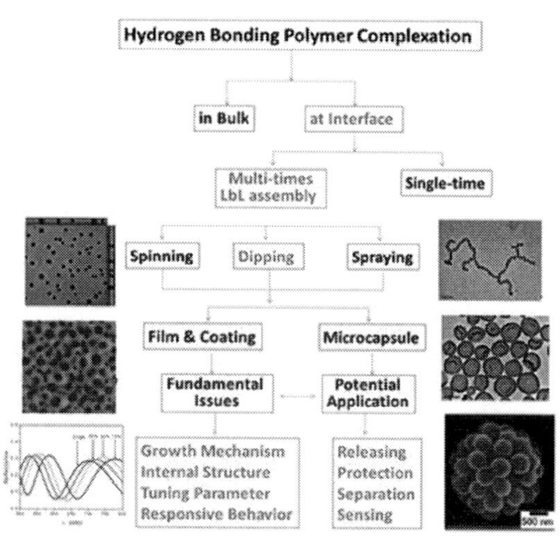

Figure 1. Applying hydrogen bonding polymer complexation to prepare thin film and microcapsule

ACKNOWLEDGMENT

S. Y. gratefully acknowledges the support from National Natural Science Foundation of China (NSFC, Grant No. 51373032), Innovation Program of Shanghai Municipal Education Commission, Fundamental Research Funds for the Central University and DHU Distinguished Young Professor Program.

REFERENCES

[1] Wang, C., Yang, S., Yu, X., Zheng, J.X., Ma, J., Xu, J., Zhu, M. *Soft Matter*, 8, 2012: 10307-10313.
[2] Ma, J., Yang, S., Li, Y., Xu, X., Xu, J. *Soft Matter*, 7, 2011: 9435-9443.
[3] Sun, J., Su, C., Zhang, X., Yin, W., Xu, J., Yang, S. *Langmuir*, 31, 2015: 5147-5154.
[4] Yang, S., Ma, S., Wang, C., Xu, J., Zhu, M. *Australian Journal of Chemistry*, 67, 2014: 11-21.
[5] Yang, S., Yu, X., Wang, L., Tu, Y., Zheng, J.X., Xu, J., Van Horn, R. M., Chen, S.Z.D. *Macromolecules*, 43, 2010: 3018-3026.
[6] Yang, S., Zhang, Y., Xu, J. *Soft Matter*, 3, 2007: 463-469.
[7] Yang, S., Zhang, Y., Guan, Y., Tan, S., Zhang, X., Cheng, S., Xu, J. *Soft Matter*, 2, 2006: 699-704.
[8] Yang, S., Zhang, Y., Wang, L., Hong, S., Xu, J., Chen, Y., Li, C. *Langmuir*, 22, 2006: 338-343.
[9] Yang, S., Zhang, Y., Yuan, G., Zhang, X., Xu, J. *Macromolecules*, 37, 2004: 10059-10062.

Directly Deposited Si NP/CNT/PVA Nanofiber Anodes for Li-Ion Battery Application

Yong Seok Kim, Ghazal Shoorideh, Yong Lak Joo
School of Chemical & Biomolecular Engineering, Cornell University, Ithaca, NY, USA
ylj2@cornell.edu

ABSTRACT

We directly prepared a novel hybrid anode of modified polyvinyl alcohol (PVA) polymer/silicon nanoparticles (Si NPs)/multiwall carbon nanotubes (CNTs) nanofibers on the current collector via water-based electrospinning. This environmental-friendly and feasible direct deposit method using water as solvent enables to eliminate using of both toxic solvents like N–Methyl–2–pyrrolidone and by-passes conventional paste preparation steps such as sonication, blending, grinding and doctor-blade coating. From modifying PVA polymer in the PVA/Si/CNTs by heat treatment under air atmosphere, the hybrid Si assemblies have very promising battery performance by solving the decay of Si nanoparticles. The half-cells using hybrid Si assemblies exhibited an outstanding initial discharge capacity of 4,100 mAh/g cycled at 0.25 A/g due to the synergy by well-dispersed Si NPs and CNTs in the PVA matrix. At a fast charging/discharging rate of 2.5 A/g, the hybrid Si assemblies showed excellent capacity retention of about 93% for 300 cycles. Such improvement of cycle life should be attributed to supporting and protecting Si NPs by modified PVA polymer, which suppresses the volume expansion or pulverization of the Si NPs and the formation of solid electrolyte interface (SEI) layers on the Si surfaces. Furthermore, the hybrid Si assemblies have very high capacities at fast charge/discharge rates such as 1,193 mAh/g at 5 A/g, 945 mAh/g at 7.5 A/g and 708 mAh/g at 12.5 A/g owing to good cooperation highly conductive carbon nanotubes in this hybrid system. The full cell using hybrid anode and $LiCoO_2$ cathode possesses a stable capacity retention of 100% during 100 cycles.

INTRODCUTION

Despite its highest theoretical gravimetric capacity, the Si anodes have still suffered from dramatic capacity fading by three kinds of critical reasons: severe volume expansion of about 400% leading to pulverization of Si anodes or electrical disconnect with copper current collectors, formation of unstable SEI by decomposing organic carbonate electrolytes on the Si surfaces, and poor intrinsic electrical conductivity of silicon itself resulting in poor rate capability. To overcome such capacity degradation and poor charge transfer, the silicon nanostructures have been hybridized with carbon nano-architectures or facile doping until now [1, 2]. Herein, we developed a novel feasible approach of water-based electrospinning to achieve both high throughput and scalable hybrid Si anodes using cost-effective PVA polymer with multiwall CNTs, as depicted in Fig. 1.

EXPERIMENTAL

0.5g of Si NPs with 100 nm in diameter (MTI Co.) were blended with the PVA/CNTs aqueous solutions which were loaded into a syringe and ejected by a pump at a feeding rate of 6 ml/min toward a vertically placed Cu foil collector that was 12 cm away from the needle of the syringe. The as-spun materials were stabilized at 200 °C to crosslink the PVA polymer in the hybrid system. SEM, TEM, TGA, and FT-IR were used to characterize the fiber morphology, distribution of Si NPs and CNTs, Si content, and degree of cross-linking of PVA. The hybrid electrodes were combined as a working electrode with Li metal as a counter electrode and a polyethylene separator to fabricate 2032 coin-type half cell batteries for examining electrochemical properties and battery performance. The homemade electrolyte composed of 1 M $LiPF_6$ with dimethylene carbonate and fluoroethylene carbonate (50:50 wt/wt%) was used for the half or full Li-ion batteries. The electrochemical properties of hybrid Si anodes were characterized by electrochemical impedance spectroscopy (PARSTAT 4000, Princeton Applied Research) and cyclic voltammetry (BST8–STAT, MTI Co.). Galvanostatic charge/discharge measurements were performed under a cutoff voltage window from 0.01 to 1.5 V versus Li/Li+ by battery cyclers (from MTI Co.). $LiCoO_2$ cathodes were obtained from MTI Co. for full cell fabrication. The full cells were tested by a cutoff capacity of 1,200 mAh/ganode, which was already obtained from the half-cell tests using hybrid Si assemblies.

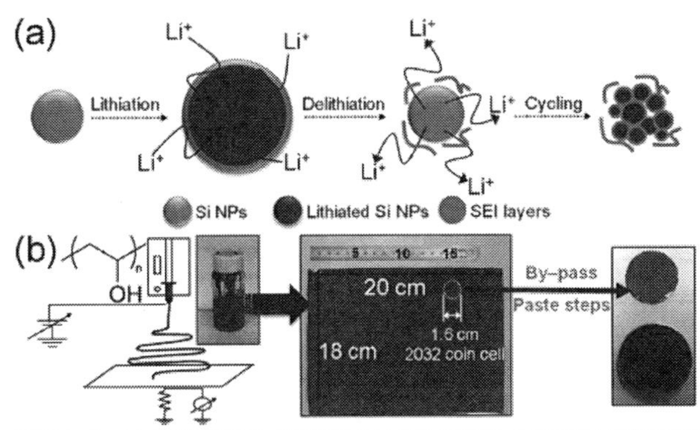

Fig. 1. Decay mechanism of Si NPs anodes during cycling; (b) procedure of direct–deposited hybrid Si assemblies via water-based electrospinning for coin-typed Li-ion batteries

RESULTS AND DISCUSSION

Characterization of Si NP/CNT/PVA nanofiber anode: As–spun materials of hybrid Si assemblies before heat treatment have 1–D morphology from PVA nanofibers incorporating Si NPs as shown in Fig. 2a, and their average diameter size is around 200 nm. After modification of PVA nanofibers, the morphology of hybrid Si anodes is greatly changed to sphere-type assemblies. In Fig. 2b of low magnification SEM image, each sphere is networked by shrunk PVA nanofibers and carbon nanotubes. Figure 2c showed the final hybrid assemblies are composed of Si NPs with CNTs having ca. 10 nm diameter size. Owing to electrospinning process and the modified PVA polymer, the Si NPs were not agglomerated. In the TEM image of Fig. 2d, we found that the layer of PVA polymer is covering and supporting the Si NPs, which are being protected from their severe volume expansion or pulverization, and formation of SEI layers on their surfaces.

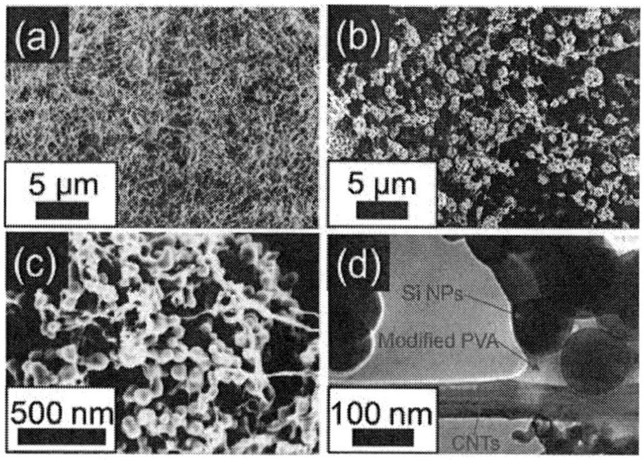

Fig 2. SEM images of as–spun hybrid Si assemblies on current collector (a) before modification; (b) low and (c) high magnification SEM images after heat treatment, (d) TEM image of hybrid Si assemblies after modification.

Battery performance and electrochemical properties: As shown in Fig. 3a, the initial lithiation capacity of hybrid Si assemblies is 4100 mAh/g as seen, while Si NPs alone have only 2500 mAh/g. Such increment would be contributed to efficiently disperse Si NPs from preparation by both electrospinning solution with ultrasonication and electrospinning process. In the investigation of battery cycle life in Fig. 3b, our hybrid Si assemblies have a reversible capacity of around 1,200 mAh/g tested at 2.5 A/g with 98% Coulombic efficiency, and retain 92% capacity retention calculated from the initial capacity to the final capacity after 300 cycles. Fig. 3c and 3d showed capacity/voltage profiles and rate capability performance at variable testing current densities from 0.25 to 12 A/g. They presented high charge/discharge capacities from slow to fast charge/discharge rates without dramatic capacity losses as seen in Fig. 3c. Especially, the capacity at 12.5 A/g for hybrid Si assemblies is ca. 650 mAh/g in Fig. 3d.

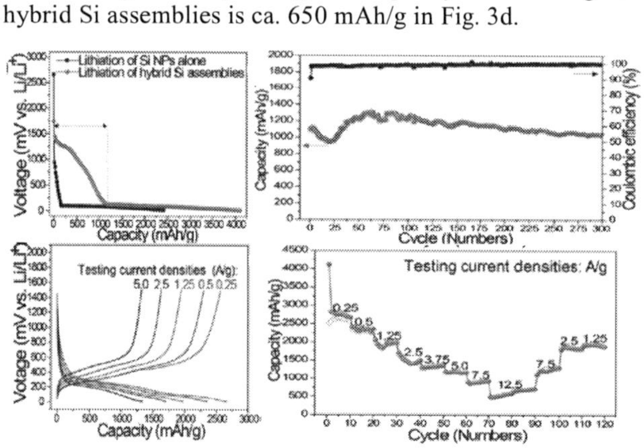

Fig 3. (a) Initial discharge curves of Si NPs alone and hybrid Si assemblies tested at 0.25 A/g; (b) cycle life of half-cell using hybrid Si assemblies tested at 2.5 A/g, (c) charge/discharge curves and (d) rate capability of half-cells using hybrid Si assemblies tested at variable current densities from 0.25 A/g to 12.5 A/g.

Full cell performance: In Fig. 4a, hybrid Si assemblies represented a constant cycle life of 100% during 100 cycles. However, the Si NPs alone in Fig. 4b exhibited stable cycle retention for only 20 cycles. Fig. 4c plotted Coulombic efficiencies (CE) of the full cells using hybrid Si assemblies and Si NPs alone. Our hybrid Si anodes have a bit better Coulombic efficiency of 98.3% than Si NPs alone (95.5%) although the hybrid anodes include non-conductive PVA polymer. It would be attributed to facilitating charge transfer feature by 1-D carbon nanotubes in the hybrid system. In conclusion, we demonstrated that the directly deposited fabrication of electrodes via water-based spinning is a quite facile and scalable approach to commercialize high-performance silicon anodes.

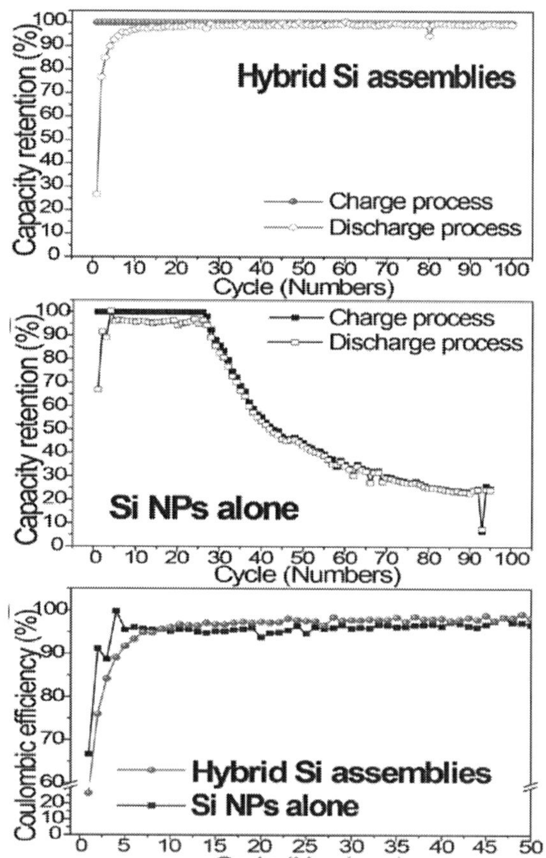

Fig 4. Full cell capacities using a) Si NP hybrid nanofibers, b) Si NPs only, and c) Coulombic efficiencies.

REFERENCES

[1] Kim, Y.S., Kim, K.W., Cho, D., Hansen, N.S., Lee, J., Loo, Y.L. *Chem Electro Chem*, 1, 2014: 220.

[2] Kim., Y.S., Kim, W.B., Joo, Y.L. *J. Mater. Chem. A*, 2, 2014: 8323.

Membranes of MnO Beading in Carbon Nanofibers as Flexible Anodes for High-Performance Lithium-Ion Batteries

Xin Zhao, Yuxuan Du, Qinghua Zhang
College of Material Science & Engineering, State Key Laboratory for Modification of Chemical Fibers and Polymer Materials, Donghua University, Shanghai, China
zhao@dhu.edu.cn

INTRODUCTION

Flexible lithium-ion batteries (LIBs) hold great promise for the next generation energy sources of future electronic devices. They have found a wide variety of promising applications including smart clothes, electronic skins and wearable sensors and so on[1-2]. Due to the low cost and easy processing, carbon nanofibers (CNFs) membrane has advantages for being used as a free-standing and flexible substrate to construct metal or metal oxides/carbon composites[3]. The electrospinning process followed by stabilization and carbonization has been proved to be a facial and controllable way for fabricating CNFs membranes[4].

Herein, we demonstrated the synthesis of MnO nanoparticles (NPs) beading in CNFs membranes by introducing MnO_2 nanowires (NWs) into the polyacrylonitrile (PAN) solution through electrospinning followed by stabilization and carbonization processes. Structural transformation from MnO_2 NWs to MnO NPs was accompanied by the simultaneous conversion of amorphous polymer to porous graphitic CNFs. The obtained membranes with MnO NPs uniformly embedded in the porous CNFs could be used directly as binder-free electrode for flexible LIBs. Effects of the loadings of MnO NPs in CNFs on the electrochemical performance of the composites were also investigated. With a preferable loading of MnO (34.6 wt%), the MnO/CNFs (MnC) membrane exhibits a high reversible capacity of 987.3 mAh g^{-1} after 150 discharge/charge cycles at 0.1 A g^{-1}, a good rate capability (406.1 mAh g^{-1} at 3 A g^{-1}) and an excellent cycling performance (655 mAh g^{-1} over 280 cycles at 0.5 A g^{-1}). Besides, the hybrid membrane shows good electrochemical performance at bending state, demonstrating a great potential as a high performance anode material for flexible LIB applications.

APPROACH

As shown in Scheme 1, the MnO_2 NWs powder was added into a certain-mass 8 wt% PAN/DMF solution as an electrospinning solution. Intensive stirring was conducted for 12 hours in order to get a homogeneously distributed solution. Then the blended solution was electrospun into MnO_2/PAN (MnP) membranes, with a controlled syringe pump of 25 μL min^{-1} and an applied voltage of 20 kV with a distance of 20 cm between the electrospinning jet and the collector. The adding amounts of MnO_2 NWs based on PAN were 10 wt%, 20 wt% and 30 wt% for MnP-1, MnP-2 and MnP-3, respectively.

The as-prepared MnP membranes were firstly performed at 280 °C in air for 2 hours for a stabilization process. Next, the temperature increased to 1000 °C with a heating rate of 3 °C/min and stayed for 1 hour in N_2 atmosphere. Then the MnC membranes were obtained after the furnace cooling down to the room temperature. During both the stabilization and carbonization processes, the membranes were sandwiched between two aluminum oxide (Al_2O_3) plates to give certain tensions against the planer dimensional shrinkage. For the controlled experiment, MnO_2 NWs powder was treated with the similar procedure in a Al_2O_3 container.

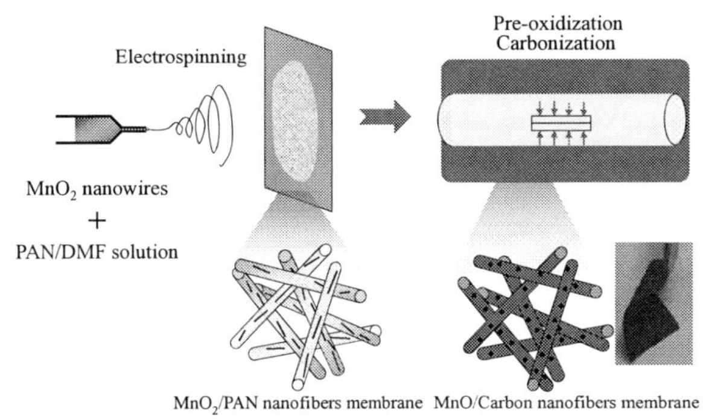

Scheme 1. Schematic illustration for the preparation process of free-standing MnO/carbon nanofiber membranes.

RESULTS AND DISCUSSION

The as-spun nanofibers in MnP membranes [Figures 1 (a) to (c)] have uniform diameters of 200-300 nm and lengths of up to several micrometers. Before thermal treatment, MnO_2 NWs are embedded in the co-axial nanofibers. Some MnO_2 NWs bundles could be observed in MnP-2 and MnP-3 [insets of Figure 1 (b) and (c)]. After carbonization, the morphologies and diameters of nanofibers in MnC membranes are well maintained. Interestingly, the MnO_2 NWs are broken into MnO NPs with a typical diameter of 20-30 nm, leading to the formation of MnO NPs embedded in the CNFs [insets of Figure 1 (d) and (e)]. Few additional particles were observed on the exterior surface of nanofibers for MnC-1 and MnC-2, but a large addition of MnO_2 led to an accumulation of MnO NPs on the external structure for MnC-3 [Figure 1 (d)].

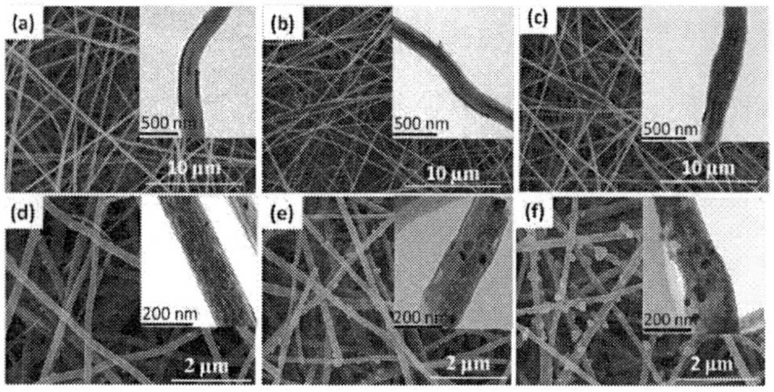

Figure 1. SEM images and TEM images (inset) for (a) MnP-1, (b) MnP-2, (c) MnP-3 and (d) MnC-1, (e) MnC-2, (f) MnC-3.

Electrochemical tests showed that with appropriate contents of MnO NPs, *i.e.*, 34.6 wt% in our work (MnC-2), a favorable porous structure with the active materials being well confined in CNFs are obtained, leading to an improvement of Li reaction kinetics. The resultant flexible MnC membrane demonstrates excellent electrochemical properties (Figure 2) with high specific capacitie of 987.3 mAh g^{-1} after 150 discharge/charge cycles at 0.1 A g^{-1}, good rate performance (406.1 mAh g^{-1} at 3 A g^{-1}) and cycling stability (655 mAh g^{-1} over 280 cycles at 0.5 A g^{-1}).

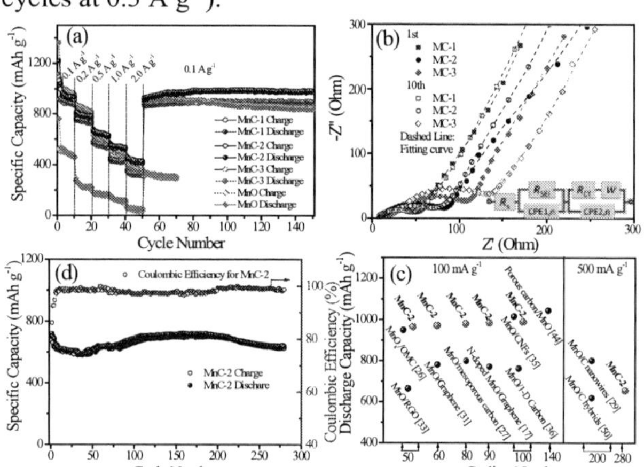

Figure 2. (a) rate capacities at different current densities and capacity at different cycles for MnC and MnO; (b) Nyquist curves of MnC samples at 1st and 10th cycle and their respective fittings with an appropriate electric equivalent circuit; (c) cycling performance of MnC-2 with columbic efficiency; (d) a comparison of the discharge capacities of MnC-2 with reported MnO-based materials at various cycles.

Furthermore, a flexible cell was assembled and tested to obtain the effect of bending on the electrochemical performance (Figure 3). It is obvious that the capacity of the battery being bent have a negligible decrease compared with that of the original flat battery. Moreover, the flexible battery showed a good cyclic stability both under flat and bent states. With respect to the original capacity, it showed a capacity retention of ~93.8% after the first 12 cycles under a flat state, and ~84% after another 20 cycles under a bent state. Such a facial and practical fabrication process of MnC membranes might be an efficient route to design and fabricate high performance flexible electrode materials for LIBs.

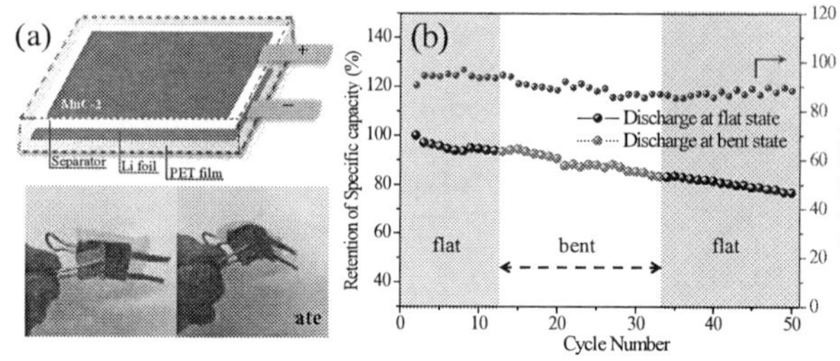

Figure 3. (a) Scheme of assembled flexible cell encapsulated by PET film and photograph of the battery tested at flat and bent states; (b) cyclic performance of the battery under flat and bent states.

CONCLUSION

In summary, we have successfully fabricated MnC membranes through incorporating MnO_2 NWs into the electrospun solution by a facial and practical electrospinning technique followed by stabilization and carbonization processes. During the thermal treatment, structural changes from MnO_2 NWs to MnO NPs occurred accompanying with the conversion from an amorphous state of polymer to a graphitic structure of CNFs. The resultant MnC architecture with MnO NPs being well embedded in the porous carbon structures not only facilitates the transport of both electrolyte ions and electrons to the electrode surface, but also enhances the utilization of the active materials. We believe that such low cost, high-performance hybrids fabricated by earth-abundant and environmentally friendly materials and scalable electrospinning techniques can offer a great promise in grid-scale electrode materials for flexible LIBs.

ACKNOWLEDGMENT

The National Natural Science Foundation of China (51403036), Shanghai Education Commission "Yangfan" Program (14YF1405300); Shanghai Education Development Foundation "Chenguang" Program (14S10656); Specialized Research Fund for the Doctoral Program of Higher Education of China (SRFDP 20130075120018); Fundamental Research Funds for the Central University (2232014D3-09).

REFERENCES

1. Gwon, H., Hong, J., Kim, H., Seo, D.-H., et al. *Energy Environ. Sci.*, 7, 2014: 538-551.
2. Zhou, G., Li, F., Cheng, H.-M. *Energy Environ. Sci.*, 7, 2014: 1307-1338.
3. Zhang, Y., Bai, W., Cheng, X., Ren, J., Weng, W., et al. *Angew. Chem. Int. Edit.*, 53, 2014: 14564-14568.
4. Jia, X., Kan, Y., Zhu, X., Ning, G., Lu, Y., Wei, F. *Nano Energy*, 10, 2014: 344-352.

Carbon Nanotube Sheet Electrodes for Anisotropic Actuation of Dielectric Elastomers

Xiaomeng Fang, Enes Cakmak, Ozkan Yildiz, Philip D. Bradford, Tushar K. Ghosh
Fiber and Polymer Science Program, Textile Engineering, Chemistry and Science Department,
North Carolina State University, Raleigh, North Carolina, USA
tghosh@ncsu.edu

OBJECTIVE

The performance of dielectric electroactive polymer (D-EAP)-based actuators depends critically on the electrode characteristics. Among the most challenging issues in the application is the device-level complexity in producing sufficient directional actuation at acceptably low electric fields. In this work, a simple carbon nanotube (CNT) based electrode for D-EAP actuators is demonstrated to vastly improve directional strain response originating from the mechanical anisotropy of the electrode material.

INTRODUCTION

Electroactive polymers (EAP) constitute a class of macromolecules that undergo a dimensional change in response to an applied electric potential[1–2] As a growing class of EAPs that shows great promise in biomimetic actuation, D-EAPs are considered well-suited for many applications including energy harvesters[3], and prosthetic devices[4]. They are inexpensive, lightweight, easily processable, and show large isotropic areal actuation strains (50-380%), large range of energy densities (0.01-3MJ m^{-3}), high electromechanical coupling efficiencies (~90%), over a broad range of electric fields (75-400 V μm^{-1})[2,5]. The real world application of D-EAPs has still been limited by a number of constraints; with the most important being the challenge of obtaining high directional electromechanical response at acceptable levels of electric field.

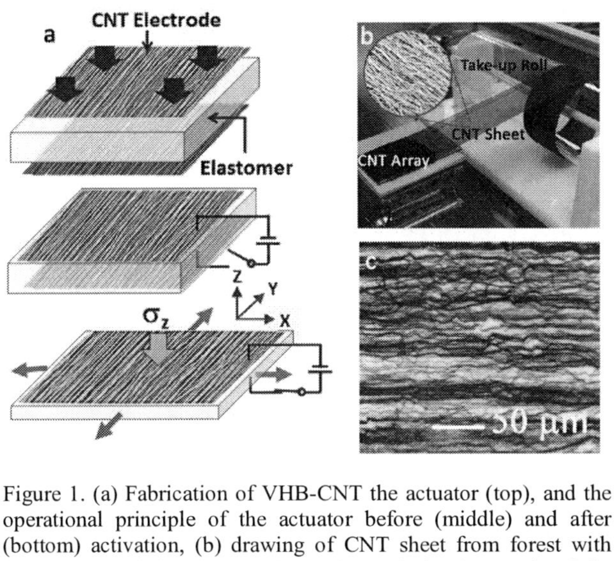

Figure 1. (a) Fabrication of VHB-CNT the actuator (top), and the operational principle of the actuator before (middle) and after (bottom) activation, (b) drawing of CNT sheet from forest with SEM image of the sheet as inset, (c) optical micrograph of the electrode area of an actuator.

Here, for the first time, we report a simple bifunctional CNT-based electrode that serves both as an electronic conductor and a mechanically anisotropic constraint to produce directional actuation. We also show that these CNT fibers penetrate the D-EAP on actuation cycling and thereby self-reconfigure toward a more stable nanolayer on the surface.

APPROACH

Sample preparation

We fabricated continuous sheets of entangled multiwall CNTs in a fibrillar network made up of mostly aligned bundles drawn out horizontally from vertically aligned CNT forests grown on quartz substrates, as shown in Figure 1b-c. The CNT sheets were then applied to uniformly prestrained acrylate adhesive films[5,6] (VHB-4905, 3M Corp.) to form the actuators (hereinafter described as "VHB-CNT") for electromechanical evaluation.

Actuator Characterization

Uniaxial tensile tests on the actuator samples were performed on a MTS-30G load frame. Low magnification morphological features of the actuator specimens were investigated using an optical microscope (Nikon, Eclipse 50i). Hig magnification images were collected on a field emission scanning electron microscope (JEOL 6400F), and a transmission electron microscope (HITACHI HF2000). Images of specimens under various levels of strain were obtained using the optical microscope, immediately after application of strain on a specially designed extension stage. Electrical properties of the electrodes on the VHB-CNT actuators, in both x- and y-directions were measured using a customized four-point probe connected to a current source and a voltmeter. Field-induced actuation response characteristics of the VHB-CNT test samples were evaluated by connecting the actuator electrodes to a high-voltage power supply with a computer controlled automated voltage trigger.

RESULTS AND DISCUSSION

Mechanical Behavior

To explore the electrode induced mechanical anisotropy of the VHB-CNT actuators as well as their ability to withstand cyclic deformation, we performed multi-cycle uniaxial tensile tests. The initial low modulus of the CNT fibers along the fiber (y) direction quickly increases to ~28.0 MPa because the aligned CNT fibers contribute to the load before significant lowering of modulus is noted due to failure of CNT fibers, see Figure 2a. The stress-strain behavior in the cross fiber direction (x), seen in Figure 2b, is unremarkable and the modulus (2.3MPa) is slightly higher than that of the neat VHB film.

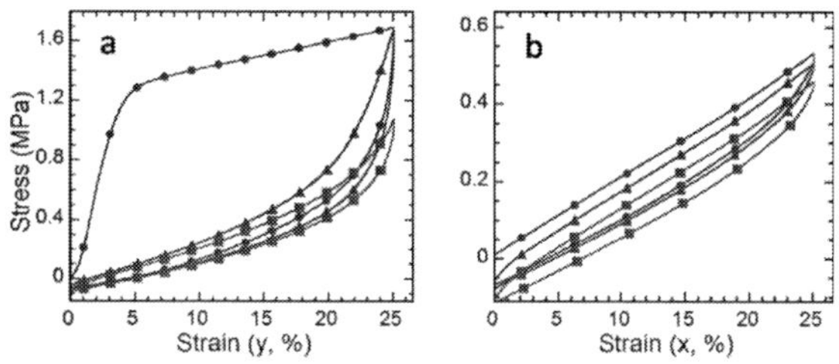

Figure 2. Cyclic tensile stress-strain curves of VHB-CNT actuators, (a) along the fiber (y) direction, (b) along the cross (x) direction. For clarity, data for 1st (●), 2nd(▲), and 25th(■) cycles are shown.

Morphological Characteristics

Morphological changes of VHB-CNT actuators upon extension cycling are presented in Figure 3. Optical microscopy images of the CNT sheet electrodes at various strain levels in the y-direction reveal lines of fracture, visible at 10% strain (Figure 3a). Close-up view of the fracture boundaries shows bridging CNT fibers across the fracture. On recovery, the CNT fibers form distinct kink bands in the direction normal to the applied strain (Figure 3b). We also note that the highly drawn-out surviving CNT fibers, near the fracture, that form distinct clusters of more tightly folded undulations and kink bands (Figure 3c). The result of strain cycling in the x-direction produced somewhat similar changes in the electrode morphology, without any evidence of fiber failure (Figure 3d).

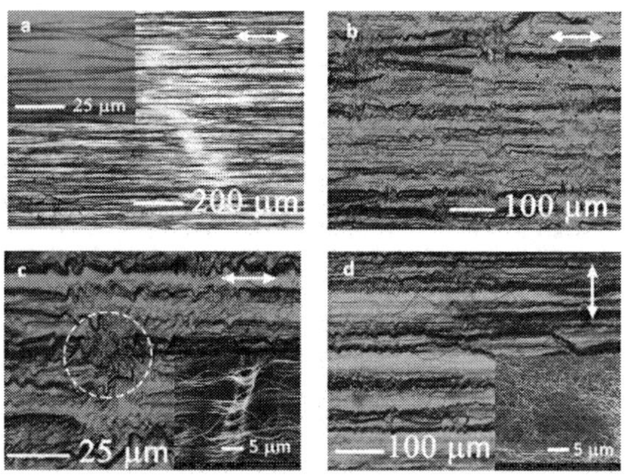

Figure 3. Optical images of electrodes showing morphological changes of CNT sheet electrodes, (a) sample under 10% tensile strain in the y-direction, (b) sample after 25 cycles of 25% strain along the y-direction, (c) CNT fiber clusters of tightly folded undulations, (d) sample after 25 strain cycles of 25% strain in the x-direction. The arrows show direction of applied starin.

Electromechanical Properties

The linear actuation strain in x-direction for the VHB-CNT system, presented in Figure 4a, is consistently greater than that in y-direction. The resulting actuation anisotropy (defined as the strain ratio Sx/Sy) values of about 7.5 compares well with the measured mechanical anisotropy of about 10. On actuation cycling for more than 100 cycles, the actuation strain increased after the first cycle and remained at approximately the same level, strongly suggesting no significant change in the electrode characteristics during actuation (Figure 4c). Figures 4d-f illustrate the anisotropic deformation of the actuator.

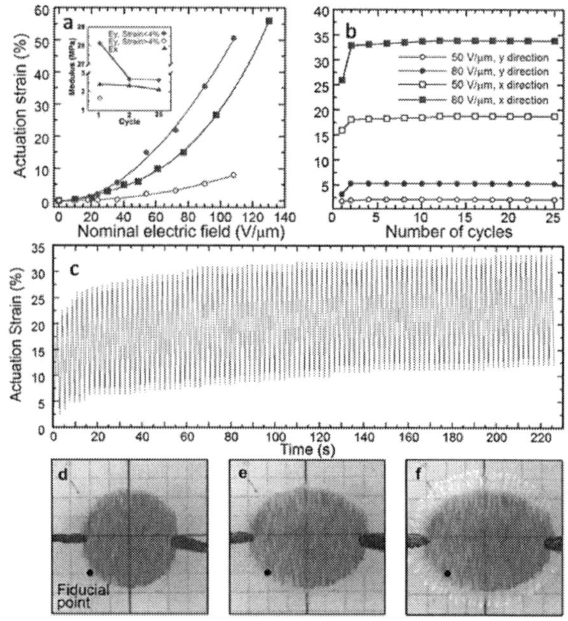

Figure 4. Linear actuation strain plotted, (a) in the fiber (○) and cross-fiber (●) directions together with directionally uniform response of a similar actuator with carbon-grease electrode (■), (b) actuation strain as a function of actuation cycles, (c) actuation strain in the cross direction (x) for the initial 100 cycles of the 370 cycles. Video capture images illustrate a test specimen, (d) before actuation, (e) at 80 V μm^{-1}, and (f) at 100 V μm^{-1}.

CONCLUSION

The electromechanical properties of the VHB-CNT actuators establish that the highly oriented CNT sheets provide an attractive alternative to conventional electrodes for D-EAPs applications. The synergistic property development results in high directional actuation strain.

FUTURE WORK

Optimization of electrical and mechanical properties of the CNT sheet is expected to improve the overall performance.

ACKNOWLEDGMENT

The authors acknowledge the U.S. Department of Commerce through the National Textile Center, and Analytical Instrumentation Facility (AIF) at North Carolina State University, which is supported by the State of North Carolina and the National Science Foundation.

REFERENCES

[1] Bar-Cohen, Y. *Electroactive Polymer (EAP) Actuators as Artificial Muscles*, SPIE Press, 2004.
[2] Pelrine, R., Kornbluh, R., Pei, Q., Joseph, J. *Science*, 287, 2000: 836.
[3] Ahnert, K., Abel, M., Kollosche, M., Jørgensen, P.J., Kofod, G. *J. Mater. Chem.*, 21, 2011: 14492.
[4] Biddiss, E., Chau, T. *Med. Eng. Phys.*, 30, 2008: 403.
[5] Shankar, R., Ghosh, T.K., Spontak, R.J. *Soft Matter*, 3, 2007: 1116.
[6] Shankar, R., Ghosh, T.K., Spontak, R.J., *Adv. Mater.*, 9, 2007: 2218.

Design of Electrospun Hybrid Nanofibrous Materials for Biomedical Applications

Shige Wang[1], Fuyin Zheng[2], Yili Zhao[3], Yu Luo[2], Xiangyang Shi[1,2,3]
[1]College of Materials Science and Engineering, [2]College of Chemistry, Chemical Engineering and Biotechnology,
[3]College of Textiles: Donghua University, Shanghai, China
xshi@dhu.edu.cn

Electrospinning is a powerful technique to produce fibers with a diameter ranging from tens of nanometers to several micrometers. Compared with single-component nanofibers, composite or hybrid nanofibers are promising due to the unique properties possessed by both the host and the guest materials. Doping nanoparticles (NPs) or nanotubes (NTs) having excellent optical, mechanical, electrical, or catalytic properties within polymer nanofibers makes it possible to produce functional nanofibers with promising applications, particularly in the fields of biomedical applications. This talk presents the results obtained in our group in the past several years in the formation of various organic/inorganic hybrid electrospun nanofibers for biomedical applications. We used electrospun poly(lactic-co-glycolic acid) nanofibers incorporated with different NPs (halloysite nanotubes, nanohydroxyapatite, or laponite nanodisks) for tissue engineering and drug delivery applications. The incorporation of these inorganic components within PLGA nanofibers is helpful to promote protein adsorption onto the nanofibers for enhanced cellular response, to stimulate osteogenic differentiation of stem cells, and to provide a double carrier system to significantly avoid the burst release profile of the encapsulated drug molecules. Likewise, via a facile crosslinking strategy or an assembly method, nanofibers can be modified with specific ligands, rendering them with an ability to capture cancer cells with high specificity.

KEYWORDS
Electrospun nanofibers; Hybrid nanomaterials; Tissue engineering scaffolds; Stem cell differentiation; Cancer cell capture.

ACKNOWLEDGMENT
This research is financially supported by the High-Tech Research and Development Program of China (2012AA030309) and the Program for Professor of Special Appointment (Eastern Scholar) at Shanghai Institutions of Higher Learning. Y. Zhao thanks the Innovation Funds of Donghua University Doctorate Dissertation of Excellence (101-06-0019014).

REFERENCES

(1) Wang, Shige, Cao, Xueyan, Shen, Mingwu, Guo, Rui, Bányai, István, Shi, Xiangyang. "Fabrication and morphology control of electrospun poly(γ-glutamic acid) nanofibers for biomedical applications." *Colloid Surf. B: Biointerfaces*, 89, 2012: 254-264.

(2) Wang, Shige, Castro, Rita, Song, Chenlei, Luo, Yu, Shen, Mingwu, Tomás, Helena, Zhu, Meifang, Shi, Xiangyang. "Electrospun laponite-doped poly(lactic-co-glycolic acid) nanofibers for osteogenic differentiation of human mesenchymal stem cells." *J. Mater. Chem.*, 22, 2012: 23357-23367.

(3) Wang, Shige, Zheng, Fuyin, Huang, Yunpeng, Fang, Yuting, Shen, Mingwu, Zhu, Meifang, Shi, Xiangyang. "Encapsulation of Amoxicillin within Laponite-Doped Poly(lactic-co-glycolic acid) Nanofibers: Preparation, Characterization and Antibacterial Activity." *ACS Appl. Mater. Interfaces*, 4, 2012: 6393-6401.

(4) Zheng, Fuyin, Wang, Shige, Wen, Shihui, Shen, Mingwu, Zhu, Meifang, Shi, Xiangyang. "Characterization and antibacterial activity of amoxicillin-loaded electrospun nano-hydroxyapatite/poly(lactic-co-glycolic acid) composite nanofibers." *Biomaterials*, 34, 2013: 1402-1412.

(5) Zheng, Fuyin, Wang, Shige, Shen, Mingwu, Zhu, Meifang, Shi, Xiangyang. "Antitumor efficacy of doxorubicin-loaded electrospun nano-hydroxyapatite/poly(lactic-co-glycolic acid) composite nanofibers." *Polym. Chem.*, 4, 4, 2013: 933-941.

(6) Wang, Shige, Zhu, Jingyi, Shen, Mingwu, Zhu, Meifang, Shi, Xiangyang. "Poly(amidoamine) Dendrimer-Enabled Simultaneous Stabilization and Functionalization of Electrospun Poly(γ-glutamic acid) Nanofibers." *ACS Appl. Mater. Interfaces*, 6, 2014: 2153-2161.

(7) Zhao, Yili, Zhu, Xiaoyue, Liu, Hui, Luo, Yu, Wang, Shige, Shen, Mingwu, Zhu, Meifang, Shi, Xiangyang. "Dendrimer-functionalized electrospun cellulose acetate nanofibers for targeted cancer cell capture applications." *J. Mater. Chem. B*, 2, 2014: 7384-7393.

(8) Wang, Zhe, Zhao, Yili, Luo, Yu, Wang, Shige, Shen, Mingwu, Tomás, Helena, Zhu, Meifang, Shi, Xiangyang. "Attapulgite-doped electrospun poly(lactic-co-glycolic acid) nanofibers enable enhanced osteogenic differentiation of human mesenchymal stem cells." *RSC Adv.*, 5, 2015: 2383-2391.

(9) Luo, Yu, Shen, Fang, Yongxiang, Cao, Yuhua, Huang, Jie, Zhang, Mengxin, Dai, Jianwu, Shi, Xiangyang, Zhang, Zhijun. "Enhanced proliferation and osteogenic differentiation of mesenchymal stem cells on graphene oxide-incorporated electrospun poly (lactic-co-glycolic acid) nanofibrous mats." *ACS Appl. Mater. Interfaces*, 7, 11, 2015: 6331-6339.

Efficient Dissolution of Cellulose in Novel Complex Solvent and Its Spinning

Yue Zhang, Xinda Li, Ya Deng, Muhuo Yu
State Key Laboratory for Modification of Chemical Fibers and Polymer Materials,
College of Materials Science and Engineering, Donghua University
zhangyue@dhu.edu.cn

ABSTRACT

Cellulose is the major carbohydrate produced by photosynthesis of plants, and is therefore the most abundant and widely distributed nature polymer in the world, whose biodegradability and renewability matches the concept of recycling economy and environmental protection. However, cellulose is hardly soluble in conventional solvents because of a large amount of hydrogen bonds[1]. Current solvent system like NMMO and ionic liquid were all excellent solvent[2, 3], but they still suffer some drawbacks like high dissolving temperature, slow rate of dissolution and high viscosity of the solutions obtained, not to mention the high costs. Thus, advances towards improved solvents for dissolving cellulose are still imperative. In this study, in order to achieve desired advantages like nonderivatization, low viscosity, high dissolution speed and low dissolution temperature, we used polar aprotic solvents with addition of few ionic liquid to dissolve cellulose, effects of type and ratio of polar aprotic solvents were investigated by means of conductivity measurement and microscope observation, and the possible mechanism of this dissolving effect was discussed.

KEYWORDS

Cellulose, Ionic liquid, Aprotic solvent, Spinning.

RESULTS AND DISCUSSION

Polarized light microscope was used to observe the dissolving process of cellulose in the complex solvent, as seen in Fig.1. The higher ω_{DMSO} is, the shorter time it needs for dissolving cellulose, and complex solvent with 60 wt% DMSO showed the shortest time of dissolving cellulose at 50 °C.

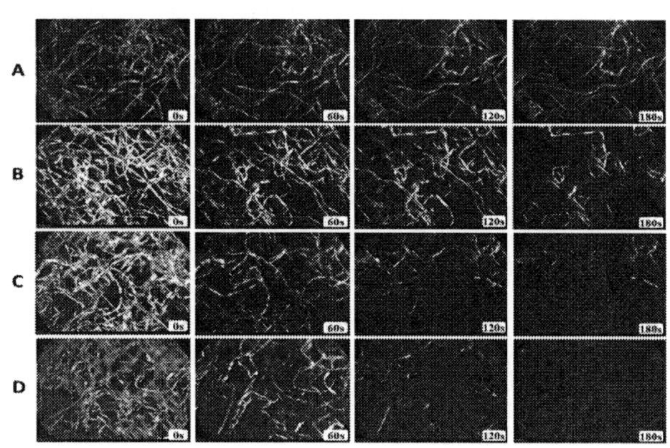

Fig. 1. PLM images of cellulose dissolved in complex solvent of DMSO/IL at 50 oC. (A) ωDMSO=0%, (B) ωDMSO=20%, (C) ωDMSO=40%, (D) ωDMSO=60%.

By means of testing the turbidity of mixture system through turbidimeter, we can easily find out the solvent of best dissolving property, as seen in Fig. 2. With the increasing of ω_{DMSO}, the complex solvent dissolving property shows a regular trend of first increase and then decrease, and the complex solvent with 70% DMSO has the best performance.

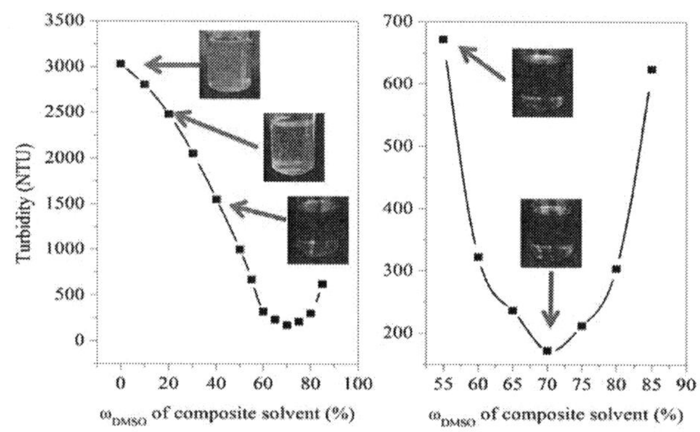

Fig. 2. Relationships between ω_{DMSO} of complex solvent and the turbidity of partly dissolved cellulose solution.

Conductivity results in Fig. 3 showed that the complex solvent with DMSO has the highest conductivity, and the equilibrium time of conductivity for three spinning solution was about 30 minutes.

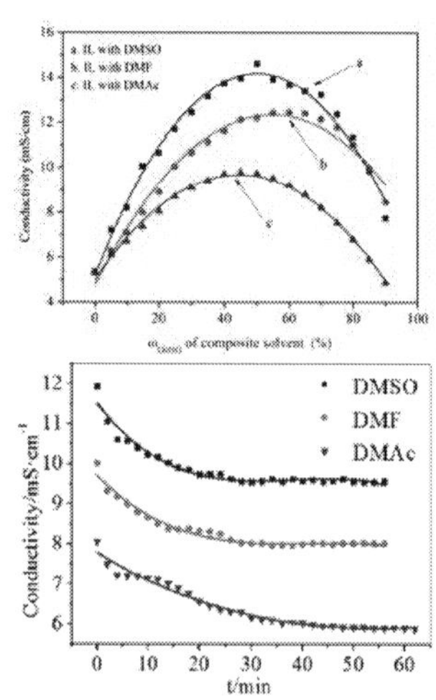

Fig. 3. Conductivity of complex solvents as a function of polar aprotic solvents content and dissolving time.

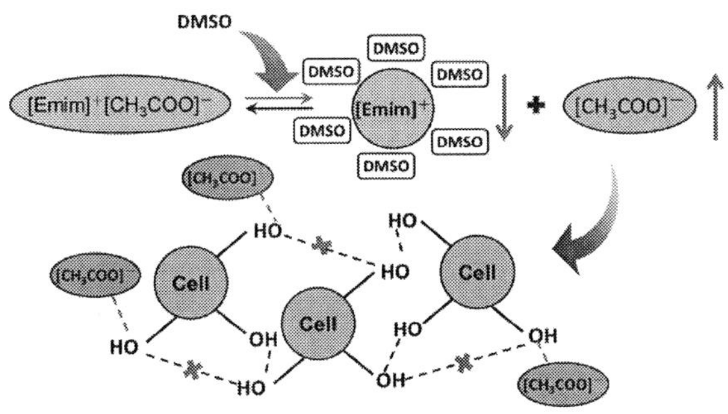

Fig. 4. Efficient dissolution mechanism of cellulose in DMSO/ionic liquid complex solvent.

Ionic liquid can ionize into cation and anion and anion plays a key role of destroying hydrogen bonds of cellulose. When adding additives like DMSO, DMF and DMAc into IL, anion is solvated by these polar aprotic solvents, and this solvation effects lead IL to ionize more amount of anion, so the dissolution speed and the solubility of cellulose both increase obviously.

CONCLUSION

1. Polar aprotic solvents have the ability of improving cellulose-dissolving property in ionic liquid.
2. Polar aprotic solvents can solvate cation of ionic liquid and increase the concentration of "free" anion.
3. This finding opens up new horizons in the design of more efficient cellulose solvents and practical use of cellulose spinning.

ACKNOWLEDGMENT

This work was supported by Shanghai Yangfan Program (14YF1405200), the Fundamental Research Funds for the Central Universities (2232015D3-08) and Textile Vision Science & Education Fund of China.

REFERENCES

[1] Heinze, T., Liebert, T. *Adv. Polym. Sci.*, 10, 2012: 83-152.
[2] Zhang, Y., Li, X., Yang, Y., Lan, A., He, X., Yu, M. *RSC Adv.*, 4, 2014: 34584-34590.
[3] Zhang, Y., Li, X., Li, H., Gibril, M.E., Han, K. Yu, M. *RSC Adv.*, 3, 2013: 11732-11737.

Cellulose Multifilament Prepared from Cellulose Carbamates in NaOH/ZnO Aqueous Solutions on a Pilot Scale

Jinping Zhou, Feiya Fu, Lina Zhang
Department of Chemistry, Wuhan University, Wuhan, China
zhoujp325@whu.edu.cn

Man-made cellulosic fibers have outstanding properties in certain respects, such as strength in sorption behavior, permeability properties, wearing properties and so on, which also meets modern environmental protection requirements[1]. As a promising substitute for the traditional polluting viscose process, cellulose carbamate (CC) process is a novel technology to produce cellulose fibers and films, which has the advantage of a better ecological compatibility[2]. In this work, novel regenerated cellulose fibers have been successfully spun from a novel CC process on a pilot plant which included the microwave-assisted synthesis of CC, dissolution of CC in a NaOH/ZnO aqueous solution, and wet fiber-spinning[3-7]. With the assist of microwave irradiation, white or slight yellow CCs with a preferable nitrogen content (1.042-1.785 *N*%) were efficiently synthesized on a large scale under solvent-free and catalyst-free conditions. Through a freezing-thawing process, the prepared CC could be well dissolved in 7-8 wt% NaOH solution with addition of a small amount of ZnO (0.4-2.0 wt%), and the CC solution could keep stable for more than 48 h at room temperature. Most recently, we founded that cellulose modified with a small amount of urea (2.1-4.1 wt%) could be well dissolved in NaOH/ZnO, and the concentration of the spinning dope reached to 8-9 wt%. High-quality cellulose filaments (Figure 1a) were successfully spun from the spinning dope on pilot equipment through an acid/salt coagulant bath. The fibers exhibited cellulose II characteristics and displayed a circular cross section with a homogeneous structure. Following multi-drawing and drying processes, the small voids on the surface of the newly shaped fibers gradually disappeared, and a much denser structure was shown in the fibers from the take-up device. An increased drawing ratio provided improved tenacity and degree of orientation in the fibers, reaching 2.36 cN/dtex and 85%, respectively. Besides, the novel fibers are safe for the excluding of sulfur and zinc, and the nitrogen content is lower than 0.3%. In addition, the novel filaments demonstrated excellent dye properties (Figure 1b) with different reactive dyes, *K/S* values were determined in the range of 50-53 for the reactive blue dye. The novel method was intentionally similar to the existing viscose system, therefore allowing for cost-effective implementation in industry. In particular, the total cost for producing a ton of novel cellulose filaments was estimated to save 15-20 % when compared to the traditional viscose method. This new and environmentally friendly method has great potential for easy application in the current cellulose industries.

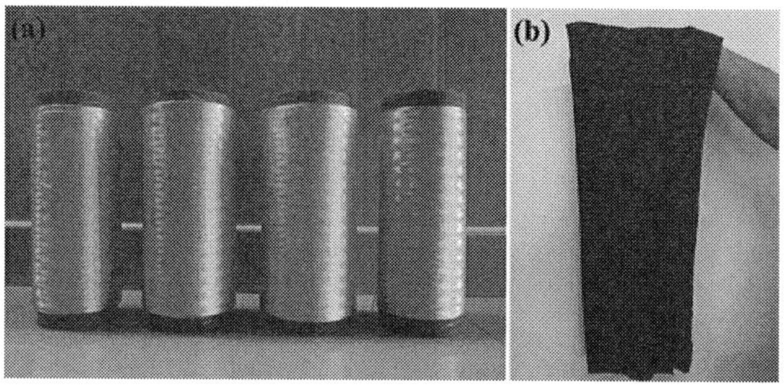

Figure 1. Photographs of the novel cellulose multifilament produced by a pilot machine (a) and the dyed yarns (b).

KEYWORDS
Cellulose filaments, Cellulose carbamate, Pilot-scale, Mechanical properties, Dye properties

ACKNOWLEDGMENT
This work was financially supported by the National Natural Science Foundation of China (51273151) and the Program for New Century Excellent Talents in University (NCET-11-0415).

REFERENCES

[1] Fink, H.P., Ganster, J., Lehmann, A. *Cellulose*, 21, 2014: 31-51.
[2] Klemm, D., Heublein, B., Fink, H.P., Bohn, A. *Angew. Chem., Int. Ed.*, 44, 2005: 3358-3393.
[3] Guo, Y., Zhou, J., Song, Y., Zhange, L., *Macromol. Rapid Commun.*, 30, 2009: 1504-1508.
[4] Guo, Y., Zhou, J., Zhang, L. *Biomacromolecules*, 12, 2011: 1927-1934.
[5] Guo, Y., Zhou, J., Wang, Y., Zhang, L., Lin, X. *Cellulose*, 17, 2010: 1115-1125.
[6] Fu, F., Zhou, J., Zhou, X, Zhang, L., Li, D., Kongo, T. *ACS Sustainable Chem. Eng.*, 2, 2014: 2363-2370.
[7] Fu, F., Yang, Q., Zhou, J., Hu, H., Jia, B., Zhang, L. *ACS Sustainable Chem. Eng.*, 2, 2014: 2604-2612.

Lower Modulus Research on Cellulose Diacetate Fiber

Xiu Du, Jiajun Wu, Xiaohong Qin
Shanghai Key Laboratory of Advanced Micro&Nano Textile Materials,
College of Textiles, Donghua University, Shanghai, China
duxiu199114@163.com; wjjdhu@126.com; xhqin@dhu.edu.cn

INTRODUCTION

Cellulose diacetate[1]has been used in cigarette filter industry for a long time[2-4]. Crisp is one of the most important procedure in production process. While after crimping[5], there are many short fibers and fiber crumbs exfoliated from the fiber body, which can be called flying[6], because crimp action makes damage to fiber, such as local damage, wrinkles or other physical defects. Also as cellulose diacetate fiber own skin-core structure, it will come to more flying. Flying can increase the burden of filaments opening and filter stick formation and have great influence on cellulose diacetate fiber quality. Thus to reduce flying is important. There is research found that lower fiber modulus before crimping can decrease fiber damage which come from crisp procedure. However, previous research in this respect is very few, so this paper will conduct a comprehensive study on that.

APPROACH

For the micro-structure concerns, fiber is high-molecular polymer. So, heating and humidification can lower fiber modulus. This is because heat can strengthen macromolecular thermal motion[7]. The thermal motion can weaken or break the bonding force which is between the molecules. And humidification, water molecules will get into the fibers, they will also weaken or break the bonding force between the macromolecules, for example, the H-bond. So heating and humidification can reduce fiber modulus. This research is to study fiber modulus variation due to heating, humidification or both methods before fiber crimping procedure.

Material

Cellulose diacetate fiber, fineness is 3.9den.

Heating treatment conditions

Temperature 40°C, 50°C, 60°C, 70°C, 80°C,90°C, treatment time for 5 min, 10 min, 15 min.

Humidification conditions

Under room temperature for 5s, 10s, 15s, 20s, 25s. Atomizing water in 40°C, 50°C, 60°C, 70°C, 80°C temperature. Treatment time for 5s, 10s, 15s. Humidifying equipment, ultrasonic nebulizer, spray power is 1.5 kg/h.

RESULTS AND DISCUSSION

Heating treatment

Fig. 1 showed the fiber modulus variation treated with different heating temperature and time. Heating treatment intensifies the fiber macromolecular thermal motion, makes the molecular chain more flexible, so we can see fiber modulus decreases apparently. As temperature going up, the molecular thermal motion is getting more and more severe, when it exceeded 80°C, fiber modulus will fall sharply.

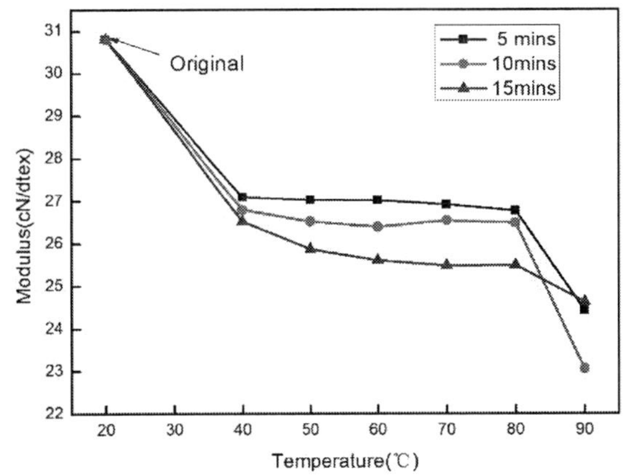

Fig. 1. Modulus-Temperature based on different heating time

Room temperature humidification treatment

Fig. 2 showed the fiber modulus variation treated with different time under room temperature. From the figure, we can know that the fiber modulus decreased with the humidifying time was increasing. And between 5~15s, it declined obviously. After 15s, modulus changes gently.

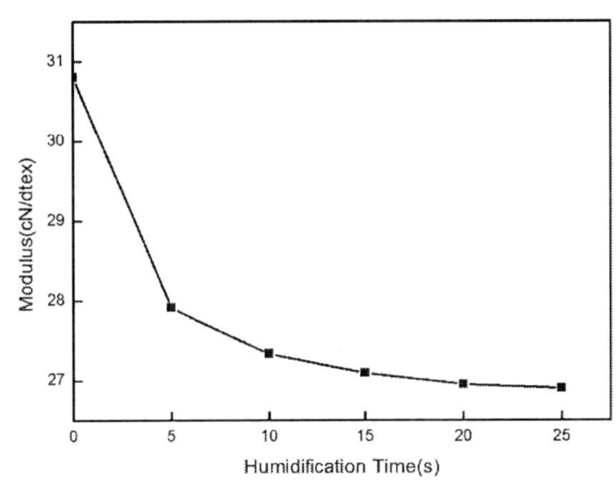

Fig. 2. Modulus-Humidification time

During humidification, some water molecules will get into the fiber, while most of them will attach to fiber surface. When fiber is all covered by water molecules, adding more atomization water will not bring obvious change of the modulus. This is why after 15s, fiber modulus was basically unchanged.

Heating and humidification treatment

Fig. 3 showed the fiber modulus variation treated with atomization water in different temperature for different treating time.

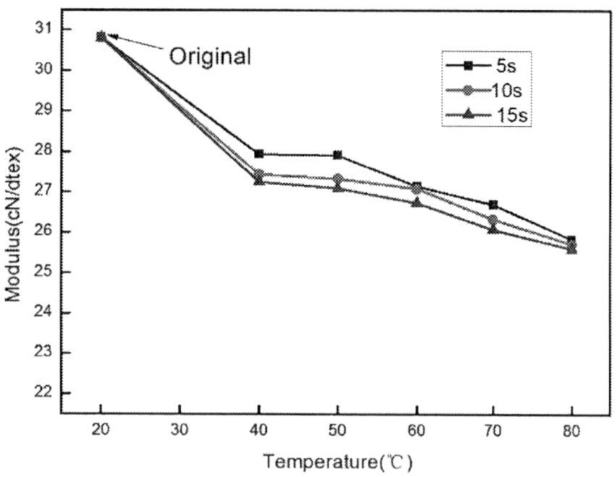

Fig. 3. Modulus-Humidification Temperature

Fiber modulus decreased apparently treated with different temperature atomization water. With the temperature of atomization water and time going up, fiber modulus declined gradually. When the condition is 80°C, 15s, modulus got its minimum 25.6cN/dtex, as the original is 30.8cN/dtex. The modulus of cellulose diacetate fiber without treatment is 30.8cN/dtex. After heating, it decreased significantly. And along with the increase of heating time, modulus is getting downward. When the temperature between 40~80°C, processing time during 5~10min, fiber modulus basically no changed. When heating longer for 15min, as temperature went up, the modulus got lower. At 90°C, the fiber modulus had a sharp decline.

CONCLUSION

With heating temperature and time increasing, fiber modulus declined. When humidifying, before 15s, fiber modulus decreased apparently, then gradual. And treated with different temperature atomization water, fiber modulus went down as temperature and time increasing.

FUTURE WORK According to stress-strain curve , we can get modulus-strain curve via differentiation. Modulus-strain curve is helpful to study fiber modulus changes during the tensile testing. Because it has almost all information that stress-strain curve has. The following work is to study the mechanism of fiber modulus changes. And for reducing flying and improve cellulose diacetate fiber quality cigarette industry, we should testing the parameters, which was got from the research, on the production line.

ACKNOWLEDGMENT

This work was partly supported by grants (51373033 and 11172064) from the National Natural Science Foundation of China to Prof. Xiaohong Qin. Furthermore, the Fundamental Research Funds for the Central Universities and "DHU Distinguished Young Professor Program" to Prof. Qin. This work was also supported by the Keygrant Project of Chinese Ministry of Education (No 113027A).

REFERENCES

[1] Sun, Xunwen, et al. "Acetone-Soluble Cellulose Acetate Extracted from Waste Blended Fabrics Via Ionic Liquid Catalyzed Acetylation." *Carbohydrote Polymers*, 98.1, 2013: 405-11.

[2] Edgar, K.J., et al. "Advances in Cellulose Ester Performance and Application." *Progress in Polymer Science*, 26.9, 2001: 1605-88.

[3] Maskos, Zofia, Khachatryan, Lavrent, Dellinger. "Role of the Filters in the Formation and Stabilization of Semiquinone Radicals Collected from Cigarette Smoke." *Energy&Fuels*, 27.9, 2013: 5506-12.

[4] Wang, H.B., et al. "Modification of Surface Properties of Cellulose Acetate Fibers with Cellulose for Cigarette Filters." Proceedings of 2005 International Conference on Advanced Fibers and Polymer Materials, 2005.

[5]Yang, Z.P., et al. "Study on Crimping Process of 2.00dtex/3.98ktex Cellulose Acetate Fiber Tow." *Quality Textiles for Quality Life*, Vols. 1-4224.

[6] Pengxiang, Li, et al. "Effect of Different Fiber Fly Quantity on Surface of Cellulose Diacetate Filament." *Chemical Fibers Interational*, 2014: 43-46.

[7] Yabuki, K. "Heat Treatment of Poly-Benzazole Fibres to Increase the Modulus Using Steam Which Moves in a Concurrent or Countercurrent Direction to Fibre." Dow Chem Co (Dowc-C) Dow Chem Co (Dowc-C) Assignee. Patent WO9412706-A; EP672203-A; US5288452-A; WO9412706-A1; AU9462262-A; ZA9309076-A; EP672203-A1; CN1097032-A; EP672203-B1; DE69312960-E; ES2105660-T3.

Synthesis, Structure, and Properties of Fiber-Related Materials

Self-Healable, Super-Tough GO/PAA Nanocomposite Hydrogels Facilitated by Dual Cross-Linking Effects

Ming Zhong, Fu-Kuan Shi, Xu-Ming Xie
Key Laboratory of Advanced Materials (MOE), Department of Chemical Engineering, Tsinghua University, Beijing, China
xxm-dce@tsinghua.edu.cn

ABSTRACT

Hydrogels are soft materials with a 3-dimensional network structure, which can retain a significant amount of water. Basically, these soft materials result from the self-assembly of hydrophilic molecules or polymers by forming a covalently or physically cross-linked network. Hydrogels can be potentially used in, e.g., tissue engineering, drug delivery, membrane separation, catalysis, and sensing technology. Although most hydrogels have good biocompatibility and stimuli-responsivity, they usually have poor mechanical strength and low deformation, which unfortunately set limitations to their high-end applications. Here we propose a facile, one-pot *in situ* free radical polymerization strategy to prepare self-healable, super tough graphene oxide (GO)/poly(acrylic acid) (PAA) nanocomposite hydrogels by using Fe^{3+} ions as a cross-linker. The 3-dimensional network structure of the GO/PAA nanocomposite hydrogels is facilitated by dual cross-linking effects through dynamic ionic interactions: (i) first cross-linking points are Fe^{3+} ions creating ionic cross-linking among PAA chains; (ii) second cross-linking points are GO nanosheets linking PAA chains through Fe^{3+} coordination. When the GO/PAA nanocomposite hydrogels are under stretching, the ionic interactions among PAA chains can dynamically break and recombine to dissipate energy, while the GO nanosheets coordinated to the PAA chains maintain the configuration of the hydrogels and work as stress transfer centers transferring the stress to the polymer matrix. In this regard, the GO/PAA nanocomposite hydrogels exhibit superior toughness (tensile strength = 777 kPa, work of extension = 11.9 MJ m^{-3}) and stretchability (elongation at break = 2980%). Furthermore, after being treated at 45 °C for 48 h, the cut-off GO/PAA nanocomposite hydrogels exhibit a good self-healing property (tensile strength = 495 kPa, elongation at break = 2470%).

ACKNOWLEDGMENT

This work was financially supported by NSFC (No. 21474058) and by State Key Laboratory for Modification of Chemical Fibers and Polymer Materials, Donghua University (Project No.LK1404).

REFERENCES

1. Shi, Fu-Kuan, Wang, Xi-Ping, Guo, Ruo-Hai, Zhong, Ming, Xie, Xu-Ming. *J. Mater. Chem. B*, 3, 2015: 1187-1192.
2. Zhong, Ming, Liu, Yi-Tao, Xie, Xu-Ming. *J. Mater. Chem. B*. In press.

Strengths of Fiber Products: Current and Historical Perspectives on Statistical Size Effects

Moon W. Suh
College of Textiles, North Carolina State University, Raleigh, NC, USA
msuh@ncsu.edu

OBJECTIVES

This presentation is aimed at examining one of the most important physical properties of all fibrous materials and products, namely, the strength, in terms of prediction, optimization, product redesign based on length, width, thickness associated with polymers, natural fibers, yarns, fabrics, cables, composites and all other forms of fiber products. While much work has been done experimentally and published, only a few theoretical papers have been widely recognized in practical applications due to the conceptual difficulty in defining the substrate geometrically, and due to the complexity involved in the derivation of statistical distributions for the wide spectrum of materials.

The paper presents a bird's eye view of the work done to date and further generalizes some of the results by relaxing the simplifying restrictive assumptions. In addition, the paper will present a new general approach for predicting the directional strength of a specimen of size k x p (width x length) based on the strength distribution of 1x1 (width x length) unit dimensions.

HISTORICAL PERSPECTIVES OF STRENGTHS IN FIBROUS PRODUCTS

F. T. Peirce [1] has written a milestone paper in 1926 to express the strength of a long fibrous specimen (fibers, yarns, fabrics) within a conceptual framework of a chain with multiple links, later called the *"weakest link theory."* This longitudinal effect, in spite of the lack of independence among chain links, has been well recognized and widely applied in materials science fields. For the widthwise effect, however, no firm theories have been firmly established until Suh [2] derived the basic probability distribution function of B_n, the strength of a bundle with *"n"* parallel filament which breaks sequentially at different breaking elongation point, by assuming that the aggregate load at each breakage point is non-decreasing and shared equally by the surviving filaments. Under the assumption, Suh [2] also established a bundle strength inequality

$$B_n/n \leq B_{n-1}/(n-1) \quad \text{for } n \geq 2,$$

and named it "The Law of Diminishing Return in Bundle Strengths."

In the complete absence of a distribution function for B_n, Peirce [1] was able to show, without a proof, that the average bundle strength of B_n converges to an asymptotic value when bundle load $L(e)$ at extension e is expressed as

$$L(e) = NKe\int_e^{\infty}\phi(x)dx\,.$$

Where, N is the number of filaments in the bundle, K the modulus of the single fiber, and $\phi(x)$ the probability density function of a single fiber breaking elongation.

H. E. Daniels [3] attempted to obtain the distribution of B_n up to n = 3 but later proven to be in error. Other significant work by Platt [5], Coleman [6], Phoenix [7], Cui, Suh and Sasser [7] dealt with various aspects of bundle strength but not on the size effects.

"THE LAW OF DIMINISHING RETURN" IN BUNDLE STRENGTH –AN ULTIMATE PROOF

This paper will show a proof for the first time that Suh's bundle strength inequality shown above holds good under a most general situation even when the bundle load at every elongation point during a breakage process is not equally shared by the surviving filaments as long as there exists no decay in the load-extension diagrams of individual filaments. This remarkable mathematical proof can be easily demonstrated by *Monte Carlo* simulations.

GENERAL STRENGTH MODEL WITH "*n*" FILAMENTS OF LENGH = ℓ

This paper will present a general model for the probability distribution function of bundle strength, when the bundle consists of n parallel filaments of length ℓ, as a function of the probability distribution $F(X)$ of the single filament strength X of unit length. In the modeling process, one may try numerous options for defining the "unit length" for X, thus providing an ample room for simulation for finding the best fit. In addition, the bundle strength B_n can be modeled as strength realized from "k" links of bundles of length "m" serially connected as long as $\ell = k \cdot m$.

MODELING OF MATERIAL STRENGTH WITH AN ARBITRAY WIDTH AND LENGTH

For a specimen of size W x L as width and length, the breaking strength can be modeled by mimicking that the specimen is an assembly of unit width and unit length arbitrarily defined. When the width W is considered as W filaments split at the borders (see the figure below), the bundle strength to be realized can be considered as the lowest limit for the strength of the uncut specimen. This simple observation presents many avenues for modeling material strength in general based on the actual test results of the unit length, unit width specimen.

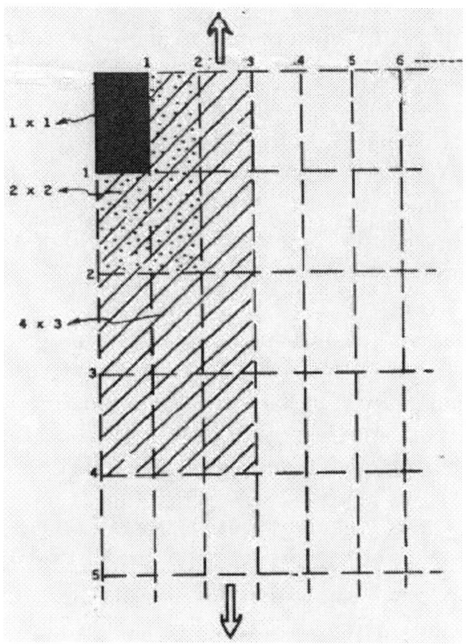

By recognizing certain fracture mechanisms that are known, these generalized bundle strength model can be quite useful. Much like this idea, Suh, Koo and Woo [8, 9] applied a new concept *"Effective Gauge Length"* in search of a bundle strength model that fits the actual spun yarn strengths best by using the individual cotton fiber strengths.

KEYWORDS

Bundle strength, weakest-link theory, statistical size effects, the Law of Diminishing Return.

REFERENCES

[1] Peirce, F.T. "Theorems on the Strength of Long and of Composite Specimen." *Journal of Textile Institute*, 17, 1926: 355-368.

[2] Suh, MW., Bhattacharyya, B.B,. Grandage, A. "On the Distribution and Moments of the Strength of a Bundle of Filaments." *J. Appl. Prob.*, 7, 7, 1970: 12-720.

[3] Daniels, H.E. "The Statistical Theory of the Strength of Bundles of Threads." *Proc. Roy. Society*, A183, 1945: 405-435.

[4] Platt, M.M., Klein, W.G., Hamburger, W.J. "Mechanics of Elastic Performance of Textile Materials Part IX." *Textile Res. J.*, 22, 1952: 641-667.

[5] Coleman, B.D. "On the Strength of Classical Fibers and Fiber Bundles." *J. Mech. Phys. Solids*, 7, 1958: 60-70.

[6] Phoenix, S.L., Taylor, H.M. "The Asymptotic Strength Distribution of a Generalized Fiber Bundle." *Adv. Appl. Prob.*, 5, 1973: 200-216.

[7] Cui, X., Suh, M.W., Sasser, P.E. "Tensile Behaviors of Slack Fiber Bundles–Theory and Application to Hvi Testing." *Textile Research Journal*, 69, 1999: 497-502.

[8] Koo, H.-J., Suh, M.W., Woo, J.L. "Variance Tolerancing and Decomposition in Short Staple Spinning Processes, Part I–Theories." *Textile Research Journal*, 71, 1, 2001: 1-7.

[9] Suh, M.W., Koo, H.-J., Woo, J.L. "Variance Tolerancing and Decomposition in Short Staple Spinning Processes, Part I: Simulation and Application to Ring Carded, Ring Combed and O-E Spun Yarns." *Textile Research Journal*, 71, 2, 2001: 105-111.

Prediction of the Tensile Strength of a Two-Ply Polyester Filament Yarn

Huang Wei, Wang Jun
College of Textiles, Donghua University, Shanghai, China
jalywei@qq.com

PURPOSE

Currently, the researches on ply yarns strength and its relationship with yarn structure and single yarn properties mostly stay in using experimental data to explore the laws among parameter variables. The lack of quantitative analysis cannot well explain the strength enhancement effects of the ply yarn, existence of critical twist factor and interactions effects between single yarn (fiber) on the tensile properties of the ply yarn, that is, the theory and experiment failed to properly be combined together to quantitatively tell these phenomena. Our group found that the parameters like yarn twist, tensile & shear modulus, elongation, fineness, single yarn interactions were important indicators affecting strength and elongation of the ply yarn. Thus, a mathematical model with these variables is established, and can be used to predict the tensile modulus, strength and tensile behavior of the finished ply yarns under different twist levels.

INTRODUCTION

In the previous researches, the approaches (models) used for studying the relationship among yarn quality, single yarn (fiber) properties and yarn structure can be briefly classified into mathematical (mechanistic), empirical (regression equation), computer simulation and neural network. Among them, Mathematical models have their basis in applied physics, providing a better understanding of the complex interrelationships of the different parameters that determine yarn properties [1,2]. Previously, Hearle, Zurek, Postle established its basic theoretical model based on force deformation analysis, while Sullivan was the first researcher trying to associate yarn strength, fiber properties and yarn twist by assuming the yarn with perfect geometry. In the 1990s, Pan in a series of papers showed that how to predict the entire stress-strain curve of yarn from fiber properties. As part of their achievements, Bogdan has established an empirical relationship between fiber properties and yarn strength, a formula for yarn breaking strength is derived, and only the intrinsic yarn strength parameter needs to be determined from experiments. The achievements made by them have contributed enormously to this area, can well predict and reflect yarn quality under certain specific conditions, but still there are three following deficiencies: 1) the relationship between introduced variables and parameters is lack of theoretical basis; 2) it is not clear enough for physical meaning of certain variables, which is always determined by empiricism; 3) the range of model applications is not wide enough, only applies to the specific conditions of yarn types or spinning conditions. For these reasons, the model of this study will avoid these disadvantages, with a clear mechanical relationship among the parameter variables and has a broad applicability.

APPROACH

An untwisted polyester filament yarn with stable performance is used for doubling and twisting by changing the twisting degrees, then followed by the tensile tests, the ply yarn strength and elongation indexes during the stretching process are measured at different twist levels.

The diameter of the ply yarn is measured by the observation platform composed of camera, PC and monitor. After obtaining the images of projective diameters in the static, image denoising and binarization are under processed with the image processing software. Then a prepared corresponding program written by Matlab is used to get the diameter data.

It is considered that the strength of ply yarns consisting of continuous filaments mainly depends on the properties of single yarns and the cohesion force by twisting. Therefore, in previous studies [3], a mathematical model expression based on geometric model for analyzing the relationship between ply yarn strength and parameter variables in the case of no single yarn slippage is derived under certain assumptions, as follows:

$$E_y = \frac{Y}{1+\frac{Y}{K}tan^2\theta} \qquad (1)$$

$$T_t = \frac{1}{4}\pi d_y^2 \varepsilon_y E_y \,/\, (1-\frac{\alpha}{2}\pi sin\theta cos\theta) \qquad (2)$$

Where E_y is the tensile modulus of the ply yarn, T_t is total strength, d_y is yarn diameter, ε_y is tensile strain, θ is twist angle, Y and K are tensile and shear modulus of single yarn, respectively, α is cohesion coefficient as an empirical parameter.

The relationship among yarn diameter d_y, twist angle θ and twist degree T can be expressed by:

$$d_y = \frac{\tan\theta}{\pi T}\sqrt{\frac{2}{\cos\theta}} \qquad (3)$$

RESULTS

Experimental data of single yarn and ply yarn: The experimental data of breaking strength and elongation at different twist levels is listed in Table I. Twist angles can be determined by Equation (3).

Table I: Experiment results on strength, elongation, diameter and twist angle

Yarn Type	Strength (cN)	Fracture elongation	Diameter (mm)	Twist angle (°)
Single yarn	1137.8	30.9%	-	-
Ply yarn 1	2605.1	45.9%	0.3388	4.2981
Ply yarn 2	2726.7	54.5%	0.3282	12.2017
Ply yarn 3	2527.8	53.6%	0.3198	19.0520
Ply yarn 4	2191.9	54.5%	0.3233	25.5275
Ply yarn 5	1732.7	48.8%	0.3100	29.9779

Note: Single yarn with linear density of 33.3 tex. Ply yarn 1, 2, 3, 4, 5 are at the twist factor of 82, 245, 408, 571 and 734.

Tensile modulus: In Equation (1), the tensile modulus of the ply yarn is determined by the twist angle, tensile and shear modulus of single yarn. Assume the single filament yarn is incompressible and isotropic, which means the Poisson's ratio μ is 0.5, and an essential relationship among Y, K and μ is $Y=2K(1+\mu)$. Y is confirmed by the tensile curve of the single yarn which is simplified by an exponential curve ($Y_2=8846*(100*\varepsilon)^{-1.142}+548$ (MPa), ε is strain), shown in Figure 1. The values exceeding the fracture strain of the single yarn are denoted by Y_2.

Tensile strength: As the diameters, twist angles and tensile modulus of the ply yarns are all known, the predicting tensile strength at different twist levels are plotted in Figure 2, with a comparison with the curves obtained by tensile equipment in Table I.

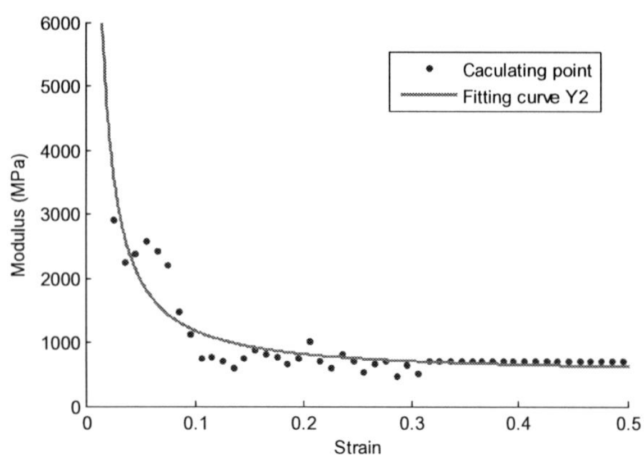

Figure 1: The tensile modulus of the single yarn and the fitting curve

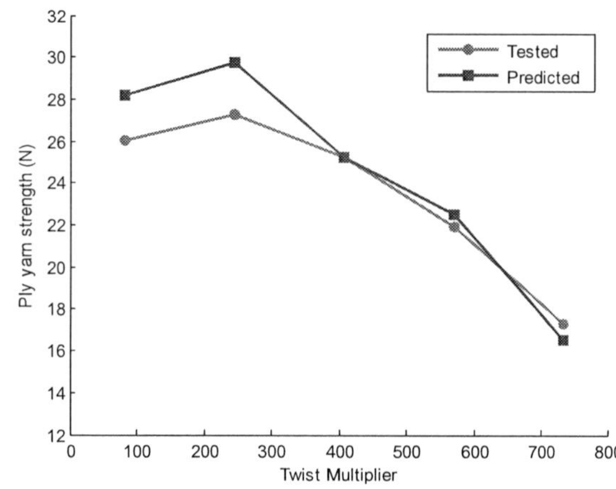

Figure 2: Ply yarn tensile strength comparison

The results show that the predicted values are quite close to the tested values with the same trend. A critical twist factor is sure to occur between the twist factor of 82~245. If possible, the whole tensile curve of the ply yarn can be plotted as well. And this would be one of further research directions of our group.

CONCLUSION

Equation (2) permits a simple deduction of the relations between the ply yarns strength and the twist factor. Then by doubling tests, observation platform, measuring equipment, analysis of the single yarn tensile curve, the model is verified quantitatively. Experimental results showed that this model can well predict the strength of finished strand yarns in actual productions and a guidance to find the critical twist factor of the yarn without the tensile tests.

FUTURE WORK

In the future, the model will be optimized for further study. Experiments can obtain more accurate parameters (such as the diameter and twist angle of the yarn) by setting up a new observing platform in order to make prediction more accurate and useful. The prediction of whole tensile curve and a comprehensive consideration of the parameters variations are also under study.

REFERENCES

[1] Ghosh, Anidya, et al. "Predictive Models for Strength of Spun Yarns: an Overview." *AUTEX Research Journal*, 5, 1, 2005: 20-29.

[2] Guha, Anirban, Chattopadhyay, R., Jayadeva, R. "Predicting Yarn Tenacity: A Comparison of Mechanistic, Statistical and Neural Network Models." *J. Text. Inst.*, 92, 2, 2001: 139-145.

[3] Huang, W., Wang, Jun. "Investigation on the Mechanical Properties of Two-ply Yarn Based on Geometric Model." *Journal of Textile Research*, 36, 2, 2015: 19-22.

Improvement of the Test Instrument of Fabric-Skin Adhesion Properties

Feng Ji[1], Jianfei Xie[2], Yan Wang[1], Ruiyun Zhang[1], Yiping Qiu[1]
[1]Key Laboratory of Textile Science & Technology, Ministry of Education, College of Textiles, Donghua University, Shanghai, China; [2]Guangzhou Fibre Product Testing and Research Institute, Guangzhou, China
jifeng@dhu.edu.cn

INTRODUCTION

Fabrics that are wet with perspiration or rain easily get adhered with skin and cause uncomfortableness for the wearer. Hence, different types of fabrics have been developed to furthest reduce the attachment of wet fabric to skins. Corresponding test methods of the wet fabric-skin adhesion properties are needed to measure and evaluate how far these fabrics have reached to get separated from human skins in wet situations.

Jou. G.[1] suggested a method to measure the cling index between a fabric and a perspired skin, however, the objective cling force was more like the skin-fabric friction in wet conditions than the adhesion force perpendicular to skin surface, thus could not effectively reflect the practical feelings that people wearing wet clothes.

Our team[2-4] developed a new method to test and evaluate the fabric-skin adhesion behaviors, and the test instrument, XAD-1A fabric-skin adhesion testing apparatus was made. In this test, the development of the adhesion force perpendicular to skin surface was measured timely when the wet fabric got separated from the skin. The adhesion force F, the maximum displacement L_m (mm), and adhesion work W (mJ) are measured or calculated so as to properly evaluate the adhesion property of the fabric to skin.

In this paper, the XAD-1A fabric-skin adhesion test instrument was improved through a liquid control system. With this device, the amount of liquid that reach the fabric sample is controllable or know, thus a controllable wet condition of fabric sample was reached and corresponding adhesion force was measured.

Three different types of fabric samples were selected and the fabric-skin adhesion test was conducted respectively when the fabric sample had absorbed 5, 10, 15ml synthetic perspiration, and totally wet with the synthetic perspiration respectively.

APPROACH

The fabric-skin adhesion test was implemented by using the XAD-1A apparatus[2]. The adhesion force-displacement curve was obtained during the process that the wet fabric was separated from the artificial skin, and the maximum adhesion force F_m (unit: cN), the maximum adhesion length L_m (mm), and adhesion work W (mJ) were calculated[3] to quantitively describe the adhesion properties of each fabric sample.

The liquid control system was constructed mainly by a thermostatic water tank and a fluid control meter, through which certain amount of liquid was allowed to reach the fabric. The fabric sample clung to the "skin" when it got wet with the liquid as was often found in daily life, and the adhesion test started as soon as the force value turned from negative (compression) to positive (adhesion).

EXPERIMENT

Three types of knit fabrics F1, F2, and F3 were selected; all were typical materials of T shirt and casual clothes. Table I presents characteristics of the samples, and Fig. 1 displays pictures of fabric samples.

Table I. Physical properties of the fabric samples.

Fabric	Fiber material	Structure	Fabric count (/5cm)	Weight (g/m^2)	Thickness (mm)
F1	100% Polyester	Plain Stitch	98.8*81.0	1.478	0.98
F2	100% Polyester	Meshwork stitch	105.2*78.8	1.279	1.19
F3	50.2%Polyester /49.8% cotton	Plain stitch	95.2*63.6	2.517	0.98

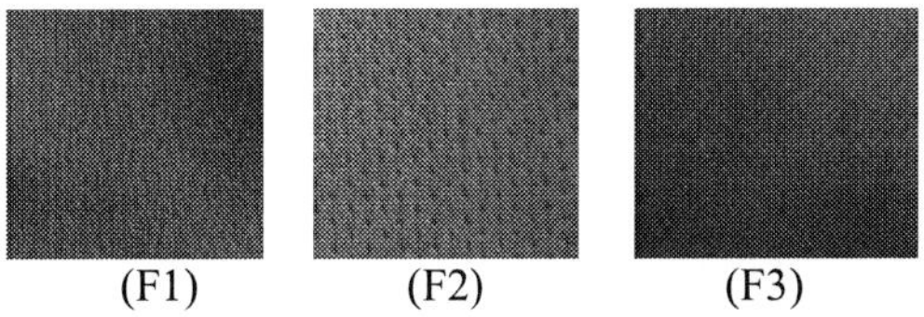

Figure 1. Pictures of three types of fabric samples.

Artificial perspiration was prepared according to GB/T 3922-2013 "Textiles-testing method for color fastness to Perspiration", and the skin was made of natural rubber. Each sample piece was cut into 8.0 cm×15.0 cm rectangular sizes and the skin area was 4.0cm×6.0 cm. The separating speed was set 30 mm/min in this research. Certain amount of artificial perspiration, 5, 10, 15 ml respectively came out from the "skin" and get to the fabric sample. Till all the liquid was absorbed the test began.

RESULTS AND DISCUSSION

The adhesion force-displacement curve was obtained when a piece of fabric sample was tested, and the corresponding F_m , L_m, and W were calculated as well. Each experiment was repeated six times. Fig. 2-4 present the typical force-displacement curves of F1, F2, F3 respectively; Table II-IV display the average value of F_m , L_m, and W of the samples respectively.

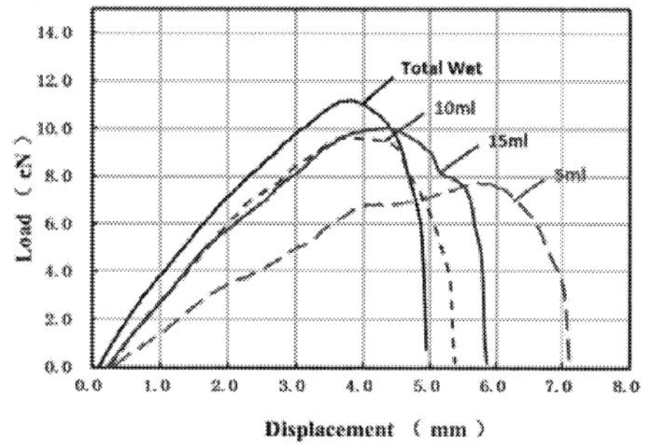

Figure 2. Development of adhesion force vs displacement of sample F1 in conditions of having absorbed 5, 10, 15 ml artificial perspiration and totally wet respectively.

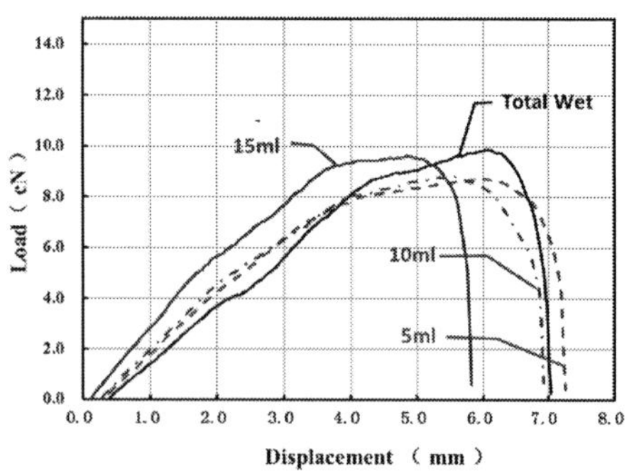

Figure 3. Adhesion force-displacement curve of sample F2 having absorbed 5, 10, 15ml artificial perspiration and totally wet respectively.

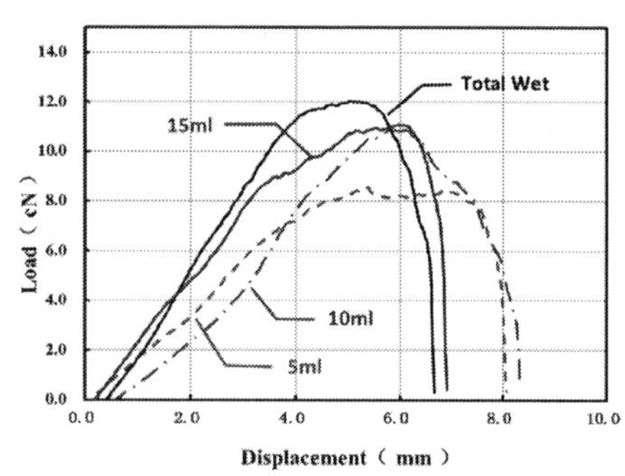

Figure 4. Adhesion force-displacement curve of sample F3 having absorbed 5, 10, 15ml artificial perspiration and totally wet respectively.

Table II. The maximum adhesion force F_m (unit: cN)

Fabric	5ml	10ml	15ml	Total Wet
F1	7.75	9.59	9.80	11.21
F2	8.82	8.14	9.84	10.09
F3	8.79	10.51	11.19	11.98

Table III. The maximum adhesion length L_m (unit: mm)

Fabric	5ml	10ml	15ml	Total Wet
F1	5.15	3.59	4.1	3.71
F2	5.42	5.06	5.5	5.75
F3	5.94	5.12	5.33	4.85

Table IV. The adhesion work W (unit: mJ)

Fabric	5ml	10ml	15ml	Total Wet
F1	0.34	0.31	0.35	0.35
F2	0.4	0.37	0.38	0.41
F3	0.51	0.48	0.48	0.49

It was found that for all the samples the adhesion force grows with the increase of absorbed liquid amount. It can be explained that more liquid creates more adhesion area, and thus larger force is needed to separate the fabric from the skin.

However, the result of L_m displayed the opposite, see Fig. 2-4 and Table III. The measured L_m obtained the largest value generally when only 5ml liquid was absorbed by the fabric. This is mostly caused by the fact that under lower liquid absorption more dry areas were remained on the fabric and dry fabric elongate more than its wet condition.

It was also found that with the growth of liquid content, the increase of adhesion force and decrease of displacement are obviously different among three fabric samples. F2 presents less variance of adhesion force and L_m with the growth of liquid content than F1 and F3. This is due to the characteristics of the fabrics. The hydrophobic fiber material and relatively looser structure of F2 make it easily to get saturated; and once the liquid amount exceeds the its maximum capacity, the increase of liquid amount has little influence on the adhesion behavior of the wet fabric.

CONCLUSION

And effective method to objectively test and evaluate the wet fabric-skin adhesion properties was proposed. In this paper, the fabric-skin adhesion tester was improved through a liquid control system. The test results of three knit fabrics show that the fabric samples present similar tendency of adhesion behavior with the growth of liquid absorption; whereas, fabric characteristics plays an important role determining the development of the adhesion behavior with growth of liquid absorption.

ACKNOWLEDGEMENT

This work was jointly supported by Shanghai Municipal Natural Science Foundation (No.14ZR1401000), the National Natural Science Foundation of China for the Youth (No.60904056), and the Fundamental Research Funds for the Central Universities (No. KLTST201504).

REFERENCES

1. Jou, G., Gau, R., Su, Y., Lin, Y. Patent (TW565697-A, 2004).
2. Ji, Feng, Qiu, Yiping, Li, Ruqin. Patent (ZL2012 1 0163310.6 P.R. China).
3. Ji, Feng, Qiu, Yiping, Xie, Jianfei, Sun, Shuyao. Advanced Materials Research, 627, 2013: 476-479.
4. Sun, Shuyao, Ji, Feng, Wang, Yan, Qiu, Yiping, Xie, Jianfei, Lou, Lin. Journal of Textile Research (Chinese), 35, 8, 2014: 17-31.

Development of Novel Poly(*p*-phenylene terephthalamide)s

Wenbin Wang, Xiang Qi, Fei Zhang, Xinhua Wan
Beijing National Laboratory for Molecular Sciences, Key Laboratory of Polymer Chemistry and Physics of Ministry of Education, Center for Soft Matter Science & Engineering, College of Chemistry and Molecular Engineering, Peking University, Beijing, China
xhwan@pku.edu.cn

ABSTRACT

Poly(*p*-phenylene terephthalamide) (PPTA) is a famous aramid polymer. Its totally aromatic, rod-like structure facilitates the fabrication of fibers (e.g., Kevlar) with a very high orientation from a liquid crystalline solution.[1] These fibers have ultrahigh tensile strength and modulus, exceptional heat and flame resistance as well as chemical, solvent and environmental stabilities, and have been used as reinforcing materials of composites in wide applications such as military, aeronautics, astronautics, automobiles, and sports.[2-6]

One main challenge in manufacturing high performance composites reinforced by Kevlar fibers is their low affinity for interfacial adhesion towards resin matrix due to the highly oriented, extended-chain morphology and the lack of reactable substituent. It is a common practice to introduce substituents on PPTA chains through either amide groups or phenyl rings to overcome this difficulty. The *N*-substitution of amide group works at the cost of damaging interchain hydrogen bondings.[7] The substitution on phenyl ring does not have such a drawback and therefore has been extensively investigated. Herein, we report our recent results on the design and synthesis of novel PPTAs.

Firstly, two PPTA-based macroinitiators with random and alternate sequences, i.e., poly(*p*-phenylene terephthalamide)-*ran*-poly[*p*-phenylene (2,2,6,6-tetramethylpiperidinyl-1-oxy)terephthalamide)] (CPPTA-*ran*) and poly(*p*-phenylene terephthalamide)-*alt*-poly[*p*-phenylene (2,2,6,6-tetramethyl-piperidinyl-1-oxy)terephthalamide)] (CPPTA-*alt*), were prepared via copolycondensation of terephthaloyl chloride, 2,2,6,6-tetramethylpiperidinyl-1-oxy (TEMPO)-functionalized terephthaloyl chloride, and *p*-phenylenediamine. The graft copolymers consisting of rigid PPTA backbones and polystyrene side chains were obtained by nitroxide-mediated radical polymerization. Both macroinitiators and graft copolymers were characterized by thermal gravimetric analysis, differential scanning calorimetry, wide-angle X-ray diffraction, and polarized optical microscopy. The regular incorporation of the TEMPO-containing co-unit imparted remarkable effects on the thermal stability, lyotropic liquid crystallinity, and macromolecular packing in bulk. CPPTA-*alt* showed better thermal stability and more ordered intermolecular structure than CPPTA-*ran*. The former generated a nematic phase at a concentration of 18 wt-% in concentrated sulfuric acid, whereas the latter did so at a concentration of 12 wt-%. For the graft copolymers, the alternative main chains exhibited sharper diffractions than the random ones. However, the sequence change exerted no discernible effect on other properties.[8]

Secondly, a facile and efficient method to prepare a novel PPTA bearing both apolar allyl and polar hydroxyl groups was developed by a combination of the step-growth polymerization and subsequent Claisen rearrangement reaction (CR). The rearrangement processes of the model compound and the precursor of target polymer were completed in 6 h at 170 °C for the former while at 190 °C for the latter without addition of any catalyst as monitored by NMR and FTIR. This polymer showed a remarkably improved solubility and hydrophilicity, and may find applications in rubber or epoxy resin-based high-performance composite acting as a reinforcing material alone or a sizing agent of Kevlar fiber. The strategy of introducing simultaneously two kinds of reactive groups to polymer backbone, established in the present work, should also be applicable to other aromatic engineering polymers by improving the interface structures and interfacial adhesions.[9]

Finally, a new cyclization reaction was found to prepare polyimides or polyimide-amides. This method may find application in aromatic polyimides with well-defined microstructures.

KEYWORDS

Poly(*p*-phenylene terephthalamide), TEMPO, nitroxide-mediated radical polymerization (NMP), Claisen rearrangement reaction, aromatic polyimide, sizing agent.

ACKNOWLEDGMENT

This work was supported by National Basic Research Program of China (2011CB606106).

REFERENCES

1) Fink, J.K. *High-performance Polymers.* New York: William Andrew, Inc., 2008.
2) García, J.M., García, F.C., Serna, F., de la Peña, J.L. *Prog. Polym. Sci.*, 35, 2010: 623.
3) García, J.M., García, F.C., Serna, F., de la Peña, J.L. *Handbook of Engineering and Specialty Thermoplastics.* John Wiley & Sons, Inc., 2011: 141-181.
4) Agarwal, U.S., Khakhar, D.V. *Nature*, 360, 1992: 53.
5) Du, S., Zhang, J., Guan, Y., Wan, X. *Aust. J. Chem.*, 2013: 67, 39.
6) Guan, Y., Zheng, Y., Cui, J., Wan, X.H. *Chinese J. Polym. Sci.*, 28, 2010: 257.
7) Takayanagi, M., Katayose, T. *J. Polym. Sci.: Polym. Chem. Ed.*, 19, 1981: 1133.
8) Dun, S.M., Zhang, J., Guan, Y., Wan, X.H. *Aust. J. Chem.*, 67, 2014: 39.
9) Du, S.M., Wang, W.B., Yan, Y., Zhang, J., Tian, M., Zhang, L.Q., Wan, X.H. *Chem. Commun.*, 50, 2014: 9929.

Solubility, Discoloration and ^{13}C NMR Spectra of Stereoregular Poly(vinyl chloride) Prepared by Urea Clathrate Polymerization at Low Temperatures

Masatomo Minagawa[1], Takayuki Kato[2], Jun Yatabe[3], Nobuhiro Sato[4], Tomochika Matsuyama[4]

[1]NPO, Dream-Create-Laboratories, Yonezawa, Japan
[2]Department of Polymer Science & Engineering, Yamagata University, Yonezawa, Japan
[3]Department of Life & Environmental Sciences, Teikyo University of Science, Uenohara, Japan
[4]Research Reactor Institute, Kyoto University, Kumatori, Japan
gakusai-minagawa@memoad.jp

ABSTRACT

Stereoregular poly(vinyl chloride) (PVC) was prepared by solid state polymerization. Its solubility into organic solvents and the thermal properties were studied in comparison with those of ordinary free radically prepared sample. Urea canal PVC was obtained in a form of white powder, which was thermally unstable as evidenced by easy discoloration in the powder state heating up to 80°C to be pale purple. In contract, free radical PVC didn't causediscoloration in the solid state under the same conditions. Easy discoloration of urea canal PVC is probably due to the absence of termination reaction, and the insolubility into organic solvents is due to the stiffness of polymer chain. An elimination reaction was studied by FT-IR and ^{13}C NMR spectra were measured. The elimination reaction was more rapid in free radical PVC (in solution) than in urea canal one (in suspension). The resulting poyene structure $-(CH=CH)_n-$ was essentially trans-form not cis-one.

^{13}C NMR spectra showed clearly that a large amount of polyene is formed in the free radical sample and very little in the urea canal one. Quite complexchemical inertness of the urea canal PVC was clarified in this way.

INTRODUCTION

Stereoregular polymer is obtained by urea clathrate polymerization at low temperatures[1]. Monomers align regularly in one dimensional narrow urea canal, and external irradiation of gamma-rays or electronbeams permits highly stereoregular polymers such as canal PVC and canal PAN (poly(acrylonitrile)). In spite of the fact that these are equally typical polar vinyl polymers, PVC has syndiotactic configuration, whereas PAN does isotactic one[2]. The reason for such different results has not been clarified. Here we studied the molecular structure and thermal properties of the former polymer.

RESULTS AND DISCUSSION

Urea canal PVC showed distinct difference in FT –IR spectra in the region of below $1400cm^{-1}$ (KBr disk). In addition to the observation by Krimm[3], specific Fermi resonance ($2933cm^{-1}$) was observed (Fig.1). The results can be reasonably explained in terms of the difference of crystalline region of syndiotactic sequence. In fact, WAXD pattern showed sixth co-axial rings as reported by Sakurada[4] in contrast with the amorphous hallow in free radical PVC. The canal PVC was totally insoluble to DMF and DMSO, whereas the free radical sample did these solvents at room temperature and provided transparent solutions. These samples were heated in DMF (solution or in suspension). In FT-IR spectra, the $1080cm^{-1}$ band (assigned to be trans double bond) appeared strongly in free radical PVC, but it did slightly in urea canal one. Fig. 2 showed the summarized results.

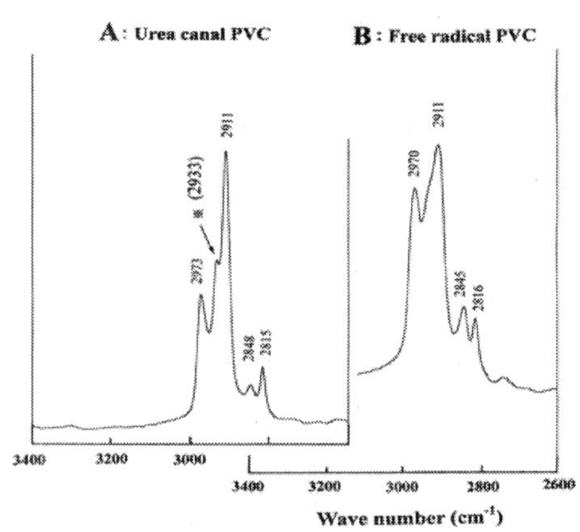

Fig. 1. Expanded FT-IR spectra of CH_2 region

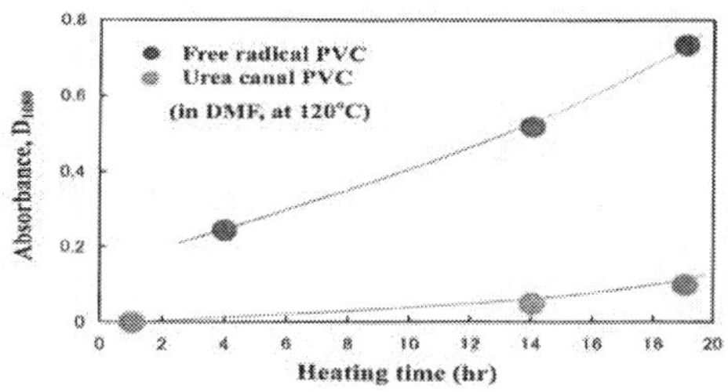

Fig. 2. Variation of intensity of $1080cm^{-1}$ band

Fig. 3 showed TG curves of two kinds of PVC. When compared with the weight loss value at 300°C, free radical PVC showed large weight loss than urea canal one.It is obvious that although initial weight loss is started at lower temperature, substantial weight loss is rather delayed in the canal PVC.

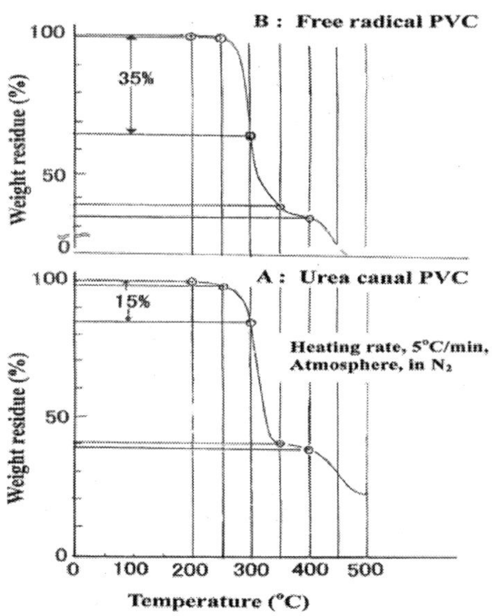

Fig. 3. Comparison of TG curves of two kinds of PVC

^{13}C NMR spectra of PVC heated in the liquid state were measured. In the original PVC, only two peaks (CH and CH_2) were observed[5]. Heated canal PVC showed almost identical spectra to that of original one except in the region of near 130ppm. In contrast-free radical one showed very broad strong peak at 130ppm, which can be assigned to polyene structure, –(CH=CH)n–. This is also supported by Fig. 4. Large spectral change was observed in free radical PVC than urea canal one. CH_2 peak became broad and split into several minor components. When compared with the NMR results heated in the liquid state and in the solid state, chemical inertness of urea canal PVC is clearly indicated, and the significance of morphological syndiotactic configuration on the thermal properties of PVC must be emphasized.

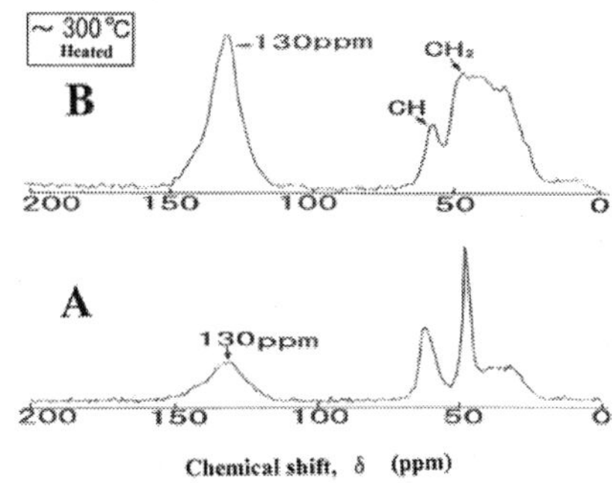

Fig. 4. Solid state NMR spectra of PVC heated in solid

CONCLUSION

Chemical inertness of urea canal PAN was discussedin comparison with that of free radical one. Easily discoloration of canal polymer was explained in terms of the absence of termination reaction, β-elimination.Rigid molecular structure of planer zigzag syndiotactic configuration was suggested based on FT-IR (Fermi resonance), TG and WAXD rings. ^{13}C NMR spectra of the residue prepared in different states (liquid, solid) was critically discussed.

REFERENCES

(1) White, D.M. J. *Am. Chem. Soc.*, 82, 1960: 5678.
(2) Minagawa, M., Utek, K., Kitayama, T., Hatada, K. *Macromolecules*, 27, 1994: 3669.
(3) Krimm, S. *Chem. Ind.*, 1959: 433.
(4) Sakurda, I., Nambu, K. *Nippon Kagaku Zasshi*, 80, 1959: 307.
(5) Minagawa, M., Narisawa, I., Sugisawa, H., Hasegawa, S., Yoshii, F. *J. Appl. Polym. Sci.*, 74, 1999: 2820.

Hyperbranched-Linear Block co-Polybenzimidazole for Non-Covalent Functionalization of Carbon Nanomaterials and the Effective Reinforcement of Their Composites

Yan Wang
College of Material Science and Engineering, Donghua University, Shanghai, China
wy@dhu.edu.cn

Carbon nanomaterials (CNMs), including carbon nanotube (CNT) and graphene, have gained extensive attention since their discovery due their excellent mechanical properties, conductivity, thermal and chemical stability, which make them promising materials for reinforcement of polymer composites. When incorporated into polymer matrix, the key factors for high-performance polymer composites are generally considered to be the homogeneous dispersion of CNMs in matrix, and the interfacial compatibility between them. Therefore, surface modifications on CNMs are necessary to address the above issues. Covalent modification of CNMs can make strong interfaces in CNMs/polymer composites, but the disruption of conjugated structures would compromise the intrinsic mechanical and electrical properties of CNMs. In contrast, non-covalent surface modifications could not only allow the homogenous dispersion of CNMs, but also preserve their integrated structures. As a result, many common surfactants or conjugated molecules have been used for functionalization of CNMs. However, when incorporated into matrix, the compatibility between modifier and matrix polymers should be delicately designed. The use of matrix polymer for direct functionalization of CNMs could thus avoid this problem, and the composites could be obtained by a simple dispersion-casting process. Aromatic polymers have long been used for this purpose because of the π-π interactions between polymer chains and CNMs. However, because of their relatively rigid polymer backbones and the poor solubility in common solvents, the polymer chains could not wrap on CNMs very well, as a result, the concentration of dispersed CNMs in solvents or polymer matrix were generally lower than flexible polymers dispersed CNMs[1,2]. On the other hand, hyperbranched polymers (HBPs) have also been used for non-covalent modification of CNMs, the extended configuration of HBPs were suggested to be helpful for adsorption of polymers on CNMs, which could amplified the interactions between polymers and CNMs[3]. However, because of the global structure and the lack of chain entanglements in HBPs, it is difficult to direct formation of macroscopic composites from the CNMs/HBPs mixture.

Based on the above considerations, we reported the use of hyperbrached-linear block copolymers for the modification and dispersion of CNMs. The hyperbranched blocks could enhance the interactions with CNMs, while the linear blocks could endow the polymers formability. As a result, homogenous composites with high concentrations of CNMs could be facile fabricated by a simple dispersion-casting process. Polybenzimidazole (PBI) was used as model polymer because of its high mechanical and thermal properties. In addition, the π-π interactions between PBI and CNT[1] or graphene[2] have been demonstrated, thus, the different dispersion efficiency between our copolymers and linear PBIs could be readily attributed to the structural changes.

The copolymers were synthesized by the copolymerization of hyperbranched PBIs (HBPBIs)[4] and linear PBIs with different weight ratios. We initially found that when the weight ratio of HBPBIs and PBIs was 1:15, the copolymers have reasonable high dispersion effect to multi-walled CNTs (MWNTs), and the concentration of dispersion of MWNTs was much higher than that of linear PBIs dispersed MWNTs. Moreover, the mechanical properties of copolymers (1:15) were similar to that of linear PBIs, while the pure HBPBIs could not form free-standing films, which confirmed the advantages of our copolymers.

Then we systematically investigated the dispersion effect of copolymers to single-walled CNTs (SWNTs), MWNTs and graphene. Raw SWNTs, MWNTs, and graphite (1 mg/mL) were sonicated in DMAc for 1 h separately in the presence of copolymers (1mg/mL), then the mixture were subjected to centrifugation at 3000 rpm for 30 min to remove the non-dispersed aggregates. CNMs were also sonicated and centrifugated in DMAc/PBIs solutions for comparison. The resultant concentrations of dispersions of CNMs were compared in Figure 1. As seen, the concentrations of all CNMs in DMAc/copolymers solutions were much higher than those in corresponding DMAc/PBIs solutions, demonstrating the effective surface modification of our copolymers to CNMs.

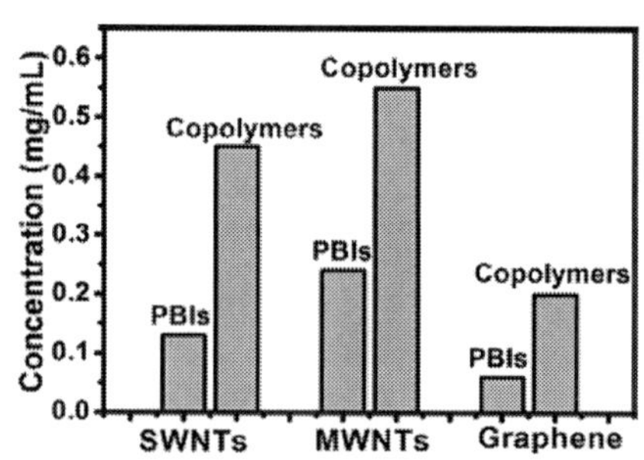

Figure 1. Comparison of the concentrations of dispersed CNMs in DMAc with the assistant of PBIs and copolymers.

The RBM bands of SWNTs and 2D bands of MWNTs and graphene in Raman spectra were used to investigate the interactions between copolymers and CNMs, the interactions between PBIs and CNMs were also probed by Raman spectra for comparison, as shown in Figure 2. For SWNTs, the peak at 165 cm^{-1} red-shifted when functionalized with PBIs and copolymers, demonstrating the π-π interactions between nanotube walls and polymers. Moreover, the shifts in SWNT/copolymers was larger than

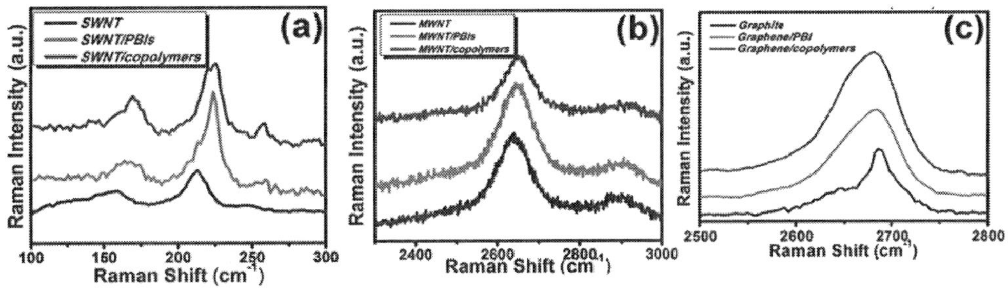

Figure 2. Raman spectra of SWNTs (a), MWNTs (b) and Graphene (c) functionalized with PBIs and copolymers.

that in SWNT/PBIs, demonstrating the stronger interactions between SWNTs and copolymers. For MWNTs, the 2D bands shifted from 2638 cm^{-1} in raw MWNTs to 2646 cm^{-1} in MWNT/PBIs and to 2651 cm^{-1} in MWNT/copolymers, conforming again the stronger interactions between nanotube walls and copolymers. For graphene, the shape of 2D bands could provide information on exfoliation state. From Figure 2c, we can see that the shapes of 2D bands of polymer exfoliated graphene were totally different from that of graphite, demonstrating the well-exfoliation state of graphene. In addition, the 2D band in graphene/copolymers were more symmetrical than that of graphene/PBIs, suggesting the more efficient exfoliation of graphene by copolymers, which was also attributed to the stronger interactions between carbon plane and copolymers.

After confirming the good dispersion state of CNMs by the assistant of copolymers and the stronger interactions between them, we next evaluated the reinforcement effects of CNMs to copolymers, CNMs reinforced PBIs were also prepared for comparison. All composites were fabricated by dispersion of CNMs in matrix and then direct solution-casting. Here, we only described the composites containing MWNTs because they were already commercialized and were more cost-effective than the other two CNMs in terms of preparing polymer composites. The typical stress-strain curves of MWNT/PBIs and MWNT/copolymers composites with different amount of MWNTs were shown in Figure 3a and 3b, the Young's modulus and tensile strength these composites were compared in Figure 3c and 3d. As can be seen, as the loadings of MWNTs were low, the mechanical properties of both MWNT/PBIs and MWNT/copolymers were improved, demonstrating the good reinforcement effects of homogeneously dispersed MWNTs in both polymers. However, as the loading of MWNTs approached 1 wt%, the modulus and strength of MWNT/PBIs were decreased, suggesting the deteriorated dispersion state of MWNTs in PBIs. In contrast, the mechanical properties of MWNT/copolymers were still improved as the content of MWNTs reached 1.5 wt%, and started to decrease as the content of MWNTs was 2 wt%, demonstrating the more efficient of MWNTs for reinforcement of copolymers, which might be resulted from the better dispersion state and the stronger interfacial interactions. The increased ratio of modulus and strength of MWNT/PBIs and MWNT/copolymers were also compared. For MWNT/PBIs composites, the highest increased ratio of modulus and strength were 24.9% and 11.4% when the MWNT content was 1 wt%. For MWNT/copolymers composites, the highest increased ratio of modulus and strength were 46.8% and 25.7%, respectively, as the MWNT content was 1.5 wt%, demonstrating again the more effective reinforcement of MWNTs to copolymers.

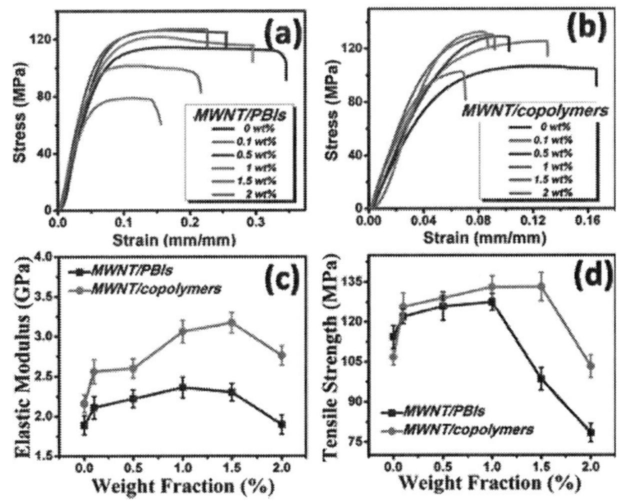

Figure 3. Typical stress-strain curves of MWNT/PBIs (a) and MWNT/copolymers (b). The comparison of Young's modulus (c) and tensile strength (d) of MWNT/PBIs and MWNT/copolymers composites.

In summary, we presented the synthesis of a hyperbranched-linear block co-polybenzimidazole for dispersion of CNMs. These copolymers were more efficient in dispersion of CNMs in solvents because of the strong interfacial interactions to CNMs as compared to that of linear polymers. Because of the better dispersion state and stronger interactions, the reinforcement effect of CNMs were also better than that of CNMs reinforced linear polymers. This simple method was expected to extend to other polymers for facile fabrication of high-performance polymer composites.

ACKNOWLEDGMENT

The author thanks the National Natural Science Foundation of China (no. 51203019) for the support.

REFERENCE

1. Okamoto, M., Fujigaya, T., Nakashima, N. *Adv. Funct. Mater.*, 18, 2008: 1776-1782.
2. Fujigaya, T., Hirata, S., Nakashima, N. *J. Mater. Chem. A*, *2*, 2014: 3888-3893.
3. Xu, L., McGraw, J.-W., Gao, F., Grundy, M., Ye, Z., Gu, Z., Shepherd, J.L. *J. Phys. Chem. C*, 117, 2013: 10730-10742.
4. Xu, H., Chen, K., Guo, X., Fang, J., Yin, J. *J. Membr. Sci.*, 288, 2007: 255-260.

Synthesis and Characterization of Melt Processable Acrylonitrile-N-Vinylimidazole Copolymers

Na Han, Hui-hui Li, Xing-xiang Zhang

State Key Laboratory of Hollow Fiber Membrane Materials and Membrane Process; Tianjin Municipal Key Lab of Fiber Modification and Functional Fiber, School of Materials Science and Engineering, Tianjin Polytechnic University, Tianjin, China

hannapolyu@163.com, zhangpolyu@aliyun.com

ABSTRACT

Polyacrylonitrile (PAN) and acrylonitrile (AN)-N-vinylimidazole (VIM) copolymers (abbreviated as P (AN-co-VIM)) with various feed molar ratios were synthesized by free-radical solution copolymerization in dimethyl formamide (DMF).

FTIR and ^{1}H NMR results revealed that P(AN-co-VIM) could be synthesized successfully by free-radical solution copolymerization. DSC results showed that the decomposition temperature of P(AN-co-VIM) shifts continuously to higher temperatures with VIM comonomer contents increasing. XRD results showed that the crystallinity and crystallite size decreased obviously with the increase of VIM contents. P (AN-co-VIM) is melt processable as VIM content is higher than 10 mol%.

KEYWORDS: Acrylonitrile-N-vinylimidazole copolymer, feeding molar ratio, melt processable

INTRODUCTION

Polyacrylonitrile (PAN) fibers which have excellent characteristics used in textiles and carbon fibers have been widely investigated [1~3]. It is well known that PAN tends to degrade (~220°C) before its melting point (~320°C) [2]. It is due to the fact that the adjacent nitrile groups on the polymer chains cyclize to form six-membered rings, and PAN undergoes a degradation reaction prior to melting at 320–330°C upon heating [4].

It was postulated that suitable comonomer could be used to reduce the long-range order of PAN. It has been demonstrated that solvent-free melt spinning can be conducted successfully [5~7]. VIM has a C=C double bond in the main chain and a big pendant imidazole group which can disrupt PAN crystallinity to make it melt-processable. Imidazole-containing macromolecules have been suggested as carrying an active moiety of several electrolytic enzymes [8, 9], which can make it has more brilliant properties and wider applied realm.

In this study, a series of P(AN-co-VIM) copolymers were synthesized by free-radical solution copolymerization. The structure and thermal property of the copolymers were studied to find a set of copolymers suitable for melt spinning.

APPROACH

A 250mL 4-necked flask was equipped with a condenser, a dynamoelectric stirrer, a dropping funnel and a nitrogen inlet. Firstly, the flask was charged with 50mL of DMF and purged with dry nitrogen for 30 min. The mixture of AN, MA, RSH and AIBN were added into the flask over 2h[10]. The polymerization reaction was carried out at 70°C with continuous stirring for 12h. The product was filtered and washed by ethyl alcohol and de-ionized water, and then dried in vacuum oven at 50°C for 48h till constant weight was obtained. A series of P(AN-co-VIM) prepared with various feed molar ratios are listed in table 1.

Table 1. P(AN-co-VIM) samples with various feed molar ratios.

P(AN-co-VIM)	Feed molar ratio (AN/VIM)	m_{AN} /g	m_{VIM} /g
A	100/0	15.918	0
B	93/7	14.804	1.977
C	90/10	14.326	2.824
D	88/12	14.008	3.388
E	85/15	13.530	4.235
F	82/18	13.053	5.082
G	80/20	12.734	5.647

RESULTS AND DISCUSSION

^{1}H NMR

A typical ^{1}H NMR spectrum of 85/15 P (AN-co-VIM) copolymer is shown in Fig. 1. The signal at δ3.1corresponds to –CH– on the backbone of AN units; the signal at around δ4.5corresponds to –CH– on the backbone of VIM units. These two characteristic peaks were used to calculate the compositions. The calculated compositions of AN/VIM copolymers were listed in table 2.

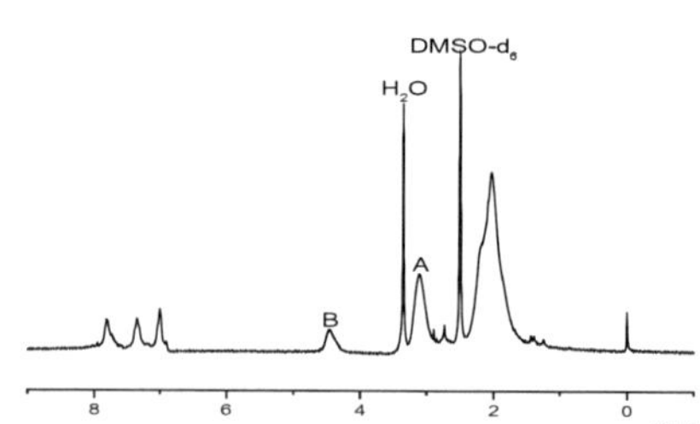

Fig. 1. ^{1}H NMR spectrum of 85/15 P(AN-co-VIM) copolymer.

DSC

It is believed that the melting peak of PAN and its copolymers could be observed clearly only at very high heating rates (i.e., 60°C/min or above) under a nitrogen atmosphere [11]. The melting points of PAN and P (AN-co-VIM) copolymers were obtained by rising the heating rate to 60°C /min as Fig. 2.

There was not a melting peak on the curve of PAN. It is noted that T_m of 93/7, 90/10 and 88/12 P(AN-co-VIM) could be observed at 232.2°C, 199.4°C and 200.2°C. However, there was no clear melting peak for 85/15, 82/18 and 80/20 P(AN-co-VIM) due to the crystalline regions decreased with the further increasing of VIM content.

Table 2. Conversion rate, intrinsic viscosity and Molar ratios of P(AN-co-VIM) copolymers.

Sample no.	Conversion /%	Intrinsic viscosity dL/g	Composition of P(AN-co-VIM) mol%
B	66.21	0.32	91.05/8.95
C	68.92	0.42	84.66/15.34
D	78.01	0.46	83.08/17.92
E	83.54	0.55	80.08/19.92
F	72.90	0.39	73.83/26.17
G	84.87	0.39	72.30/27.70

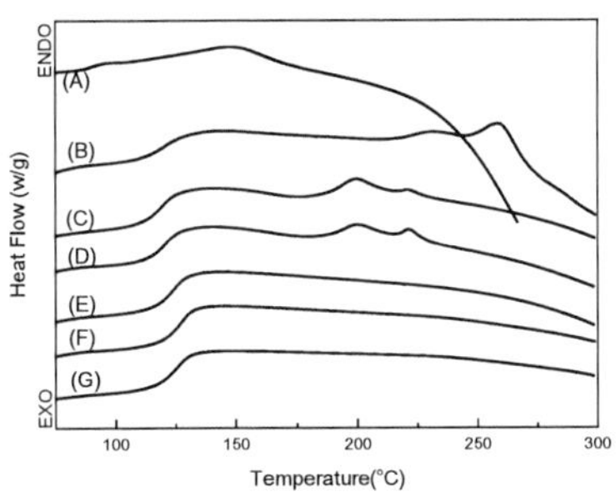

Fig. 2. The DSC heating curves (60°C/min) of PAN and P(AN-co-VIM) copolymers with varous feed molar ratios: (A) PAN; (B) 93/7; (C) 90/10; (D) 88/12; (E)85/15; (F) 82/18; (G) 80/20.

POM

A series of POM photos of 85/15 P(AN-co-VIM) were shown in Fig. 3. It starts to melt at the temperature of 170°C. The transparent areas (melting part) connect with each other and the edge goes on shrinking at 180°C. The melting part is further enlarged and the edge becomes smooth at 190°C. The copolymer is almost completely melt at 20°C. It was observed obviously that the copolymer is melt processable as the VIM content is higher than 10 mol%, which is corresponding to DSC curves. It is furtherly evidenced that melt processable polyacrylonitrile based copolymer has been successfully synthesized by incorporating VIM units into PAN molecular chains.

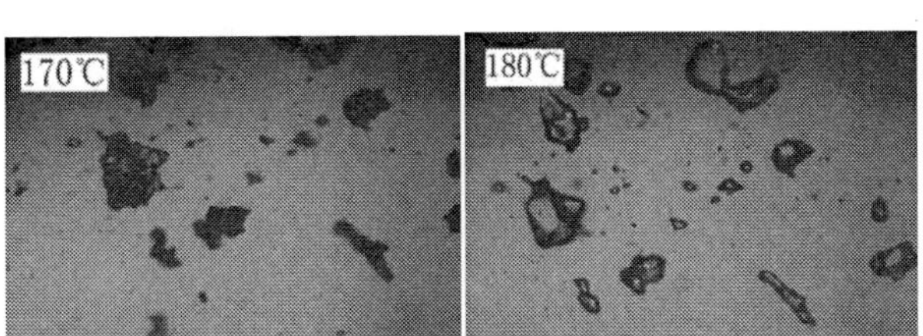

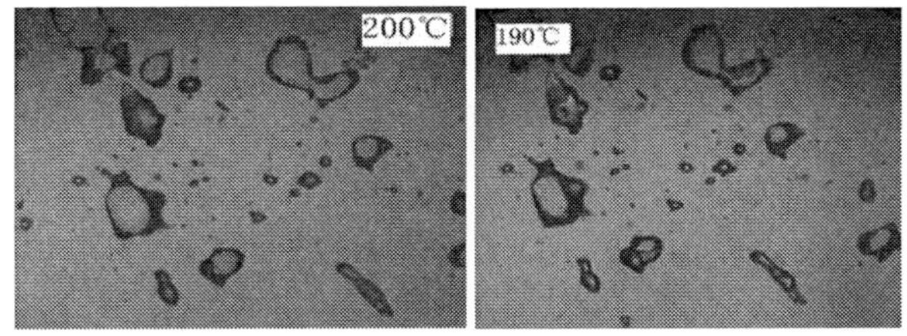

Fig. 3. Hot stage microscope photo of P(AN-co-VIM) 85/15 copolymer.

CONCLUSIONS

Polyacrylonitrile and acrylonitrile-N-vinylimidazole copolymers (P(AN-co-VIM)) with various feed molar ratios were synthesized successfully. P (AN-co-VIM) copolymer is melt processable when the VIM content is higher than 10 mol%. With VIM comonomer content increasing, the glass transition temperature of P(AN-co-VIM) decreases and the decomposition temperature increases. The melting temperature can be depressed to about 180°C and the initial decomposition temperature can be raised to 321.2°C. It is concluded that VIM is a promising comonomer for preparing PAN fiber by melting spinning, which will be applied in the carbon fiber or commercial fiber.

ACKNOWLEDGMENT

The authors are thankful to the National Natural Science Fund of China (21206123); Doctoral Program of Higher Education of China (20111201120001); Postdoctoral Program projects (2014M551026, 201402011); Tianjin Municipal Applied Basic Research and Frontier Technology Research Programs (13JCQNJC02300) for financial support.

REFERENCES

[1] Coxe, C.D. US, 2585444, 1952.
[2] Masson, James C., Chen, G.K., Shen, X.Y., Yao, L. *Acrylic Fiber Technology and Applications*. Beijing: China Textile and Apparel Press, 2004.
[3] Fitzer, E. *Carbon*, 27, 5, 1989: 621.
[4] Clarke, A. J., Bailey, J.E. *Nature*, 243, 1973: 146.
[5] Bajaj, P., Sreekumar, T.V., Sen, K.J. *Appl. Polym. Sci.*, 86, 3, 2002: 773.
[6] Han, N., Zhang, X.X., Wang, X.C. *Iranian Polym. J.*, 19, 4,2010: 243.
[7] Mukundan, T., Bhanu, V.A., Wiles, K.B., et al. *Polymer*, 47, 2006: 4163.
[8] Overberger, C.G. *Bioorg. Chem.*, 3, 1, 1974: 35.
[9] Alpert, J.A., Regnier, F.E. *J. Chromatoger*, 185, 1979: 375-92.
[10] Wang, X.H., Qin, X.Y., Lv, Y.G. *Mater. Sci. & Engin.*, 27, 4, 2011: 85.
[11] Gupta, A. K., Paliwal, D. K., Bajaj, P. *J. Appl. Polym. Sci.*, 70, 1998: 2703.

Posters

Engineering of Fluorescent Emission Silk Fibroin Composite Materials by Material Assembly

Zhang Fayin[1], Xu Hongyao[1], Lin Naibo[2], Meng Zhaohui[2], Liu Xiangyang[3]
[1]College of Material Science and Engineering & State Key Laboratory for Modification of Chemical Fibers and Polymer Materials, Donghua University, Shanghai, China
[2]Research Institute for Soft Matter and Biomimetics, College of Materials & Department of Physics, Xiamen University, Xiamen, China
[3]Department of Physics, National University of Singapore, Singapore
hongyaoxu@dhu.edu.cn; zhangfy0516@163.com

STATEMENT OF PURPOSE

The materials assembly technology allows the designated materials to acquire additional/ enhanced performance by fixing "functional components" into the materials. To fabricate such multifunctional materials, we should be able to assemble functional nanoparticles/molecules into the "substrate" materials on a chemically diverse range. In this research, we will explore a novel approach of material assembly to fabricate fluorescent emission silk fibroin bio composite materials.[1] The dominating hydrophobic peptide repeats self-organize into the β-sheet structures, mediated by hydrophobic hydration. [2] Recently, Bombyx mori silk fibers, silk regenerated fibers, films and sponges have been adsorbing great interests in the fields of biomedicine and high performance materials, [3]due to their intrinsic functions, ie. in vitro and in vivo biocompatibility,[4] robust mechanical properties,[5] and relatively slow biodegradation. [6] The organic styryl-fluorene molecules and QDs have been widely investigated in the last two decades because of the unique photonluminescence and TPF properties in photonic devices.[7] The silk fibroin biomaterials with efficient TPF would be of great scientific and technological relevancy in the areas of 3D fluorescence bio-imaging,[11] 3D optical data storage,[8] up-converted lasing, [9] and photodynamic therapy.[10]

INTRODUCTION

Nowadays, advanced multifunctional materials are of great importance in astronomical science & technology, military, biomedical and domestic applications, which will have a great impact in our lives in the near future.[11] A wide range of environment-friendly bio-composites have demonstrated their various applications in drug/gene delivery, tissue engineering, cosmetic orthodontics, agriculture, modern construction, electronics.[12] As a branch of multifunctional biomaterials, silk fibroin biomaterials have attracted more and more attention. In order to obtain multifunctional silk, silk materials can be doped with specific function materials, such as organic material and nanomaterial. When specific function materials are embedded in organic polymer matrices, the obtained materials not only exhibit the combined properties of the original components but also exhibit improved performances over the original components.[13] The new trend and challenge for nano sciences and technologies are to assemble functional nanoparticles in a control manner so as to acquire some robust multifunction and performance.[14]

APPROACH

The nitro groups in the TPF styryl-fluorene molecules and the amide groups in silk fibroin molecules are supported to recognized and interacted. The silk fibroin solution and CdTe QDs solution were mixed to form stable and uniform protein/QDs hybrid solution in the alkaline condition by the hydrogen-bonding interaction between NH groups of silk fibroin and carboxylic group (COO-) of QDs.

RESULTS AND DISCUSSION

TPF styryl-fluorene molecules are supposed to be incorporated into the amorphous region of silk fibers at molecular levels by molecular recognition and interaction. NM in the silk fibers exhibits the bright-yellow emission with an elevated fluorescence efficiency of about 22 %, which is significantly higher than that (1%) of NM in the solid state. This confirms that the TPF emission enhancement is caused by the -NO2 terminal groups of NM molecules, which can interact effectively with the amide groups of silk fibroins. The occurrence of the hydrogen -bonding interaction between NH groups of silk fibroin and carboxylic group (COO-) of QDs is key factor in controlling the QDs dispersion. The TPF properties of CdTe QDs/silk fibroin composite films shows that the fluorescence emission is from both the surface and the centers of the CdTe QDs/silk fibroin composite film, indicating the uniform distribution of CdTe QDs in CdTe QDs/silk fibroin composite films.

CONCLUSION

We report on a new and versatile material assembly method to obtain fluorescent emission silk fibroin bio composite materials, and it is applied in non-invasive scaffold imaging. Preliminary results show that this material assembly method endows the resulting silk fibroin composite materials with TPF properties. The molecular recognition between the nitro groups of the NM molecules and amide groups of the silk fibroin greatly enhances the TPF emission. CdTe QDs with tunable TPF was obtianed, which can interact with silk fibroin by hydrogen bond.

REFERENCES

1. Ghosh, S., Parker, S.T., Kaplan, D.L., Lewis, J.A. *Adv. Funct. Mater.*, 18, 13, 2008: 1883-1889;
2. Du, N., Liu, X. Y., Narayanan, J., Li, L. A., Li, D.Q., *Biophys. J.*, 91, 12, 2006: 4528-4535;
3. Place, E.S., Evans, N.D., Stevens, M.M., *Nat. Mater.*, 8, 6, 2009: 457-470;
4. Jiang, C., Wang, X., Naik, R.R., Tsukruk, V.V. *Adv. Funct. Mater.*, 17,13, 2007: 2229-2237.
5. Vasconcelos, A., Freddi, G., Cavaco-Paulo, A. *Biomacromolecules*, 9, 4, 2008: 1299-1305;
6. Jacek, J., Signorini, S., Bozio, R., Martucci, A., Mulvaney, P. *Adv. Mater.*, 20, 1, 2008: 69-73;
7. Denk, W., Strickler, J.H., Webb, W.W. *Science*, 248, 4951, 1990: 73-76;
8. Jordan, G., Kobayashi, T., Blau, W.J., Horhold, H.H. *Adv. Funct. Mater.*, 13, 10, 2003: 751-754;
9. Dichtel, W.R., Frechet, J.M.J., Prasad, P. N., *J. Am. Chem. Soc.*, 126, 17, 2004: 5380-5381;
10. Qi, Z.D., Li, D.W., Jiang, P., Cheah, K. W. *J. Mater. Chem.*, 21, 8, 2011: 2455-2458.
11. Benabid, F., Knight, J.C., Antonopoulos, G., Russell, P.S.J. *Science*, 298, 5592, 2002: 399-402.
12. Shao, Z., Vollrath, F. *Nature*, 418, 6899, 2002: 741-741;
13. Shallcross, R.C., Pyun, J., Armstrong, N. R. *J. Am. Chem. Soc.*, 132, 8, 2010: 2622-2632;
14. Vercelli, B., Zotti, G., Berlin, A., Natali, M. *Chem. Mater.*, 22, 6, 2010: 2001-2009.

Knitting Technologies of Novel, Curved, Flat-Knitted Three-Dimensional Spacer Fabrics

Xiaoying Li[1,2], Gaoming Jiang[1], Pibo Ma[1]
[1]Engineering Research Center for Knitting Technology, Jiangnan University, Wuxi, China
[2]School of Art and Design, Guangdong University of Technology, Guangzhou, China
mapibo@jiangnan.edu.cn

ABSTRACT

This paper introduces a knitting technique for making innovative curved three-dimensional (3D) spacer fabrics by the computer flat-knitting machine. During manufacturing, a number of reinforcement yarns made of aramid fibers are inserted into 3D spacer fabrics along the weft direction to enhance the fabric tensile properties. This research can help the further study of 3D spacer fabric when application to composites.

KEY WORDS

Three-dimensional spacer fabric, aramid fiber, computer flat-knitting machine, knitting technology

INTRODUCTION

New textile structures have attracted more and more researchers' attention in exploration of textile composites. Recent years, flat-knitted three-dimensional (3D) spacer fabrics have been becoming attractive for making composites due to the specific properties. The structure of flat-knitted 3D spacer fabric is novel, which is made of two separate fabric layers connected vertically by pile yarns. These 3D spacer fabrics normally have high thickness and excellent performance, such as good structural integrity, low weight and density, and high impact resistance, heat resistance and sound absorption. These structural 3D fabrics have been used for transport vehicles, aircraft, storage, bridge, marine applications and others [1-4]. The separated layers of flat-knitted 3D spacer fabrics can be connected by pile yarns or fabric layers. The spacer fabrics bound with fabric layers have greater potential applications and development value because of the fabric without limited thickness from the distance between two needle beds. Flat-knitted 3D spacer fabrics could be manufactured using the advanced modern flat knitting machine, which may be a good solution to develop such innovative 3D structures. One goal of the current study is to develop curvilinear flat-knitted 3D spacer fabrics by using aramid fibers. These fabrics can be used as reinforced materials for different composites with improved mechanical properties. This can meet the high requirements of industrial production. In order to achieve this objective, a few actual spacer fabrics were manufactured by using aramid fibers.

APPROACH

Fat-knitted innovative 3D spacer fabrics

Materials and computer flat knitting machine

Aramid fibers (440Dtex/100f×3) are used in knitting. Knitting is carried out by a modern computer flat knitting machine (CMS530 of company STOLL from Germany), Gauge is 3.5 mm.

Flat-knitted 3D spacer fabrics

Computer flat knitting technique is a better way in developing innovative structures by using unique technical features such as holding down sinkers, transfer, racking, etc. Spacer fabrics show complex 3D constructions with two separate fabric layers bonded each other. The yarns experience a long way from the yarn feeder to the knitting area which produces large friction and big input tension.

The flat-knitted 3D spacer fabrics are developed based on the above mentioned innovative knitting skill. In our previous work, rib-course has been done only once in connecting layer. However, single rib has been founded that it is too easy bent to meet the requirement after experiment. So the program is improved to keep the space shape better by knitting rib-course twice.

Development of flat-knitted 3D spacer fabrics with reinforcement yarns

In order to enhance the tensile properties of flat-knitted 3D spacer fabrics, the reinforcement yarns are integrated into 3D spacer fabrics in surface layers or in fabric layers (in the weft direction). If the reinforcement yarns are replaced by functional or conductive yarns (such as carbon filament, copper wire) in flat-knitted 3D spacer fabrics. If the reinforcement yarns are replaced by other high-functioning fibers, the flat-knitted 3D spacer fabrics could be used as high-functioning fabrics. The way that reinforcement yarns are inserted into 3D spacer fabrics in surface layers is shown in Figure 1.

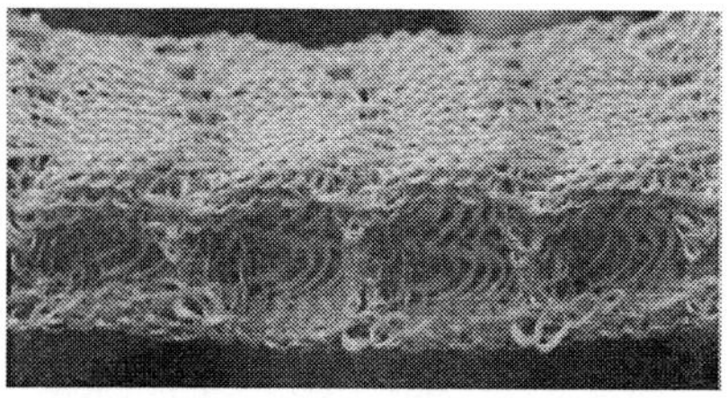

With reinforced yarns

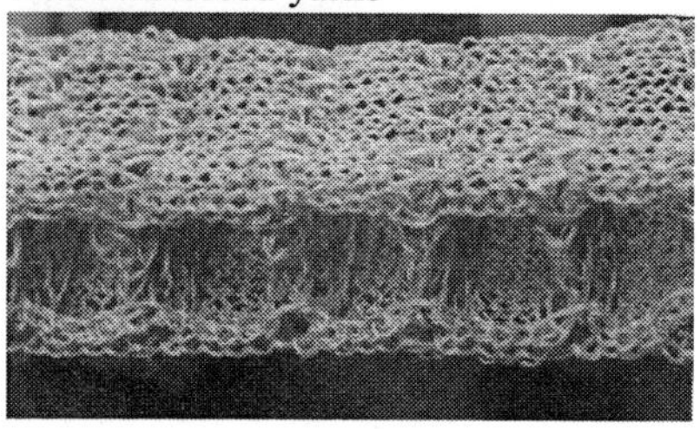

Without reinforced yarns

Figure 1. Flat-knitting 3D spacer fabric with reinforced yarns and without reinforced yarns

Flat-knitted 3D curvilinrear spacer fabrics

The advanced flat-knitting technique for 3D spacer fabrics also permits the various curvilinear shapes in the warp direction. The curvilinear spacer is innovative structure, which can be developed out of flat-knitted 3D spacer fabrics. To maintain the relative balance of comb-down, the length of H is changed, and the L should be adjusted accordingly to ensure knit smoothly. For instance, as it is shown in Figure 2.

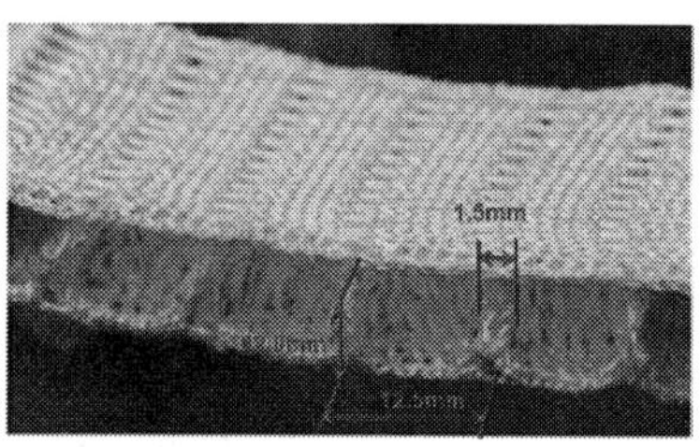

Flat-knitting 3D spacer fabric with reinforced yarns

Spacer fabric with 180° curvature

Figure 2. Fabric samples

RESULTS AND DISCUSSION

Innovative flat-knitted 3D spacer fabrics and curvilinrear effects

Shape stability is simply achieved by knitting the rib twice in fabric layers. This innovative flat-knitted 3D spacer fabrics show the better mechanical properties obviously. And then in order to enhance the tensile properties of such spacer fabrics, the reinforcement yarns are integrated in weft direction(Figure 2-a). The 3D curved spacer fabrics of arc angle in warp direction from 0° to 360° were realized successfully (Figure 2-b). The distance between the surface layers and fabric layers is designed to 15mm. Because of inclined joining by the fabric layers, it is dissimilar for both surface layers of the spacer fabrics with the curvature angles of 90° and 360°.

CONCLUSION

Three-dimensional spacer fabrics are produced innovatively and successfully by knitting rib-course twice in fabric layers and inserting reinforcement yarns in surface layers (weft direction), and then with different angles (warp direction). All these fabrics were manufactured using aramid fiber by computer flat-knitting machine.

Hence, beyond the traditional application trends in the fields of lightweight composite, integrated sandwich composites using this textile-reinforced structure can potentially be used in some industrial fields by filling in the wax, wires, stacking electronic chips and embedded electronic components.

ACKNOWLEDGMENT

The authors acknowledge the financial support from the National Science Foundation of China (No.11302085 and 51403080), and the Innovation fund project of Cooperation among Industries, Universities & Research Institutes of Jiangsu Province (No. BY2014023-34 and BY2014023-20).

REFERENCES

[1] Abounaim, M.D., Chokri, C. Flat-knitted innovative three-dimensional spacer fabrics: a competitive solution for lightweight composite applications." *Textile Research Journal*, 82, 3: 2012: 288-298.

[2] Abounaim, M.D., Olaf, D., Gerald, H., Chokri, C. "Thermoplastic composites from curvilinear 3D multi-layer spacer fabrics." *Journal of Reinforced Plastics and Composites*, 29, 24, 2010: 3554-3565.

[3] Abounaim, M.D., Gerald, H., Olaf, D., Chokri C. "Thermoplastic composites from curvilinear flat knitted 3D multi-layer spacer fabric using hybrid yarn and the study of 2D mechanical properties." *Composites Science and Technology*, 70, 2010: 363-370.

[4] Unal, A., Hoffmann. G., Cherif, C. "Development of weft knitted spacer fabrics for composite materials." *Melliand Textile Berichte*, 4, 224-226, 2006: E49-50.

Folic Acid-Modified Electrospun Polyvinyl Alcohol/Polyethyleneimine Nanofibers for Cancer Cell Capture Applications

Zhangyu Fan[1], Yili Zhao[2], Xiangyang Shi[1,2]
[1]College of Chemistry, Chemical Engineering and Biotechnology, Donghua University, Shanghai, China
[2]College of Textiles, Donghua University, Shanghai, China
xshi@dhu.edu.cn

INTRODUCTION

Electrospun nanofibers can provide a large number of cell contact sites and be used for cell capture applications (Zhang et al.). Folate receptor (FR) is a single-chain glycoprotein with high specific affinity for folic acid (FA) (Wang et al.; Zhao et al.). In this study, PEGylated FA was conjugated to the surface of the PVA/PEI electrospun nanofibers for the specific capture of cancer cells.

EXPERIMENTAL

Synthesis of FA-PEG-COOH

FA was first activated with EDC and NHS in DMSO at room temperature for 3 h. The resulting solution of activated FA was dropwise added into the DMSO solution of NH_2-PEG-COOH (Mw = 2000) under vigorous magnetic stirring at room temperature for 3 days. Finally, the reaction mixture was dialyzed against phosphate buffered saline (PBS) and water for 3 days using a dialysis membrane with MWCO of 1000, followed by lyophilization to obtain the product FA-PEG-COOH.

Preparation of FA-modified PVA/PEI nanofibers

12 wt% PVA/PEI (w/w = 3:1) solution was electrospun at the following parameters: 0.3 mL/h, 25 cm, 18.6 kV to obtain PVA/PEI nanofibrous mats according to our previous work (Fang et al.). To render the water stability of the mats, they were exposed to GA vapor under vacuum for 24 h. After that, the prepared FA-PEG-COOH was activated using EDC/NHS under stirring for 3 h. A circular crosslinked PVA/PEI nanofibrous mat was exposed to the activated FA-PEG-COOH solution under gentle shaking at room temperature for 3 days. Then triethylamine with 5 molar equivalents of the PEI amines was first added and shaken for 30 min, acetic anhydride with similar molar equivalents of triethylamine was then added and shaken for 24 h. For comparison, non-targeted PVA/PEI nanofibers (*m*PEG-PVA/PEI-Ac) were also prepared.

Cell capture assay

U87MG cells were seeded onto the FA-PEG-PVA/PEI-Ac or *m*PEG-PVA/PEI-Ac nanofibrous mats pre-fixed in 24-well cell culture plates at a density of 10^5 cells/well with the medium volume of 400 μL/well, and incubated for 10, 20, 40, 60, and 120 min, respectively. At each pre-determined time point, the cell culture medium and the PBS washing solutions containing suspended cells were collected, followed by counting the number of cells using Scepter 2.0 Handheld Automated Cell Counter. To further prove the targeted cancer cell capture ability, L929 cells were also seeded onto the targeted and non-targeted PVA/PEI nanofibrous mats, treated, and analyzed under similar experimental conditions. Confocal microscopy was also used to check the cancer cell capture ability of FA-PEG-PVA/PEI-Ac nanofibrous mats. At each pre-determined time point, the cells were washed with PBS for 3 times, fixed with glutaraldehyde (2.5%) for 15 min at 4 °C, and counterstained with Hoechst 33342 (1mg/mL, 20 μL) for 20 min at 37 °C using a standard procedure. The cells attached onto the nanofibrous mats were imaged using a 10× objective lens.

RESULTS AND DISCUSSION

Characterization of FA-PEG-COOH

Figure 1 shows that the peak between 3 and 4 ppm is assigned to the $-CH_2-$ protons of the NH_2-PEG-COOH. The appearance of aromatic protons associated with FA at 6.62, 7.70, and 8.64 ppm in the ^1H NMR spectrum of FA-PEG-COOH clearly indicates that FA has been successfully conjugated to the NH_2-PEG-COOH. Based on NMR integration, the average number of FA coupled to each PEG was estimated to be 0.85.

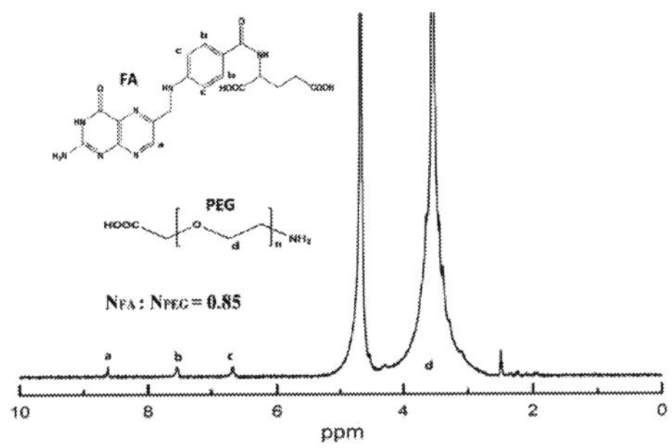

Figure 1. ^1H NMR spectrum of FA-PEG-COOH conjugate.

Formation of FA-modified PVA/PEI nanofibers

SEM (Figure 2) shows that the diameter of the formed PVA/PEI nanofibers is 425 ± 60.8 nm. After exposure of the nanofibers to the GA vapor, the nanofibrous mats became water-insoluble and still kept excellent porous fiber structure, the mean diameter increased to 493 ± 65 nm, possibly due to the swelling of the nanofibers during GA vapor crosslinking. After surface modification, the morphology become curly and the mean diameter of *m*PEG-PVA/PEI-Ac (599 ± 55.4 nm) and FA-PEG-PVA/PEI-Ac (625 ± 55.4 nm) nanofibers increases, which should be attributed to the swelling of the fibers during the process of modification.

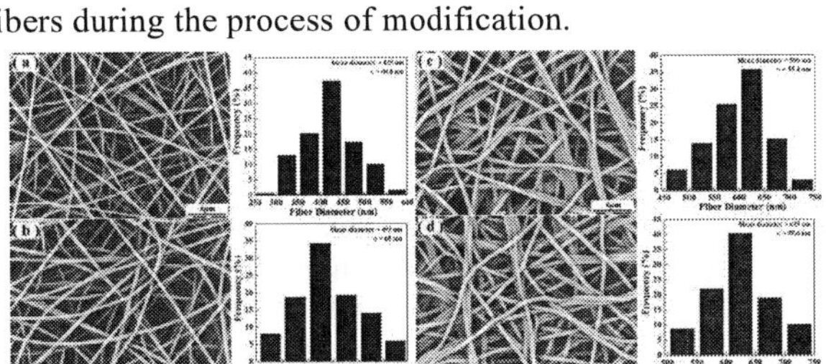

Figure 2. SEM micrographs and diameter distribution histograms of (a) PVA/PEI nanofibers, (b) Crosslinked PVA/PEI nanofibers, (c) *m*PEG-PVA/PEI-Ac nanofibers, and (d) FA-PEG-PVA/PEI-Ac nanofibers, respectively.

As shown in Figure 3, a new peak at 1650 cm^{-1} represents the formation of aldimine linkage after crosslinking, which demonstrated the successful crosslinking reaction. Both FA-modified nanofibers and FA-PEG-COOH have three peaks near 1500 cm^{-1}, which can be assigned to the benzene ring of FA, further confirming the successful modification of FA-PEG-COOH onto the fiber surfaces.

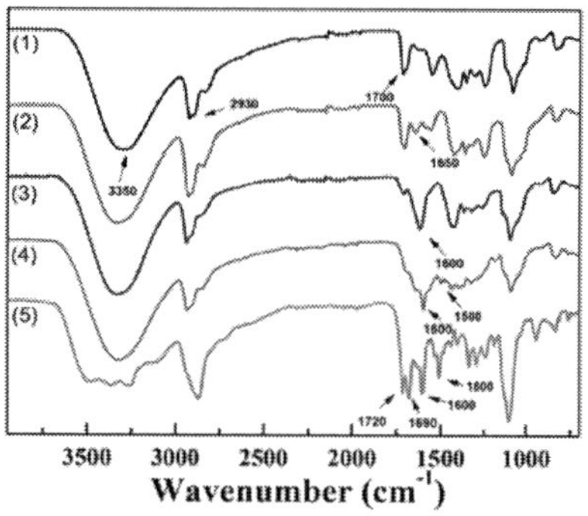

Figure 3. FTIR spectra of (1) PVA/PEI nanofibers, (2) crosslinked PVA/PEI nanofibers, (3) *m*PEG-PVA/PEI-Ac nanofibers, (4) FA-PEG-PVA/PEI-Ac nanofibers, and (5) FA-PEG-COOH.

Cell capture assy

The cell capture efficiency of FA-PEG-PVA/PEI-Ac nanofibrous mat at different time points was investigated (Figure 4). It can be seen that with the capture time, both FA-PEG-PVA/PEI-Ac and *m*PEG-PVA/PEI-Ac nanofibrous mats have increased efficiency to capture U87MG and L929 cells, which is due to the nonspecific adsorption of the materials. But starting from the 40 min time point, the U87MG cells capture efficiency of FA-PEG-PVA/PEI-Ac nanofibrous mat is much higher than that of the *m*PEG-PVA/PEI-Ac nanofibrous mat, but the efficiencies to capture L929 cells using FA-PEG-PVA/PEI-Ac nanofibrous mat is almost equivalent to those using *m*PEG-PVA/PEI-Ac nanofibrous mat at a given time point, which indicates that the FA-PEG-PVA/PEI-Ac nanofibrous mat show great specificity to capture U87MG cells.

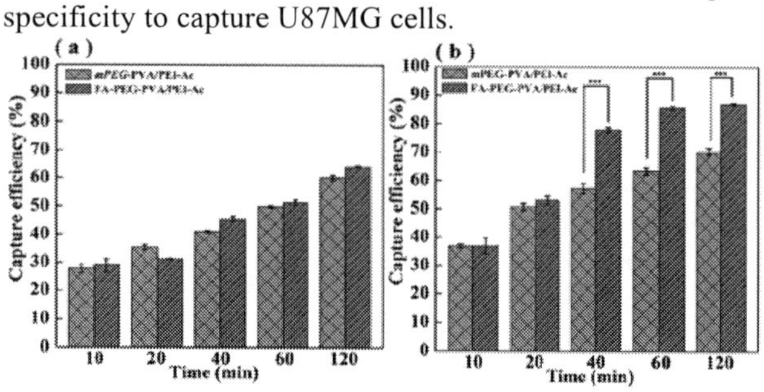

Figure 4. The capture efficiency of L929 cells (a) and U87MG cells (b) using FA-PEG-PVA/PEI-Ac and *m*PEG-PVA/PEI-Ac nanofibrous mats at different time points.

The specific cancer cell capture ability of the FA-PEG-PVA/PEI-Ac nanofibrous mat was also confirmed by confocal microscopic observation (Figure 5). In the confocal microscopic images, Hoechst 33342-labeled blue fluorescent U87MG cells can be clearly seen. At a given time point, much more U87MG cells can be captured onto the FA-PEG-PVA/PEI-Ac nanofibrous mat than onto *m*PEG-PVA/PEI-Ac nanofibrous mat, especially at the time points of 40, 60, and 120 min.

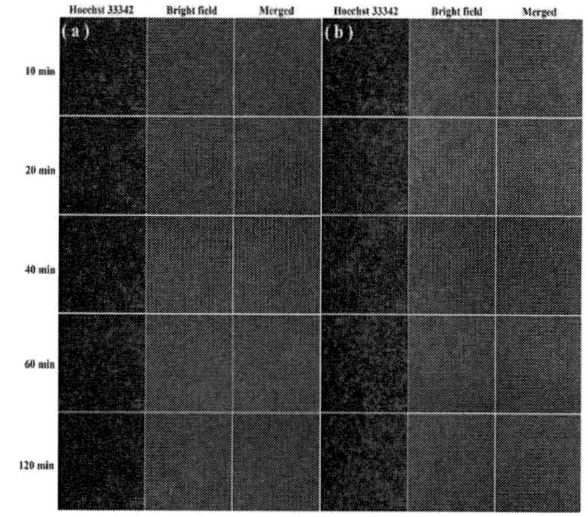

Figure 5. CLSM images of U87MG cells captured on *m*PEG-PVA/PEI-Ac (a) and FA-PEG-PVA/PEI-Ac (b) nanofibrous mats at different time points.

CONCLUSION

In summary, we developed a convenient approach to fabricating FA-modified electrospun PVA/PEI nanofibrous mats for targeted cancer cell capture applications. The FA-modified nanofibers exhibit superior capability to capture U87MG cells overexpressing high-affinity FA receptors. The formed FA-modified electrospun PVA/PEI nanofibrous mats hold great promise to be used for detection of circulating tumor cells in lymphatic system or bloodstream.

AKNOWLEDGMENT

This research is financially supported by the High-Tech Research and Development Program of China (2012AA030309) and the Innovation Funds of Donghua University Doctorate Dissertation of Excellence (101-06-0019014).

REFERENCES

Fang, Xu, et al. "Fabrication and Characterization of Water-Stable Electrospun Polyethyleneimine/Polyvinyl Alcohol Nanofibers with Super Dye Sorption Capability." *New Journal of Chemistry*, 35.2, 2011: 360-68.

Wang, Su He, et al. "Dendrimer-Functionalized Iron Oxide Nanoparticles for Specific Targeting and Imaging of Cancer Cells." *Advanced Functional Materials*, 17.16, 2007: 3043-50.

Zhang, Nangang, et al. "Electrospun Tio2 Nanofiber-Based Cell Capture Assay for Detecting Circulating Tumor Cells from Colorectal and Gastric Cancer Patients." *Advanced Materials*, 24.20, 2012: 2756-60.

Zhao, Yili, et al. "Dendrimer-Functionalized Electrospun Cellulose Acetate Nanofibers for Targeted Cancer Cell Capture Applications." *Journal of Materials Chemistry B*, 2.42, 2014: 7384-93.

Improvement of Dentist Gowns Comfort and Protection by Increasing the Adhesion of a Nanofiber Layer on Woven Fabric

A.V. Apetrei[1], L. Schacher[2], A. Curteza[1], D. Adolphe[2], N. Khenoussi[2]
[1]Technichal University "Gheorghe Asachi" of Iasi – Romania
[2]Université de Haute-Alsace – ENSISA – Laboratoire de Physique et Mécanique Textiles Mulhouse – France
apetrei.a@gmail.com; laurence.schacher@uha.fr

OBJECTIVE
The goal of our work is to attach a layer of nanofibers to a chosen woven fabric in order to improve the protection that the dentist gown provides by shielding the dentist from external contamination factors [1].

INTRODUCTION
Today the design of the dentist gowns for daily use takes into consideration aspects such as protection, comfort and appearance. The main purpose of the fabric used for manufacturing the dentist gown is to prevent the penetration of microbes and liquids (blood and saliva included), as well as providing the required comfort (the garment should allow freedom of movement, prevent heat build-up and create an aseptic barrier) [2].

According to the EN 13795 (European Standards for Surgical Drapes and Gowns) [3], the two critical areas that require a special attention are the chest and the sleeves (as shown in figure 1).

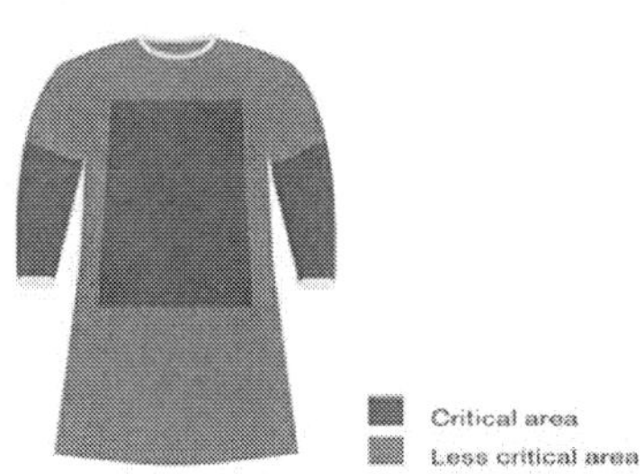

Fig. 1 – Critical areas according to EN 13795 (European Standards for Surgical Drapes and Gowns)

The studies conducted so far in the field focused either on waterproof fabrics[4] or non-woven layers (especially for the single use drapes and gowns) in order to protect the critical areas pointed out in the standards.
By using the electrospinning process [5] a nanofiber layer was attached to the fabric, thus enabling us to improve the resistance of the gowns to external contamination factors.

The electrospinning process uses an electrical field in order to create a charged jet of polymer solution. The jet travels through air while the solvent evaporates leaving behind a charged fiber that can be collected on different kind of collectors, in our case the woven fabric. Fibers with a variety of sizes and shapes were produced from PA 6/6 polymer. The diameter of these fibers was in the range of 0.05 - 0.150 microns. PA 6/6 (Nylon-6.6) is a semi-crystalline polyamide most commonly used in fibres applications. It is also used as an engineering material due to its good abrasion resistance and self-lubricating properties [6].

It was proven that in the case of daily use gowns if the fabrics used are made of cotton or of a mixture of cotton and polyester they only fulfill some requirements (such as comfort) but they are not too efficient in preventing the infiltration of microbes and liquids [7].
However a survey conducted in France [8] shows that the medical staffs prefer using a mixture of cotton and polyester.

APPROACH
For this reason, the fabric we opted for is a blend of 60% cotton and 40% polyester (table 1). Using a wire-base ELMARCO NS 1WS500U electrospinning device, we took into consideration the following process set of parameters: air-flow, wire distance from the collector and wire speed.

Table I – Fabric parameters of the test samples

Properties	60% cotton and 40% PES fabric
Fabric structure	Plain wave (Basket wave) Cotton weft – 2 treads PES warp – 2 treads
Fabric weight	210 g/m^2
Fabric thickness	0.48 ± 0,10mm

We prepared a solution of PA 6/6 pallets and formic acid solvent with a 20 wt. % which allowed us to test the adhesion of a nanofiber layer on the woven fabric. The mixture was stirred at 50°C for 24h in order to obtain a homogenous solution. During these preliminary tests we also used a nonwoven fabric in order to improve the grip between the nanofibers and fabric without any chemical treatment [9]. The reason why we chose this type of tests was to avoid the direct contact between the skin and the chemically treated fabric (the dentist gowns are usually worn directly on the skin, which results in direct exposure).

RESULTS AND DISCUSSION
In order to improve the adhesion between the nanofiber layer and the woven fabric, we subjected the woven fabric samples to the electrospinning process. This enabled us to create a bond between the fabric and the nanofiber layer. The electrospinning parameters used for all samples were modified one by one: the air flow, wire distance from the collector and wire speed which varied from minimum to maximum values possible with the used set-up enabling us to observe the adhesion process in its different stages.

During the next stage, the samples were examined on the Scanning Electron Microscope (SEM) at different magnifications in order to collect the right sizes of the fibers.

The parameter 1 – wire distance from the collector, shown in Fig. 2 shows us that the best distance to work with during the tests in the future is 200 mm. In figure 3 we can observe that the air flow influences the deposition and although the best values are with low air flow, the machine would be damaged in long term usage; as a consequence we chose the next values which are around 100 m^3/h. We tested the parameter 3 - wire speed as seen in figure 4 - to see if there is any effect on the samples during the electrospinning process.

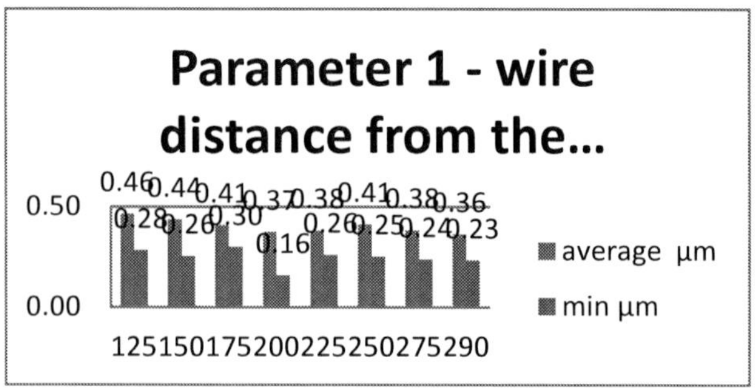

Fig. 2 – Parameter 1 - Wire distance from the collector

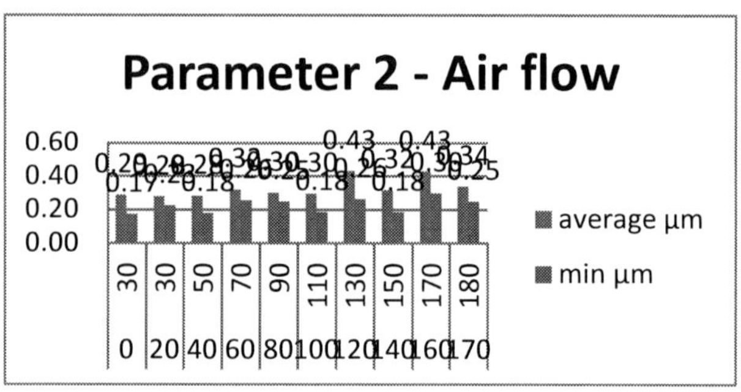

Fig. 3 – Parameter 2 - Air flow

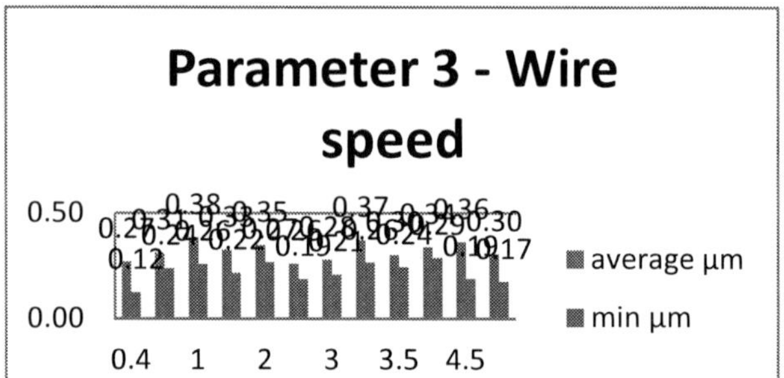

Fig. 4 – Parameter 3 - Wire speed

CONCLUSION

The paper presents a simple solution that is still under development with a final objective of attaching the nanofiber layer directly to the fabric. These preliminary tests were carried out on ELMARCO NS 1WS500U electrospinning device, which allowed us so far to set the optimum parameters in order to obtain a good deposition with a good adhesion.

FUTURE WORK

During the next phase of this research, the main objective is to use the obtained samples in creating a new fabric for dentists gowns. The gown produced will address protection issues and, at the same time, guarantee comfort according to the established standards and therefore create a blouse that is more appropriate to the needs of the final customer.

KEYWORDS: gown, dentist, protection, nanofibers.

ACKNOWLEDGMENT

This work is supported by the training program with the host organization Université de Haute-Alsace – ENSISA – Laboratoire de Physique et Mécanique Textiles Mulhouse – France.

REFERENCES

[1] Belkin, Nathan L. Belkin. "Surgical gowns and drapes as antiseptic barriers." *American Journal of Infection Control*, 16, 1, February 1988: 14-18.

[2] Abreu, M.J., Silva, M.E., Schacher, L., Adolphe, D. "Designing surgical clothing and drapes according to the new technical standards." *International Journal of Clothing Science and Technology*, 15,1, 2003: 69-74.

[3] EN 13795. *European Standards for Surgical Drapes and Gowns*.

[4] Schacher, Marie, Haïkel, Youssef, Berger, Stéphane, Schacher, Laurence, Adolphe, Dominique. "Improvement of dentist gowns–new constraints and new risks." *International Journal of Clothing Science and Technology*, 21, 4, 2009: 180-192.

[5] Ramakrishna, Seeram, Fujihara, Kazutoshi, Teo, Wee-Eong, Lim, Teik-Cheng, Ma, Zuwei. "An introduction to electrospinning and nanofibers." *World Scientific Publishing Co. Pte.*

[6] Deopura, B.L., Alagirusamy, R., Joshi, M., Gupta, B. *Polyester and Polyamides*. Cambridge: England, 2008.

[7] Abreu, M.J., Silva, M.E., Schacher, L., Adolphe, D. "Designing surgical clothing and drapes according to the new technical standards." *International Journal of Clothing Science and Technology*, 15, 1, 2003: 69-74.

[8] Arnold, Matthieu. "Development of a professional outfit for specialized dental care." Thesis of State Diploma of Doctor of Dental Surgery.

[9] Varesanoa, Alessio, Aluigia, Annalisa, Florio, Luca, Fabris, Riccardo. "Multifunctional cotton fabrics." *Synthetic Metals*, 159, 2009: 1082-1089.

Acrylic Fibers with Designed Cationic Phthalocyanine

Hak-Soo Kim, Jong-Woo Jeong, Hee-Soo So, Sehoon Lee, Kyosun Ku, Chang-Kyem Kim, Young-A Son
Department of Advanced Organic Materials and Textile System Engineering, Chungnam National University, Daejeon, Korea
yason@cnu.ac.kr

INTRODUCTION

Acrylic fabrics have wide applications in the textile industry, and there is always a continuous increase in the production and consumption of these fabrics. Acrylic fabrics obtained by using at least 85% acrylonitrile as a monomer are available in cationic dyeable or anionic dyeable forms. Acrylic polymers containing small amount of anionic centers, such as sulfonic acid or carboxylic acid groups, can be dyed easily with dyes bearing a positive charge, viz. cationic dyes. Cationic dyeable fabrics contain negatively charged dye sites and therefore can be dyed with cationic (i.e. basic) dyes, on the other hand the anionic dyeable fabrics contain positively charged dye sites and can be dyed with anionic dyes. Of these two forms, the cationic dyeable fabrics are more common in commercial use. Cationic dyes have good solubility in water, high colored value and are amongst the brightest dyes available. The mechanism of dyeing of acrylic fabrics with cationic dyes involves electrostatic attraction between the anionic sites on the fabric and the colored cations of the ionized dye. The experiments were carried out using C.I. Basic Red 26, C.I. Basic Yellow 67, Kayacryl Blue TL-ED and cationic phthalocyanine at different temperature, pH condition and liquor ratio.

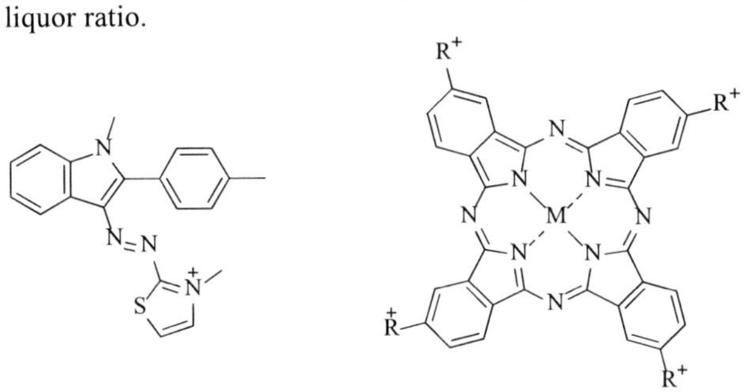

C.I. Basic Red 26 Cationic phthalocyanine

Figure 1. Structure of cationic dyes

RESULTS AND DISCUSSION

Effect of liquor ratio

Important variable to be considered in the processes is liquor/fabric ratio. Selection of appropriate liquor ratio contributes to producing desired results. The experiments were carried out at the six different liquor ratios 60:1, 70:1, 80:1, 90:1, 100:1, and 110:1. Fig. 2 shows the effects of liquid ratio on the color strength (K/S). It was observed that lower liquor ratio resulted in higher color strength (K/S) and increase in liquor ratio resulted in decrease in color strength (K/S). Maximum color strength (K/S) was achieved at the liquor/acrylic fabric ratio of 60:1. This may be attributed to the fact that lowering the liquor ratio results in concentrated solution with maximum probable molecular collisions between acrylic fabric and cationic dyes and thus making more number of cationic molecules to interact with acrylic fabric.

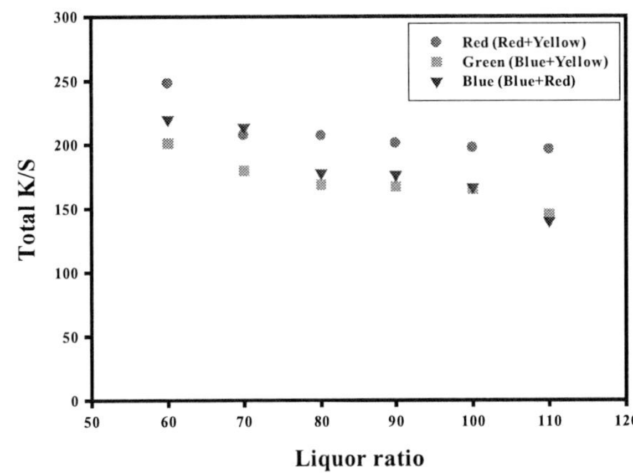

Figure 2. Total K/S of cationic dyes on acrylic fabric at various liquor ratios.

Effect of pH condition

Fig. 3 shows that the pH conditions of the dye bath have substantial effect on the dyeability of acrylic fabric with cationic dyes. The effect of pH condition can be attributed to the correlation between dye structure and acrylic fabric. When cationic dyes are bound on acrylic fabric, further kinds of interactions take place together with ionic forces.

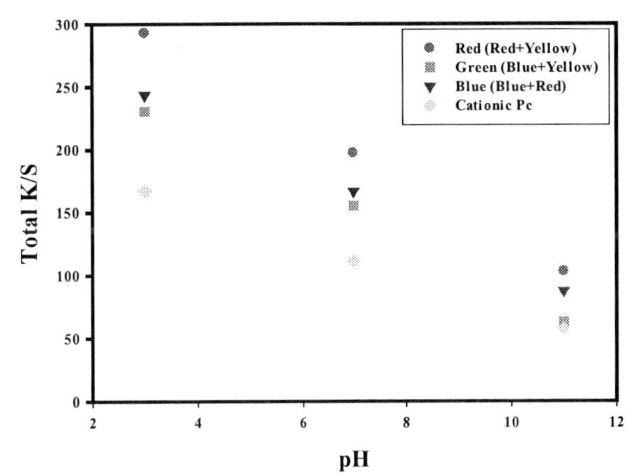

Figure 3. Total K/S of cationic dyes on acrylic fabric at various pH condition.

Effect of dyeing temperature

In order to monitor the effect of temperature on acrylic fabrics were dyed with cationic dyes at various temperatures (60-100 degrees Celsius). Generally, the color strength (K/S) of the dyeing increased with increasing application temperature. The adsorption was more well-disposed at higher temperature than at lower temperature. Within the lower range of dyeing temperature (60-80 degrees Celsius), these temperatures had little effect on the dye uptake to the substrates. Notably, satisfactory dye uptake was obtained in the range of temperatures of 90-100 degrees Celsius.

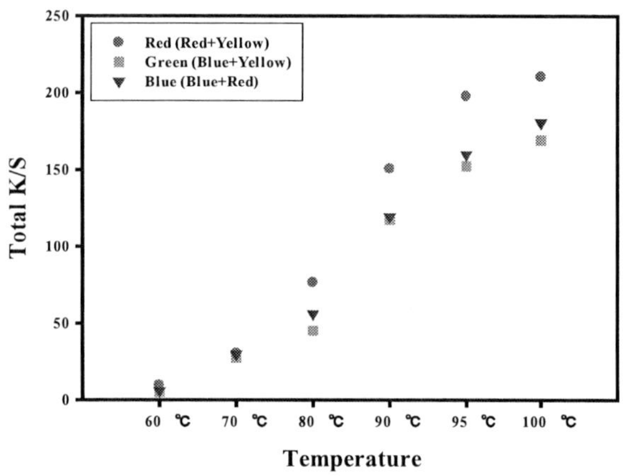

Figure 4. Total K/S of cationic dyes on acrylic fabric at various temperature.

CONCLUSION

Acrylic fabric was dyed with four cationic dyes in this work. As expected, four cationic dyes could be introduced into acrylic fabric by following traditional cationic dyeing process. The adsorption of cationic dye onto acrylic fabric was found to follow a Langmuir mechanism, indicating that the dye molecules interacted with the anionic sites of the acrylic fabric in the form of ionic linkage.

KEYWORDS

Cationic dye; Phthalocyanine; Dyeing; Acrylic fabric

ACKNOWLEDGMENT

This work was supported by Ministry of Trade, Industry & Energy (Global Expertise Development Project No. 10048453)

REFERENCES

1. Liu, L., Qin, C., Tang, R.-C., Chen, G. "Dyeing Properties and Colour Characteristics of a Novel Fluorescent Dye Applied to Acrylic Fabric." *Fibers & Textiles in Eastern Europe*, 4, 2013: 144-147.

2. Ma, M., Sun, G. "Antimicrobial cationic dyes. Part 3: simultaneous dyeing and antimicrobial finishing of acrylic fabrics." *Dyes and Pigments*, 66, 2005: 33-41.

3. Kamel, M.M., Helmy, H.M., Mashaly, H.M., Kafafy, H.H. "Ultrasonic Assisted Dyeing, Dyeing of Acrylic Fabrics C.I. Astrazon Basic Red 5BL 200%." *Ultrasonics Sonochemistry*, 17, 2010: 92-97.

Wearable Sensors Based on Optical Fibers and Textiles

Luciano F. Boesel, Brit Maike Quandt, Marek Krehel, Susanne Widmer, Markus B. Bannwarth, René M. Rossi
Empa, Lerchenfeldstrasse 5, CH-9014 St. Gallen, Switzerland
luciano.boesel@empa.ch; rene.rossi@empa.ch

INTRODUCTION

Flexible sensors based on incorporating polymer optical fibers (POF) or fluorescent moieties into textiles possess several advantages in the medical and chemical fields. POFs have become more and more popular for sensing applications due to increased homogeneity. E.g., in the medical area, POFs have been used for measuring various vital parameters, such as heart rate, oxygen saturation, and respiratory rate. [1] Optical fibers show great potential for usage in combination with other medical diagnostic machinery as they are immune to electromagnetic interference, are resilient to harsh environmental conditions, and show high sensitivity. [1,2]

Another application which may profit from the use of POFs is the chemical detection of aggressive and/or toxic gases. As an example, several optical ammonia gas sensors have been reported, based on spectrophotometric ammonia detection by monitoring a change in fluorescence of some fluorescent dye or dye pair. [3] The development of a similar gas sensor system based on flexible POFs with flexible sensing coatings, which finally can be integrated into textiles, would increase sensor wearability. Moreover, the same or similar dyes could be used to coat conventional medical textiles (bandages, wound pads, etc.), creating specialty products able to sense the pH of the wound or any metabolite present in it.

The flexibility has often been a bottleneck for integration of POFs into textiles. To that aim, POFs more flexible than the current gold-standard of PMMA are needed, since high deformation and deformation rates are present during textile integration, e. g, stitching. [2]

In this work we report our recent developments on 3 different sensors based on optical fibers or fluorescent textiles: 1) POF-based sensors for oxygen saturation (photopletismography); 2) POF-based chemical sensors for ammonia; 3) textile sensor for pH of the wound.

APPROACH

Photopletismograph: The optical fibres used in this study were produced using a melt-spinning process. The core of the optical fibres was made of cycloolefin polymer, (Zeonor 1020R, Zeon, Düsseldorf, Germany) and as cladding material a fluorinated polymer (THVP 2030GX, Dyneon, Burgkirchen, Germany) was used. The optical yarn was embroidered on the carrier fabric using an embroidery machine ERA TM 0625 (Saurer AG, Switzerland). Pulse wave signals were analysed with a near infrared tissue spectrometer Oxiplex provided by ISS (Champaign, USA), the same apparatus providing the light sources. The measurements were performed by locating the PPG-fabric on the middle finger of the right hand. Further details on the production of the optical fibres and the final PPG patch, and on the PPG measurements are found in [2]

Chemical sensors: commercial PMMA fibers (i-Fiberoptics, USA) were coated with a sol-gel based coating containing a fluorescent dye pair (fluorescein/rhodamin). Fibers were coated by dip-coating in a solution of the dyes and then dried in air at high-temperature (70oC). Sol-gels were prepared with organically modified silanes (Ormosils), on which solution the 2 dyes were dissolved and let "age". After drying, the fibers were connected to a home-made gas-detection chamber and subjected to different concentrations of ammonia between 50-10000ppm. The change in fluorescence of the dyes due to absorption of ammonia was converted with a photodetector to a change in voltage and recorded electronically. For more details refer to [3].

Textile pH sensors: the same sol-gel system as for the chemical sensor was also used to coat wound pads in order to develop textile pH sensors. In this case, only one dye (fluorescein or fluorescein derivative) was incorporated into the xerogel. Alternatively, the dye was directly bound to the wound pad material. The system comprises the use of a biocompatible alginate matrix with the pH-sensitive dye. The dye was chosen to be especially sensitive in the pH-range 6-8 that occurs in chronic wounds. We then measured the fluorescence of the dye both in solution as well as immobilized in the pad with solutions of different pHs.

RESULTS AND DISCUSSION

Photopletismograph: The stitched textile PPG-setup contains 4 lines of stitched optical fibers, 2 for outcoupling the incoming red (692 nm) and infra-red (834 nm) light, and two for incoupling the light reflected back by the tissue. Fig. 1 shows the calculated heart rate and blood oxygenation values. These values were also compared with a standard commercial product (Nellcor). It is clear that the values of our experimental textile PPG-setup agree with those of a commercial product. Another important property of our setup is the high comfort levels, as demonstrated by friction measurements against synthetic skin models (PUR).

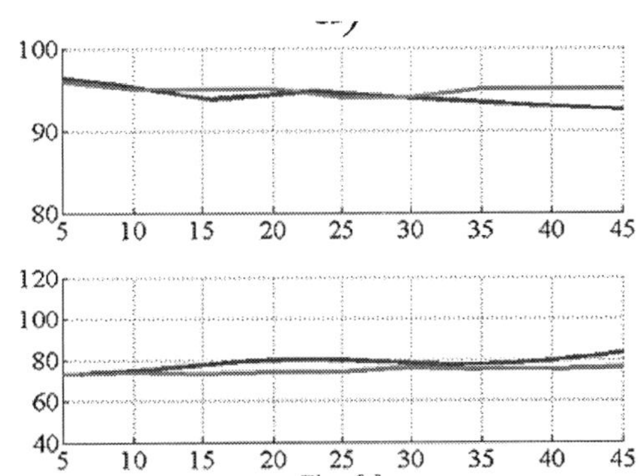

Figure 1. Heart rate (top) and blood oxygenation (bottom) of PPG-setup (blue line) and commercial product (red line).

As shown in Fig. 2, the coefficient of friction of the setup is well below those of typical hospital bedsheets, either cotton or synthetic material. Therefore, we do not expect any discomfort when the setups will be used for extended periods of time.

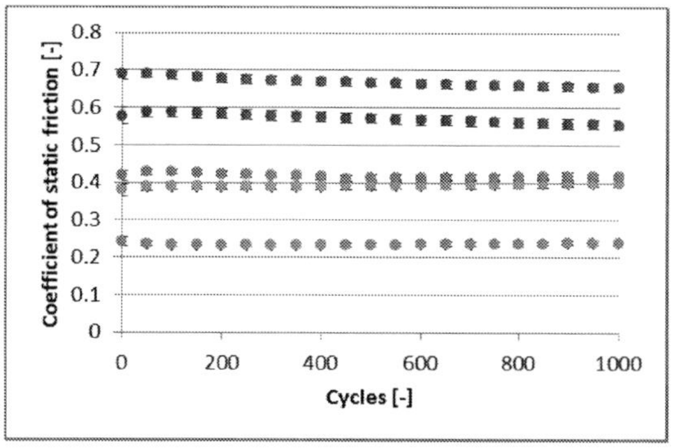

Figure 2. Friction of PPG-setup (orange/blue) with a skin model compared to hospital bedsheet (red/violet) and low-friction textile (green).

Chemical sensors: Fig. 3 shows the gas sensor response of fluorescent fibers, while Fig. 4 displays an image of a coated fiber. The red color derives from the high amounts of rhodamin B which were coated. As seen in Fig. 3, the system responded quickly to ammonia exposure, while the recovery took longer times. The reason being the strong absorption of the ammonia to any free OH groups in the xerogel structure, making it hard to remove the ammonia from the coatings. In any case, both response as well as recovery times in the range of hundreds of seconds were obtained. These fiber-based sensors could be easily integrated into textiles, by simultaneously miniaturizing the electronics (LED and photoreceptor) and allowing for a wireless transmission of the response to a remote control unit.

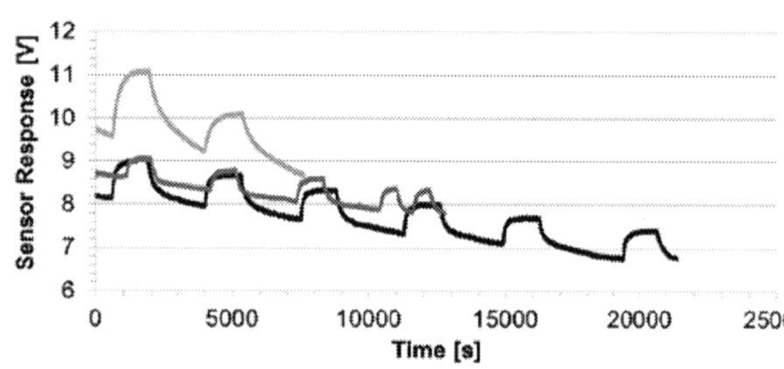

Figure 3. Response of coated POF to ammonia and recovery with nitrogen.

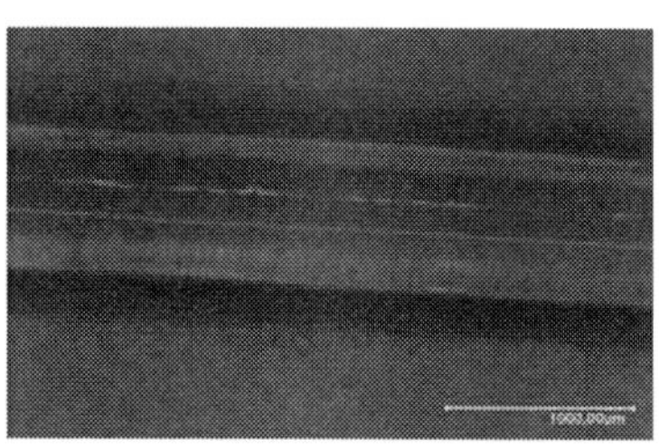

Figure 4. POF coated with sol-gel containing fluorescein and rhodamin B dyes.

Textile pH sensors: Fig. 5 shows the response of the functional textile to the external pH. Fluorescein is a pH sensitive dye and, Depending on the ambient pH value, fluorescein exists in cationic, neutral, mono-ionic or dianionic forms or as a mixture of thereof.[3] Emission spectra are dependent on the presence of species in different protonation states, leading to pH-dependent emission over the range 5 to 9. This could be clearly shown in Fig. 5, where we achieved a very high change of fluorescence in the pH range of interest to chronic wounds. The dyes could be easily incorporated in wound pads by either directly coupling or coupling to an intermediate matrix which is then adsorbed on textile pads.

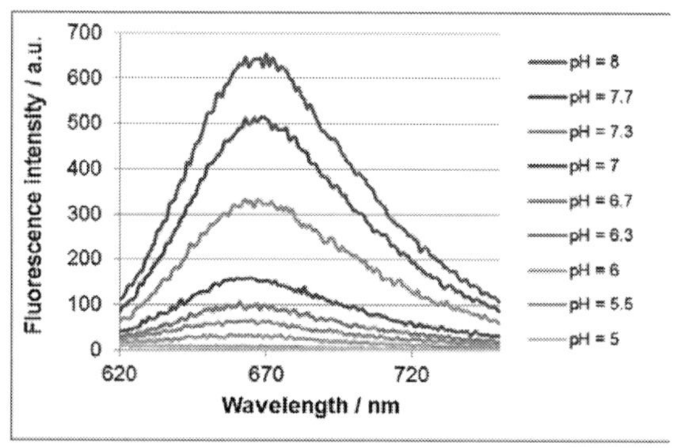

Figure 5. Fluorescence response of textile sensors exposed to solutions of different pHs.

CONCLUSION and OUTLOOK

In this work, we presented three examples of the use of optical fibers and fluorescent textiles as wearable sensors. These photonic/fluorescent textiles are very versatile for wearable long-term monitoring and allow the measurements in different parts of the body like forehead, chest, arms, etc. Furthermore, the presented fabrics have the haptic and comfort of conventional fabrics (or even better) and thus enhances the acceptance of the wearer, which is necessary for long-term monitoring purposes. Future work will concentrate on increasing the signal response and signal/noise ratio of the different systems, as well as adapting the developed textile pH sensors to the detection of metabolites (biosensors). Generally speaking, the real potential of wearable photonic sensors is seen in applications and fields in which electronic counterparts are unsuitable, such as diagnostic means (e.g., MRI) but also other biomedical appliances (e.g., pace maker). Wearable and comfortable monitoring devices will allow for an improvement in the quality of life of elderly patients.

REFERENCES

[1] Quandt, B.M., et al. *Advanced Healthcare Materials*, 4, 2015: 330.

[2] Krehel, M., et al. *Biomedical Optics Express*, 5, 2014: 2537.

[3] Widmer, S. et al. *Analyst*, 139, 2014: 4335.

Preparation and Characterization of a Novel Anti-Dripping Flame Retardant PET/Phosphate Glass Nanocomposite

Huichao Liu[1], Yilong Huang[1], Jinghong Ma[1], Jinghua Gong[1], Shouyun Zhang[1], Jian Xu[1,2]

[1]State Key Laboratory for Modification of Chemical Fibers and Polymer Materials, College of Material Science and Engineering, Donghua University, Shanghai, China
[2]Beijing National Laboratory for Molecular Sciences, Laboratory of Polymer Physics and Chemistry, Institute of Chemistry, Chinese Academy of Sciences, Beijing, China
huichaoliu@126.com

ABSTRACT

In the present study, the effect of phosphorus flame retardation (FR) agent, 2-carboxyl ethyl (phenyl) phosphinic acid (CEPPA), incorporating phosphate glass (Pglass) on the flame retardation and melt dripping behavior of poly(ethylene terephthalate) (PET) were investigated. The Pglass/FR-PET nanocomposites (P/FR-PET) were prepared by melt blending. The morphology, thermal degradation behavior and flamability of the nanocomposites were investigated by means of scanning electron microscopy (SEM), thermogravimetric analysis (TGA), limiting oxygen index test (LOI) and vertical combustion test. SEM micrographs showed as the increasing of Pglass contents the morphologies of Pglass in the FR-PET matrix can change from nanoparticles (50 nm to 100 nm) to network structures. The thermal stability of nanocomposites at high temperature increased with the addition of Pglass contents. Meanwhile, the LOI values of the nanocomposites increased from 33.6 for neat FR-PET to 36.1 for 50% P/FR-PET nanocomposites. The vertical combustion test revealed the P/FR-PET nanocomposites had excellent anti-dripping flame retardant property which can achieve UL-94 V-0 level. The formation of the Pglass network-structured protective layer, which acts as a supporting and heat shield layer, must be crucial for the large reduction in melt dripping. In a word, the experimental results showed that the CEPPA in combination with Pglass has excellent flame retardation and anti-dripping effect.

KEYWORDS: PET; phosphate glass; nanocomposite; flame retardation; anti-dripping

Optimal Design of Firefighters' Protective Clothing Based on Ergonomics

Yun Su[1,2], Yunyi Wang[1,2,3], Jun Li[1,2,3]
[1]Fashion Institute, Donghua University, Shanghai, China
[2]Protective Clothing Research Center, Donghua University, Shangha, China
[3]Key Laboratory of Clothing Design & Technology, Donghua University, Ministry of Education, Shanghai, China
lijun@dhu.edu.cn

INTRODUCTION

Fire-fighters' protective clothing is used to protect firefighter from thermal exposures and other life threatening risks. It is essential and one of the most effective way for firefighting, emergency rescue and anti-terrorism in varieties of hazards. Meantime, ergonomics design of fire-fighters' protective clothing have greatly impact on work efficiency and body safety. According to the developing status of firefighters' protective clothing in past decade years, the standard protective clothing is a multilayer construction, which is heavy, bulky and discomfort to provide the desired thermal protection. In order to optimize fire-fighters' protective clothing, the correlation of human-garment-environment is analyzed based on ergonomics. The commonly fabric types and the essential performance of each layer for fire-fighters' protective clothing is summarized. Therefore, optimizing method of fire-fighters' protective clothing is put forward in terms of style design, pattern design and the use of smart textiles. Hence, the next generation of protective clothing will be provided for improve function and comfort of fire-fighters.

HUMAN ERGONOMICS

Ergonomics mainly study the relations among human, environment and clothing. According to different working conditions and functional demanding, design of fire-fighters' protective clothing is essential to be fulfilled from garments' three elements such as style, pattern and fabrics.

FABRICS SELECTION AND PERFORMANCE

The fire-fighters' protective clothing is usually comprised of multiple layers. It consists of a flame resistant outer shell and a thermal liner comprised of a moisture barrier, a thermal barrier and a lining material (figure 1). According to the concept of human ergonomics, the fire-fighters' protective clothing should provide for protection, comfort, mobility and aesthetic feeling. Therefore, each layer in multi-layer clothing assemblies is essential to fulfill the following requirements (figure 1).

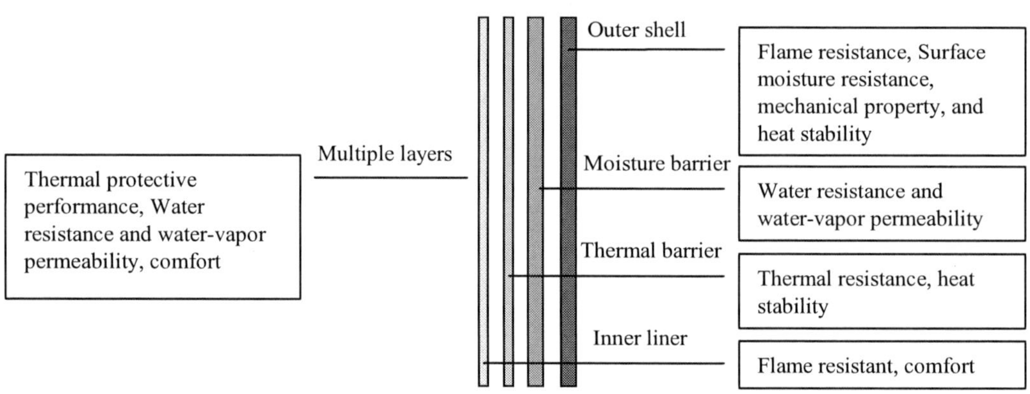

Figure 1. Multi-layer clothing assemblies and performance requirements

OPTIMAL DESIGN PROTOCOLS

Style design

Style design in fire-fighters' protective clothing plays an important role. Fire-fighters' protective clothing should be designed by taking into consideration the factors such as fire-fighting conditions, ergonomics of movement, heat stress, anthropometry and thermo-physiological behavior in addition to the protection. It includes design of collar, sleeves, shoulders, waist, crotch, pant legs and cooling methods by changing clothing shape and attaching some functional enclosures.

Pattern design

The clothing pattern design can improve the local mobility and enhance workers' comfort in varieties of heat flux exposures. And it affects the air gap between the wearer's body and the clothing, which in turn affects the protective performance. It is indispensable to optimize the garment's pattern in terms of the chest's ease, hip structure, knee structure, clothing sizes and fit.

Use of smart textiles

The next generation of fire-fighters' protective clothing will be smart protective clothing which interact with the human and environment. This new generation of protective clothing can sense and react to environmental conditions and be active in many fields. The smart suits can be divided into three subgroups: passive (only able to sense), active (reactive sensing to stimuli) and very smart (able to sense, react and adapt). The smart protective clothing will include many hi-tech functions such as physiological monitoring, locating fire-fighters, predicting the stress, communication and environmental monitoring for using of smart textiles, wireless sensor, actuators and controlling unit. It can increase work safety and firefighting efficiency through sensing, reacting and adapting.

CONCLUSION

On the basis of the performance requirements and using status of fire-fighters' protective clothing, the paper apply for style design in collar, sleeves, shoulders and like that as well as pattern design by ease of chest, clothing sizes and fit, pleats and darts according to human and ergonomics. Meantime, The use of electronic textiles can monitor the physiological conditions of the wearer and the environmental conditions. These studies can provide proper guidelines for the application of fire-fighters' protective clothing so as to improve protective performance, comfort and ergonomics, thus meet the technological requirements on safety development and public security prevention.

KEYWORDS: fire-fighters' protective clothing; ergonomics; design; smart textiles

ACKNOWLEDGMENT

This work was financially supported by the Fundamental Research Funds for the Central Universities.

REFERENCES

[1] Barr, David, Gregson, Warren, Reilly, Thomas. "The thermal ergonomics of firefighting reviewed." *Applied Ergonomics*, 41.1, 2010: 161-172.

[2] Lu, Yehu, Li, Jun, Wang, Yunyi. "Research and developing trend of heat and moisture transfermodel of thermal protective clothing." *Journal of Textile Research* 33.1, 2012: 151-156.

[3] *NFPA 1971. Protective Ensemble for Structural Fire-Fighting*. Quincy, MA: National Fire Protection Association, 2007.

[4] Li, .J, Wang, Y., Zhang, X., et al. "Properties and Combination of Multilayered Fabrics of Firefighter Protective Clothing." *Journal of Donghua University (Natural Science Ed.)*, 34.4, 2008: 410-415.

[5] Wang, Y., Zong, Y., Li, J., et al. "Evaluating the moisture transfer property of the multi-layered fabric system in firefighter turnout clothing." *Fibres & Textiles in Eastern Europe*, 19.6, 2011: 101-105.

[6] Wang, Y., Zhang, Z., Li, J., et al. "Effects of inner and outer clothing combinations on firefighter ensembles' thermal-and moisture-related comfort levels." *Journal of the Textile Institute*, 104.5, 2013: 530-540.

[7] Lu Yehu, Li Jun, and Wang Yunyi." Research on functional design pattern of protective clothing." *Journal of Textile Research* 32.11 (2011): 119-125.

[8] Tian, M., Wang, Y., Zhang, X. "Comfort-Ergonomics Evaluation and Optimization of Thermal Protective Clothing." *Journal of Donghua University (Natural Science Ed.)*, 39.6, 2013: 754-759.

[9] Nayak, Rajkishore, Houshyar, Shadi, Padhye, Rajiv. "Recent trends and future scope in the protection and comfort of fire-fighters' personal protective clothing." *Fire Science Reviews*, 3.1, 2014: 1-19.

[10] Salim, Flora, et al. "Design and evaluation of smart wearable undergarment for monitoring physiological extremes in firefighting." *Proceedings of the 2014 ACM International Symposium on Wearable Computers: Adjunct Program*, 2014.

[11] Florea, G., et al. "Wearable system for heat stress monitoring in firefighting applications: Recent advances in computer and net-working." *Proceedings of the 2nt International Conference on Information Technology and Computer Networks (ITCN'13), Antalya*. 2013.

Influence of Thermal Imidization Conditions on Structure and Mechanical Properties of BPDA-PDA Polyimide Fibers

Wenke Yang[1,3], Fangfang Liu[2], Xueming Dai[2], Zhixin Dong[2], Xuepeng Qiu[2], Xiangling Ji[1]
[1]State Key Laboratory of Polymer Physics and Chemistry, Changchun Institute of Applied Chemistry, Chinese Academy of Sciences, Changchun, China
[2]Laboratory of Polymer Composites and Engineering, Changchun Institute of Applied Chemistry, Chinese Academy of Sciences, Changchun, China
[3]University of Chinese Academy of Science, Beijing, China
wkyang@ciac.ac.cn; xp_q@ciac.ac.cn; xlji@ciac.ac.cn

INTRODUCTION

Aromatic polyimide (PI) fibers have a great potential application due to their good mechanical properties, thermal stability and excellent chemical resistance[1, 2]. Most polyimide fibers are fabricated by two-step method because of their poor solubility. Firstly, polyamic acid (PAA) fibers were synthesized. Then the PAA fibers were thermally or chemically treated into PI fibers. Definitely, the process of imidization is of great importance to influence the final mechanical properties of the PI fibers. Here, the PAA fibers were prepared by means of dry-jet wet spinning with typical monomers of 3,3',4,4'-biphenyltetra-carboxylic dianhydride (BPDA) and p-phenylenediamine (PDA), as shown in Figure 1. As-prepared PAA fibers were treated into PI fibers under different temperatures and time in air and the variation of structure and properties have been discussed.

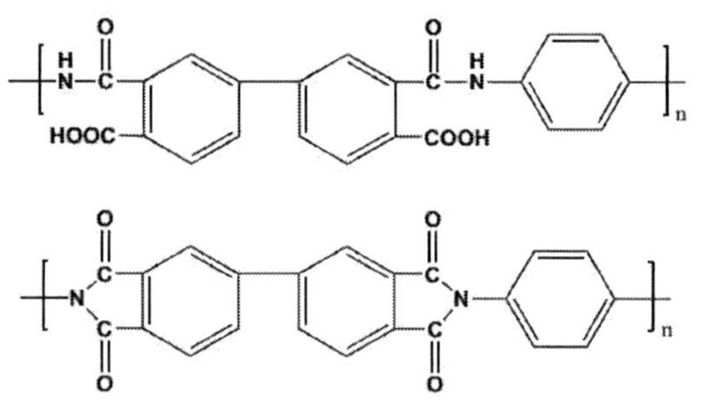

Figure 1.Chemical structure of BPDA-PDA polyamic acid(a) and polyimide(b).

RESULTS AND DISCUSSION

Characterization of imidization degree

Figure 2 shows the isothermal TGA results for PAA fibers treated at different temperatures. The weight loss reflects the stage of the imidization process. The PAA fibers show an immediate decrease as a start. As increasing temperature, the weight retention levels off in a short time and keeps at specific values constantly until to the end. The inset in Figure 2 shows that the finial imidization degree of each sample grows from 79.57% to almost 100%. It demonstrates that, comparing with treating time, temperature is a more decisive factor to affect imidization degree.

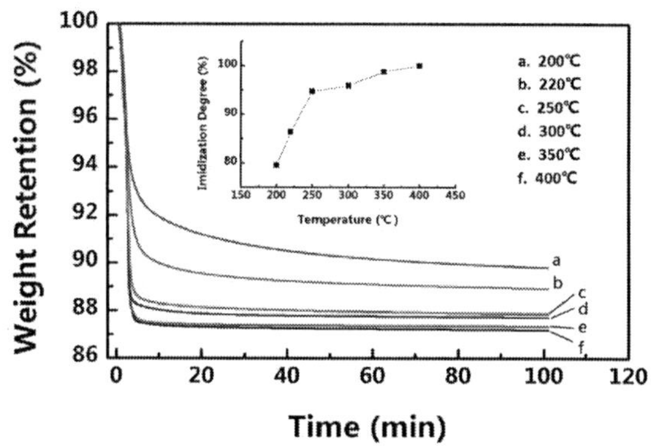

Figure 2. Isothermal TGA curves of PAA fibers under different temperatures.

The H_{1772}/H_{1552} ratios, which were used to characterize the ID [3, 4], calculated from the height of absorption bands at 1772 cm^{-1} and 1552 cm^{-1} of FTIR spectra are consistent with the TGA results, as shown in Figure 3. Generally, longer treating time can make higher imidization degrees, even within a very short time. But treating at low temperature, even prolonging treating time, it can only increase the imidization degree to some extent. This is because, under a constant temperature, the imidization reaction would be halted as imidization reaction proceeds, as the ability of chain movement is reduced. So, the imidization will stay at a certain value if the temperature is not raised.

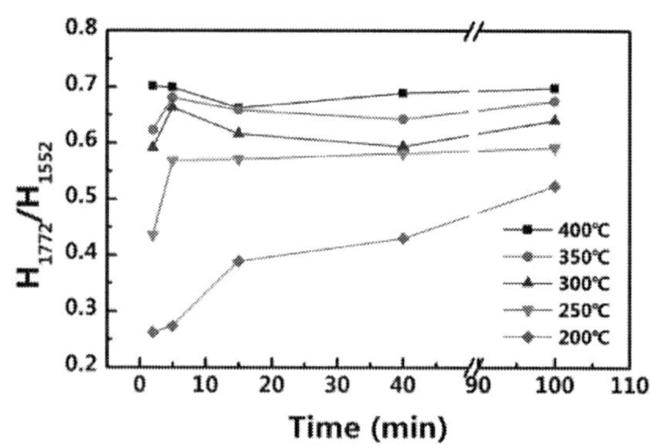

Figure 3. H_{1772}/H_{1552} ratios of PI fibers calculated from FTIR spectra.

Formation of crystals during imidization process

Figure 4 shows the XRD patterns along fiber axis of PAA fibers cured at 250 °C and 350 °C for different time. The diffraction peaks at 11.15°, 17.07°, 27.87° and 28.00° demonstrate the formation of crystals and orientation of some certain crystal planes. Treating temperature also plays an important role. As increasing temperature from 250 °C to 350 °C, there is a significant increase of the peak intensity. However, extending treating time seems do not have a very obvious promotion for samples treated at 250 °C. Only the samples treated at 350 °C exhibit an increasing trend with lasting time. And this could also be explained by the ability of chain movement as discussed above.

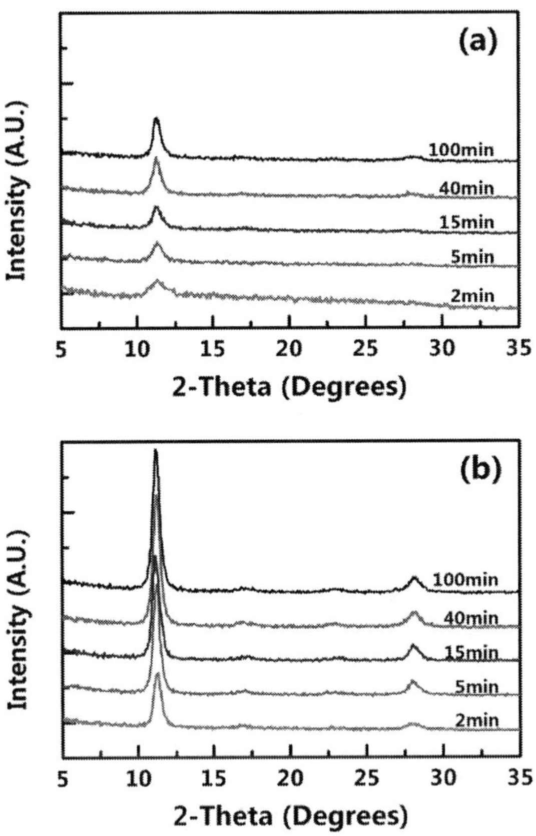

Figure 4. XRD patterns along fiber axis of PI fibers cured under 250 °C(a) and 350 °C(b).

Mechanical properties

Tensile strength of fibers treated at different time and temperature are shown in Figure 5. From a general aspect, higher temperature can make fibers have better tensile strength that ranges from 2 cN/dtex to 4 cN/dtex because they could get higher imidization and crystallinity as discussed above. Definitely, the initial moduli of these fibers also increase from 100 cN/dtex to 300 cN/dtex with rising treating temperature as the tensile strength. Regarding tensile strength, it shows a depression after treating time exceeding 15min at higher temperature. And this depression is probably caused by the oxidation on fiber surface at high-temperature. Defects may be produced and make the fibers rupture at lower stresses.

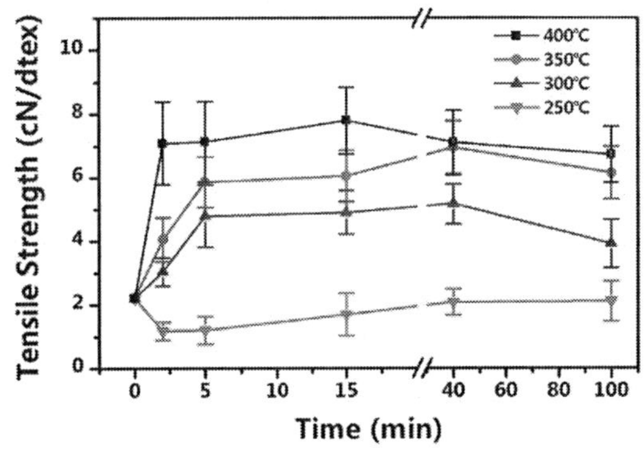

Figure 5. Tensile strength of as-prepared fibers treated at different temperatures.

CONCLUSIONS

The imidization process of PAA (BPDA-PDA) fibers has been characterized by TGA and FTIR techniques. And the WAXD patterns along fiber axis are applied to characterize the formation of crystals during imidization process. Increasing temperature shows a more remarkable promotion than prolonging treating time to make fibers get higher imidization degree and crystallinity. This is because higher temperature could enhance the ability of chain movement as PAA chains imidized into PI chains. High mechanical properties are available when PAA fibers are treated at relatively high temperatures.

REFERENCES

[1] Zhang, C., Zhang, Q., Xue, Y., et al.. *Chem. Res. Chin. Univ.*, 30, 2014: 163-167.

[2] Huang, S., Ma, X., Guo, H., et al. *Chin. J. App. Chem.*, 29, 2012: 863-867.

[3] Snyder, R.W., Thomson, B., Bartges, B., et al. *Macromolecules*, 22, 1989: 4166-4172.

[4] Brekner, M.-J., Feger, C. *J. Polym. Sci. Part A: Polym. Chem.*, 25, 1987: 2005-2020.

3-D Simulation of the Dynamic of Weft-Knitted Fabric Based on Improved Mass-Spring System

Sha Sha, Gaoming Jiang, Pibo Ma, Xinxin Li
Engineering Research Center for Knitting Technology, Ministry of Education, Jiangnan University, Wuxi, Jiangsu, China
sarah0323@sina.com

ABSTRACT

In order to solve the problem in simulating of the complicated organization of weft knitted fabric and lack sense of reality, mass-spring model was improved and three-dimensional mass-spring model was proposed. The complicated organization of weft-knitted fabric could be well simulated via the improved mass-spring model. The phenomenon of super-elastic in fabric dynamic simulation was also solved by non-linear spring force. The spatial displacement of particles was calculated using velocity-Verlet numerical integration to realize the 3D dynamic simulation of weft knitted fabric. The algorithm showed its advantages on computing speed and real-time performance.

INTRODUCTION

Because of the ubiquity of weft knitted fabric in everyday life, realistic simulating of such articles has been an active area of computer graphics[1-3]. Simulating the dynamic behavior of weft knitted fabric object in interactive application requires two difficult issues to be solved: the realistic effects of intricate knitted surfaces and the stable simulation scheme for real-time animation.

Since soft object animation was characterized by a deformable surface problem, various physical based methods have been proposed. Because the normal mass-spring model is simple and practical, the simulation of the model has been widely used.

To simulate the dynamic fabric, numerous techniques have been carried out. Explicit Euler integration method[12] was widely used for numerical integration. But it could not be applied for large time steps. One of the most significant advancement was implicit integration. However, since the implicit integration is a linear solving system, it cannot be computed easily. Therefore, some efficient approximate approaches have been proposed[2-4].

PARTICLE DYNAMICS

Compared with woven fabric, the physical performance of weft knitted fabric is substantially similar except for obvious thickness and weave. In addition, the traditional model is simplicity and effective. Therefore, the physical properties of mass-spring model was still used for simulating the dynamic of weft knitted fabric. To implement volumetric performance and complex stitch structure of weft knitted fabric, we used springs to link the cuboid particle system. Every face of the cuboid is modeled by stretching, shearing spring. The coefficient of bending springs is minimal which is ignored in this paper.

As shown in Fig. 1, particles were uniformly distributed on the cuboid grid, the model was constructed i×j×k (k=2) particles. To distinguish the particles on different surfaces, the value of subscript 's' was used to denote the particles on which surfaces. Particles distribute on the cuboid surface could be denoted as $P(t) = \{h_s(t, r, s) \mid t \in [0, i-1], r \in [0, j-1], s \in [0, k-1]\}$, where i represents the number of column, j represents the number of row, s represents the number of surface. Each prism of the cuboid was used as the stretching springs to maintaining the distance of particles in warp and weft directions. Each diagonal of the cuboid was used as the shearing springs to maintain distance of particles in oblique directions. After building the model, the stretching springs was defined by different nodes with only one index different from its respective counterpart, and there is only one difference in value among these relevant indexes. And similarly, one shearing spring was defined by different nodes with two sub-indexes different from its respective counterpart, and there are two differences in value among these relevant indexes.

Because particles are moved by external forces and internal forces, bonding points are moved at each time step. The moved bonding points resulted in the change of loop shape. These particles belong to the dynamics system. Each particle moves according to the Newton's law:

$$f = f_{ext}(s, t) + f_{int}(s, t) = m \cdot a = m \cdot \frac{\partial^2 s}{\partial t^2} \qquad (1)$$

To compute the spring force, the vector was denoted from the particle $h_s(t, r, s)$ to the particle , its normalized vector was denoted as $h_s(t + \Delta t, r + \Delta r, s + \Delta s)$, particles in arbitrary positions of the same surface were connected by stretching springs whose vector could be denoted as $\hat{\varpi}_s^m(t, r, s, \Delta t, \Delta r, \Delta s)$, such that:

$$\hat{\varpi}_s^m(t, r, \Delta s, \Delta r) = \frac{l_s^m(t + \Delta t, r + \Delta r, s + \Delta s) - l_s^m(t, r, s)}{\left| l_s^m(t + \Delta t, r + \Delta r, s + \Delta s) - l_s^m(t, r, s) \right|}$$

(2)

$$\varpi_s^m(t, r, s, \Delta t, \Delta r, \Delta s) = l_s^m(t + \Delta t, r + \Delta r, s + \Delta s) - l_s^m(t, r, s)$$

(3)

$$\eta_s^m(t, r, s, \Delta t, \Delta r, \Delta s, \alpha_\lambda, \beta_\lambda) = -a_\lambda \begin{pmatrix} \left(\left| \varpi_s^m(t, r, s, \Delta t, \Delta r, \Delta s) \right| - \left| \varpi_s^0(t, r, s, \Delta t, \Delta r, \Delta s) \right| \right) \\ \hat{\varpi}_s^m(t, r, s, \Delta t, \Delta r, \Delta s) \end{pmatrix}$$

$$+ b_\lambda \begin{pmatrix} v_s^m(t + \Delta t, r + \Delta r, s + \Delta s) \\ - v_s^m(t, r, s) \end{pmatrix} \hat{\varpi}_s^m(t, r, s, \Delta t, \Delta r, \Delta s)$$

(4)

The connected particles were on the same surface, in equation (4), a_λ is the spring coefficient, b_λ is the damping coefficient, $v_s^m(t, r, s)$ and $v_s^m(t + \Delta t, r + \Delta r, s + \Delta s)$ are the current velocity of two adjacent particles.

When a user applies a force to the weft knitted fabric, the fabric will deform. According to the principle that the force spread and diffusion is from near and far, the force particle will spread the force to the adjacent particles[5]. The external forces will not cause all particles move, the displacement only occurred in adjacent particles. Therefore, just consider the appropriate particles in the corresponding range for improve efficiency.

The force is spread layer by layer until a certain depth n, and the layer behind n is not affected. This method is not only ensure the accuracy of the calculation, but also reduce amount of computation. In 3-D space, if the particle subjected to the force and the propagation depth is limited n, forces on the total number of particles is N, such that:

$$N = (n + 2)(2n^2 + 8n + 9) \tag{5}$$

IMPLEMENTATION

The queue template "deque" of STL was applied to building the bidirectional queue. The bidirectional queue could be added and deleted elements in any position. It could traversal, access and read elements randomly, which are the traditional dull list can't accomplished. Compared with ordinary list, by using this method to initialize the particles, springs and bonding points, the program is more flexible and efficient. As show in Fig.2, the simulation of weft knitted fabric forces by the procedures. By defining the physical properties and computing the velocity and displacement of particles, the dynamic simulation of particles was achieved.

RESULTS AND DISCUSSION

The cuboid particle system used for simulating the dynamic of weft knitted fabric is shown in FIG.3. We implemented the technique on as system with Intel XEON Processor E3-123v3 LGA1150, which equipped with cores on 3.30 GHz, 16 GB RAM, and GTX 690 running on Windows 7 OS.

Fig..3 exhibits a range of stitching patterns that can be implemented by the cuboid particle system. Comparison of the two column, the cuboid particle system could simulate actual fabric vividly. A plane particle system was also been implemented for demonstrating the accuracy of cuboid particle system. However, the velocity-Verlet method used as the numerical integrator in this paper produces stable and reasonable behavior. The cuboid particle system could reflect the accurate dimensional relationship of loop between the front needle bed and back needle bed.

CONCLUSION

In this paper, an efficient and stable method was proposed for establishing a dynamics system for weft knitted fabric simulation. The cuboid particle system was introduced and the bonding points were interpolated from coordinates of the particles, which could flexible represent the shape of actual weft knitted fabric. The STL was introduced to create an efficient particle system. When simulated fabric force, limiting the spread of the depth to ensure the accuracy of the calculation, and reducing the amount of computation.

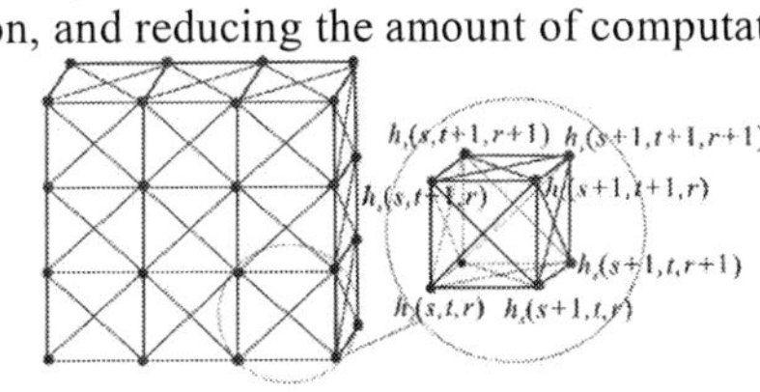

Fig. 1 Improved spring-mass model.

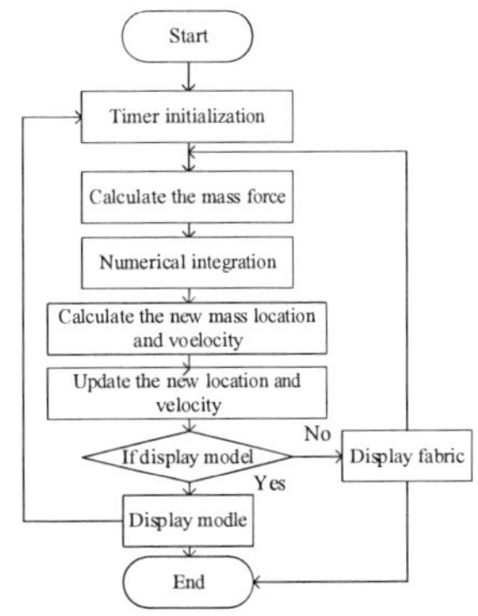

Fig. 2 Work flowchart.

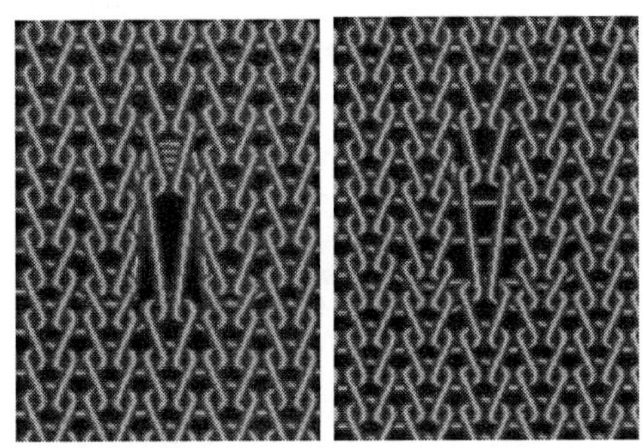

Fig. 3. Exhibits a range of stitching patterns that can be implemented by the cuboid particle system.

REFERENCES

[1] Breen, D.E., House, D.H., Wozny, M.J. "A particle-based model for simulating the draping behavior of woven cloth." *Textile Research Joural*, 64, 11, 1994: 663-685.

[2] Eberhardt, B., Weber, A., Straßer, W. "A Fast, Flexible, Particle-System Model for Cloth Draping." *IEEE Computer Graphics & Applications*. 16, 5, 1996: 52-60.

[3] House, D.H., Breen, D.E. "Cloth Modeling and Animation." Natick, Massachusetts: A.K. Peters, 2000.

[4] Kaldor, Jonathan M., James, Doug L., Marschner, Steve. "Simulating Knitted Cloth at the Yarn Level." ACM SIGGRAPH Conference Proceedings.

[5] Eißner, M., Eberhardt, B. "The Art of Knitted Fabric, Realistic and Physically Based Modelling of Knitted Patterns." *Eurographics*, 17, 3.

Ultralight 3D Electrospun Nanofiber Sponge for Tissue Engineering

Gaigai Duan[1], Valérie Jérôme[2], Ruth Freitag[2], Seema Agarwal,[1] Andreas Greiner[1]
[1]Macromolecular Chemistry II and [2]Process Biotechnology, University of Bayreuth, Germany
greiner@uni-bayreuth.de

Natural sponges are a unique class of complex materials and represent the simplest form of multi-cellular creatures. For example, a sponge with a density of 15 mg / cm^3 had a fiber interconnected structure with a fiber diameter of 20 μm. Electrospinning is known as an excellent method to prepare fibers with fiber diameters in the submicrometer range.[1] Here we report on ultra-light 3D polymer sponges, which were prepared by freeze-drying of dispersions of short electrospun nanofibers (Figure 1).[2] With the use of dioxane as solvent we discovered the formation of a complex hierarchical network of pores. As a consequence open cell soft polymer sponges with extremely low densities were obtained with high mechanical integrity.

Figure 1. Ultralight sponge composed of short electrospun fibers on a feather.

In contrast to other ultra-light sponges or foams the polymer sponges are compressible and soft, which make them particular suitable for novel complex cell tissue design as shown here (Figure 2). Details of the cell culture the cell culture experiment, including hydrophilization and sterilization of the sponges, cell seeding, cell growth, and survival tests will be discussed on the poster.

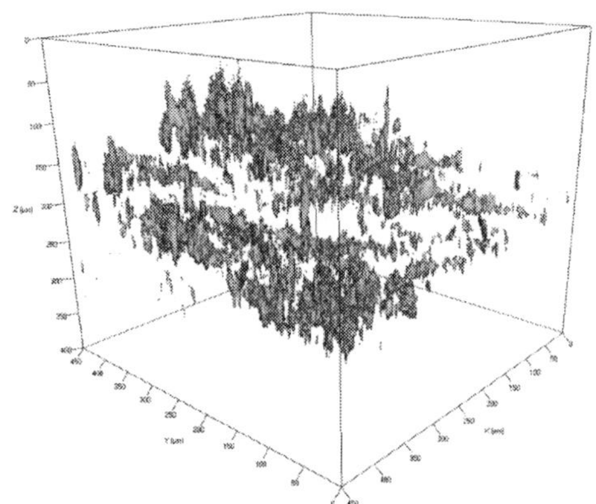

Figure 2. 3D confocal microscopy of live (green) and dead (red) Jurkat cells in sponge after 20 days (x × y × z = 450 × 450 × 400) of cells incubation.

REFERENCES

[1] Agarwal, S., et al. *Prog. Polym. Sci.*, 38, 2013: 963.
[2] Duan, G., et al. *Adv. Funct. Mater.*, Accepted.

.

Effect of the Exposure Time on Transmitted and Stored Energy Through Protective Fabric Systems

Jiazhen He[1,2], Jun Li[1,2]
[1]Protective Clothing Research Center of Fashion Institute , Donghua University, Shanghai, China
[2]Key Laboratory of Clothing Design & Technology , Donghua University, Shanghai, China
lijun@dhu.edu.cn

INTRODUCTION

When exposed to the heat and flam, fabrics will be heated up to a considerably high temperature, consequently thermal energy can be stored within the fabrics. The stored thermal energy released to the skin after exposure can also contribute to skin burn injuries. [1-3] Comprehensive studies regarding the transmitted energy during exposure and stored energy after exposure are demanded to be further investigated.

Moreover most of the studies that explored the influencing factors on the thermal protective performance of clothing focused on the physical properties of the fabric, the entrapped air layer and the presence of moisture. Only a few of studies were carried out to examine the effect of environmental parameters on the thermal insulation of protective materials. These studies concentrated only on the intensity and the constitution of the heat source,[4-5] however there is lack of study concerning the duration of exposure.

The aim of this laboratory study was to assess the effect of heat exposure time on the transmitted energy and the stored energy through multilayer protective fabric systems. The research finding will provide better understanding of the mechanisms associated with thermal stored energy in protective clothing.

APPROACH

Fabric materials

This study was conducted on two protective fabric systems made up with thermal liner selected to have different weight. For this experiment series, the thermal liner layers (TLA and TLB) were combined with the same outer shell fabric (OS), moisture barrier (MB) and inner layer (IL) to make up the fabric systems S1 and S2, as shown in Table I.

Table I. Properties and TPP values of the fabric systems

Code	Assembly description	Thickness (mm)	Weight (g/m^2)
S1	OS+MB+TLA+IL	3.75	578
S2	OS+MB+TLB+IL	2.61	512

Apparatus and procedures

The testing apparatus employed in this study was the widely used thermal protective performance (TPP) tester. The nominal heat source of the tester is $84kW/m^2$. The temperature rise at the back of the specimen was measured by the thermal sensor. The sensor assembly can be in contact with the specimen or a spacer can introduce a 6.4mm air gap between the fabric and the sensor. The heat exposure time from 6 seconds to 27 seconds with increments of 3 seconds was selected in this study. After heat exposure with specific duration, the specimen was terminated by inserting the blocking shutter. Continue to record the data after terminating the heat exposure to the specimen and until the total data acquisition time was 90 seconds.

Transmitted energy (Q_t), stored energy (Q_s) and the new index of decrement factor to second degree burn (f)

The details of the method used for cumulative thermal energy calculation are described in ASTM F2703.[6] The total cumulative energy (Q_{max}) throughout the test was found to be the maximum thermal energy detected by the sensor, while the transmitted energy (Q_t) was defined as the thermal energy measured by the sensor during exposure. Therefore, the difference between the total cumulative energy (Q_{max}) and the transmitted energy (Q_t) was calculated as the stored energy (Q_s) discharged to the sensor after the thermal exposure, as shown in Figure 1.

In order to examine the intersections occurred between the sensor response (Qi) and the "Stoll" criterial (Si), a new index f was proposed to describe the decrement factor to second degree burn. This new index was calculated as:

$$f=\frac{(S_i-Q_i)_{min}}{S_{i@min}}$$

Where $(S_i - Q_i)_{min}$ stands for the minimum difference between the "Stoll" criteria and the sensor response after the exposure (J/cm^2); $S_{i@min}$ stands for the value in the "Stoll" criteria at the time when the minimum difference is detected (J/cm^2).

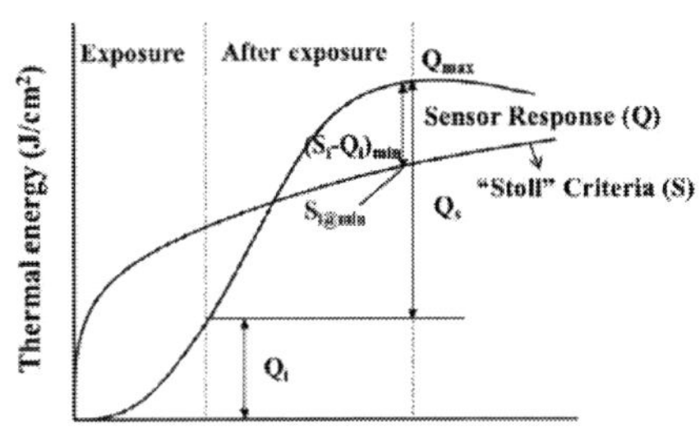

Figure 1. The schematic representation of transmitted energy (Q_t), stored energy (Q_s) and the determination of the decrement factor to second degree burn (f)

RESULTS AND DISCUSSION

Effect of the exposure time on transmitted energy (Q_t)

The transmitted energy (Q_t) through two fabric systems exposed to eight levels of heat duration was compared in Figure 2. It was inescapably clear that the transmitted energy through both of the fabric systems changed largely as the exposure time increased. For the configuration without a spacer, the transmitted energy increased sharply after 15s exposure, whereas for the configuration with a spacer the transmitted energy increased sharply after 21s exposure. In both configurations with and without a spacer, the transmitted energy obtained through fabric system S2 was higher than that from S1. Moreover, the introduction of a spacer led to a decrease in the transmitted energy.

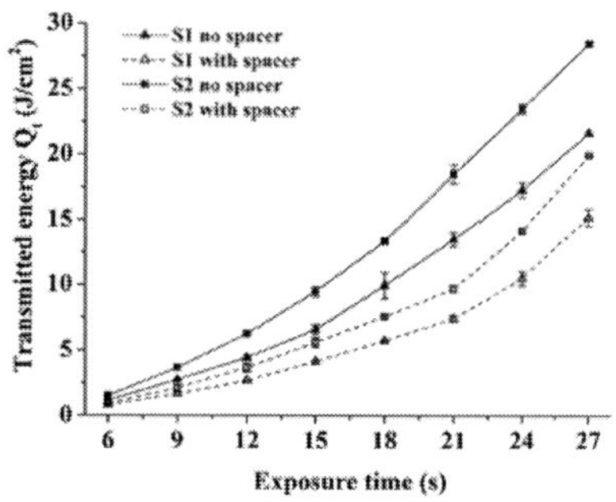

Figure 2. Comparison of transmitted energy (Qt) through two fabric systems with and without spacer

Effect of the exposure time on the stored energy (Q_s)

The relationship between stored energy released from fabric systems and the exposure time was illustrated in Figure 3. The values of the stored energy showed a relatively smooth change to that observed in the transmitted energy; however the variations of stored energy became more complicated as exposure time altered. As the heat exposing duration was ascended from 6 seconds to 27 seconds, the stored energy discharged from the fabric systems increased firstly and then decreased. The overall trend of the stored energy demonstrated that thermal storage capacity of the fabric seemed to be disturbed if the fabric system was heated for a rather long time. The decline of the stored energy that occurred after long heat exposures would be caused by the thermal degradation of the fabric.

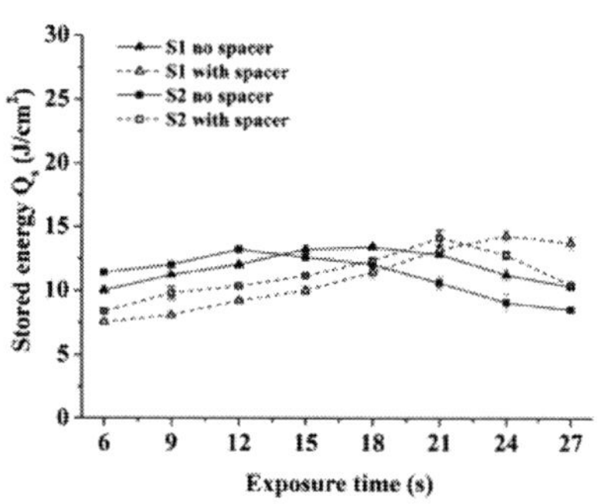

Figure 3. Comparison of stored energy (Qs) from two fabric systems with and without spacer

Effect of the exposure time on the decrement factor (f)

Figure 4 showed the changes of decrement factor to second degree burn obtained from two fabric systems. It was clear that the decrement factors calculated from fabric system S1 with a spacer stayed in the highest position, followed by the fabric system S2 with a spacer, S1 without a spacer and then S2 without a spacer, indicating that the better thermal protection was provided by thicker fabric system with the air layer even though stored energy effect was taken into account. The trend of the decrement factor demonstrated that the values in the decrement factor fell with the increase of heat exposure time. It revealed the fact that thermal protection provided by the protective ensembles was greatly reduced with the increasing of the exposure time. The index of decrement factor introduced in this study could be used to distinct the thermal protection provided by different fabric systems even with various exposure time.

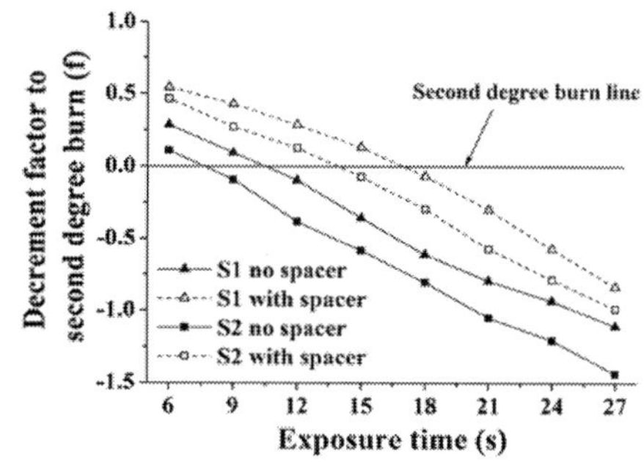

Figure 4. The change of the decrement factor to second degree burn

CONCLUSION

The analyses demonstrated that the amount of transmitted energy went up continuously with the increasing of exposure time; however the amount of stored energy approached a peak value, which indicated that the stored energy released to the skin was limited in the process of exposure time adding. This study also proposed a new index named the decrement factor to second degree burn to assess the protective performance of fabric systems. This new index integrated the effect of the transmitted energy and also the stored energy. The decrement factor to second degree burn fell with the increase of heat exposure time. The data produced by this research clearly showed that, besides the improvement of personal protective clothing properties, the safety of firefighters was essentially related to good control of the exposure times to these high intensity fluxes.

REFERENCES

1. Torvi, D.A., Eng, P., Threlfall, T.G. "Heat transfer model of flame resistant fabrics during cooling after exposure to fire." *Fire Technol.*, 42: 2006: 27-48.
2. Barker, R.L., Guerth, C., Behnke, W.P., et al. "Measuring the thermal energy stored in firefighter protective clothing. *Performance of Protective Clothing: Issues and Priorities for the 21st Century.* West Conshohocken PA: American Society for Testing and Materials, ASTM STP 1386, 2000: 33-44.
3. Song, G.W., Paskaluk, S., Sati, R., et al. "Thermal protective performance of protective clothing used for low radiant heat protection." *Textile Res. J.*, 81, 3, 2010: 311-323.
4. Lee, Y.M., Barker, R.L. "Thermal protective performance of heat-resistance fabrics in various high intensity heat exposures." *Textile Res. J.*, 57, 3, 1987: 123.
5. Lawson, .LK., Crown, E.M., Ackerman, M.Y., et al. "Moisture effects in heat transfer through clothing systems for wildland firefighters." *International Journal of Occupational Safety and Ergonomics*, 10, 2004: 227-238.
6. "Standard Test Method for Unsteady-State Heat Transfer Evaluation of Flame Resistant Materials for Clothing with Burn Injury Prediction." West Conshohocken, PA: American Society for Testing and Materials, ASTM F2703-08, 2008.

Conductive Polyaniline/Viscose Fiber Composites

Wang Ning[1], Li Guodong[1], Zhang Xingxiang[1], Liu Dongqi[2]

[1]Tianjin Municipal Key Lab of Fiber Modification and Functional Fiber, School of Material Science and Engineering, Tianjin Polytechnic University, Tianjin, China; [2]Xinxiang Bailu Hua Xian Group Co., LTD., Xinxiang, China

wangntjpu@hotmail.com

INTRODUCTION

The discovery of conducting polymers has opened up a new promising field in material science and engineering. Among them, polyaniline (PANI) is one of the most promising conducting polymers because of its unique properties, its ease of preparation, and excellent environmental stability. [1] PANI has the utilization potential in a large number of applications. [2] Recent research on conductive textiles has already been attracted much attentions due to the potential applications on clothing, biomedicine and the covering of the equipments. Many approaches have been used such as the insertion of metallic wires inside the yarns, coating fibers with metals or conductive films and the incorporation of conductive fillers. [3] Viscose fiber (VF) is a linear β-1, 4-glycosidically linked polyglucan containing a mass of intermolecular and intramolecular hydrogen bonds, therefore VF exhibits unique properties such as chemical stability, biocompatibility and biodegradation [4]. In this work, the conductive PANI/VF composites are fabricated by polymerization using ammonium persulfate as the oxidant, HCl and dodecylbenzene sulfonic acid (DBSA) as dopant respectively. The resultant PANI/VF composites are characterized by FTIR, the field-emission scanning electron microscopy. The properties of PANI/VF composites such as electrical conductivity, tensile strength and washing resistance are also investigated.

APPROACH

VF was soaked in acetone solution to remove the oil on fibers and dried in vacuum. The above mentioned VF was dispensed in HCl and DBSA solution respectively. Aniline monomer was introduced, and then the mixture was stirred to facilitate the adsorption of aniline onto the fiber. Then $(NH_4)_2S_2O_8$ solution was added dropwise into the above system. The mixture was stirred for different time. The reaction products were filtrated and washed successively with acetone and deionized water until the filtrate become colorless, and then dried in vacuum to obtain dark green fibers.

RESULTS AND DISCUSSION

The FTIR spectra of VF, PANI and PANI/VF with different dopants were shown in Fig. 1. The bands at 3443, 2917, 1642, 1439, 1378, 1157, and 1062 cm^{-1} in VF were associated with VF [5]. The typical feature of pure PANI was also exhibited. The peak at 801 cm^{-1} corresponding to the out-of-plane bending vibration of the C-H band of p-disubstituted benzene ring appeared. The peaks at 1303 and 1138 cm^{-1} due to the stretching of the C-N band and vibration of C-H in the benzene ring were observed. The peaks around 1559 and 1495 cm^{-1} were respectively assigned to stretching vibration of N-B-N and N=Q=N structures (B and Q represent benzenoid and quinoid moieties in the PANI chains) [6]. Fig. 1c and 1d show the spectra of the PANI/VF compsoties dopant with HCl and DBSA respectively. The peaks at 801, 1303 and 1495 cm^{-1} indicated the components of PANI. And the absorption peak at 3447 cm^{-1} in the spectrum of VF shifted blue in the spectrum of the PANI/VF composites. It revealed that the interaction formed duo to the intermolecular hydrogen bands between VF and PANI.

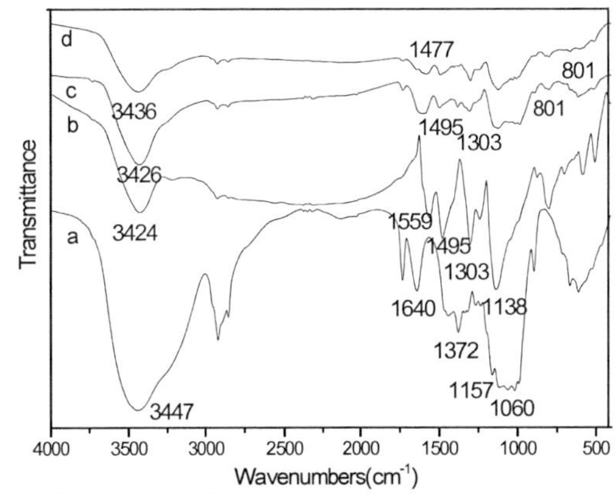

Fig. 1. FTIR spectra of VF (a) PANI (b) PANI/VF dopant.

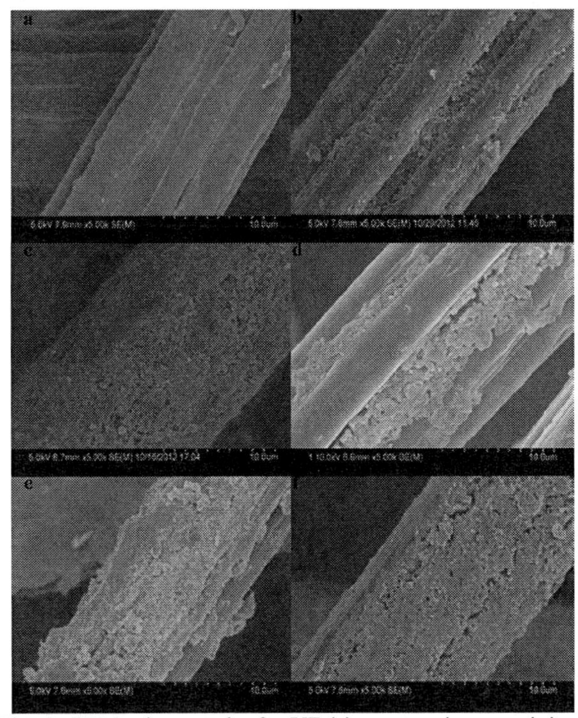

Fig. 2. SEM micrographs for VF (a), composite containing 4.0 wt % PANI doped with HCl (b), composite containing 11.8 wt % PANI doped with HCl (c), composite containing 4.8 wt % PANI doped with DBSA (d), composite containing 10.5 wt % PANI doped with DBSA (e), and composite containing 11.2 wt % PANI doped with DBSA (f).

Fig. 2 depicts the morphology of VF and PANI/VF composites containing the different PANI contents. Several longitudinal grooves were observed clearly on the smooth surface of the VF in Fig. 2a. At the low PANI content (as shown in Fig. 2b and 2d), some disconnected PANI particles appeared on the VF. It was obvious that the addition of PANI to the matrix resulted in the rougher bulk as compared to the pure VF. As revealed in Fig. 2c and 2e, the PANI particles deposited on the surface of fibers connected to form a continuous layers and filled grooves along VF fiber templates with the improving content of PANI. The hydrogen bands between hydroxyl groups of VF and amine groups of aniline might serve as a traction force to assist the growing of the continuous layers of PANI over cellulose [7].

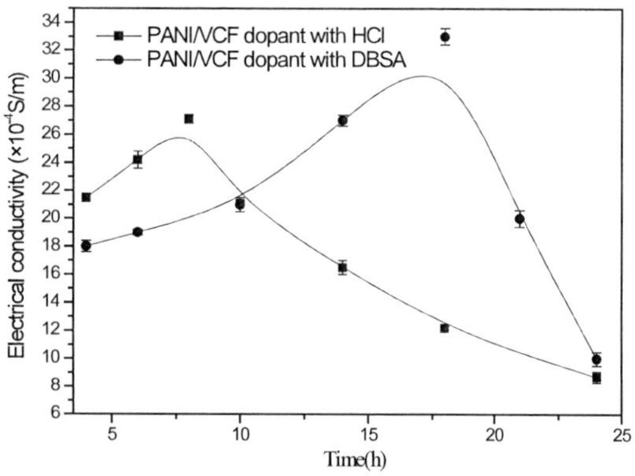

Fig. 3. Influence of reaction time on room temperature electrical conductivity of PANI/VF composites.

Fig. 3 shows the effect of the reaction time on the electrical conductivity of PANI/VF composites. There was a general trend of initial increase and subsequent decrease in electrical conductivity of PANI/VF composites with the increasing of reaction time. It could be explained that a large number of aniline monomers were loaded on the cellulose surface because of the strong hydrogen band interaction between aniline and hydrogen groups of cellulose, which resulted in the improvement of PANI contents in adducts. The disconnected PANI particles gradually grew into a continuous layer along the VF templates. However, the longer reaction time could result in the precipitation or aggregation of PANI particles and excessive oxidation of the aniline monomer with the fracture of the PANI conjugated chain. It is similar to the results of the reference [7]. At the same time, the doping system may be undermined to some extent, and the insoluble PANI may generate in the composite fibers which decrease the conductive component, thus reducing the corresponding conductivity of composite fibers [8].

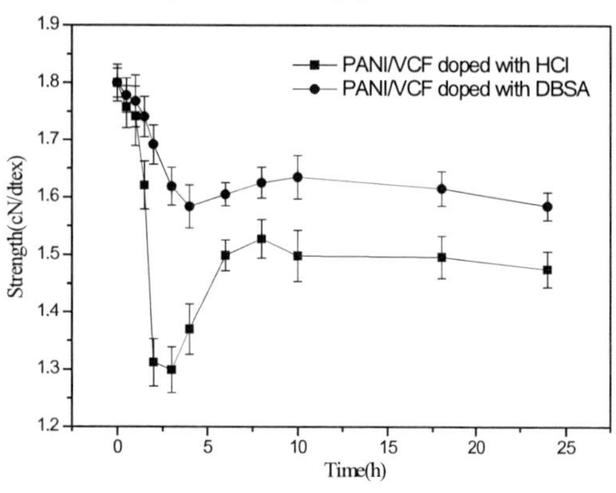

Fig. 4. Influence of reaction time on tensile strength of PANI/VF composites.

Influences of reaction time on tensile strength of PANI/VF composites were revealed in Fig. 4. With the increasing of reaction time, the tensile strength initially decreased and subsequently increased in different protonic acids. The initial reduce could be attributed to the corrosion of hydrochloric acid; Subsequent improvement might be related to the PANI particles deposited on the surface of fibers connected to form a continuous layers. When the reaction time was more than 8h, the strength reduced slowly.

In Fig. 5, PANI/VF composites containing11.8 wt % and 10.5 wt % PANI had a slight variation of the conductivity, after they were repeatedly washed with water. A small amount of PANI, poorly combining with VF was washed, the conductivity was basically unchanged. When the adduct was washed 50 times, it was stable after more washing. It indicated that PANI has good adhesion on VF.

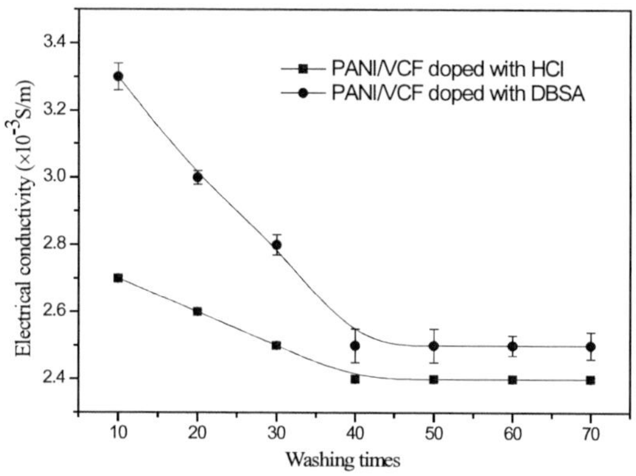

Fig. 5. Washing resistance of PANI/VF composites.

CONCLUSION

In this work, PANI/VF composites were synthesized using in situ polymerization. PANI particles deposited on the surface of VF connected to form a continuous network along the VF template. The introduction of PANI in the composite provides high conductivity as well as a relatively high tensile strength. PANI has good adhesion on VF after washing 50 times. These intriguing features make it quite a suitable and promising material for conductive textiles, applying on clothing, biomedical uses and the covering for outside equipment.

ACKNOWLEDGMENT

The authors thank science funds of aviation (201329Q2001) for financial support

REFERENCES

1) Mu, S. *Synth. Met.*, 143, 2004: 269.
2) Tahir, Z.M., Alocilja, E.C., Grooms, D.L. *Biosens Bioelectron*, 20, 2005: 1690.
3) Gurjanov, O.P., Ibragimova, N.N., Gnezdilov, O.I., et al. *Carbohydr Polym.*, 72, 2008: 719.
4) Madani, M., Aly, S.S., S.M. El-Sayed, S.M. *H. Perform Polym.*, 22, 2010: 515.
5) Hu, W.L., Liu, S.P., Chen, S.Y., et al. *Cellulose*, 18, 2011: 655.
6) Dhibar, S., Sahoo, S., Das, C.K. *Polym Compos.*, 34, 2013: 517.
7) Mo, Z.L. Zhao, Z.L., Chen, H., et al. *Carbohydr. Polym.*, 75, 2009: 660.
8) Hu, W.L., Chen, S.Y., Yang, Z.H., et al. *J. Phys. Chem.*, 115, 2011: 8453.

High-Performance Polyimide Nanofiber

Yu Zhuo[1], Ning Wang[1], Guodong Li[1], Xingxiang Zhang[1], Dongqi Liu[2]
[1]Tianjin Municipal Key Lab of Fiber Modiication and Functional Fiber, School of Material Science and Engineering, Tianjin Polytechnic University, Tianjin, China; [2]Xinxiang Bailu Hua Xian Group Co., LTD., Xinxiang, China
wangntjpu@hotmail.com

INTRODUCTION

In the last years, polymeric nanofibrous materials have received considerable attentions on their potential applications in science, technology and industry for their unique properties of higher specific surface area, higher porosity, smaller pore size, easy post processing than the conventional fibers[1]. As a high-performance polymer, poly(pyromellitic dianhydride-co-4,4-oxydianiline) (PI) have been widely used for a variety of interesting applications, owing to its superior mechanical properties, thermal stability, high glass transition temperature, chemical resistance [2,3]. Solution blow spinning as a new nanofiber fabricating method with productivity several times higher than electrospinning was recently proposed [4]. In the present study, PMDA/ODA PI nanofibers were successfully solution blown and several nanofibers of different spinning parameters were investigated. The resultant polyimide nanofibers are characterized.

APPROACH

PAA precursors for spinning were prepared by co-polymerizing PMDA and ODA in N,N-dimethyl-formamide (DMF) solvent. And then the PAA solution was spun into fiber web using an solution blow spinning apparatus. It consisted of a high velocity compressed air supply, a spinning die, a spinning cabinet and a collector. After the solvent of the PAA nanofiber mat was removed, next stage was the process of thermal imidization. At last, the PMDA/ODA PI nanofibers mat can be obtained.

RESULTS AND DISCUSSION

The FTIR spectra of PAA and PI were shown in Fig. 1. 1778 cm^{-1}(symmetric C=O stretching, imide I), 1716 cm^{-1}(asymmetric C=O stretching, imide I), and 1378 cm^{-1}(C–N–C stretching, imide II), 725 cm^{-1}(C=O bending) were the characteristic bands of PI. At the same time, the typical polyamic acid absorption bands at 1660 cm^{-1}(C=O stretching, amide I) and 1550 cm^{-1}(C–Nstretching, amide II), and 1400 cm^{-1}(C–Nstretching, acid amide III) are totally absent indicating the polyamic acid was completely transformed to polyimide under thermal imidization[2,5].

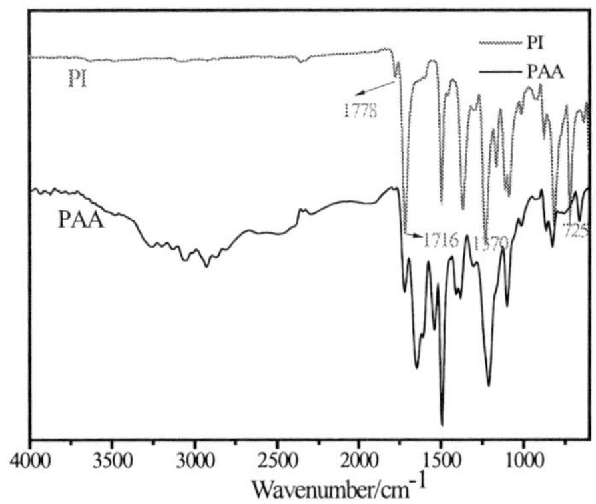

Fig. 1. ATR-FTIR spectrum of the PAA and PI nanofiber.

Fig. 2 shows the thermal gravimetric behavior of PI mats in nitrogen atmospheres. From the Fig. 2, it was seen that the temperatures of the PI web at a mass loss of 5% was 561 °C, whereas at a mass loss of 10% it was 579 °C that could be attributed to solvent evaporation and imidization reactions. When heated in nitrogen to temperatures above 800℃, the weight of PI web was nearly constant at 53%, indicating a relatively high thermal stability.

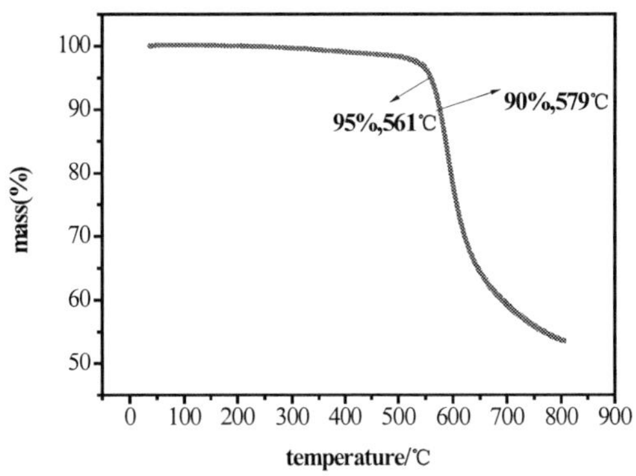

Fig. 2. TGA thermograms of PI nanofiber mat under nitrogen atmosphere.

In the process of solution blow spinning, several nanofibers of different spinning parameters were successfully solution blown. From the Fig. 3(a), it was seen that at feeding rate(v) of 5ml/h, solution concentration(c) of 3003mpa.s, gas pressure(p) of 0.16MPa, the most fiber diameter ranged from 63.16 to 157.91 nm with the average diameter of 112.15 nm. From the Fig. 3(b), on conditions that c=3868mpa.s;p=0.1 MPa; v=8ml/h, the most fiber diameter ranged from 116.47 to 231.65 nm with the average diameter of 165.81 nm. From the Fig. 3(c), it was seen that on conditions that c=7058mpa.s;p=0.16MPa; v=10ml/h, the fiber diameter distribution ranged from 126.33 to 219.72nm with the average diameter of 168.61nm. From the Fig. 3(d), it was seen that on conditions that c=3003mpa.s; p=0.16MPa; v=12ml/h, the fiber diameter distributions was at the range of113.87 to 233.78 nm with the average diameter of 155.30nm. With the gas pressure increasing, the nanofibers mat became more curled or twisted loop. With the solution concentration and feeding rate increasing, the average diameter would be thicker.

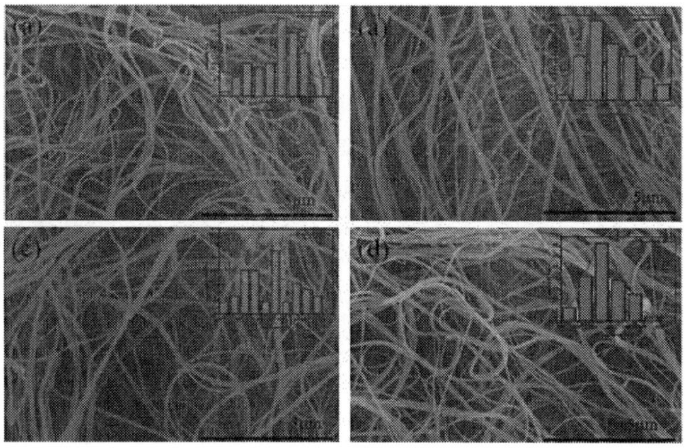

Fig. 3. SEM micrographs for:
(a) c=3003mpa.s;p=0.16MPa;v=5ml/h
(b) c=3868mpa.s;p=0.16MPa;v=8ml/h
(c) c=7058mpa.s;p=0.16MPa;v=10ml/h
(d) c=3003mpa.s;p=0.16MPa;v=12ml/h

CONCLUSION

In the present work, the PMDA/ODA PI nanofibers of different spinning parameters were successfully solution blown. PI nofibers have more advantages due to the combination of nanofibers and polyimide. PMDA/ODA PI nanofibrous mats possess unique properties of high specific surface area, high porosity, small pore size which will be a promising candidate for widely use in filtration filed.

FUTURE WORK

In consideration of the advantages of PI nanofibers, We expected that this novel process could be extended to fabricate polyimide nanofibers with higher productivity for filtration application. Therefore, the next step will be that filtration efficiency will be examined to evaluate the filtration performance of the PI nanofibers mats. Furthermore, the carbonization of PI nanofibers mats will be interesting. Of course, high temperature resistance will be promoted.

ACKNOWLEDGMENT

The authors thank science funds of aviation (201329Q2001) for financial support

REFERENCES

1) Shi, Lei, Zhuang, Xupin, Cheng, Bo-wen. "Solution Blowing of Poly(dimethylsiloxane) / Nylon 6 Nanofiber Mats for Protective Applications." *Chinese Journal of Polymer Science*, 32, 2014: 786-792.

2) Ba, Chaoyi, Economy, James. "Preparation of PMDA/ODA polyimide membrane for use as substrate in a thermally stable composite reverse osmosis membrane." *Journal of Membrane Science*, 363, 2010: 140-148.

3) Yang, K.S., Edie, Dan D., Lim, D.Y. "Preparation of carbon fiber web from electrostatic spinning of PMDA-ODA poly(amic acid) solution." *Carbon*, 41, 2003: 2039-2046.

4) Zhang, Xupin, Shi, Lei, Jia, Kaifei. "Solution blown nanofibrous membrane for microfiltration." *Journal of Membrane Science*, 429, 2013: 66-70.

5) Han, Enlin, Wu, Dezhen, Qi, Shengli. "Incorporation of Silver Nanoparticles into the Bulk of the Electrospun Ultrafine Polyimide Nanofibers via a Direct Ion Exchange Self-Metallization Process." *Appl. Mater. Interfaces*, 4, 2012: 2583-2590.

Characterization of the Crystalline Structure of Polyimide Fiber by Using Synchrotron Two-Dimensional X-Ray Diffraction

Jidong Zhang[1], Fang Yang[1], Ru Xiao[2]
[1]Changchun Institute of Applied Chemistry, CAS, China; [2] State Key Laboratory for Modification of Chemical Fibers and Polymer Materials, College of Materials Science and Engineering, Donghua University, Shanghai, China
jdzhang@ciac.ac.cn

Polyimide (PI) fibers have attracted more and more attention due to their properties such as outstanding thermal stability, chemical resistance, excellent mechanical properties and good irradiation resistance. Here we characterize the crystalline structure of a commercial polyimide fiber by using synchrotron two-dimensional X-ray diffraction. The measurements were carried out at Beamline 14BL in Shanghai Synchrotron Radiation Facility (SSRF) with a MarCCD detector. The X-ray wavelength is 0.124nm. All the patterns were taken with a bundle of fibers oriented horizontally. The 2DXRD pattern was processed with the software Fit2D. To eliminate the influence of background noise, the data were treated by subtracting 2DXRD pattern without fibers.

Figure 1 and 2 are the original and treated 2DXRD pattern of PI fiber. After subtracting background signals, the diffraction signals become more clear. In the meridian direction there are four obvious diffraction corresponding to (00l) planes. In the equator direction there are four weak diffraction arcs corresponding to molecular chain packing. These results indicate that the molecular chains orient along fiber axial direction. Using integration to achieved one-dimensional diffraction patterns, the effect of background subtraction is more obviously, especially in equator direction. Integrating from equator to meridian direction in 90° range, we can achieve diffraction pattern (Figure 5) equivalent to truly powder diffraction, which contains a very strong (001) diffraction peak and other weak diffraction peaks that demonstrate the high degree of crystallinity.

In conclusion, the crystalline structure of PI fiber was characterized by using synchrotron 2DXRD. More detailed work will be carried out to study the relationship between crystalline structure and properties.

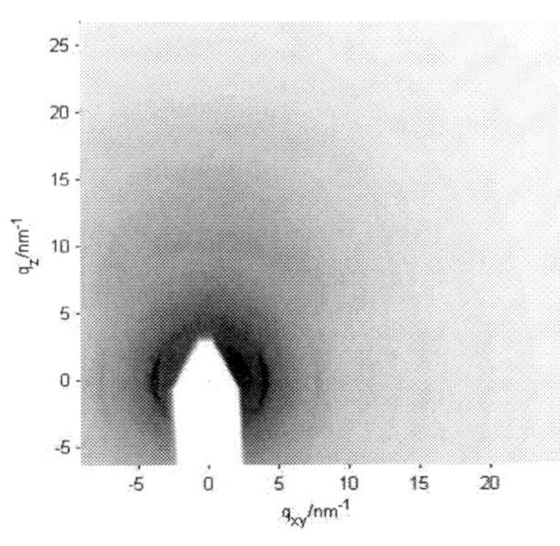

(1)

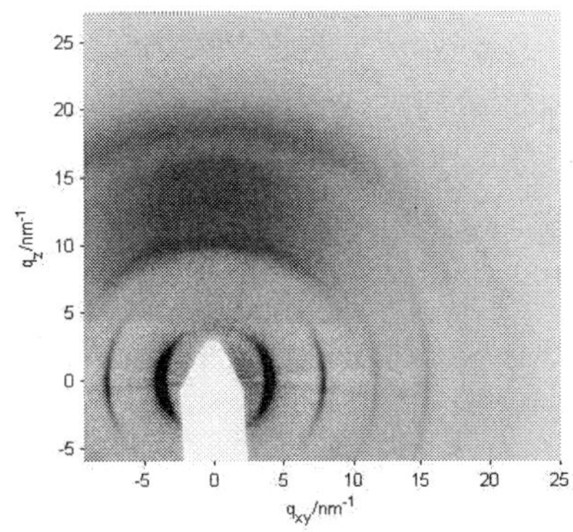

(2)

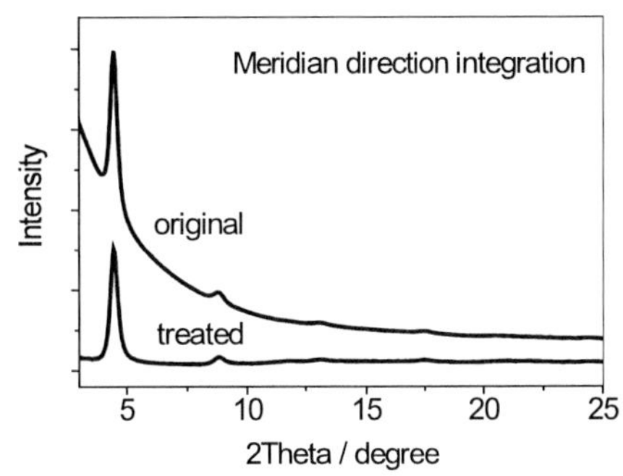

(3)

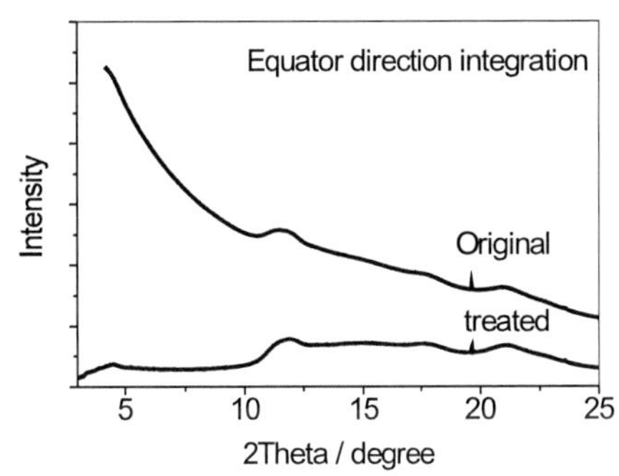

(4)

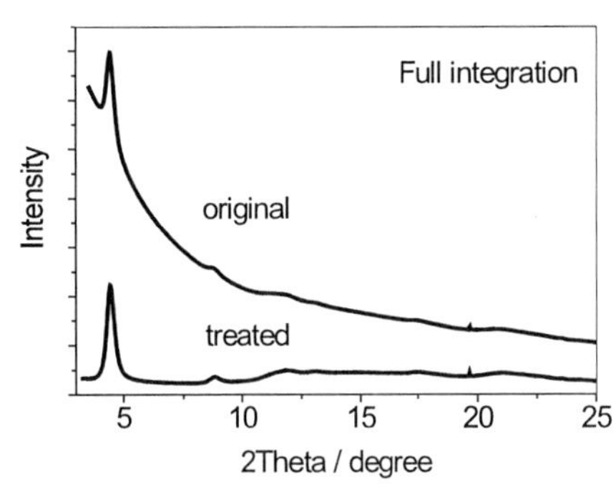

(5)

Heat-Resistance and Mechanical Properties of Injected Molded Poly(L-lactic acid)/ Poly(D-lactic acid) Blends with Different Molecular Weight

Yu-zeng Li, Ge-sheng Yang, Rui-hao Ming, Qiao Li, Hui-hui Zhang, Hui-li Shao
State Key Laboratory for Modification of Chemical Fibers and Polymer Materials, College of Material Science and Engineering, Donghua University, Shanghai, China
hhzhang@dhu.edu.cn; 15316738186@163.com

INTRODUCTION

Poly(lactic acid) (PLA) is a typical biodegradable and bio-based thermoplastic polyester with high modulus and strength. However, the widespread application of PLA is still limited because of its poor heat-resistance and low crystallization rate. In 1987, Ikada et al [1] firstly reported stereo-complexation formation between enantiomeric poly(L-lactide)(PLLA) and poly(D-lactide) (PDLA). Stereocomplex polylactide (sc-PLA) shows a high melting temperature around 230°C, 50°C higher than homo-crystallized polylactides [2]. Therefore, it has become a new approach to prepare high heat resistant PLA in recent years. Melt stereocomplexation of PLLA/PDLA blend depends on different parameters, such as the molecular weight and optical purity of the raw material, the blending ratio as well as the melt temperature and processing time [3].

In this work, two kinds of PLLA and three kinds of PDLA pellets with different molecular weight were used as the raw materials, and equivalent proportion of PLLA/PDLA were melt blended and then followed by injection molding to form the stereocomplex, then the heat-resistance and mechanical properties of PLLA/PDLA blends with different molecular weight were compared.

APPROACH

PLLA-1 and PLLA-2 were supplied by Nature Works LLC USA, and Zhejiang Hisun Biomaterials Co., Ltd, China, respectively. PDLA-1 and PDLA-2 were supplied by Changchun Institute of Applied Chemistry Chinese Academy of Science, and Shanghai Academy Biomaterials Corporation, China. PDLA-3 was purchased from Purac Corbion Biochem., Singapore. The detailed data of PLLA and PDLA pellets are shown in Table I.

Table I. Basic data of PLLA and PDLA pellets.

Samples	T_m(°C)	M_η(g/mol)*
PLLA-1	169.4	3.7×10^4
PLLA-2	179.9	8.6×10^4
PDLA-1	176.5	3.0×10^4
PDLA-2	157.6	5.1×10^4
PDLA-3	193.7	9.6×10^4

* M_η was calculated according to the Mark-Howink equation: $[\eta]=5.45\times10^{-4}M_\eta^{0.73}$

The equivalent proportion of PLLA and PDLA with different molecular weight was dried in a vacuum oven at 60°C for 12h, and then they were melting blended in a Rheocord90 torque rheometer (Haake Co, Germany) with a rotating rate of 50rpm at 220°C for 5min. At last, the blend was pelletized and then injecting molded into samples. In order to make a comparison, pure PLLA were also prepared by this way, while PLLA was processed at the temperature of 190°C.

DSC was made on a Modulated DSC 2910 (TA Instruments Ltd, USA) at a heating rate of 10°C /min from ambient temperature to 250°C in the presence of nitrogen gas (40ml/min). Crystallinity degree of stereocomplex crystallite (X_{sc}) was calculated according to reference [4].
Vicat softening temperature (VST) was carried out in a HDT & Vicat 6921 (Ceast Co., Italy). The sample was tested with a load of 10N and a heating rate of 2°C/min.
Tensile testing was performed on WDW3100 universal testing machine (Changchun Kexin Testing Machine Co. Ltd, China) at 2mm/min.

RESULTS AND DISCUSSION

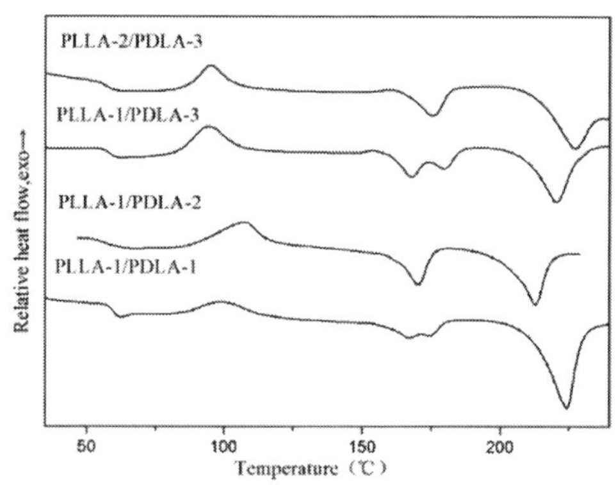

Figure 1. The DSC thermograms of different PLLA/PDLA blends.

Table II. The thermal properties of different PLLA/ PDLA blends

Samples	$T_{m,hc}$(°C)	$T_{m,sc}$(°C)	X_{sc}(%)
PLLA-1/PDLA-1	169.6	226.4	24.8
PLLA-1/PDLA-2	170.2	212.5	11.4
PLLA-1/PDLA-3	167.3	220.5	17.3
PLLA-2/PDLA-3	176.5	228.0	23.2

Fig. 1 shows the DSC thermograms of different PLLA/PDLA blends, and the thermal properties of PLLA/PDLA blends are shown in Table II. It can be seen from Fig.1 that all PLLA/PDLA blends showed two melting peaks at 160~190°C, and ~220°C, which can be assigned to the melting of PLLA or PDLA homo-crystallites(hc) and stereo- complex crystallites(sc), respectively. As can be seen, the $T_{m,sc}$ is ~50℃ higher than $T_{m,hc}$.

The molecularweightof PLLA or PDLA pellets strongly influences the $T_{m,sc}$ and crystallinity of the blends. When keeping PLLA-1 constant, the PLLA-1/PDLA-1 blend showed highest $T_{m,sc}$ and X_{sc}, compared with PLLA-1/PDLA-2 blend and PLLA-1/PDLA-3 blend. This is because low molecular weight of PLLA or PDLA pellet can facilitate the formation of stereocomplex crystallite. On the other hand, PLLA-1 and PDLA-1 have similar molecular weight, which maybe favoring the formation of hydrogen bonds between PLLA and PDLA, since this type of interaction between L and D-units occur pairwise, each oxygen atom of carbonyl groups being hydrogenbonded with two hydrogen atoms of its enantiomeric chain [5]. In addition, PLLA-2 and PDLA-3 also have similar molecular weight, and $T_{m,sc}$ of PLLA-2/PDLA-3 blend reached 228°C, which was higher slightly than PLLA-1/PDLA-1, whereas X_{sc} of PLLA-2/PDLA-3 blend was lower than PLLA-1/PDLA-1 blend because the stereoselective association occurred preferably to PDLA and PLLA with lower molecular weight.

Table III. The Vicat softening temperature of PLLA and PLLA/PDLA blends

Samples	Vicat softening temperature (°C)
PLLA-1	58.3
PLLA-1/PDLA-1	161.1
PLLA-1/PDLA-2	74.7
PLLA-1/PDLA-3	157.4
PLLA-2/PDLA-3	167.5

It can be seen from the Table III that PLLA-2/PDLA-3 blend exhibited a highest VST at 167.5°C, which is about 110°C higher than that of pure PLLA (58.3°C). The VST of PLLA-1/PDLA-2 blend was only 74.7°C, which maybe result from the low X_{sc}(11.4%) of blend, and the high content of hc in PLLA-1/PDLA-2 blend caused a phase separation between hc and sc.

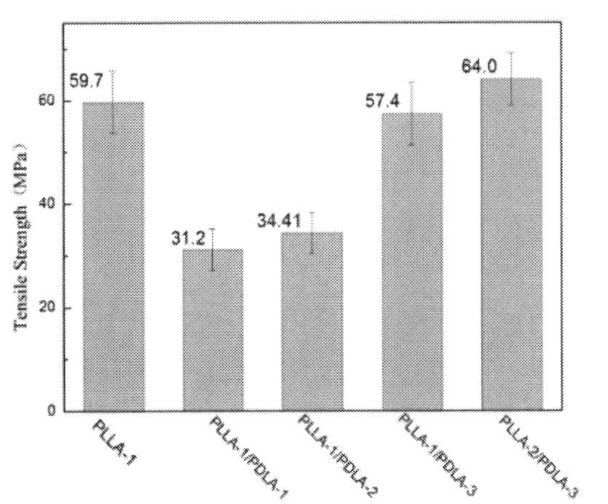

Figure 2. The tensile strength of PLLA and PLLA/PDLA blends.

As can be seen from Fig.2, the tensile strength of PLLA/PDLA blend was lower than the pure PLLA when the molecular weight of pellet was too low. The possible reason is that PLLA/PDLA blend was processed at the comparatively high temperature (220°C), which may lead to the thermal degradation of PLA molecular chains to some extent. However, the tensile strength of the PLLA/PDLA blend increased with the increasing of molecular weight of PLLA or PDLA pellet. It can be noticed that the tensile strength of PLLA-2/PDLA-3 blend reached 64.0MPa, which was higher than pure PLLA.

CONCLUSIONS

The effect of molecular weight of PLLA and PDLA pellets on the heat-resistance and mechanical properties of injection molded PLLA/ PDLA blend was investigated. It could be found that the homo-crystallites ($T_{m,hc}$=160 ~190 °C) and stereocomplex crystallites ($T_{m,sc}$≈220°C) coexisted in PLLA/PDLA blend. PLLA and PDLA with low and similar molecular weight could facilitate the formation of stereo-complex crystallite and improvement of VST in PLLA/PDLA blend.VST of PLLA-2/PDLA-3 blend reached 167.5°C, which is around 110°C higher than pure PLLA. VST of PLLA-2/PDLA-3 blend reached 167.5 °C, which is around 110 °C higher than that of pure PLLA. The tensile strength of the PLLA/PDLA blend increased with the increasing of molecular weight of PLLA or PDLA pellet, and it can exceed pure PLLA if the PLLA or PDLA pellets with high molecular weight were used.

REFERENCES

[1] Ikada, Y., Jamshidi, K., Tsuji, H., et al. "Stereocomplex formation between enantiomeric poly(lactides)." *Macromolecules*, 20, 1987: 904-906.

[2] Bertin, A. "Emergence of polymer sterecomplexes for biomedical applications." *Macromol. Chem. Phys.*, 213, 22, 2012: 2329-2352.

[3] Rui, Y.B., Yang, W., Jiang, W.R., et al. "Stereocomplex formation of high molecular weight polylactides: A low temperature approach." *Polymer*, 53, 2012: 5449-5454.

[4] Sarasua, J.R., Arraiza, A.L. "Crystallinity and mechanical properties of optically pure poly- lactides and their blends." *Polym. Eng. Sci.*, 45, 5, 2005: 745-753.

[5] López-Rodríguez, N., deArenaza, I.M., Meaurio, E., et al. "Improvement of toughness by stereocomplex crystal formation in optically pure polylactides of high molecular weight." *J. Mech. Behav. Biomed. Mater.*, 37, 2014: 219-225.

A Practical Development of Superhydrophobic Polyester Fabric by Alkaline Hydrolysis

Mi Seon Han, Chung Hee Park, Seon Young Youn
Department of Clothing & Textiles, Seoul National University, Seoul, Korea
junghee@snu.ac.kr; syyoun@snu.ac.kr

OBJECTIVE

The objective of this paper was to obtain the binary roughness of the polyester fabric surface through alkali hydrolysis and treatment with fluorine-based chemical. In order to make the various levels of roughness, alkaline treatment time was varied. In addition, Poly (vinylidene fluoride-co-hexafluoropropylene) and 97% 1H, 1H, 2H, 2H-Perfluorodecyltriethoxysilane was used for the superhydrophobic surface. Surface morphology, chemical composition, thickness, weight, fiber diameter and tensile strength were evaluated to examine the changes of the fabric performance. Static contact angles and shedding angles were measured as well.

RESULTS AND DISCUSSION

As the alkali treatment time increased, the surface roughness increased creating lots of micro and nano pores on the fiber surface. Forming the air pocket between the pits on the surface of the treated fabric caused to increase the static contact angle and decrease the shedding angle. Within this experiment, the fabric treated with alkaline for 20 minutes exhibited the most superior superhydrophobic surface. The treatment time of 15minutes which showed the static contact angle of 162.2±2.7° and shedding angle of 8.8±0.2° was considered as the optimum condition without too much tensile strength loss.

CONCLUSION

This research developed superhydrophobic polyester fabric that would allow the mass production by relatively simple and widely used alkaline hydrolysis treatment.

KEYWORDS

Alkaline hydrolysis, superhydrophobicity, surface roughness, contact angle, shedding angle, roll-off

ACKNOWLEDGMENT

This study was supported by the 'BK21 Plus' Project funded by the National Research Foundation of Korea in South Korea.

Polyaniline Nanostructure Tuning with Oxidants in Interfacial Polymerization System

Fanxin Zeng, Zongyi Qin, Banglei Liang, Na Liu, Zhe Zhou, Meifang Zhu
State Key Laboratory for Modification of Chemical Fibers and Polymer Materials, College of Materials Science and Engineering, Donghua University, Shanghai 201620, China
656087782@qq.com; phqin@dhu.edu.cn

INTRODUCTION

Polyaniline (PANI) as a fine conducting polymer has attracted more and more attention due to its advantages, such as simple synthetic process, cheap and accessible raw material and well environmental stability. Recently, PANI has been designed to a variety of nano-structure materials, such as nanofibers, nanospheres, nanorods and nanotubes. These nano-structured PANI exhibited great potential applications in supercapacitors, metallic anti-corrosion coatings, sensors and conducting materials [1]. In this work, PANI was prepared in different oxidation system at room temperature, and the influences of oxidation system on the PANI structure and properties were studied through field emission scanning electron microscopy (FE-SEM), UV-visible spectrometry (UV-vis), Fourier infrared spectrometry (FTIR), laser Raman spectrometry, X-ray diffraction spectrometry (XRD) and two-electrode method, respectively.

APPROACH

Materials and reagents

Aniline (An) monomers with an average molecular weight of 93 g/mol, APS, $FeCl_3$, $K_2Cr_2O_7$, carbon tetrachloride (CCl_4) and hydrochloric acid (HCl) were purchased by Sinopharm Chemical Reagent Co., and used without further purificaition. The deionized water was used for all the experiment.

Synthesis of polyanline

The aniline monomers (0.3 M) were dissolved in 50 mL carbon tetrachloride, and the hydrochloric acid aqueous solution (1 M) was dissolved in the oxidant system of APS, APS/$FeCl_3$, APS/$K_2Cr_2O_7$, respectively. Then oxidant solution was carefully poured into the organic solution to form a static interface. After reaction for 12 h, PANIs in aqueous phase were extracted, washed with deionized water, and the product was dried under vacuum at 60 ℃ for 24 h. The illustration for interfacial polymerization of polyanline was shown in Fig. 1.

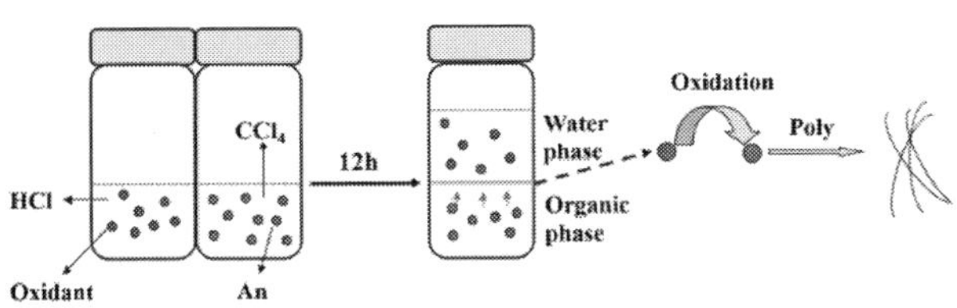

Fig. 1. Illustration for interfacial polymerization of polyaniline

RESULTS AND DISCUSSION

Morphology

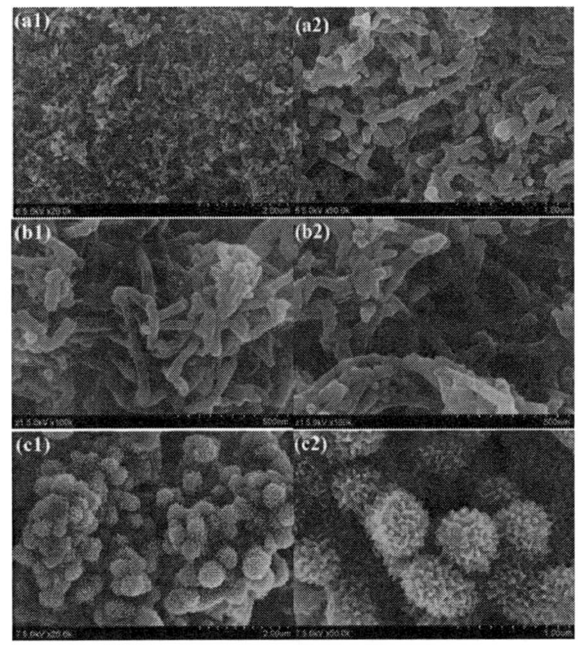

Fig. 2. SEM images of PANI prepared with different oxidants. (a1)(a2) APS ; (b1)(b2) APS/$FeCl_3$; (c1)(c2) APS/$K_2Cr_2O_7$

Fig. 2 shows the SEM images of PANI synthesized via interfacial polymerization with different oxidants. When useing composite oxidants of APS/$FeCl_3$, PANI nanofibers with the diameters of 45-70 nm can be obtained, thinner than those by using pure APS as a single oxidant. However, when using the composite oxidants of APS/$K_2Cr_2O_7$, PANI exhibited petal-like morphology. Detailed data on the diameter, conductivity and yield of the samples were shown in Table I.

Table I. Diameter, conductivity and yield of PANI prepared with different oxidants

	APS	APS/ $FeCl_3$	APS/ $K_2Cr_2O_7$
Diameter (nm)	60-80	45-70	430-570
Conductivity (10^{-3} S/m)	1.27	1.46	1.02
Yield (%)	9.02	33.32	57.35

Chemical structure

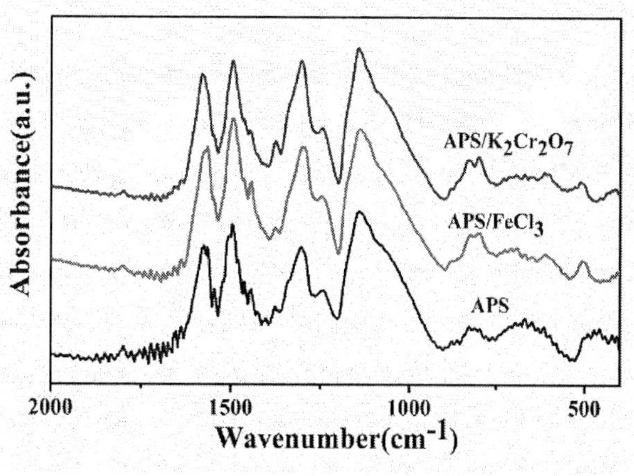

Fig. 3. FTIR spectra for PANI prepared with different oxidants

Fig. 3 shows the FTIR spectra of PANI synthesized with different oxidants through interfacial polymerization. As can be seen, the characteristic absorption peaks at 1577 cm^{-1} and 1492 cm^{-1} can be assigned to the C=C stretching vibration of the quinoid ring and benzenoid ring, respectively [2]. The peak at 1140 cm^{-1} is ascribed to quinonoid unit of doped PANI. Furthermore, among all the samples, the peak intensity of PANI synthesized with composite oxidants of $APS/K_2Cr_2O_7$ is the strongest, illustrating its highest electron delocalization.

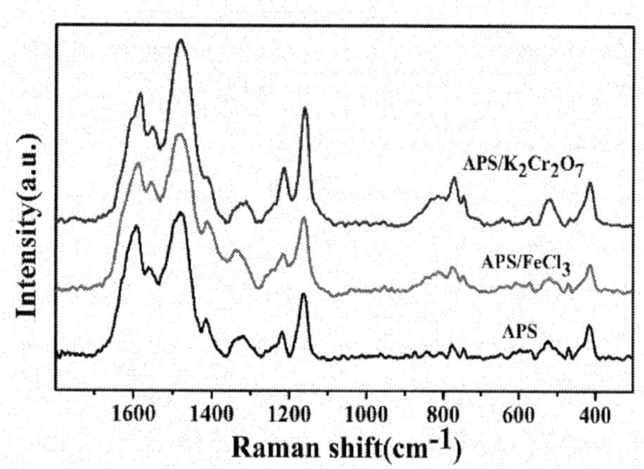

Fig. 4. Raman spectra for PANI prepared with different oxidants

Raman spectra of the PANI synthesized by using different oxidants via interfacial polymerization is shown in Fig. 4. Clearly, there are seven characteristic absorption peaks at 1590 cm^{-1}, 1480 cm^{-1}, 1330 cm^{-1}, 1160 cm^{-1}, 775 cm^{-1}, 520 cm^{-1} and 414 cm^{-1}. The bands at 1590 cm^{-1} (νC=C) and 1480 cm^{-1} (νC=N) are due to bipolaronic units. The band at about 1160 cm^{-1} (βC-H) can be assigned to polaronic units.

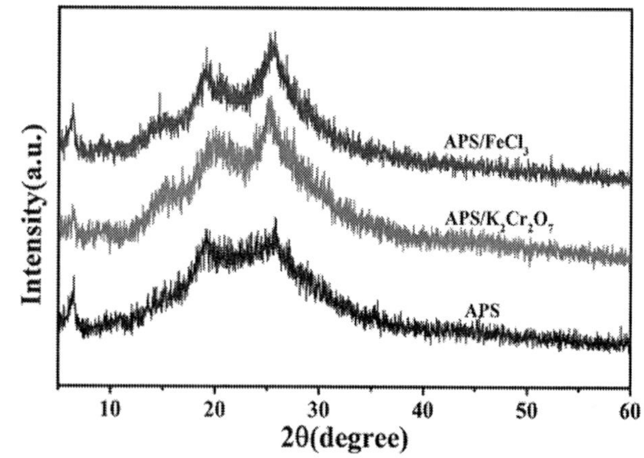

Fig. 5. XRD patterns for PANI prepared with different oxidants

As observed in Fig. 5, the diffraction characteristic peaks at $2\theta = 20°$ and 26° showing the existence of high crystallinity. Moreover, the two sharp peaks at $2\theta = 20°$ and 26° are ascribed to periodicity parallel and periodicity perpendicular to the polymer chain, respectively [3]. What's more, the crystallinity is related to conductivity of polyanline, the higher the crystallinity degree, the bigger the conductivity. Clearly, the diffraction characteristic peaks of polyanline synthesized with composite oxidants of $APS/FeCl_3$ are very sharp, illustrating its high conductivity. And this experimental result is accord with the result in Table I.

CONCLUSION

The morphology and diameter of PANI nanofibers synthesized by $APS/FeCl_3$ as composite oxidants via interfacial polymerization were more uniform and thinner than those produced by the pure APS. However, petal-like morphology was observed by using of $APS/K_2Cr_2O_7$. The yield of PANI obtained from the interfacial polymerization with composite oxidants is much higher than those with single oxidant, illustrating that the composite oxidants can make the reaction proceed more completely.

ACKNOWLEDGMENT

This work has been financially supported by National Natural Science Foundation of China (21274019), Fundamental Research Funds for the Central Universities (2232013A3-02), and Program for Changjiang Scholars and Innovative Research Team in University (IRT1221)

REFERENCE

[1] Wang, J., Zhang, D. *Advances in Polymer Technology*, 32, 2013: 323-368.

[2] Ayad, M., Amer, W., Whdan, M. *Journal of Applied Polymer Science*, 125, 2012: 2695-2700.

[3] Zhang, X., Zhu, J., Neel, H. *Polymer*, 53, 2012: 2109-2120.

Electrospun Nanofibrous Composite Membrane for Air Filtration

Ce Wang, Xuefen Wang
State Key Lab for Modification of Chemical Fibers and Polymer Material,
Donghua University, Shanghai, China
wangce91@163.com

OBJECTIVE

To improve the air filtration performance of nanofibrous membrane by the incorporation of titania(TiO2) nanoparticles via electrospinning.

INTRODUCTION

Population growth, together with the expansion of industrial and agricultural production, are creating unprecedented demands on clean air supplies all over the world. Fine particles, especially particulate matter with an aerodynamic diameter smaller than 2.5 μm, have been proved to be a major cause of adverse health effects ranging from the human respiratory tract to extrapulmonary organs.[1] However, the most fine particle cannot be removed by the present nonwoven filter media because the size of pores formed with micron-scale fibers is considerably larger.[2] To enhance the filtration efficiency of the filter media, it is necessary to fabricate much thicker media for smaller pore sizes. Owing to the tunable fiber diameter, high porosity, remarkable specific surface area, and interconnected porous structure, electrospun fibrous membrane is considered as an effective medium for air filtration.[1]

APPROACH

Polyvinylidene fluoride (PVDF) solutions (15wt%) containing 0, 4 and 8 wt% of TiO_2 NPs (nanoparticles) were prepared by using DMF with stirring for at least 24 h at 70 °C. The dope solutions were then cooled down at room temperature and sonicated for 1 h before electrospinning.

The solution was placed in a 5-mL syringe and ejected at a controllable feed rate (5μl/min) through a blunt metal needle with 0.37 mm inner diameter operated by a syringe pump. The PVDF/TiO_2 nanofibrous membranes were deposited on a grounded drum covered with a piece of PP (polypropylene) nonwoven, which was positioned 15 cm from the tip of the needle. The voltage was set at 30 kV, and the electrospinning conditions were maintained at 30±2 °C and 50±5 % relative humidity. The obtained electrospun samples were denoted as PVDF/TiO_2-X, where X is the weight percent of TiO_2 NPs with respect to the total solid weight (PVDF+TiO_2 NPs).

The surface morphology of the electrospun samples were investigated by field emission scanning electron microscopy. An image analyzer, namely, ImageJ 2X software were used to measure the PVDF/TiO_2 fiber diameters together with its distribution. The filtration performance was measured by an automated filter tester.

RESULTS AND DISCUSSION

Figure 1(A)-(C) demonstrated the typical FE-SEM images of the as-prepared PVDF and PVDF/TiO_2 membranes with different weight percent of TiO_2, which revealed the randomly oriented nonwoven membranes with uniform nanofiber morphology. With the increase of TiO_2 NPs, the nano-protrusions were clearly visible on the nanofibers surface of PVDF/TiO_2-4 and PVDF/TiO_2-8, which constructed the hierarchical structure. Owing to this multi-scaled morphology, not only the surface roughness of PVDF fibers was enhanced, but also its effective surface area was also increased, which significantly boosted the ability of the fibers in capturing particles in air.[3] Moreover, with increasing of the TiO_2 contents, the fiber diameters of the PVDF/TiO_2-0, PVDF/TiO_2-4 and PVDF/TiO_2-8 decreased gradually and became more uniform, as shown in Figure 1(D)-(F), which were related to the weakened conductivity of the composite solutions[4].

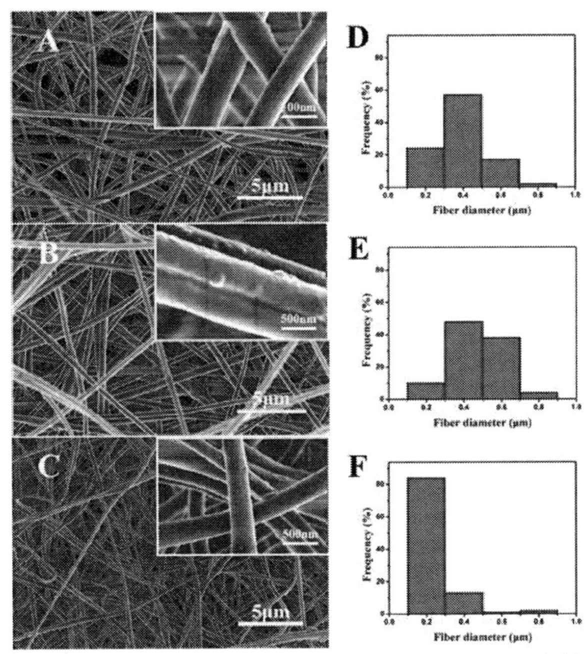

Figure 1. SEM images of (A) PVDF/TiO_2-0, (B) PVDF/TiO_2-4 and (C) PVDF/TiO_2-8,and Fiber together with respective distribution of (D) PVDF/TiO_2-0, (E) PVDF/TiO_2-4 and (F) PVDF/TiO_2-8.

The filtration performance of PVDF/TiO_2-0 and PVDF/TiO_2-8 were investigated by using the charged neutralized NaCl particles with the size range from 300 to 500 nm under the air flow of 85 L/min. As shown in Figure 2(A), the filtration efficiency of PVDF/TiO_2-8 has been improved 6-10% compared to that of PVDF/TiO_2-0, which could be attributed to the hierarchical structure and decreased fiber diameter (Figure 1). Meanwhile, the pressure drop was nearly invariable. To evaluate the filtration performance of a given medium, a quality factor (QF) is often used:

$$QF = \frac{-\ln(1-\eta)}{\Delta P} \qquad (1)$$

Where η and ΔP are the filtration efficiency and pressure drop across the membrane, respectively. This equation clearly indicates that large QF values require high filtration efficiencies and/or low pressure drops, and a large quality factor indicates good filter quality.[5] As shown in Figure 2(B), the QF of both membrane were reduced as the basis weight was increased, and the measured QF of PVDF/TiO_2-8 was higher than that of PVDF/TiO_2-0, indicating that an enhanced filtration performance (filtration efficiency of 92 %, pressure drop of 180.8 Pa) was obtained.

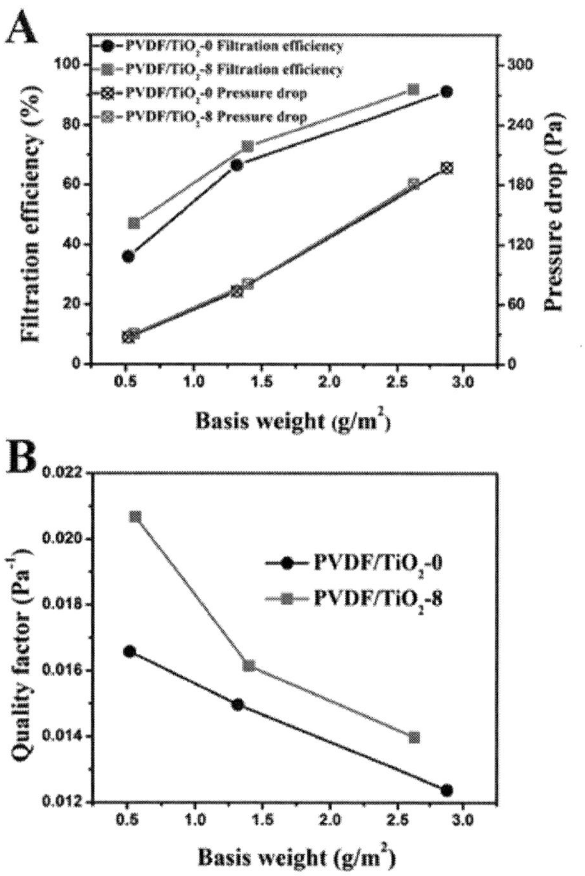

Figure 2. (A)Filtration performance and (B)quality factor of PVDF/TiO_2-0 and PVDF/TiO_2-8 with varied basis weight.

CONCLUSION

Electrospun PVDF/TiO_2 composite nanofibers with hierarchical structure have been prepared for air filtration. The results have shown that nano-protrusions and smaller fiber diameter of PVDF/TiO_2-8 enhanced its filtration performance, which demonstrated the feasibility of designing and manufacturing novel PVDF/TiO_2 nanofibrous membranes for high efficiency/low pressure drop air filtration.

KEYWORD

Electrospinning, PVDF/TiO_2 nanofibers, multi-scaled morphology, air filtration performance,

REFERENCES

[1] Wang, N., Zhu, Z., Sheng, J., Al-Deyab, S.S., Yu, J., Ding, B. *J. Colloid Interface Sci.* 428, 2014: 41-48.

[2] Cho, D., Naydich, A., Frey, M.W., Joo, Y. L. *Polymer* 54, 2013: 2364-2372.

[3] Wang, N., Si, Y., Wang, N., Sun, G., El-Newehy, M., Al-Deyab, S.S., Ding, B. *Sep. Purif. Technol.*, 126, 2014: 44-51.

[4] Wan, H., Wang, N., Yang, J., Si, Y., Chen, K., Ding, B., Sun, G., El-Newehy, M., Al-Deyab, S.S., Yu, J. *J. Colloid Interface* Sci., 417, 2014: 18-26.

[5] Wang, Z., Zhao, C., Pan, Z. *J. Colloid Interface Sci.*, 441, 2015:121-129.

The Effect of Binary Structure on Self-Cleaning Fabric via Plasma Treatment

Sohyun Park, Hey-sang Kim, Ho Bin Kim, Chung Hee Park
Department of textile, Merchandising and Fashion design, Seoul National University, Seoul, Korea
junghee@snu.ac.kr

INTRODUCTION

Extreme superhydrophobic surfaces with binary structure of micro and nano scale roughness and low surface energy can not only repel water droplet but also liquid of low surface tension[1]. And a dirt particle is removed by rolling off of water drop on these surfaces. This superomniphobic surface can repel a variety of liquids like wine, coffee, etc. and has gained significant scientific and industrial interest for its potential applications in outdoor wear and protective textiles. In this study, self-cleaning polyester fabric with binary structure was prepared via plasma etching and HMDSO plasma enhanced CVD(PECVD). And the wetting behavior and self-cleaning effect of superhydrophobic fabric was compared to that of specimens having a variety of micro and nano scale surface roughness, by controlling with number of monofilament fibers and oxygen plasma treatment time.

APPROACH

Material

Micro scale roughness was controlled by number of monofilament fiber 36, 72 and 144(Table 1)

Table I. Characteristics of the specimens.

Specimen code	Fiber composition	Thickness (mm)	Density (in inch)
F	100% Polyester Film	0.01	-
36f	100% Polyester 75d/36f Filament	0.153 (±0.002)	144 × 112
72f	100% Polyester 75d/72f Filament	0.149 (±0.002)	144 × 112
144f	100% Polyester 75d/144f filament	0.144 (±0.002)	144 × 112

Plasma process

The plasma process was consisted of two steps in order to modify surface roughness and surface free energy of PET fiber: (1) plasma etching in the presence of oxygen to etch a nano-pillar on the fiber, and (2) subsequent plasma enhanced deposition of hexamethyldisiloxane(HMDSO). For the plasma process, specimens of 8 cm × 8 cm in size were placed on a stainless steel cathode plate, and the reactor was evacuated to a base pressure lower than 20 mTorr. Oxygen gas flow was initiated at a flow rate of 20sccm. Plasma etching was started with a radio frequency glow discharge of oxygen gas at a bias voltage of −400 V. The time of oxygen plasma etching was varied as 1, 2, 3, 5, 7, 10, 15 minutes to varied nano-scale surface roughness. After the plasma etching, PECVD of HMDSO was carried for 30 seconds to lower surface free energy of the roughened PET fiber. HMDSO precursor gas was supplied at a flow rate of 10 sccm.

EXPERIMENTAL

Surface morphological changes after the plasma treatment were observed by field-emission scanning electron microscopy (FE-SEM, NanoSEM200, USA).

In order to evaluate superomniphobicity, contact angle(CA) and shedding angle(SHA) were measured using a contact angle goniometer(Attension® Theta Lite, Sweden). For the CA, 4.0 ± 0.2 μl drop of deionized water and 10% isopropyl alcohol was dropped at different spots on the fabric surface and was measured 5 times. For the SHA, the sliding angle at which a 12.5 ± 0.2 μl size water droplet rolls off at a distance of 2 cm or greater on the substrate was measured. The distance between the syringe tip and the fabric surface was fixed at 1 cm.

In order to evaluate self-cleaning effect, The hydrophilic silicon carbide and hydrophobic Sudan black B particle were scattered on the fabric specimens and 4.0 ± 0.2 μl drops of water and 10% isopropyl alcohol were dropped. The self-cleaning effect was determined confirming the removal of particles that were eliminated by roll-off liquid drops.

RESULTS AND DISCUSSION

Dual scale roughness with nano-pillar was achieved with a uniform size and distribution after plasma etching. After 2 minute of oxygen plasma etching, nanometer scale pillar were appeared densely on the surface of the PET fibers, as shown in Figure. 1. The nano-pillar grew longer and appeared to aggregate with increased plasma etching treatment after an etching time of 3 minutes. Thereby, a number of nano-pillar were placed on the top of the micrometer scale PET fiber(8~15μm), resulting in a dual scale surface roughness.

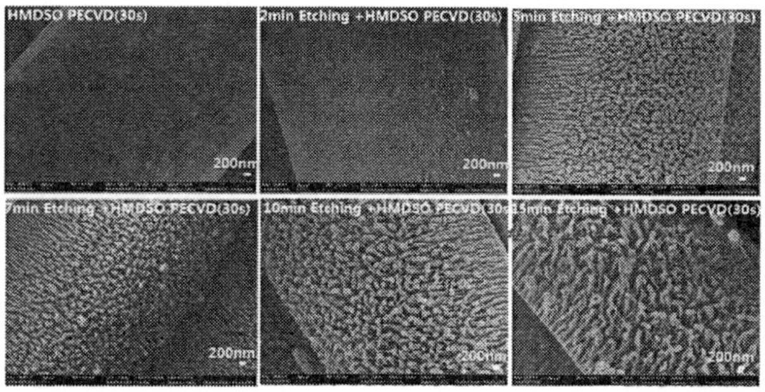

Figure 1. FE-SEM images of PET fiber treated by oxygen plasma etching and HMDSO PECVD. The scale bar of insets is 200 nm.

The omniphobic polyester fabric was developed having contact angle greater than 160° and shedding angle less than 1° for water, whereas contact angle greater than 160° and shedding angle less than 1.5° for 10% isopropyl alcohol solution. The surface structure of PET fabric had two levels of roughness that enable the trapping of air under liquid droplets, thereby contributing to the rolling off liquid drops. The effect of the surface roughness by monofilament strains was insignificant for achieving superhydrophobic surfaces, but the micro-structure of a particular fabric noticeably influenced the hydrophobicity in that a contact angle difference between the film and fabric deposited only HMDSO was about 41°.

The result of the self-cleaning effect test demonstrated that both hydrophilic silicon carbide and hydrophobic Sudan Black B particles from the slope of 1°degree were removed from 144f fabric specimens etched for 3 minutes. This finding indicated that a hydrophobic particle that simply attached to the fabric specimen was able to be removed. However, liquid droplet of 10% isopropyl alcohol was rolled off within 1cm distance and stopped on the 144f fabric etched for 7 minutes, and both particles were not eliminated.

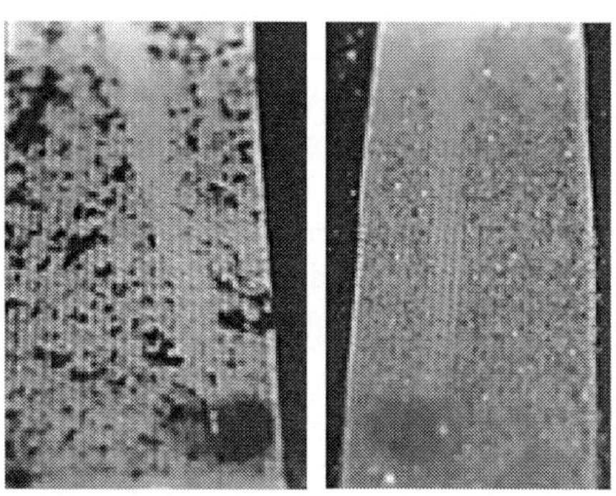

Figure 2. Image of self-cleaning effect by water on 144f etched for 3 minutes. Removal of Sudan black B(left), removal of silicon carbide(right).

CONCLUSION

An superhydrophobic and omniphobic PET fabric was fabricated via by plasma etching and the subsequent deposition of HMDSO, resulting in a dual scale surface roughness on the fabric surface, having an high CA of water and 10% isopropyl alcohol and low SHA of water and 10% isopropyl alcohol. The highest water CA and lowest SA (167° and 1.5°, respectively) were achieved under the condition of 2 minutes of oxygen etching and 30 seconds of HMDSO deposition.

For 10% isopropyl alcohol, the highest CA and lowest SA (166.5° and 2.5°, respectively) were achieved under the condition of 5 minutes of oxygen etching and 30 seconds of HMDSO deposition. Sudan black B and silicon carbide was successfully removed by rolling water drop on 144f specimen etched for 3 minutes under condition of the inclination of 1°.

KEYWORDS

Superhydrophobic, self-cleaning, omniphobicity, binary structure roughness, polyester, HMDSO

ACKNOWLEDGMENT

This study was supported by the 'BK21 Plus' Project funded by the National Research Foundation of Korea in South Korea.

Assessment of Temperature Adaptability of PCM Patches Containing Octadecane and Eicosane

Yeji Choi, Eunae Kim
Functional Textile System Research Lab, Department of Clothing and Textiles, Yonsei University, Korea
cyeji1074@yonsei.ac.kr; eakim@yonsei.ac.kr

INTRODUCTION

'Phase change' is the process of going from one physical state to another i.e. from liquid to solid and vice versa. During the phase change, the materials absorb or release a lot of energy as phase change take place. By absorbing or releasing heat, the materials show the temperature adaptability. Textile products which have automatic acclimatizing properties have attracted more attention recently.

OBJECTIVE

The temperature adaptability of PCM is closely related to the heat storage capacity(J/g) of the material, the more quantity of PCM is used, the higher temperature adaptability is shown. As the quantity of PCM incorporated in the textile product is, however, very small, whether the effect of the PCM product is sensible by human or not is in controversy. Moreover, once PCMs provide the temperature adaptability by changing its phase, it terminates when the phase change in all of the PCMs would complete.

In order to apply PCMs to textile product effectively, it is important to assess the microclimate changes due to their heat capacity by weight. We investigated the relationship between the material and the microclimate temperature, we used bulk of PCMs, not microcapsulated PCMs. In textile field, Eicosane and Octadecane are commonly used due to their melting and crystallization temperature close to human body temperature. In this study, by making and investigating PCM patches containing different weight of Eicosane or Octadecane, temperature adaptability of PCM depending on weight will be discussed.

EXPERIMENTAL

Materials

The patches containing 1g, 5g, 10g of Octadecane(n-Octadecane, Sigma-Aldrich Korea) and 1g, 5g, 10g of Eicosane(n-Eicosane, Sigma-Aldrich Korea) were prepared. The pouch(Hwasung, Korea) used for patches was made of 3ply of Al(Aluminum), PET(Polyethylene terephthalate), and CPP(Cast Polypropylene) and the size of the patches was 5cm×10cm.

Table I. Characteristics of Octadecane and Eicosane

PCM	Formula	Melting Temperature (℃)	Crystallization Temperature (℃)	Heat Storage Capacity (J/g)
Octadecane	$C_{18}H_{38}$	28.2	25.4	244
Eicosane	$C_{20}H_{42}$	36.1	30.6	247

Test apparatus

To test the temperature adaptability of the PCM patches, the dynamic change of the environment which simulates the actual wear condition is important. The Human-Clothing-Environment (HCE) Simulator (Korea patent 0483996, 2003; US patent 7,680,638 B2, 2010) has been reported as a useful apparatus for evaluating fabric samples in terms of accuracy, time and expenses. It can simulate the actual wear situations like, shuttling between the inside and the outside, sweating conditions, and layering effects. The HCE simulator can measure the micro climate changes in clothing layers and heat flux (W/m^2) by the environmental condition change automatically.

Measurement

The specimens were composed of two layers. The first layer, closest to the hot plate, was the PCM patch, and the second one was quick-dry knit fabric which has excellent moisture management property. Every layer, sensors were located and collected data every minute.

Test protocol

The temperature needed to be above 36.1℃ for the high temperature, or below 28.2℃ for the low temperature in order to make it possible for all specimens to react. The test protocol of this study is shown in Figure 1. All the data were collected 3 times repeatedly and the average value of data was used for the analysis.

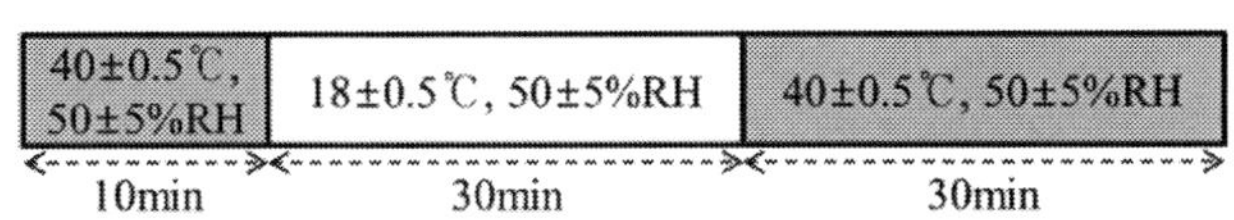

Figure 1. Test protocol

Data collection and calculation

By measuring microclimate temperature at dynamic change in environment condition, temperature buffering index(B_t) is calculated.

$$B_t = \frac{C}{tan_\alpha \cdot \Delta T_{i-f}}$$

Where,
tan α : initial decrease rate of microclimate temperature
ΔT_{i-f}: temperature difference between initial and final state
C : constant(1000)

RESULTS AND DISCUSSION

Temperature difference

To investigate the temperature difference between with and without PCM, temperature difference was calculated by subtracting the value of control specimen. As shown in Figure 2, the patches containing Eicosane were higher than those of Octadecane. The temperature differences increased when the amount of PCM contents increased.

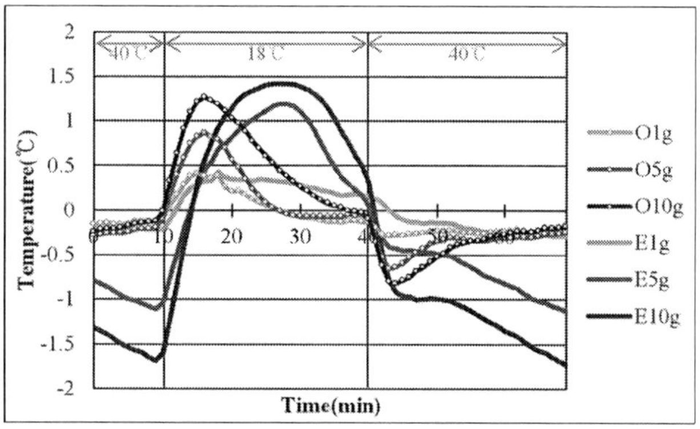

Figure 2. Temperature difference between with and without PCM of 6 specimens

At the period of exposing to low temperature, Eicosane started reacting slowly, after exposing in 19 minutes, the data showed the maximum temperature difference. On the other hands, Octadecane started reacting quickly, within 7 minutes after exposing. At the period of low temperature exposure, Eicosane reacted slowly and showed the more differences consistently. Octadecane reacted fast, it took 5 minutes till the maximum temperature difference. Figure 3 shows the maximum temperature difference between with and without PCM in accordance with the type and weight of PCMs.

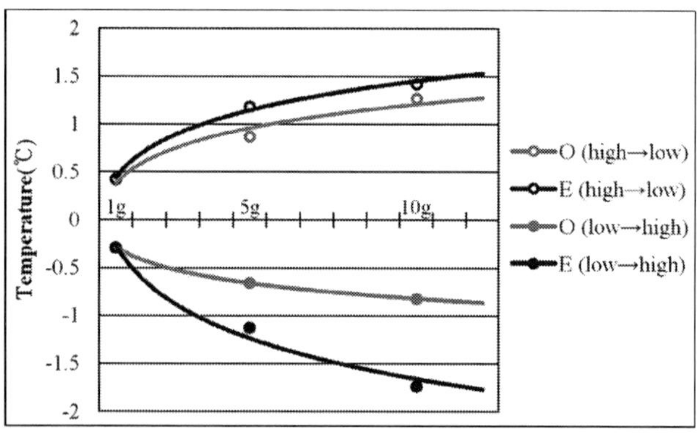

Figure 3. The maximum temperature difference of 6 specimens

Depending on the type and amount of PCM content, the degree of difference existed, orderly. Heating effect in hot temperature and cooling effect in cold temperature increased according to the amount of PCMs. Eicosane showed higher temperature difference and time delay than that of Octadecane. These results can be explained by the heat capacity. The heat storage capacity of Eicosane is 247J/g, and that of Octadecane is 244J/g. The higher heat storage capacity contributed to the more temperature difference, which might lead to the better temperature adaptable performance.

Temperature buffering capacity

After the specimens were stabilized at 40℃ for 30 minutes, it is shown that temperature difference appeared in sequence of Octadecane < Eicosane. Eicosane specimens showed lower temperature at initial state and the higher temperature at final state than those of Octadecane specimens. Those effects were reflected in the temperature buffering index(B_t). As shown in table II, B_t value showed in sequence of Octadecane < Eicosane as well corresponding the heat storage capacity. In case of environmental temperature change from low to high, some of the values reversed, but they followed on the overall sequence.

Table II. Temperature buffering capacity of the specimens

Temperature Change(℃)	Octadecane			Eicosane		
	1g	5g	10g	1g	5g	10g
high → low	286	323	381	323	315	403
low → high	389	439	431	447	412	458

It can be concluded that the temperature buffering capacity is consistent with the heat storage capacity and the amount of PCM content.

CONCLUSIONS

The heat storage capacity is an important factor of the temperature regulating effects of PCMs, the amount of PCMs used corresponds with the temperature adaptable performance. Even though the changes in microclimate and buffering capacity are observed by the quantity of the PCMs, the effect is relatively small. There are some controversial issues about the effects, because amount of PCM incorporated in the textile product is small to be perceived by human. We may need further study how we can facilitate the maximum effect of PCMs.

KEYWORDS

Phase change materials, Octadecane, Eicosane, Temperature adaptability

ACKNOWLEDGMENT

This work was supported by the National Research Foundation of Korea (NRF) grant funded by the Korea government (MEST) (No. NRF-2012R1A2A2A03047265).

REFERENCES

1. Kim, Eunae, Yoo, Shinjung. "Human-clothing-environment simulator." U.S. Patent No. 7,680,638. 16 Mar. 2010.

2. Mondal, S. "Phase change materials for smart textiles–an overview." *Applied Thermal Engineering*, 28.11, 2008: 1536-1550.

3. Yoo, Hwasook, Lim, Jihye, Kim, Eunae. "Effects of the number and position of phase-change material-treated fabrics on the thermo-regulating properties of phase-change material garments." Textile Research Journal, 83.7, 2013: 671-682.

Fabric Movement Analysis of 3.25kg Laundry Load in a Front-Loading Washer

Yoonkyung Cho, Changsang Yun, Jungyoon Kim, Chunghee Park
Department of Textiles, Merchandising and Fashion Design, Seoul National University
junghee@snu.ac.kr

Fabric movements in a front-loading washer were analyzed and the correlation between fabric movements and detergency was investigated. With 3.25kg laundry load, 25% of maximum load of the drum type washer used in this study, various fabric movements were induced with different drum rotating speed. Fabric movements were observed in respect of macro and micro dimension and they were digitized and visualized. Finally, the influence of fabric movements on detergency was investigated.

As for the macro fabric movements, the occupied area ratios by laundry loads in the washer were varied according to the drum rotating speed. In terms of the micro fabric movements, the movements of tracer fabric were divided into 12 movements. The tendency of micro fabric movements as well as the degree of interactions between fabrics at each rotating speed were investigated. The characteristics of macro and micro fabric movement also influenced on detergency and resulted in the maximum detergency at 46rpm.

Among ten fabric movement indexes, five indexes (occupied area of laundry batch, maximum and minimum value ratio of occupied area of laundry batch, distance between the center of drum and indicator fabric, speed difference between drum and indicator fabric, appearance frequency of indicator fabric) came out to have a statistically significant correlation with detergency. Also, a regression equation of detergency was obtained with R^2 value of 0.707 and p-value below 0.01. The two independent variables of the regression equation were 'average value of distance between center of the drum and tracer fabric' and 'appearance frequency of tracer fabric', with an inverse correlation with detergency.

Therefore, it was concluded that not only mechanical force like falling movement but also interactions between fabrics like abrasion and mixing were needed to improve detergency for practical laundry loads.

KEYWORDS
Fabric movement, Mechanical action, Detergency, Practical laundry load, Front-loading washer

ACKNOWLEDGMENT
This study was supported by the 'BK21 Plus' Project funded by the National Research Foundation of Korea in South Korea.

REFERENCE
Yun, Changsang, Park, Chunghee. "The effect of fabric movement on washing performance in a front-loading washer II: under various physical washing conditions." *Textile Research Journal*, 85.3, 2015: 251-261.

The Study on Market Demand of Technical Jackets in Shanghai

Manhao Guan[1,2], Jun Li[1,2,3]
[1]Fashion Institute, Donghua University, Shanghai, China
[2]Protective Clothing Research Center, Donghua University, Shanghai, China
[3]Key Laboratory of Clothing Design & Technology, Donghua University, Ministry of Education, Shanghai, China
lijun@dhu.edu.cn; gmhouse92@163.com

STATEMENT OF PURPOSE

The study investigates and analysis the market demand of technical jacket in Shanghai. The overall development situation of technical jacket has been presented. The scale and demographic characteristics of consumers in three consumption levels and the factors influencing the purchase have been analyzed. We also estimate the market size of technical jacket in Shanghai.

INTRODUCTION

With the increasing scale of outdoor industry in China [1], the demand for technical jackets as one of the basic equipment is expanding. In fact, the targeted consumers of technical jacket include the general public for daily wear as well as the professional outdoor athletes and amateur outdoor sports enthusiasts. However, national standards on the related technical requirements have not yet introduced [2] and few researches on the factors of technical jackets' market, especially those from clothing itself have done.

The study is aimed at investigating and analyzing the market demand of technical jacket in Shanghai, mainly focusing on the clothing itself.

APPROACH

Investigation on market condition

The overall market situation should be grasped firstly, including the market shares of brands, product categories, price distribution, distribution channels, publicity and sales performance, etc. In addition, the factors which might influence market performance should be considered, such as economic development, spending power [3] and popular culture, etc. The investigation consists of references of secondary data and the online and offline survey [4]. The major part of online and offline survey is the study of network shops and entities shops, mainly including product categories, features, prices and sales.

Survey of consumer demand

The study on consumer demand for technical jackets mainly includes three parts: the demographic characteristics of consumers [5] in different consumption levels, the purchase behavior and willingness of consumers [6] in different consumption levels and the factors influencing the purchase of different consumption levels.

The survey is mainly based on two kinds of questionnaires which have been issued through the Internet and field. The first questionnaire is aimed at identifying the current market size of technical jackets and future trend. Meantime, the consumer proportion and the overall consumption attitude toward technical jackets could be analyzed. The second questionnaire is designed for the targeted consumers. The basic information and the structure distribution of targeted consumers have been identified by the grasp of the demographic environment. The analyses of the consumption attitudes and behaviors have been done with the 5W1H analysis method [7]. The five-point scale [8] is designed for the evaluation of factors influencing the purchasing of technical jackets.

RESULTS AND DISCUSSION

Market overview

The brands of technical jackets in domestic market can be divided into four levels: top brands, high-end brands, middle brands and low-end brands. The vast majority of foreign brands are high-end brands and above, which have strong ability of research and design and good reputation, but the price is relatively high. Meantime, there are large numbers of domestic brands most of which are middle or low-end brands, cost-effective and of strong channel expansion capability [9].

Estimation of market size

The per capita expense on technical jackets of highly educated population in Shanghai is about 317.14 yuan. The consumers of technical jackets account for 13.78% in the general population in Shanghai [10, 11]. The expense spent on technical jackets in Shanghai accounts for 4.26% in the expense of all categories of clothing.

Consumer purchase behavior

The most well-known brands and the brands which consumers buy most have been concluded, as well as the main purchase season and purchase channels. Besides, the summary of purposes of purchasing the technical jackets has shown that the majority consumers buy jackets for general sports and daily wear and only a few buy them for professional outdoor sports.

Consumer characteristics of different levels

The consumers who have purchased technical jackets have high levels of education. 87.5% of consumers have Bachelor's degrees or above. The consumers owning master degrees or higher account for 25% of all the consumers.

The income of low level consumers is low and has a wide age range. The need of the kind of consumers is not stable and continuous. They often use technical jackets for long-haul outdoor sports. They pay more attention to the functionality and durability of clothing than appearance and publicity.

The income of middle level consumers is the highest whose age range is from 25 to 40. This kind of consumers has a relative stable and continuous demand for technical jackets. They often wear jackets for daily life or tours. They lay emphasis on functionality and quality of jackets, as well as price. Compared with the low level consumes, they focus more on the appearance of clothing. Similarly, they pay no much attention to the publicity.

The income of high level consumer is in the middle. Their age range is from 25 to 35, which is the youngest group. This kind of consumers has a most stable and continuous demand for technical jackets. The frequency of long-haul outdoor sports is the highest as well. They pay most attention to the quality and functionality of clothing instead of publicity too.

CONCLUSION

The study investigates and analysis the market demand of technical jacket in Shanghai through online and offline survey. Compared with the domestic brands, the foreign brands is more well-known and of higher market shares. The high-end and middle brands are mostly the foreign brands while the domestic brands occupy the low-end market for the price advantage. The current sales channels are still the physical stores, such as the shopping malls and outdoor shops. The main part of online sales is the domestic brand's flagship stores. The market of technical jackets is expanding from the professional field to the general public. The overall development situation and market size of technical jackets has been estimated.

In addition, three aspects about targeted consumers have been analyzed, including the scale and demographic characteristics of consumers in three different consumption levels, their purchase behavior and willingness and the factors influencing the purchase of the three levels. The consumers of technical jackets are mainly the highly educated young population. The proportion of male consumers is higher, which is almost the twice of the female. The consumers pay more attention on factors, such as functionality, quality and price. Consumers expected prices are mainly concentrated in less than 3,000 yuan while higher prices have a limited market demand higher.

FUTURE WORK

The investigation has shown that the aged is an important consumer group of technical jackets which cannot be neglected. However, during this period of survey, the sample size of old people is not sufficient because of the limit of time. More samples of aged people and a more reasonable of sample distribution would be required in the following study.

KEYWORDS

Technical Jackets, Consumer Demand, Functionality, Shanghai

ACKNOWLEDGMENT

This work was financially supported by the Fundamental Research Funds for the Central Universities.

REFERENCES

[1] Qian, Baoshan, Ren, Huiying, Guan, Chengxue. "Demand for Outdoor Sports Industry and Value Innovation Strategy." *Science and Technology Innovation Herald*, 20, 2013: 235-235.

[2] Hu, Hao. "Market Situation and Standard of Outdoor Jackets." *Knitting Industries*, 8, 2012: 66-67.

[3] http://www.stats-sh.gov.cn/data/toTjnj.xhtml?y=2013.

[4] Liu, Jian. "Discussion on The Impact of Strategy on Outdoor Industry." *Contemporary Sports Technology*, 4.19, 2014: 193-193.

[5] http://www.stats-sh.gov.cn/tjnj/nj13.htm?d1=2013tjnj/C0202.htm.

[6] Solomon, Michael, Russell-Bennett, Rebekah, Previte, Josephine. *Consumer Behaviour*. Pearson Higher Education AU, 2012.

[7] Li, Xiaolong, et al. "Innovative Product Design Based on Comprehensive Customer Requirements of Different Cognitive Levels." *The Scientific World Journal*, 2014.

[8] Zhang, Weiyuan. *Clothing Comfort and Function*. Donghua University Press, 2011.

[9] Wang, Bo. "Analyses of Brand Competition Pattern of Chinese Outdoor Market." *China Business & Trade*, 35, 2011: 013.

[10] http://www.stats-sh.gov.cn/tjnj/nj13.htm?d1=2013tjnj/C0202.html.

[11] http://www.stats-sh.gov.cn/sjfb/201203/239823.html.

Effects of Carbon Nanotubes on Mechanical Properties of Polyimide Fibers

Chao Zheng[1], Xuepeng Qiu[2], Guomin Li[2], Zhixin Dong[2]

[1]University of Chinese Academy of Sciences, Changchun, China

[2]Changchun Institute of Applied Chemistry, Chinese Academy of Sciences, Changchun, China

xp_q@ciac.ac.cn; zhchao@ciac.ac.cn

STATEMENT OF PURPOSE

The purpose of this research is to improve mechanical properties of polyimide (PI) fibers. Herein, the effects of addition of carboxylic functionalized carbon nanotubes (CNTs) into PI fibers on their mechanical properties were investigated in detail.

INTRODUCTION

CNTs, as an excellent material for reinforcement of polymer materials due to its excellent mechanical, thermal, and electrical properties etc.[1-3], are widely used to improve polymer materials' performances [4]. PI fiber is a high performance fiber with great mechanical properties and extraordinary thermal performance. Heretofore, there are plenty of work on PI/CNTs composite materials [5-6], and Yin *et. al.* [7] fabricated PI/CNTs composite fibers via wet spinning, which showed an improvement on mechanical properties. In this study, CNTs' change after carboxylic functionalization and the influences on PI fibers' mechanical properties were fully investigated.

APPROACH

Multi-walled carbon nanotubes were modified with carboxylic groups on surface by mixed sulfuric acid and nitric acid treatment for different hours. Then they were added into the mixture of monomers (4, 4'-oxydianilline, 4, 4'-biphitalic dianhydride, and pyromellitic dianhydride), and the polymerization was conducted with stirring at room temperature under nitrogen flow for 24 hours before spinning. The as-spun polyamic acid/CNT composite fibers were prepared via dry-jet-wet-spinning process. The final PI/CNT composite fibers were acquired after thermal imidization and hot-drawn.

The contents of carboxylic groups on CNTs' surface were characterized by thermogravimetric analysis (TGA) and X-ray photoelectron spectroscopy (XPS). The morphologies of CNTs and PI/CNT composite fibers were observed on scanning electron microscope (SEM). The tensile tests of PI/CNT composite fibers were performed on a XQ-1 tensile tester.

RESULTS AND DISCUSSION

Figure 1 depicts the change of carboxylic group contents on CNTs' surface with the increase of treatment time. The content of carboxylic group increases sharply from 1.14% to 12.4% in the beginning 4 hours and then grows slowly to 19.6% in the end of 48-hour treatment according to the TGA curve (the XPS curve shows a similar trend), which means that the oxidation of CNTs reacts intensely at first and reaches a saturate state ultimately. Meanwhile, Figure 2 describes the lengths of modified CNTs. Obviously the average length of CNTs falls drastically in the beginning of the oxidation process, and then decrease gradually: CNTs were shortened from an average length of 10μm to 396nm in the first 2 hours, and finally they are even shorter than 100nm. CNTs are shortened due to the oxidation of mixed acid, and the reduce rate is consistent with the increase rate of carboxylic group.

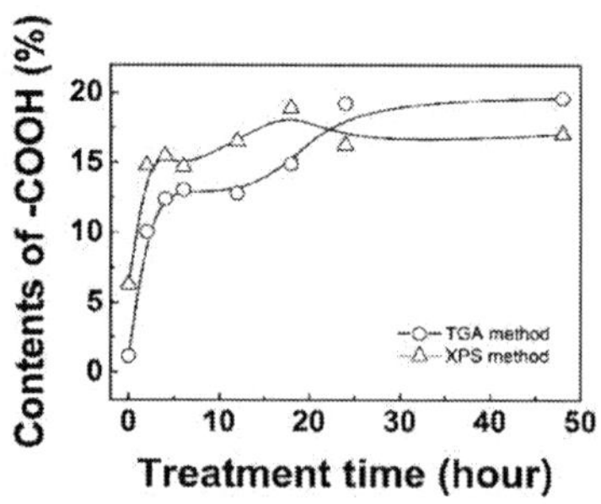

Figure 1. Contents of carboxylic groups on CNTs' surface

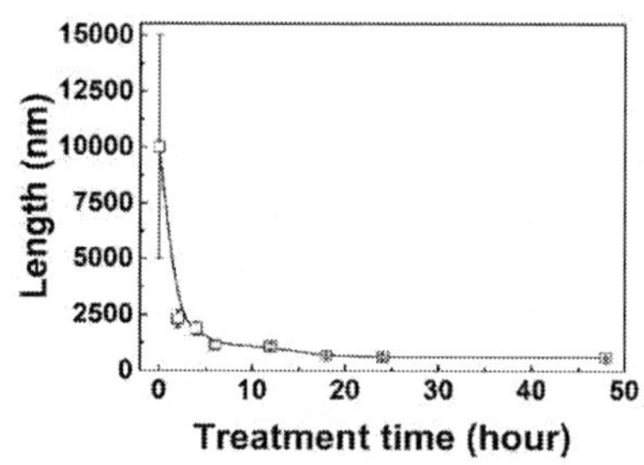

Figure 2. Lengths of carboxylic functionalized CNTs

The morphology of PI/CNT composite fibers is illustrated in Figure 3, which shows that composite fibers have a dense inner structure (Figure 3a) and smooth surface (Figure 3b). CNTs can be clearly observed after that the composite fibers were etched by KOH and they distribute evenly within the PI matrix (Figure 3c), which can be attributed to the CNTs' surface modification.

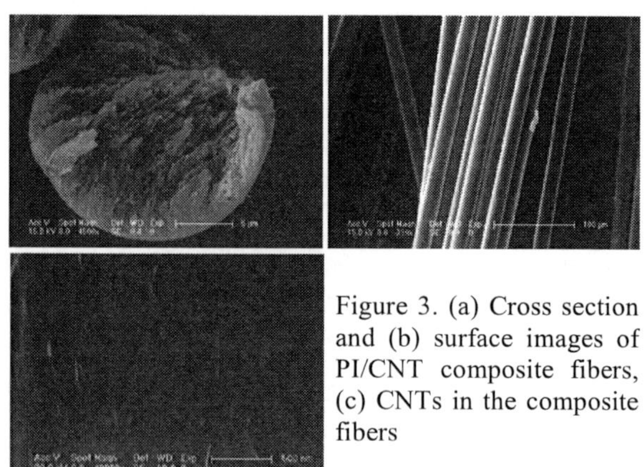

Figure 3. (a) Cross section and (b) surface images of PI/CNT composite fibers, (c) CNTs in the composite fibers

Figure 4 shows the mechanical properties of PI/CNT composite fibers containing CNTs with different treatment time. The tensile strength and elongation at break have similar trend with the increase of CNTs' treatment time: they increase steadily at first, reach a peak value, and then decrease gradually. As depicted in Figure 4, the tensile strength of composite fibers rises from 0.93GPa (pure PI fiber) to 1.40GPa when CNTs are treated for 4 hours, and then decreases to 0.99GPa when CNTs are oxidized for 24 hours. So does the elongation of composite fibers: it rises from 10.8% to 14.3%, and then declines to 11.8%. The reasons are as follows: the carboxylic groups on CNTs' surface and the carbonyl groups in the PI can form hydrogen bonds, thus, they disperse evenly in the matrix and the interface between CNTs and the matrix is well-built [8]. As a result, the composite fibers become stronger and tougher. However, with the increase of CNTs' treatment time, they become shorter, leading to fewer link points between CNTs and PI molecular chains per CNT. Consequently, the binding force between each other is likely to reduce, which causes the decrease of composite fibers' tensile strength and elongation. Overall, there is a significant improvement in the mechanical properties of composite fibers when CNTs are not over treated by mixed acid. Unfortunately, it is difficult to identify the contribution of both effects to composite fibers' mechanical properties. Furthermore, the modulus of composite fibers barely changes and maintains around 10.0GPa, as shown in Table I.

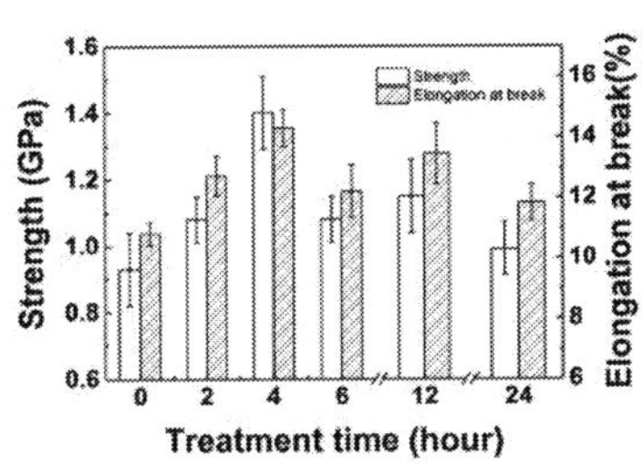

Figure 4 Tensile strength and elongation at break of PI/CNT composite fibers

Table I. Modulus of PI/CNT composite fibers

Treatment time (hour)	Pure PI	2	4	6	12	24
Modulus (GPa)	10.1 ±1.1	8.1 ±0.9	10.0 ±0.9	8.5 ±0.8	9.1 ±0.7	10.1 ±0.7

CONCLUSION

It is concluded that the mechanical properties of PI fibers are influenced by both the content of carboxylic groups on modified CNTs' surface and the length of CNTs: on one hand, the more carboxylic groups there are, the stronger the interface binding between CNTs and PI matrix is, so the better the mechanical properties of composite fibers are; on the other hand, the shorter the CNTs are, the fewer PI molecular chains with which they can link there are, leading to an decrease in the mechanical properties of composite fibers.

KEYWORDS

Polyimide Fiber, Carbon Nanotubes, Composite

ACKNOWLEDGMENT

The authors would like to acknowledge the National Basic Research Program (2014CB643604) and National Natural Science Foundation of China (51173178) for sponsoring this research.

REFERENCES

[1] Treacy, M.M.J., Ebbesen, T.W., Gibson, J.M. "Exceptionally high Young's modulus observed for individual carbon nanotubes." *Nature*, 381, 1996: 678-680.

[2] Berber, Savas, Kwon, Young-Kyun, Tománek, David. "Unusually high thermal conductivity of carbon nanotubes." *Physical Review Letters*, 84, 2000: 4613-4616.

[3] Ebbesen, T.W., Lezec, H.J., Hiura, H., et al. "Electrical conductivity of individual carbon nanotubes." *Nature*, 382, 1996: 54-56.

[4] Honglei, Wang, Xingui, Zhou, Haijiao, Yu, et al. "Research and development of carbon nanotubes reinforced composites." *Materials Review*, 22, 2008: 146-149.

[5] Zhu, Bao-Ku, Xie, Shu-Hui, Xu, Zhi-Kang, et al. "Preparation and properties of the PI/multi-walled carbon nanotubes (MWNTs) nanocomposites." *Composites Science and Technology*, 66, 2006: 548-554.

[6] Guo-wen, He, Ling, Xie, Kai-yuan, Tan, et al. "Preparation of MWNTs/PI nanocomposite materials and their dynamic mechanical properties and dielectric properties." *The Chinese Journal of Nonferrous Metals*, 21, 5, 2011: 1123-1130.

[7] Yin, et al. "Large fabrication of PI fibres containing functionalized multiwalled carbon nanotubes via wet spinning." *Composites: Part B*, 58: 2014: 430-437.

[8] Hwa So, Hyang, Cho, Jae Whan, Sahoo, Nanda Gopal. "Effect of carbon nanotubes on mechanical and electrical properties of PI/carbon nanotubes nanocomposites." European Polymer Journal, 43, 2007:3750-3756.

Preparation of Novel, Aromatic, Polyimide Fiber Containing Phosphorous

Zhao Yong[1], Xuepeng Qiu[2], Guomin Li[2], Zhixin Dong[2]
[1]University of Chinese Academy of Sciences, Beijing, China
[2]Polymer Composites Engineering Laboratory, Changchun Institute of Applied Chemistry, Chinese Academy of Sciences, Changchun, China
xp_q@ciac.ac.cn; zhaoyong@ciac.ac.cn

INTRODUCTION

Polyimide (PI) was one of performance polymers since the mid-twentieth century. This type of polymer in which the main chain consisted of aromatic or heterocyclic rings quickly drew many people attentions because of their good dimensional stability, lower dielectric property, excellent mechanical properties, temperature resistance and resistance to radiation capability[1]. In 1960s, DuPont firstly developed commercial PI product, named Kapton. From then on, this material has attracted many people to research it and been growing rapidly. At present, PI have been widely used in industries and manufacturing of spaceships and air engines.

While several different polyimides were available as polymers, films and resins. The fiber formation of PI polymer was very limited. One of commercial fibers was the P84 which was a semi-commercial PI fibers produced by Evonik company right now. Conventionally, polyimide fibers were used in applications such as high temperature filtration, braided packing and protective clothing[2]. Recently, with the development of industries and aerospace applications, PI fibers have been partly used as spacesuits, aircraft bag materials and satellite antenna tension ropes. It can be predicted that polyimide fibers will be more potential in aerospace industries in the future. However, there is a big problem when polyimide materials were serviced in the low earth orbit (LEO) environment (160-800 km in altitude), where the atomic oxygen (AO) seriously eroded the surfaces of most organic polymers. Hence, seeking for effective approaches to enhance the ability of AO resistance have been a hot subject in recent years[3]. Among the approaches to improve the AO resistance of PI, a method to prepare inherently AO-resistant PI films has been synthesized from diamines with phenylphosphine oxide (PPO) groups[4-15]. As early as 1983, P-containing PIs have been synthesized and showed high thermal stability, good adhesive property[16]. From then on, the report about polyimide fibers containing phosphorous with AO resistance capability has never been seen in past years. It can be obviously expected that PI fibers with P-containing will enhance the capability of fire-resistance and thermal properties and intrinsic AO-resistance property.

In this article, we reported our recent works on polyimide fibers by our group. A diamine monomer bis(3-aminophenyl) methyl phosphine oxide (DAMPO) which contained PPO group was introduced into copolymer with Oxydianiline (ODA) and Biphenyltetracarboxylic dianhydride (BPDA) in an attempt to modify traditional ODA/BPDA PI fibers. This type of polyimide fibers demonstrated that the introducing of DAMPO had not only improved further thermal stability properties of PI fibers, but also improved their dimensional stability. At the same time, the fibers still remained mechanical properties with increasing the diamines ratios of DAMPO and ODA. However, the improvements of the fibers properties were related with the content of DAMPO. Herein, the discussion focused on thermal stability and the characterization of AO-resistant capability will discuss in our further work. This paper described the synthesis and characterization of polymers which contained phosphorous in the main backbone. The important monomer DAMPO was synthesized by an improvement of a literature method. Synthesis and characterization of PI fibers based on DAMPO are reported in this paper.

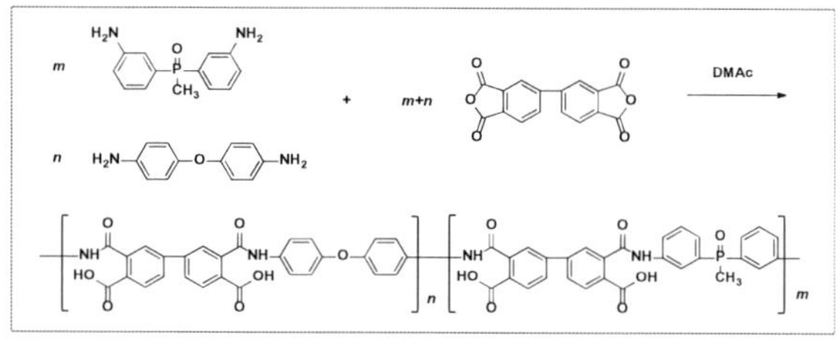

Figure 1. Synthesis route. (m/n = 0/10, 2/8, 4/6, 6/4).

APPROACH

The diamine containing phosphorous was synthesized via a modified method[17]. The polyamide acid was prepared by reaction between the diamine and dianhydride in DMAc. In our work, the dianhydrides of the polyamide dope was a specific one (s-BPDA), however, the diamine of the dope was a mixture of both ODA and DAMPO, or only ODA with adjusting the ratio of both diamines to obtain the polyamide acid dope containing different phosphorus content. More than 60% phosphorous content dope was not synthesized, due to the formation of gel was not conducive to spinning process. The inherent viscosities of the dopes were presented in Table 1, and the polymerization route of the polyamide acid (PAA-PPO) dopes was showed in Figure 1. The fibers were prepared by a dry-jet wet spinning process as shown in Figure 2.

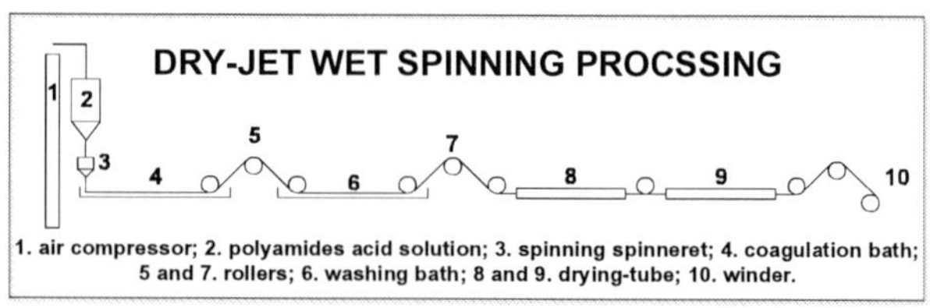

Figure 2. Preparation processing of spinning fibers.

RESULTS AND DISCUSSION

The mechanical properties of polyimide and copolyimide fibers were listed in Table 1. In the same condition of draw ratio the strength, modulus and strain of fibers ranged from 2.81 to 6.14 cN/dtex, 35.99 to 65.55 cN/dtex and 15.63 to 56.90%, respectively. The introduction of DAMPO into copolyimde fibers decreased the strength and modulus, but increased the strain of fibers at same draw ratio. It indicated that the use of DAMPO has efficiently improved the strain with increaseing of the DAMPO content. Thus, the larger DAMPO content was, the higher fibers were drawn. In the Table 1, PI-6' and PI-6" were the PI fibers which were drawn in ratio of 2.0 and 3.0, comparing with PI-6. In was shown that the strength and modulus of fibers were increased with increasing of the draw ratio. It was unexpected that the introduction of phosphorous into the structure of PI molecule chain didn't changed the mechanical properties.

Table 1. Mechanical properties of fibers.

No.	Ratio ($m:n$)	η_{inh} (dl/g)	Mechanical Properties Strength (cN/dtex)	Modulus (cN/dtex)	Strain (%)
PI-0	0:10	1.77	6.14	65.55	15.63
PI-2	2:8	1.68	5.98	67.21	18.49
PI-4	4:6	1.50	4.14	49.31	35.78
PI-6	6:4	1.27	2.81	35.99	56.90
PI-6'	6:4	1.27	3.97	71.43	22.67
PI-6"	6:4	1.27	5.61	91.92	18.82

All the glass transition temperatures of PI fibers were shown in Figure 3 from 268 °C to 295 °C. As can be seen in Figure 3, the copolyimide fibers had T_g higher than polyimide fibers. Moreover, with the increasing content of DAMPO the T_gs obviously increased. The reason of T_gs increasing was that the higher glass transition temperature of copolyimide fibers has been attributed to the increasing of the meta-structure proportion, requiring a larger sweep volume for the crank shaft motion and accordingly more energy (higher temperature) to obtain this motion.

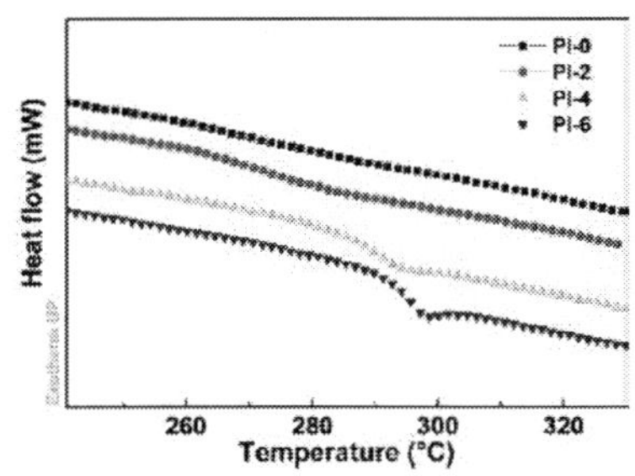

Figure 3. Glass transition of temperatures (T_g) in N_2.

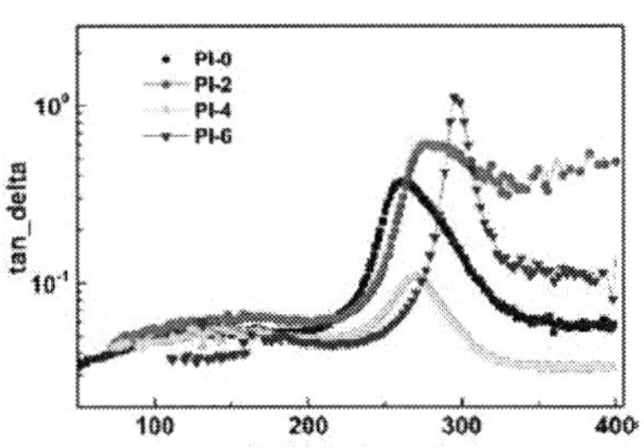

Figure 4. DMTA spectra of PI fibers.

Dynamic mechanical thermal analysis (DMTA) was performed at a heating rate 10 °C/min on all PI fibers in air atmosphere. The DMTA spectra of PI fibers was presented in Figure 4. At the temperature ranging from 250-310 °C, the α relaxations of PI fibers were observed, which was related to the glass transition. The DMTA curves indicated that the glass transition temperatures of fibers were in the range of 250-310 °C, increased with the increasing content of DAMPO. The values of PI-0, PI-2, PI-4 and PI-6 were estimated to be 262, 276, 270 and 297 °C respectively, which indicated that the introduction of DAMPO had efficiently improved the T_g of the fibers. The result of DMTA characterization consistent with the above DSC consequence.

CONCLUSION

A series of copolyamide acid solutions with various diamine ratios of DAMPO/ODA were synthesized and then spun into PAA fibers via dry-jet wet spinning process. The PI fibers were obtained by the heat treatment of PAA fibers. The tensile strength and initial modulus of fibers were practically not changed with the introduction of phosphorous into the main chain structure. The glass transition temperatures of PI fibers were in the range of 268-295 °C in DSC spectrum and 262-297 °C in DMTA spectrum.

KEYWORDS

Polyimide fiber, atomic oxygen, flame retardant, phosphorous

ACKNOWLEDGMENT

Support of this work by The National Basic Research Program (2014CB643604) and National Natural Science Foundation of China (51173178) is gratefully acknowledged.

REFERENCES

1. Zhang, Y., Shen, J., Zhang, Q., Xu, Z., Yeung, K.W.K., Yi, C. *Science of Advanced Materials*, 6, 44, 2014.
2. Bhat, G.S., Schwanke, R. *Journal of Thermal Analysis*, 49, 1997: 399.
3. Liu, B., Ji, M., Liu, J., Fan, L., Yang, S. *High Performance Polymers*, 25, 2013: 907.
4. Connell, J.W., Watson, K.A. *High Performance Polymers*, 13, 2001: 23.
5. Smith, J.G., Thompson, C.M., Watson, K.A., Connell, J.W. *High Performance Polymers*, 14, 225, 2002.
6. Watson, K.A., Palmieri, F.L., Connell, J.W. *Macromolecules*, 35, 2002: 4968.
7. Thompson, C.M., Smith, J.G., Connell, J.W. *High Performance Polymers*, 15, 2003: 181.
8. Smith Jr, J.G., Connell, J.W., Hergenrother, P.M. *Polymer*, 35, 1994: 2834.
9. Connell, J.W., Smith Jr, J.G., Hedrick, J.L. *Polymer*, 36, 1995: 13.
10. Connell, J.W., Smith Jr, J.G., Hergenrother, P.M. *Polymer*, 36, 1995: 5.
11. Jin, J., Topping, C.M., Suresh, S., Foulger, S.H., Rice, N., Mojazza, B.H., Smith Jr, D.W. *Polymer*, 46, 2005: 6923.
12. Zhu, Y., Zhao, P., Cai, X., Meng, W.-D., Qing, F.-L. *Polymer*, 48, 2007: 3116.
13. Li, Z., Liu, J., Gao, Z., Yin, Z., Fan, L., Yang, S. *European Polymer Journal*, 45, 2009: 1139.
14. Li, Z., Song, H., He, M., Liu, J., Yang, S. *Progress in Organic Coatings*, 75, 2012: 49.
15. Connell, J.W., Smith Jr, J.G., Hergenrother, P.M., Watson, K.A., Thompson, C.M. Google Patents: 2006.
16. Varma, I.K., Rao, B.S. *Journal of Applied Polymer Science*, 28, 1983: 805.
17. Tan, B., Tchatachoua, C.N., Dong, L., McGrath, J.E. *Polymers for Advanced Technologies*, 9, 1998: 84.

Wettability and Water Absorption Properties of Lignin Nanofibrous Membranes

Youjung Song, Eunsil Lee, Seungsin Lee
Department of Clothing and Textiles, Yonsei University, Seoul, Korea
yjssong@yonsei.ac.kr

INTRODUCTION

Lignin is an abundant natural polymer in the biosphere and second only to cellulose; however, it is underutilized and considered a waste. Recently, lignin has received new attention as a renewable resource, and considerable efforts are being made to develop more valuable applications of waste lignin [1].

Electrospinning is a fiber-forming technique that can yield polymer fibers with diameters in the nanometer to micron range. Nonwoven mats of electrospun nanofibers provide exceptionally high ratios of surface area to volume with very small pores, which is attractive for various applications ranging from filter media to tissue engineering scaffolds, wound dressings, sensors, and high performance apparel [2].

Previously, we successfully fabricated lignin nanofibers with electrospinning and examined the critical parameters that affected the electrospinnability of the resulting fibers [3]. The wetting behavior and hydrophilic/hydrophobic nature of nanofibrous membranes is of great interest for many applications such as wound dressings. In this study, we examine wetting behavior and water absorption properties of lignin nanofibrous membranes with a view to investigating further applications of lignin nanofibers.

EXPERIMENTAL

Lignin (low sulfonate content, Mw=~10,000, Sigma Aldrich Co., USA) was used as received and poly(vinyl alcohol) (PVA) (>99% hydrolyzed, Mw=89,000-98,000, Sigma Aldrich Co., USA) was used as a carrier polymer to facilitate the fiber formation of lignin. Distilled water was used as a solvent.

All the procedures for preparing lignin and PVA hybrid solutions as well as for the electrospinning of lignin nanofibers have been described in our previous study [3]. The mass ratios of lignin to PVA were set at 85:15, 50:50, and 0:100 (pristine PVA). The concentration of all PVA precursor solutions in this study was fixed at 5 wt%. The prepared solutions were electrospun on substrate material (polyester nonwoven) with a voltage of 18 kV at a working distance of 11.5 cm and a feed rate of 0.2 ml/hr through a 30 gauge needle. After electrospinning, an insolubilization treatment was conducted on the resulting fibers to stabilize the structures against dissolution in water.

The structure of electrospun lignin nanofibers was examined using a field emission scanning electron microscope (FE-SEM) (JSM 6701-F, JEOL Ltd., Japan). Atomic force microscopy (XE-Bio, Park systems, Korea) was used to observe the surface morphology and roughness of the materials.

The absorbency of lignin nanofibrous membranes was evaluated in accordance with AATCC 79-2010, Absorbency of Textiles, at 21°C, 30% RH. To evaluate the hydrophilic/hydrophobic properties of the lignin nanofibers, we examined the contact angle and water uptake. To determine the water uptake, lignin nanofibrous membranes were cut into pieces in 2 cm × 2 cm and completely dried using a vacuum oven. The dried samples were weighed and subsequently immersed in distilled water. After 2 h of immersion, the samples were taken and weighed after removing the surface water with filter paper. The water uptake (%) was calculated as follows:

$$\text{Water uptake (\%)} = (m_1 - m_0)/m_0 \times 100$$

where, m_0 and m_1 represent the masses of the sample before and after immersion in water, respectively. The wetting behavior of the electrospun lignin nanofibrous membranes was examined using static contact angles. Static contact angles of the lignin nanofibrous membranes were determined on a Drop Shape Analyzer (CAM 200, KSV Instrument, Finland).

RESULTS AND DISCUSSION

The SEM images of lignin nanofibers showed that ultrafine fibers with diameters ranging from 130 to 180 nm were fabricated. AFM analysis was carried out to further characterize the surface roughness of the nanofibrous membrane. The AFM imaging showed that the roughness of the surface gradually increased as the amount of lignin in the lignin/PVA hybrid solution increased.

To understand the wetting behavior and hydrophilic nature of the electrospun lignin nanofibrous membranes with different mass ratios of lignin to PVA, the water absorbency, contact angle and water uptake on each of the membrane were examined. Water drops were absorbed instantly into the PVA nanofibrous membranes. Similar phenomena happened when measuring the lignin nanofibrous membranes, but it took a little longer time for water drops getting into the specimens containing 85 wt% lignin.

Pure PVA nanofibrous membranes showed over 400% water uptake since PVA is a water-soluble polymer. The water uptake decreased as the amount of lignin in the lignin/PVA hybrid solution increased.

CONCLUSIONS

Lignin nanofibers were fabricated by electrospinning with the addition of PVA. Cylindrical fibers with diameters of approximately 150 nm without beads were obtained. Overall, water drops were quickly absorbed into the membranes with different mass ratios of lignin to PVA. As the amount of lignin in the lignin/PVA hybrid solution increased, the water uptake decreased and the water drop absorption time increased slightly. The electrospun lignin nanofibrous membranes with higher water uptake may have potential in biomedical applications such as wound dressings.

ACKNOWLEDGMENT

This research was supported by Basic Science Research Program through the National Research Foundation of Korea (NRF-2013R1A1A2009730).

REFERENCES

[1] Thakur, V.K., Thakur, M.K., Raghavan, P., Kessler, M.R. "Progress in green polymer composites from lignin for multifunctional applications: a review." *ACS Sustainable Chem. Eng.*, 2, 5, 2014: 1072-1092.

[2] Shabafrooz, V., Mozafari, M., Vashaee, D., Tayebi, L. "Electrospun nanofibers: from filtration membranes to highly specialized tissue engineering scaffolds." *J. Nanosci. Nanotechnol.*, 14, 1, 2014: 522-534.

[3] Lee, E., Lee, S. "Fabrication of lignin nanofibers using electrospinning." *J. Korean Soc. Cloth. Text.*, 38, 3, 2014: 372-385.

Determination of Wire and Spring Diameter of TWSMA Springs for Intelligent Firefighters' Turnout Gear

Eunjin Park, Eunae Kim
Functional Textile System Research Lab, Department of Clothing & Textiles, Yonsei University, Korea
blessedjina@naver.com; eakim@yonsei.ac.kr

OBJECTIVE

Firefighters' turnout gear can have intelligent thermal protection by using two-way shape memory alloy (TWSMA) springs embedded thermal liner. The springs expand at high temperature and make air gap and collapse at room temperature. The purpose of this study is to examine optimum shape memory alloy (SMA) wire and spring diameters to have effective air gaps.

INTRODUCTION

Firefighters need to wear thermal protective clothing in order to protect from hazard. However, wearing thermal protective garments may increase heat stress and diminish firefighting performance. It is necessary to examine the techniques to reduce thermal stress and enhance the performance of firefighters during firefighting. In order to achieve improvements in thermal insulation, one novel way is to use the air as the insulator. In our previous studies, we have reported the usefulness of these TWSMA springs in terms of increased thermal insulation and comfort properties.

EXPERIMENTAL

Preparation of TWSMA springs

Commercially available Nitinol alloy composed of 50.8at% Nickel and 49.2at% Titanium was used. Six different sizes of TWSMA springs were prepared as shown in Table Ⅰ. The springs are compressed at room temperature but when they exposed to high temperature above 40℃, they are actuated and expand to their latent length instantaneously.

Table Ⅰ. Sizes of the specimens

Specimen	Wire diameter (mm)	Coil diameter (mm)	Mass (g)
SMA 1-15	1mm	15mm	0.88±0.008
SMA 1-18	1mm	18mm	1.08±0.017
SMA 1-22	1mm	22mm	1.35±0.004
SMA 1.2-18	1.2mm	18mm	1.51±0.013
SMA 1.2-22	1.2mm	22mm	1.91±0.005
SMA 1.2-26	1.2mm	26mm	2.19±0.002

Hysteresis, response time and compressional resistance of SMA springs

Hysteresis was evaluated by the change in displacement according to temperature changes ascending from 40℃ to 70℃ and descending from 70℃ to 40℃ at the interval of 10℃. Response time was evaluated by measuring the average time to expand to their maximum latent length at the temperature from 40℃ to 70℃ at the interval of 10℃. Compressional resistance were calculated by adding weights(1g, 10g, 50g, 100g) until the spring is completely pressed under two temperature between 40℃and 70℃.

Thermal insulation performance test of clothing system containing TWSMA springs

Human-Clothing-Environment (HCE) simulator (US patent 7,680,638 B2, 2010) was used to investigate the thermal insulation performance of the protective fabric system with TWSMA springs under two different ambient conditions according to microclimate temperature changes. The hot plate was maintained at 35±0.5 ℃ to simulate the skin temperature of human body. The test fabrics and sensors were mounted on the hot plate with the order as shown in Figure 1. Sensors were located 1~2mm away from each layers. The Test protocol is shown in Figure 2.

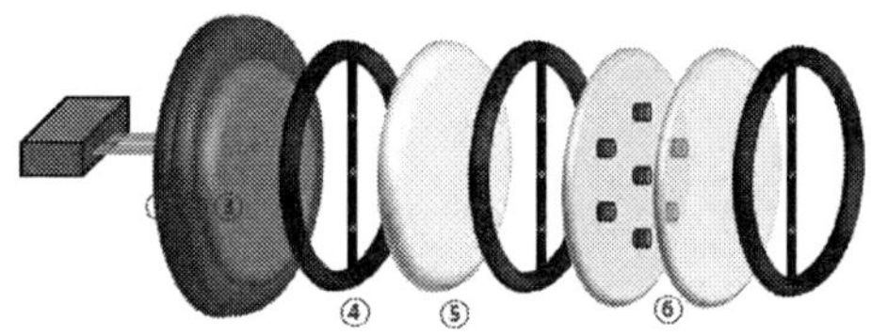

① Bottom plate
② Guard plate
③ Hot plate
④ Temperature and humidity sensors
⑤ Underwear
⑥ Thermal liner with TWSMA springs

Figure 1. Arrangement of TWSMA Springs onto the test fabrics.

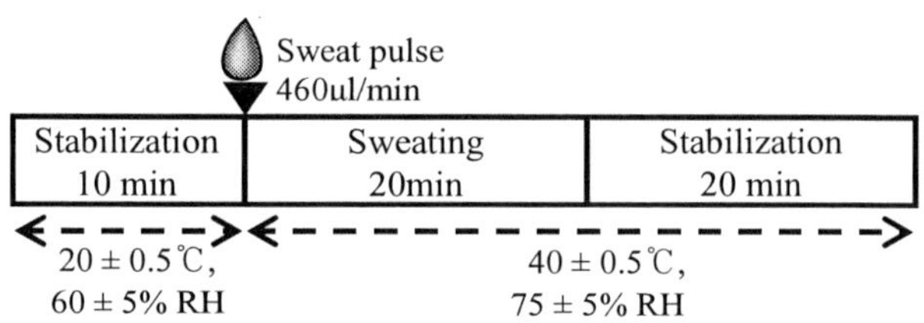

Figure 2. Test protocol.

RESULTS AND DISCUSSION

Original and latent length

The original and latent length are shown Table Ⅱ.

Table Ⅱ. Original and latent length of the specimens.

Specimen	Original length (mm)	Latent length (mm)
SMA 1-15	4.02±0.02	13.60±0.06
SMA 1-18	4.04±0.04	14.65±0.04
SMA 1-22	4.10±0.08	18.90±0.01
SMA 1.2-18	4.81±0.01	14.00±0.04
SMA 1.2-22	4.80±0.00	18.10±0.08
SMA 1.2-26	4.85±0.09	22.40±0.07

Hysteresis
Hysteresis of TWSMA springs was less than 1.5mm for all sizes (Figure 3). There was no change in the hysteresis of SMA spring in spite of its repetitive compression and expansion.

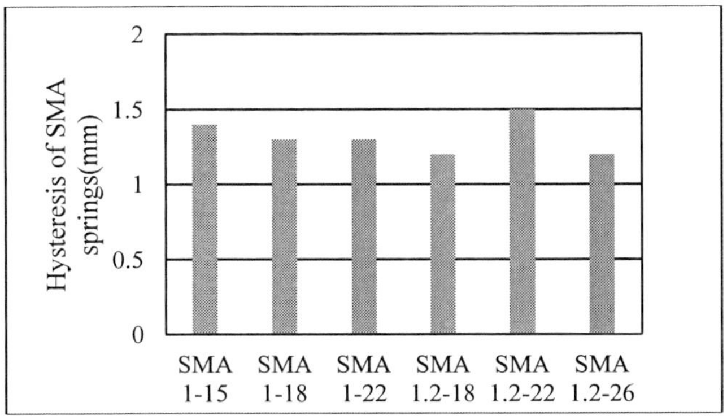

Figure 3. Hysteresis of TWSMA springs.

Response time
All six types of SMA spring have fast response time that take about 10 seconds at 70℃and 45seconds at 40℃, which means it is appropriate to use in the protective clothing.

Compressional resistance
As the temperature increases, SMA spring compressional resistance appeared significantly high as shown in Figure 4 & Figure 5.

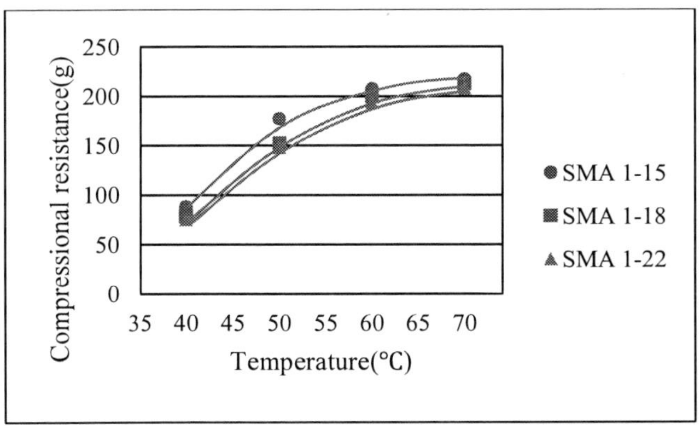

Figure 4. Compressional resistance of TWSMA springs of 1mm wire diameter.

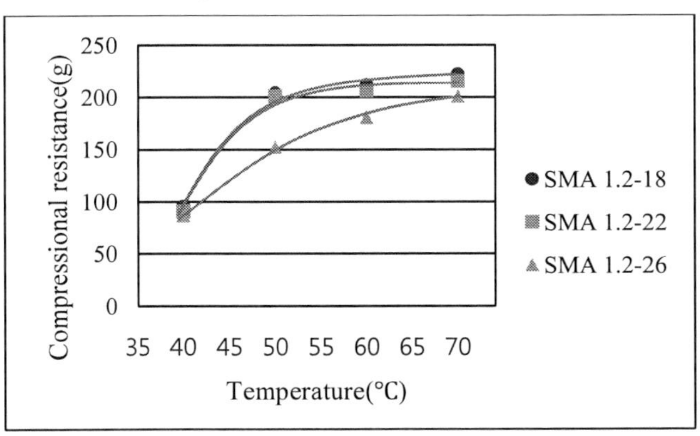

Figure 5. Compressional resistance of SMA springs with 1.2mm wire diameter.

Thermal insulation performance test of clothing system containing TWSMA springs
Figure 6 shows the results of microclimate temperature in the first air layer. Microclimate temperature changes of the clothing system with TWSMA springs were lower than commercial original thermal liner except SMA 1.2-26, whereas the clothing system with SMA 1.2-26 was almost same as original thermal liner (TL).

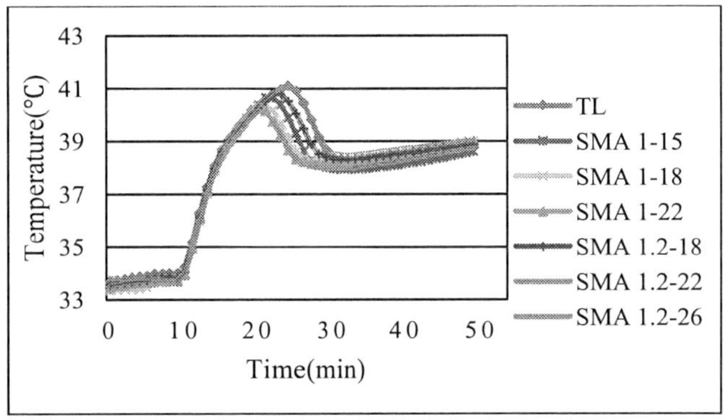

Figure 6. Changes of 1st layer microclimate temperature of SMA springs embedded Thermal liner in HCE simulator.

CONCLUSION
This research results favorably indicate that the created air layer by SMA springs expansion when firefighters exposing under extreme environmental conditions will provide thermal insulation. Among 6 types of SMA spring having different wire and spring diameter, SMA 1.2-18 is the optimum size. As results, it is possible to manufacture firefighters' turnout gear embedded SMA springs for mass production which has more effective to form air gaps than original ones.

FUTURE WORK
Shape memory alloy springs can be successfully used in the various fields of technical intelligent garments or textiles, among which medical textiles or sport wear and equipment. For more effective application, designs of TWSMA springs, actuating temperatures, number of springs in the fabric system, as well as the optimum size of diameter of SMA wire and springs should be considered in each field of clothing.

KEYWORDS
Shape memory alloy, Wire diameter, Spring diameter, Air gap, Firefighters' turnout gear

ACKNOWLEDGMENT
This work was supported by the National Emergency Management Agency of Korea (NEMA) grant funded by the Korea government (No. 2014-NEMA14-019-01010002-2014).

REFERENCES

[1] Slater, Keith. "Comfort or protection: The clothing dilemma." *ASTM Special Technical Publication* 1237, 1996: 486-497.
[2] Kim, Eunae, Yoo, Shinjung. "Human-clothing-environment simulator." U.S. Patent No. 7,680,638, 16 March, 2010.

Electromagnetic Shielding Effectiveness of Knitting Fabrics Which Consist of Polyester Yarns Doped Barium Titanate

Rumeysa Tural, Yusuf Ulcay
University of Uludağ, Textile Engineering Department, Bursa, Turkey
rumeysa@uludag.edu.tr

ABSTRACT

It was aimed to produce conductive textile surfaces to use electromagnetic shielding applications especially in home textiles (e.g. curtain, netting).

Usage of electrical and electronic devices has been increased with developing technology. These devices spread electromagnetic radiation which has harmful effects on the plants, animals and also human health especially in children. According to United Nations, electromagnetic radiation pollution is a serious problem that must be kept down [1]. Electrically conductive textile surfaces are one of the electromagnetic shielding materials. Textiles were preferred due to low density, flexibility and low cost [2].

Textiles can be gained electrical conductivity with different methods such as using conductive polymers, yarns or coatings. In this study, additive polyester yarns were produced in sheath and core form through bi-component spinning method. Three different dope ratio of barium titanate masterbatch were used at core section. Knitted fabrics were obtained from bicomponent yarns having two different fabric densities at circular knitting machine. These fabrics also dyed. Electromagnetic shielding effectiveness (EMSE) of samples (raw and dyed fabrics) was measured according to the standard ASTM D 4935. Shielding effectiveness (SE) test results were investigated in the range 0,01-1,30 GHz. The effect of dope ratio, fabric density and dyeing process on EMSE was investigated.

All fabrics showed the best EMSE results at 0,02 GHz. Fabrics with the highest content of conductive component and more fabric density showed the highest shielding efficiency. Increase in dope ratio and density effected EMSE test results as expected. Dyeing process changed EMSE test results. According to the standard Functional Technical Textile Standard (FTTS-FA-003) fabrics showed sufficient SE for home textiles.

In future work, repeated washing will be made to fabrics. The effect of washing process on EMSE will be investigated. However, weaving fabrics will obtained with these additive yarns. The effect of fabric structure on EMSE will be investigated. Weaving and knitting fabrics will be compared.

REFERENCES

[1] Yağmur, F., Bozbıyık, A., Hancı, İ.H. 2003. Elektromanyetik dalgaların insan biyokimyası üzerine etkileri. *Sted.*, 12, 8, 2003: 296-7.

[2] Kılıç, G., Örtlek, G.H., Saraçoğlu, G.Ö., 2007. "Elektromanyetik Çevre Kirliligi ve Bu Kirlilikten Korunmada Tekstil Çözümleri." *Tekstil ve Mühendis*, 14, 67, 2007.

Research on Testing Hygroscopicity and Moisture Conductivity of Chest Wader Lining

Qing, Zhan, Fengjiao He, Chongchong Yang, Xiaoxia Wu, Zheng Liu
College of Fashion and Art, Zhejiang Sci-Tech University, China
:zjxu_liu@126.com.ext

ABSTRACT
Chest wader applied in production only meets the waterproof requirement currently. Functional and comfortable design is little considered in the chest wader production. Under the large diversity work outside, chest wader can't expel the moisture originate from metabolism and sweating. The lower limbs of workers are placed in the wet environment for a long time. In order to improve the situation, this paper focuses on testing moisture permeability and moisture conductivity of the chest wader lining.

INTRODUCTION

Many researchers improve the fabric performance and the structure rationality of the chest wader. There is little study concern on the humid microclimate caused by for long-term wearing. Attaching the inner lining is a simple and effective way. Therefore, this paper studies the fabric hygroscopicity and moisture conductivity under the situation of simulating water pressure. It can be a reference to chest wader lining design.

APPROACH

Four pieces of fabrics have been selected as chest wader lining, they are: Coolmax fabric, 100% polyester; non-nap flannel, 60% cotton, 40% polyester; two-sided weft knitted fabric, 100% polyester; woven cloth, 100% polyester. This study designs a double layer fabrics absorption experiment. We simulate the water pressure in the certain weight. Fabrics and super absorbent resin material are combined as a aggregate in the test. The performance of the hygroscopicity and moisture conductivity are measured in a the specific pressure circumstance.

RESULTS AND DISCUSSION

The result of the aggregate water loss test ranking in the list of from small to large is combination of weft knitted double-sided cloth; combination of woven plain cloth; combination of non-nap flannel; combination of Coolmax. Water absorption of the corresponding high absorbent resin material ranking in the list of small to large followed by combination of weft knitted double-sided cloth; combination of woven plain cloth; combination of Coolmax; combination of non-nap flannel.

It is demonstrated that the performance in water loss and water absorption ability of the fabric and high absorbent resin material combination is different.

CONCLUSION

Compared with the Coolmax performance, the woven cloth combination has a good internal moisture transfer rate and a weak water storage capacity. Weft knitted double-sided cloth combination has a good internal moisture transfer rate and a medium water storage capacity. Non- nap flannel combination has a weak internal moisture transfer rate and a strong water storage capacity.

FUTURE WORK

According to the test result and body sweating map, we would design a composite inner lining of chest wader. Make a sample first, and take field tests several times to find the problems necessary to be solved. The improvement of moisture and temperature in dressed chest wader lining would be tested under the operating environment.

REFERENCES

[1] Liu, Changming, Fang, Ruihua "Wading clothing." *China Personal Protective Equipment*, 3, 1996:19-24.
[2] Karlsson, I.C. Marianne, Rosenblad, Elsa F.S. "Evaluating functional clothing in climatic chamber tests versus field tests: a comparison of quantitative and qualitative methods in product development." *Ergonomics*, 10, 41, 1998: 1399-1420
[3] Ingunn, Holmen, Geving, Reitan, Sandsund, et al. "Safer Work Clothing for Fishermen." *Marit. Health*, 57, 2006: 1-4.
[4] Wood, Michael Tate. "Waterfowl hunter's dual-function top garments." USP: US7,318,239 B2, 2008- 01-15.
[5] Blauer, Stephen J., Mordecai, Mark A., Atallah, Toufic G. "Unlined Waterproof Clothing." USP: US6,336,221 B1, 2001-01-08.
[6] Hussein, Shehata, Ziskin, Alan. "Waterproof and breathable microporous thermoplastic laminated fabric." USP: US2003/0215617 A1, 2003-11-20.

Influence of Cotton/Polyester Blend Level on Color Fastness

Yingjie Wu, Chunhui Xiang
Department of Apparel, Events and Hospitality Management, Iowa State University, Ames, Iowa, USA
chxiang@iastate.edu

OBJECTIVE

The purpose of this study is to investigate the influence of blend level in a fabric on the performance of color fastness. In this study, cotton and polyester blend fabrics were dyed with acid and reactive dyes. After dyeing, laundering, perspiration and crocking tests were conducted to evaluate the color fastness of the dyed fabrics.

INTRODUCTION

Dyes will be dropped under some external force such as abrasion, laundry and light[1] then have some bad feedback to human body. So, the textile color fastness has become an important indicator of the quality inspection of the internal quality assessment[2]. Evidence shows that finishing[3, 4] and pretreatment[5] process can improve the color fastness properties. Moreover, using a neutral buffer system bath[6] can also improve the color fastness performance. The study on dyeing behaviors showed that the effect of dyeing conditions such as temperature and time[7], dye depth[8], and pH value[9] would affect the color fastness. At the same time, online monitoring dyeing system [10] has been introduced to assess in the dyeing process. In this study, the color difference and color parameter were evaluated to identify the influence of cotton/polyester blend level on color fastness.

APPROACH

Six different blend level fabrics (table I) were used to investigate the dyeing and color fastness performance. The fabrics were dyed with 4% (on weight the fabric, o.w.f.) of acid red and reactive fuchsin dye with a liquor-to-goods ratio of 50:1. Vinegar and soda ash were used to adjust pH of the dyeing bath.

Table I. Fabric Information

Fabric No.	Fabric Composition	Weight (acid)	Weight (reactive)
#1	100% Cotton	25.32g	22.32g
#2	65/35 C/P	40.97g	40.76g
#3	50/50 C/P	27.78g	26.68g
#4	35/65 C/P	27.21g	25.29g
#5	20/80 C/P	31.77g	30.68g
#6	100% Polyester	15.71g	26.09g

Twelve different fabrics dyed with two different dyes were tested for colorfastness. Color fastness to washing (AATCC 15), (AATCC 8), and perspiration (AATCC 61). The color fastness and staining ratings were evaluated by comparing the AATCC Grey Scale from 5 to 1 for Evaluating Change in Color and Grey Scale for Evaluating Staining. The accurate color value were read under the CIELAB system. The three coordinates of CIELAB represent the lightness (L) of the color, its position between red and green (a*) and its position between yellow and blue (b). ΔE value is the color difference between reference and sample. Also, the relative color difference was evaluated by equation below:

$$\text{Relative color difference} = \frac{\Delta E_{dyed} - \Delta E_{lab}}{\Delta E_{dyed}} \times 100\%$$

RESULTS AND DISCUSSION

Dyeing Performance

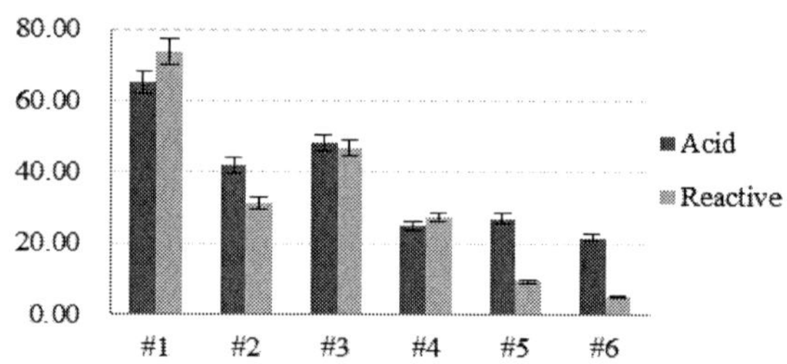

Figure 1 Color Difference of the Samples after Dyeing Process

Figure 1 shows the color difference of the fabrics dyed with acid and reactive dye. Under the same dyeing concentration, liquor-to-goods ratio, and pH value conditions, 100% cotton showed the best performance of dye uptake to both acid and reactive dye, while 100% polyester exhibited the worst dye uptake performance. Moreover, the 50/50 cotton/polyester blend level showed the best dye uptake among the tested cotton/polyester blend fabrics. As the cotton content decreased from 50% to 20% in the blend fabrics, the dye uptake performance decreased significantly, which is in agreement with our 100% cotton fabric results.

Color Fastness Performance

Table II. Average Stain Evaluation of Crocking Test

	#1	#2	#3	#4	#5	#6
Acid dry	1.7	4.8	2.7	4.3	4.3	4.0
Reactive dry	2.2	5.0	4.2	4.8	5.0	5.0
Acid wet	1.2	2.0	2.2	4.0	4.0	4.0
Reactive wet	1.5	3.7	3.5	4.2	5.0	5.0

Table II shows the results of color fastness to rubbing. The fabric crocking test showed that 100% cotton had poor fastness but it can be improve a lot after blend. Moreover, special fabric structure can also improve the dry crocking color fastness performance.

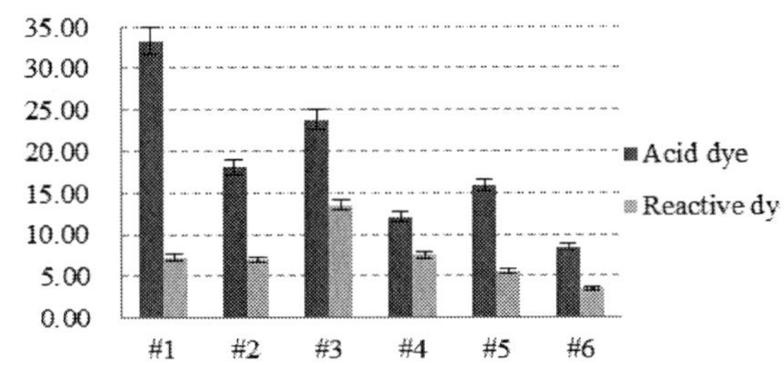

Figure 2 Color Difference after Laundry Test

Figure 2 shows the color difference after washing. After laundry test, the fabrics had the color fade and color stain problem. The absolute color fade results showed that the 100% polyester had the best laundry color fastness (acid dye and reactive dye), and the relative color fade results showed that the 100% cotton had a very good laundry fastness when using the reactive dye, but at the same time, blend did not influence the color fade under the acid dyeing process.

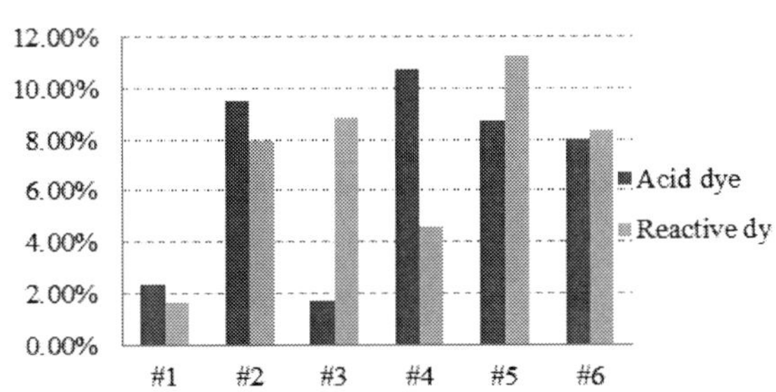

Figure 3 Color Change Percentage after Perspiration Test

Figure 3 shows the color fastness to perspiration. 50/50 cotton/polyester fabric showed the best color fastness to acid dye and 100% polyester (absolute color fade) and 100% cotton (relative color fade) were best for reactive dye. Moreover, blend can also help to improve the color stain performance under the perspiration condition.

CONCLUSION

Blend level of fabrics showed influence on color fastness of the fabrics. 100% cotton performed best under the reactive dye process. 50/50 cotton/polyester showed best color fastness to acid dye.

REFERENCES

[1] Li, Y.M. *Study on the Perspiration-Light Stability of Reactive Dyes*. Shanghai: Donghua University, 2007.

[2] Huang, L. "Comparison Among Testing Standards for Color Fastness and Analysis of Common Fastness Problems." *Dye Fabrics*, 36, 7, 2010: 36-40.

[3] Zuber, M., et al. "Modification of cellulosic fabric using polyvinyl alcohol, part-II: Colorfastness properties." *Carbohydrate Polymers*, 87, 4, 2012: 2439-2446.

[4] Zuber, M. et al. "Preparation of rich handles soft cellulosic fabric using amino silicone based softener, part II: Colorfastness properties." *International Journal of Biological Macromolecules*, 49, 1, 2011: 1-6.

[5] Blackburn, R.S., Burkinshaw, S.M., Gordon, R. "Dyeing Cotton/Polyamide Blend Fabrics with 1: 2 Metal Complex Acid Dyes." *Textile Chemist & Colorist & American Dyestuff Reporter*, 32, 12, 2000: 44-48.

[6] Anis, P., Eren, H.A. "Improving the fastness properties of one-step dyed polyester/cotton fabrics." *AATCC Review*, 3, 4, 2003: 20-24.

[7] Yang, Y., Li, S., Brown, H., Casey, P. "Dyeing of PTT/PET Blends." *AATCC Review*, 2, 8, 2002: 54-59.

[8] Chen, Q.H., Yang, P., Lu, B.T. "Factors exerting influence on light fastness of the reactive dyeings." *Dyeing and Finishing*, 1, 2007: 19-23.

[9] Li, Y.M. et al. "Color fastness to the combination of perspiration and light of the reactive dyeings." *Dyeing and Finishing*, 3, 2007: 1-4.

[10] Li, X.Y. et al. "Dyeing behaviour of THCDP and its blend with cotton at low temperature." *Materials Research Innovations*, 18, 4, 2014: 896-900.

Enhanced Mechanical Properties of Polyimide Fibers Containing Benzimidazole and Benzoxazole Units

Qinghua Zhang[1], Chaoqing Yin[1], Jie Dong[1], Jinyou Lin[2]
[1]State Key Laboratory for Modification of Chemical Fibers and Polymer Materials, College of Materials Science and Engineering, Donghua University, Shanghai, China
[2]Shanghai Synchrotron Radiation Facility, Shanghai Institute of Applied Physics, Chinese Academy of Sciences, Shanghai, China
qhzhang@dhu.edu.cn

OBJECTIVE

High-performance polymeric fibers are of great interest to researchers due to their high strength-to-weight ratio, ease of processing and availability of structural variations to control both mechanical and physical properties. In response to the need of high strength polymeric fibers for various engineering applications, scientists have long been active in researches in the area of aromatic polymer fibers with rigid heterocyclic units[1,2]. Incorporating rigid aromatic heterocyclic units has been regarded as one of the most effective approaches, as it could increase the rigidity of the polymer backbones and introduce additional intermolecular associations such as hydrogen bonding interaction[3, 4]. Among many existing functional groups, benzimidazole is a rigid-rod unit, which has strong inter- or intra-molecular hydrogen-bonding formation capability that may strengthen the mechanical properties of resulting materials. The polymers containing benzimidazole units in backbone, such as polybenzimidazoles, which generally exhibit highly ordered structure, it also has favorable adhesive properties in preparation of composites by forming complexes with the resin matrix. Besides, As a kind of high performance polymer fibers containing benzoxazole units, Zylon® fiber (PBO) possesses very high tensile strength (~5.8 GPa) and modulus (~270 GPa) and excellent thermal stability as well as good environmental resistance, moreover, the heat treated Zylon® HM fiber almost shows zero moisture sensitivity. Hence, the introduction of the benzoxadazole group into polyimide backbones may greatly enhance the mechanical properties of the resulting polymers. Taking into consideration of the above factors, we hypothesize that the introduction of these two versatile units into the polyimide backbone is expected to fabricate high performance PI fibers by two-step process with outstanding mechanical, thermal and other properties to meet the integrated performance requirements for applications.

EXPERIMENTAL APPROACH

The precursors of the co-polyimide, poly(amic acid)s (PAAs), were prepared by mixing equimolar amounts of the dianhydride and diamine dissolved in DMAc under a dried nitrogen atmosphere. With this procedure, co-PAA solutions with various BIA/BOA molar ratios (10/0, 9/1, 7/3, 5/5, 3/7, 0/10) were synthesized in the present work. The PAA precursor fibers were produced by extruding the viscous PAA solutions through a spinneret (50 holes, 80 μm in diameter) into a H_2O/DMAc coagulation bath. The PAAs were then converted to polyimides by successive heating at 100, 200 and 300 °C for 1 h at each temperature. Besides, for each type of PI fibers, efforts were made to achieve the maximum draw ratios at more than 450 °C.

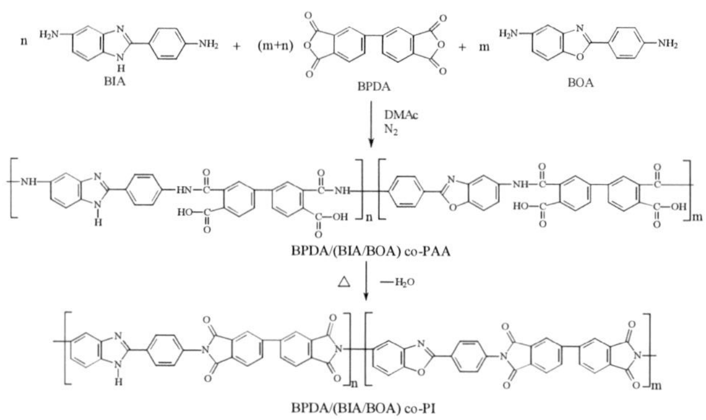

Scheme 1. Schematic diagrams for the synthesis of the BPDA/BIA/BOA co-PAA and co-PI.

RESULTS AND DISCUSSION

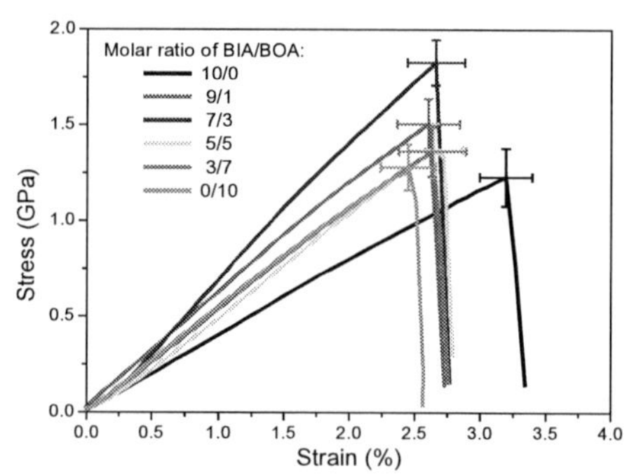

Figure 1. Stress-strain curve of polyimide fibers with various diamine ratios.

The typical stress-strain curves of co-PI fibers are shown in Figure 1, from which it can be seen that the four co-PI fibers simultaneously containing BIA and BOA units show better mechanical properties compared with the two homo-PI fibers. It is clear that homo-BPDA/BIA and homo-BPDA/BOA PI fibers possess tensile strength of 1.23 and 1.29 GPa, and modulus of 40.1 and 51.6 GPa, respectively. Along with increasing BOA content in the co-PIs, the fibers exhibit a rising trend in strength as well as initial modulus. It may be worth noting that the strongest fiber is co-PI-2, in which, the BOA molar content reaches 30%, and the corresponding tensile strength and modulus are 1.74 and 74.4 GPa, respectively. However, the strength decreases when the BOA molar content increases to 50%, which may be caused by the following two factors: (A) limited molecular weights, and (B) decreased molecular orientation and crystallinity with more BOA moieties.

As we know, orientation and crystallinity in fibers play important roles on the performance of polymeric fibers. Figure 2 shows the 2D WAXD patterns of the obtained PI fibers. Figure 2A of homo-BPDA/BIA PI fiber exhibits several clear diffraction streaks in both the region of meridian and equatorial directions, revealing a well-defined crystal structure forms in the homo-BPDA/BIA polyimide fiber. WAXD patterns of co-PI-1 and co-PI-2 fibers, which are quite similar to the homo-BPDA/BIA PI fiber, are shown in Figure 2B and 2C, from which it can be deduced that the segments of BPDA-BIA are mainly packed in the crystal phase in the three systems.

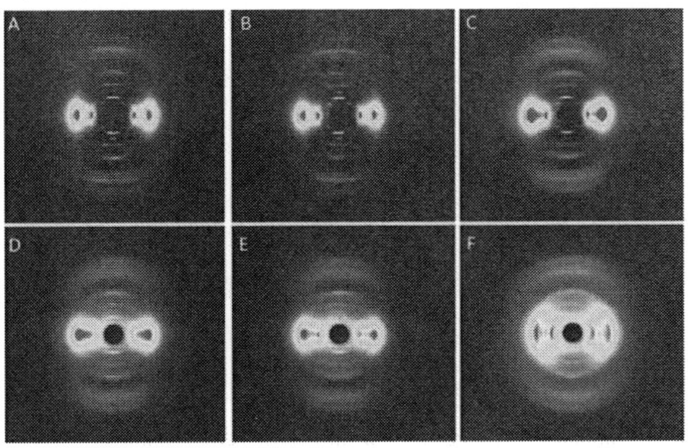

Figure 2. 2D WAXD patterns PI fibers with various diamine ratios: (A) homo-BPDA/BIA, (B) BIA/BOA=9/1, (C) BIA/BOA=7/3, (D) BIA/BOA=5/5, (E) BIA/BOA=3/7 and (F) homo-BPDA/BOA.

Dramatic changes in the 2D WAXD patterns occur when the molar content of incorporated BOA reaches 50%. As shown in Figure 2D, the equatorial arcs originated from lateral packing become obscure, which are so-called "amorphous halos", indicating a reduction in the lateral packing order along the transverse direction of fibers, which may be resulted from the more or less random distribution of BPDA/BOA segments in polymer chains. For the co-PI-4 and homo-BPDA/BOA fiber, as shown in Figure 2E and 2F, the X-ray reflections converge to a new set of diffractions, indicating the formation of a new lateral packing ordered structure. Curve-fitting deconvolution of the WAXD spectra was performed using Peak-fit Software to separate crystallization and amorphous contributions, as shown in Figure 3. It is evident that the homo-BPDA/BIA PI fiber shows the highest crystallinity with a well-defined crystal structure. With increasing BOA content, the crystallinity of the co-PI fibers decreases gradually until the BOA molar content reaches 50%, and the co-PI-3 fiber shows the lowest crystallinity (20.7%), indicating the lower lateral packing order resulted from the somehow random distribution of the BPDA-BOA units in the polymer chains. Further increasing BOA content leads to a rising trend of crystallinity for polyimide fibers, and it may be mainly attributed to the formation of new crystal structures. For polymeric fibers, the molecular orientation plays an important role on the final mechanical performance. The degree of orientation of the ordered polymer chains along the fiber axis was also studied according to the Herman's equation. Apparently, the co-PI-3 fiber with BOA molar ratio of 50% has the lowest degree of orientation along the fiber axis (72.5%), while the orientation factors for the homo-BPDA/BIA PI and co-PI-2 fibers are over 80%, indicating the molecular chains form highly ordered structures along the axial direction and thus lead to relatively higher tensile strength and modulus.

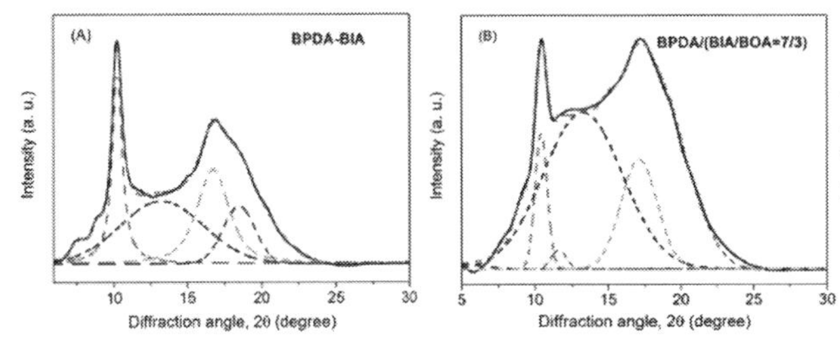

Figure 3. WAXD peak fitting results for the polyimide fibers with various diamine ratios: (A) homo-BPDA/BIA, (B) BIA/BOA=7/3.

Figure 4 shows the SAXS patterns of the co-PI fibers with various diamine ratios. It should be noted that there is a detectable scattering along the meridian scattering direction for the homo-BPDA/BIA fiber (rectangle boxes in Figure 4A), suggesting the existence of a periodic lamellar arrangement between crystalline and amorphous regions. However, as adding the BOA monomers into the polymer chains, this feature disappears in all of the other samples, further investigation may be performed on the detailed information about this phenomenon. Besides, in all polyimide fibers, the scattering pattern shows a sharp and elongated equatorial streak, which may be attributed to the existence of microvoids in the wet-spun co-PI fibers.

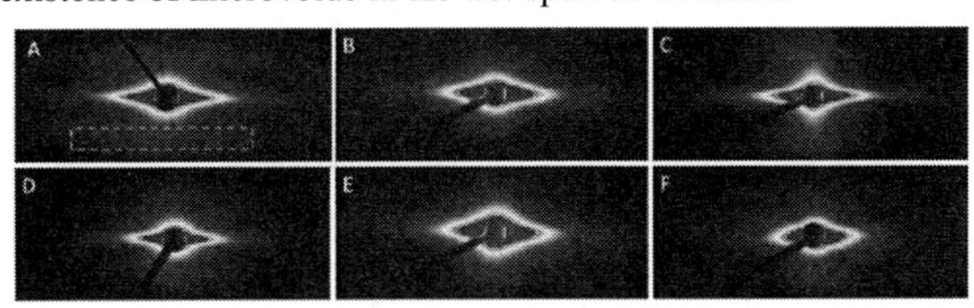

Figure 4. 2D SAXS patterns co-PI fibers with various diamine ratios: (A) homo-BPDA/BIA, (B) BIA/BOA=9/1, (C) BIA/BOA=7/3, (D) BIA/BOA=5/5, (E) BIA/BOA=3/7 and (F) homo-BPDA/BOA.

ACKNOWLEDGMENTS

Financial support of this work is provided by NSFC (51233001, 51173024), 973 Plan (2014CB643603), 863 plan (2012AA03A211), and 111 Project (111-2-04) and Chinese Universities Scientific Fund.

REFERENCES

[1] Zhang, Q., Dai, M., Ding, M., Chen, D., Gao, L. *European Polymer Journal*, 40, 2004: 2487-2493.

[2] Jin, L., Zhang, Q., Xu, Y., Chen, D. *European Polymer Journal*, 45, 2009: 2805-2811.

[3] Dong, J., Yin, C., Zhao, X., Li, Y., Zhang, Q. *Polymer*, 54, 2013: 6415-6424.

[4] Dong, J., Yin, C., Luo, W., Zhang, Q. *Journal of Materials Science*, 48, 2013: 7594-7868.

Structure and Properties of Cellulose Diacetate Fibers Using Ionic Liquid as Solvent

Na Liu, Baochun Wang, Huaping Wang
State Key Laboratory for Modification of Chemical Fibers and Polymer Materials,
College of Materials Science and Engineering, Donghua University, Shanghai, China
wanghp@dhu.edu.cn, liuna@dhu.edu.cn

ABSTRACT

Cellulose acetates (CAs) are important derivatives of cellulose as a result of substituting acetyl groups for hydroxyl group.[1] The most popular cellulose acetates are cellulose diacetate (CDA) with an average degree of substitution (DS) ranging from 2.2 to 2.7 and cellulose triacetate (CTA) having average DS above 2.7.[2] They have many applications in fibers, plastics, films, membranes and so on.[3] As one of regenerated cellulose fibers, Cellulose diacetate fiber known as "acetate", has distinctive properties combining the characteristics of synthetic fibers and cellulose fibers and had played an irreplaceable important role in the textile fibers for decades. It possess not only a pleasant feel against the skin, a softness and suppleness with a silk-like hand, but also natural crease resistance, low moisture absorption, excellent draping qualities and distinct filtering effect.[4,5] Thus it can be widely used as high-grade clothing fabric material, advanced medical material and cigarette filter material. [5]

Acetate is commercially manufactured by means of dry spinning with large quantities of acetone used as solvent, which is potentially damaging to the environment for the high volatility and flammability of acetone and lacks efficiency for its long processing flow.[6] Moreover, the acetate via dry spinning shows a low tensile strength as reported in literature with a range of 1.06-1.23cN/dtex[5], which limited its application in textile field.

Mechanical properties of regenerated cellulose fibers vary from different processing technology.[7] As non-derivatizing cellulose solvent, Ionic liquids (ILs) have been used to dissolve and process cellulose[8-9]. It was also found that the cellulose fibers from ILs technology possess good mechanical properties similar to those of Lyocell fibers[10].

In this study, a type of ionic liquids, 1-Butyl-3-methylimidazolium chloride (BMIMCl) was used as solvent of CDA and then CDA fibers were spun by dry-wet spinning. The rheological behavior of CDA/ BMIMCl was tested. The structure and properties of CDA fibers were then investigated.

Commercial CDA was ground to powder and kept in vacuum at 80 °C for at least 12 hours to remove the absorbed water as much as possible. Then dried CDA powder was added into BMIMCl at 90 °C in a stirring kneader for 1 hour, and then the mixture was put in vacuum at 85 °C to get rid of the air bubbles until the mixture turned to a homogeneous solution. Finally spinning solution of 12 wt% CDA in BMIMCl was obtained.

The rheological test was performed on a rotational rheometer, Phsica MCR 301 (Anton Paar Instrument Corporation). It was very clear that the CDA/BMIMCl solution exhibits the typical shear-thinning behavior, with the temperature increasing, the zero-viscosity tended to decrease (Fig. 1). Meanwhile, when the temperature was over 70 °C, η at different temperature was very close under the same shear rate, that is, the temperature has little influence on the η when shear rate exceeds $100s^{-1}$. Thus the spinning temperature was set as 70 °C in this study.

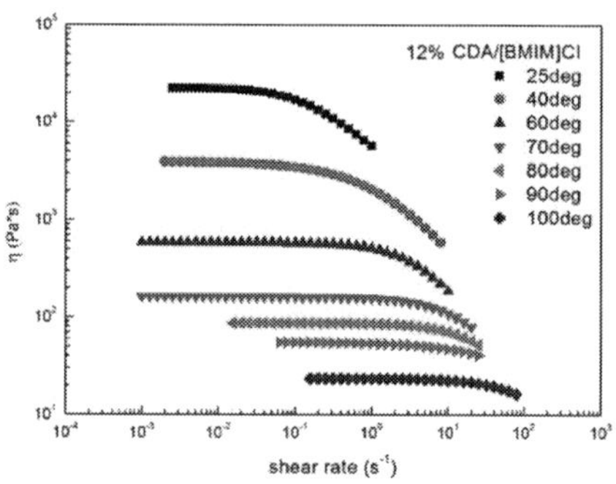

Fig. 1. viscosity-shear rate curves of 12 wt% CDA/[BMIM]Cl solution at different temperature.

Spinning solution was transferred to a spinning tank at 70 °C, and extruded through a spinneret orifice with a diameter of 0.1 mm, then passed through a 60-mm-long air gap and immersed in a coagulation bath of water to precipitate cellulose into a filament form. The filaments were washed with water, wound, and dried. Fiber section is a regular circle (Fig. 2a), the surface is very smooth (Fig. 2b).

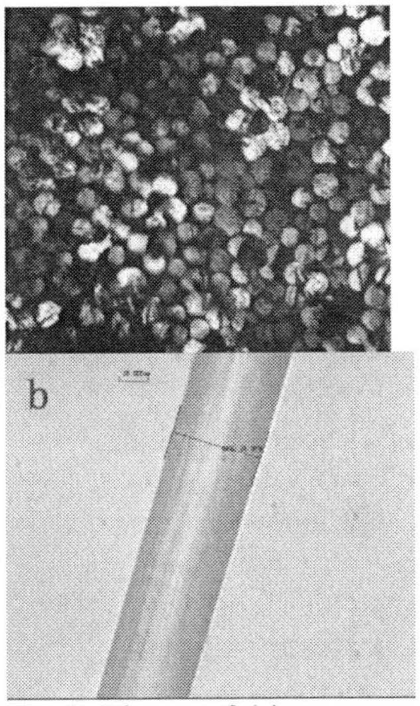

Fig. 2. Photos of (a) cross-section and (b) surface of CDA fibers.

The structure of CDA before and after dry-wet spinning was investigated by FTIR (Fig.3). It was noted that there was no significant absorption peak shift of CDA after dry-wet spinning, which indicated that no new compound was formed during fiber formation. Thus it was inferred that BMIMCl was non-derivatizing solvent of CDA, same as cellulose[9].

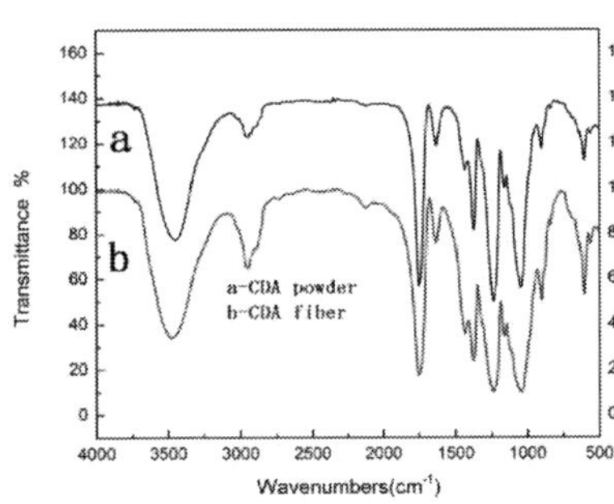

Fig. 3 IR spectra of (a) CDA powder and (b) CDA fiber.

The mechanical properties and crystallinity of CDA fibers at different draft ratio were list at table I. The tensile strength of CDA fiber was 1.24 cN/dtex when the draft ratio was 7.4, which was greater than what had been reported in literature[5]. Moreover the tensile strength of CDA fibers increased with the increase of draft ratio, so did the crystallinity. However the elongation at break decreased. As is known to all, the mechanical properties of fibers mainly rely on the crystallinity and orientation of fibers. The spinning tension in the spinning line increased when the draft ratio increased, the CDA macromolecules aligned gradually in an orderly fashion along the drawing direction, resulting in an increasing orientation. At the same time, the crystallinity of the fibers increased because of the crystallization induced by the orientation. Thus the CDA fibers showed increasing tensile strength with draft ratio increasing.

Table I. Mechanical properties and crystallinity of CDA fibers.

sample	Draft ratio	Tensile strength (cN/dtex)	Elongation at break (%)	Crystallinity (%)
1#	7.4	1.24	10.34	31.87
2#	8.4	1.74	8.94	33.66
3#	9.4	1.77	7.70	33.87
4#	10.4	1.78	10.61	37.78
5#	11.8	1.80	8.00	45.04

CONCLUSIONS

To conclude, CDA/BMIMCl solution was a typical shear-thinning fluid, and the temperature has little influence on the η of solution when shear rate exceeds $100s^{-1}$. CDA fibers with round structure and smooth surface were prepared successfully using BMIMCl as solvent. The tensile strength of CDA fibers over 1.24cN/dtex was greater than that of commercial acetate, which was reported as 1.06-1.23cN/dtex. Therefore, this study presents a novel process route of high performance cellulose diacetate fibers. The CDA fibers were then further drawn at 150 °C and it was found the heat-treated fibers exhibited a higher crystallinity and orientation. More investigation about effect of heat treatment will be done in our future work.

ACKNOWLEDGMENT

This work was financially supported by the National Natural Science Foundation of China (51273041) and the Fundamental Research Funds for the Central Universities (2232014D3-36).

REFERENCES

1. Xu, D.S. *Cellulose Derivatives.* Beijing: Chemical Industry Press, 2001: 48.
2. Heinze, T., Liebert, T. *Prog. Polym. Sci.*, 26, 2001: 1689.
3. Heinze, T., Liebert, T. *Macromol. Symp.*, 208, 2004: 167.
4. Zugenmaier, P. *Macromol. Symp.*, 208, 2004: 81.
5. Law, R.C. *Macromol. Symp.*, 208, 2004: 255.
6. Heinze, T., Liebert, T. *Macromol. Symp.*, 208, 2004: 49.
7. Kosan, B., Michels, V., Meister, F. *Cellulose*, 15, 2008: 59.
8. Swatloski, R.P., Spear, S.K., Holbrey, J.D., Rogers, R.D. *JACS*, 1, 2002: 4974.
9. Hermanutz, F., Gähr, F., Uerdingen, E., Meister, F., Kosan, B. *Macromol. Symp.*, 262, 2008: 23.
10. Cai, T., Zhang, H.H., Guo, Q.H., Shao, H.L., Hu, X.C. *J. Appl. Polym. Sci.*, 115, 2010: 1047.

Synthesis and Characterization of Surface-Modified Electrospun Poly (Acrylonitrile-co-Styrene) Nanofibers for Dye Decolorization

M.R. El-Aassar[1, 2], M.F. El-Kady[3, 4], H. Shokry Hassan[5], Salem S. Al-Deyab[6]

[1]Polymer Materials Research Department, Advanced Technology and New Materials Researches Institute, City of Scientific Research and Technological Applications, New Borg El-Arab, Alexandria, Egypt
[2]College of Chemistry, Chemical Engineering and Biotechnology, Donghua University, Shanghai, China
[3]Chemical and Petrochemical Engineering Department, Egypt-Japan University of Science and Technology, New Borg El-Arab City, Alexandria, Egypt
[4]Fabrication Technology Department, Advanced Technology and New Materials Researches Institute, City of Scientific Research and Technological Applications, New Borg El-Arab City, Alexandria, Egypt
[5]Electronic Materials Researches Department, Advanced Technology and New Materials Researches Institute, City of Scientific Researches and technological applications, New Borg El-Arab City, Alexandria, Egypt
[6]Petrochemical Research Chair, Chemistry Department, College of Science, King Saud University, Riyadh, Saudi Arabia

ABSTRACT

In the present work, Poly Acrylonitrile-co- Styrene (poly (AN-co-ST)) copolymer was prepared using solution polymerization. The synthesized copolymer was fabricated into nanofiber using electrospinning technique with an average diameter of approximately 220 nm were successfully synthesized. This nanofiber was functionalized with surface carboxylic acid groups. Characterization of the poly (AN-co-ST) nanofiber and its modified nanofibers were studied by SEM, FTIR, and TGA, and the chemical modification process has positive effect on the mechanical properties of the produced nanofiber as elucidated from the tensile stress–strain test. The chemically modified copolymer nanofiber (CMCN) was evaluated for decolorization of C.I. Basic violet 14 from aqueous solutions. The decolorization process onto the nanofiber was relatively fast and the equilibrium was achieved within 30 min. The influence of the different parameters affecting on the dye sorption process were established. The increment on both the pH and the temperature of the dye solution enhance the dye sorption onto the prepared chemically modified nanofiber. Meanwhile, the presence of cation salts accompanying with the dye ions hinder the dye sorption process onto the prepared nanofiber. The chemically modified nanofiber proved its effectiveness as adsorbent material for dye decontamination from the polluted wastewater.

KEYWORDS: Synthesis; Surface modification; Nanofiber; Dye removal

Kinetics of Catalytic Glycolysis of PET with Sodium Titanium Tris(glycolate)

Shaobo Wang, Chaosheng Wang, Huaping Wang
State Key Laboratory for Modification of Chemical Fibers and Polymer Materials,
College of Materials Science and Engineering, Donghua University, Shanghai, China
wsbharry@163.com; cswang@dhu.edu.cn

INTRODUCTION

Poly(ethylene terephthalate) (PET), as an indispensable polymer material is wildly used in the textile and packaging industry, approximately 70% of synthetic fibers are made from PET. Consequently, the recycling of PET has found its continuously growing popularity owing to economic and ecological considerations. Great attention is currently paid to chemical recycling of PET because it has unique advantages on recycling PET waste with high content of impurities and color, particularly those textile wastes. Chemical recycling can utilize processes such as hydrolysis, methanolysis, glycolysis, ammonolysis and aminolysis.[1] Among them, glycolysis with ethylene glycol (EG) is one of the most attractive method because it requires mild conditions and its product bis(2-hydroxyethyl terephthalate) (BHET) can be readily integrated into conventional production of PET or other functional co-polymers.

Glycolysis of PET is usually carried out at 175-195°C with catalyst as a heterogeneous reaction for two main reasons: 1) serious side reactions will occur at higher temperature such as formation of diethylene glycol. [2] 2) uncatalyzed process is sluggish. Kinetic studies of PET chemical recycling are mainly based on two typical models: shrinking-core model and power-law model. Shrinking-core model can well describe hydrolysis catalyzed by acid and alkali, [3, 4] because the whole reaction is clearly heterogeneous. However it fails to describe the glycolysis due to the heterogeneous reaction only occurring in the initial reaction stage. Soluble oligomer will be released from solid to liquid phase during the glycolysis and change the reaction kinetics. Lopez-Fonseca, et al [5] used power-law model to describe glycolysis of PET particles (size smaller than 0.25mm) at 165-195°C by assuming it as a homogeneous reaction, but when larger size of PET particles were applied, this model would be no longer suitable and the activation energy obtained from this model was significantly affected by omitting the phase changing effect. On the other hand, kinetic descriptions of catalyzed PET glycolysis are mainly based on reactions with zinc acetate as catalyst. Recently, we found sodium titanium tris(glycolate) is a suitable and green catalyst for glycolysis of PET.[6] Therefore, in this work we have developed a new suitable kinetic model to describe PET glycolysis at 175-195°C more accurately by considering the phase changing and equilibrium of depolymerization during the reaction. Also we use this model to investigate the kinetics of catalytic glycolysis of PET with sodium titanium tris(glycolate), which is necessary for developing the industrial process of chemical recycling of PET wastes.

KEYWORDS
PET wastes recycling, chemical recycling, catalyzed glycolysis, kinetic modeling.

APPROACH

Uncolored virgin PET granules (super bright textile grade, intrinsic viscosity: 0.650±0.015dLg^{-1}) with a mean radius of 0.5~1mm, chemical pure grade EG and synthesized sodium titanium tris(glycolate) [6] were used as raw material. The glycolysis of PET reactions were carried out in reactor at 175-195°C under atmospheric pressure with nitrogen atmosphere, stirring and EG refluxed. The molar ratio of EG/PET was settled at 8. Briefly, the extent of reaction and yield of products were determined by weighting method. The related filtrations and subsequent precipitations are schematically depicted in Fig. 1. Different fractions obtained from the glycolysis were analyzed by differential scanning calorimetry (DSC).

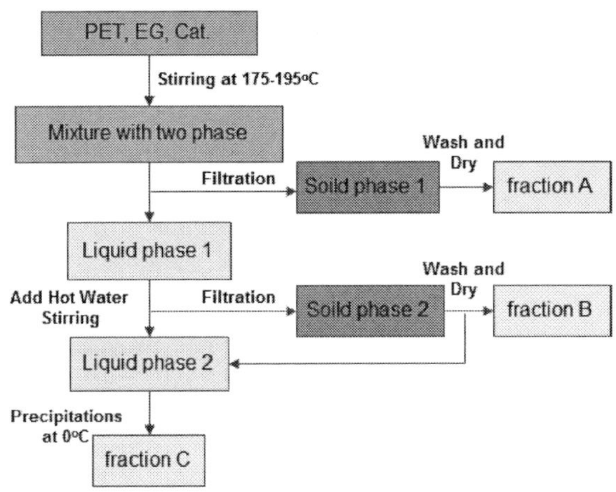

Fig.1. Schematic diagram of the experiment and procedure for the catalytic glycolysis of PET.

RESULTS AND DISCUSSION

The DSC thermograms of different fractions obtained from glycolysis experiment (185°C, 60min) are shown in Fig. 2.

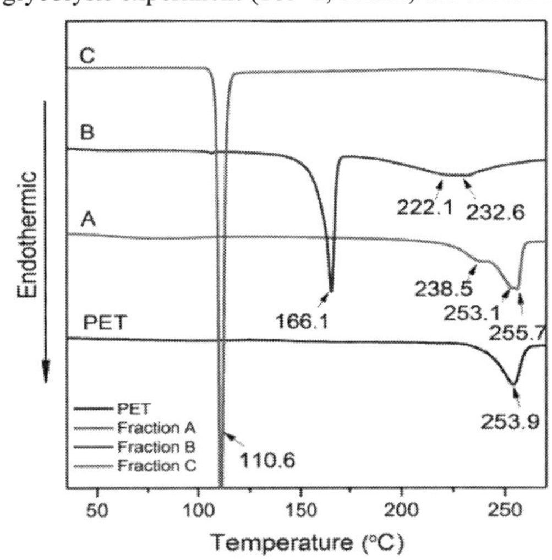

Fig. 2. DSC profiles of different fractions (185oC, 60min).

The DSC thermograms of different fractions obtained from glycolysis experiment (185°C, 60min) are shown in Fig.2. The melting endothermic peaks of these fractions clearly identify all matters from different phases during glycolysis reaction and reveal the evolution of the glycolysis. Fraction A, B, and C are unreacted PET, soluble oligomers (including dimer), and BHET, respectively. [7]

Meanwhile, The evolutions of yields of glycolysis products and reaction conversion (Fig. 4) clearly present "S-shaped". Therefore, the reaction rate must be additionally enhanced during the process. Based on these results, it is reasonable to infer that the products of glycolysis play an important role to change the phase during the reaction, which agrees with the principle 'like dissolves like'. In order to successfully simulate the heterogeneous glycolysis of PET, the acceleration of phase changing caused by depolymerization products and the equilibrium of depolymerization have been taken into account. By treating heterogeneous reaction and homogeneous reaction as two consecutive steps, the scheme evolution of heterogeneous PET glycolysis and kinetic model can be developed as shown in Fig. 3

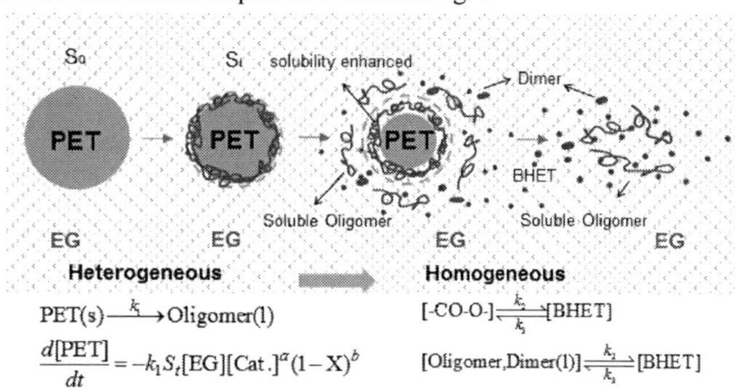

Fig. 3. Scheme evolution of heterogeneous PET glycolysis and kinetic model depiction.

The fitting outcomes of glycolysis experiment data (for example: at 185°C) by using the proposed model and by typical shrinking-core or power law model are shown in Fig.4 and Fig.5. It can be seen that our model can simulate the heterogeneous glycolysis experiment data well and much better than the other two typical models did.

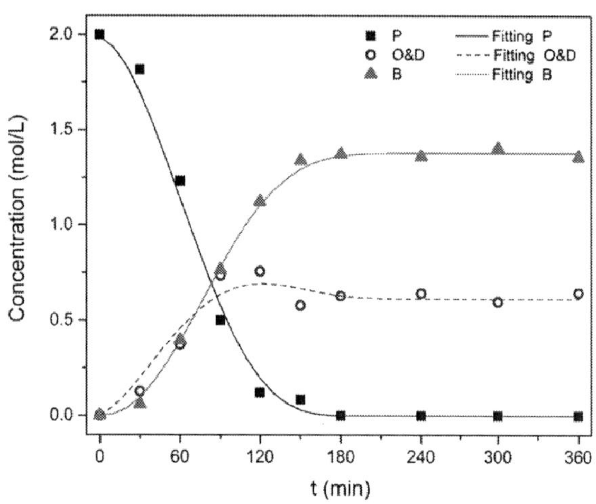

Fig. 4. Experiment data and model fitting results (185°C).

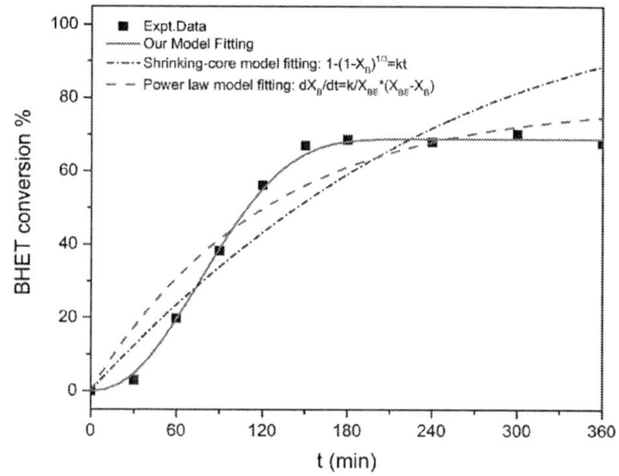

Fig. 5. Fitting results by using different kinetic models (185°C)

CONCLUSION

In this work, a new suitable kinetic model for describing heterogeneous glycolysis of PET was developed by considering the phase changing and equilibrium of depolymerization during the reaction. This model can simulate the catalytic glycolysis of PET with sodium titanium tris(glycolate) more accurately than shrinking-core model and power-law model did.

FUTURE WORK

Based on this kinetic model and the data of further catalytic glycolysis experiments to obtain the full kinetic parameters of catalytic glycolysis of PET with sodium titanium tris(glycolate).

ACKNOWLEDGMENT

This work was supported by the Fundamental Research Funds for the Central Universities (No. 2232013D3-17)

REFERENCES

[1] Achilias, Dimitris S. *Material Recycling–Trends and Perspectives.* Rijeka: Janeza Trdine, 9, 51000, 2012.

[2] Scheirs, J., Long, T.E. *Modern Polyesters: Chemistry and Technology of Polyesters and Copolyesters*. Wiley. 2005.

[3] Yoshioka, T., Okayama, N., Okuwaki, A. "Kinetics of hydrolysis of PET powder in nitric acid by a modified shrinking-core model." *Ind. Eng. Chem. Res.*, 37, 1998: 336-40.

[4] Kumar, S., Guria, C. "Alkaline hydrolysis of waste poly(ethylene terephthalate): A modified shrinking core model." *J. Macromol. Sci. Pure.* A42, 2005: 237-51.

[5] Lopez-Fonseca, R., Duque-Ingunza, I., de Rivas, B., Flores-Giraldo, L., Gutierrez-Ortiz, J.I. "Kinetics of catalytic glycolysis of PET wastes with sodium carbonate." *Chem. Eng. J.*, 168: 2011: 312-20.

[6] Wang, S., Wang, C., Wang, H., Chen, X., Wang, S. "Sodium titanium tris(glycolate) as a catalyst for the chemical recycling of poly(ethylene terephthalate) via glycolysis and repolycondensation." *Polym. Degrad. Stabil.*, 114, 2015: 105-14.

[7] Guclu, G., Kasgoz, A., Ozbudak, S., Ozgumus, S., Orbay, M. "Glycolysis of poly(ethylene terephthalate) wastes in xylene." *J. Appl. Polym. Sci.*, 69, 1998: 2311-9.

Low-Observable Radar Absorbing Composites Composed of Carbon Fiber Felt, Frequency-Selective Surfaces and Sandwich Structures

Shaoxu Xia[1], Yanni Shi[1], Shi Chen[1], Qilin Wu[1,2]
[1]State Key Laboratory for Modification of Chemical Fibres and Polymer Materials, Donghua University, Shanghai, China,
[2]College of Materials Science and Engineering, Donghua University, Shanghai, China
wql@dhu.edu.cn; dreamingrain@yeah.net

OBJECTIVE

In this work, three groups of Frequency Selective Surfaces are designed and low-observable radar absorbing materials with low Reflection Loss(RL) and wide effect bandwidth(EB) in low-frequency were fabricated.

INTRODUCTION

Low-observable radar absorbing materials should exhibit strong microwave absorption (MA) properties over a wide frequency range and need to be with thin thickness as well as low total weight, especially for aircraft[1] use. Carbon materials have attracted increasing interest because of their potential applications in ideal absorbers[2-5]. Carbon fibers(CFs) composites have been found to be promising absorbers owing to their excellent electrical properties, low cost, and wide availability. Traditional FSS is made of metals, which results in a decrease of mechanical properties or increase of the total weight of the aircraft[6]. Activated Carbon Fiber Felt Screens (ACFFS) with appropriate electromagnetic modification are used to replace the metal FSS[7]. Moreover, the MA properties of FSS mainly depend on the structure parameters of the element[8]. Therefore, outstanding absorbing composites with low RL and wide EB could be obtained if proper structure and parameters of FSS element are designed.

APPROACH

Glass fiber (GF) fabric: thickness of 0.2mm; Viscose-based ACF felt : areal density of 109.82 g/m^2, average thickness of 1.2mm, Donghua University, Shanghai; Epoxy Resin: 6002, shanghai resin factory co., LTD; Curing Agent: T-31, shanghai resin factory co., LTD.

The Epoxy Resin was blended uniformly with the curing agent in the mass ratio of 5:1. Then the ACFFS and GF-fabric were laid into the die after adequately infiltrated. And the mixture was poured into the die. Subsequently, molding was degassed and then maintained at 45 ℃ for 12 hours to obtain specimens. The RL measurement was held by free space method over 2-18GHz in the Science and Technology on Electromagnetic Scattering Laboratory in Shanghai, China, according to GJB 2038-94 in the incident electric field of horizontal polarization (HP).

RESULTS AND DISCUSSION

Three types of samples were prepared with different structure: A-type samples with circular apertures unit as Fig. 1 (a); B-type samples with circular patches unit as Fig. 1 (b); C-type samples with combination unit that each patch in the center of the aperture as Fig. 1 (c).

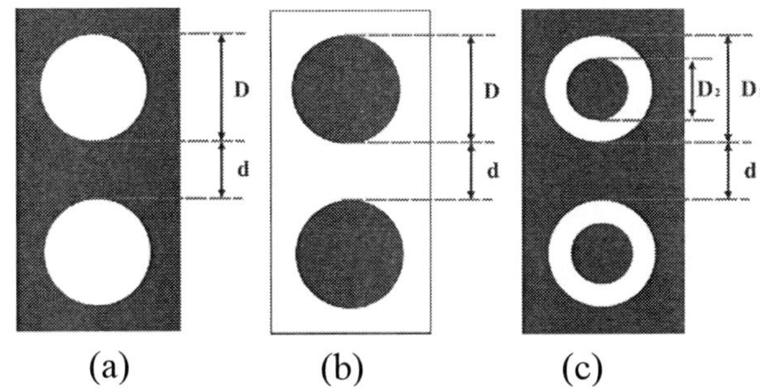

Fig. 1. Design of the ACFFS Configuration:
(a) A-Type, (b) B-Type, (c) C-Type

MA properties of the composites containing A-Type ACFFS

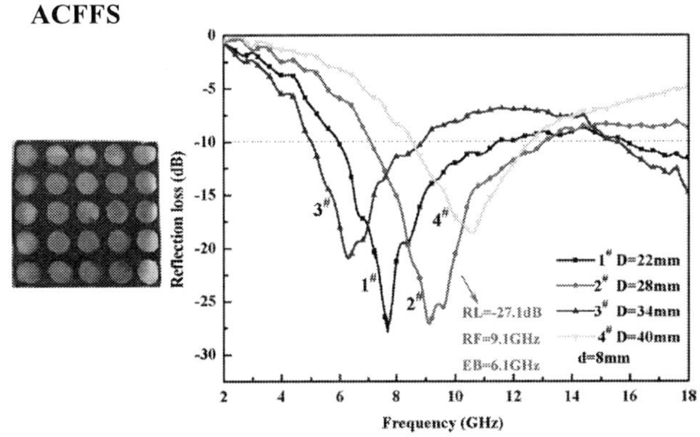

Fig. 2. The photograph of A-type FSS specimen (left) and RL characteristics with respect to D (right)

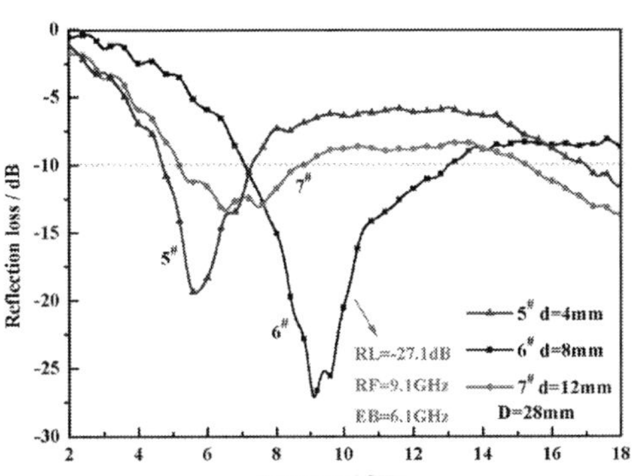

Fig. 3. The RL characteristics with respect to d

Fig. 2 shows the measured RL of A-type FSS with constant interval between circular apertures(d) of 8 mm. After introducing the A-type FSS into ACFFS, the EB increased significantly and then narrowed with the increasing D, demonstrating D is one of the crucial parameters to affect the intensity and the RF. When D=28mm, the RL at 9.1 GHz is -27.1 dB with EB of 6.1GHz from 7.1 to 13.2GHz. Specimens with different d were fabricated to measure their MA characteristics. The RL, as shown in Fig. 3, decreased to -27.1dB first and then increased to -13dB as d increased, while the EB increased from 4GHz to 6.3GHz as d increased. Moreover, we got an interesting result, the RL of Sample 5# and 7# showed double-peak absorbing.

MA properties of the composites containing B-Type ACFFS

Fig. 4 shows the dependence of absorption on D when d is 8mm. The RL and RF, increased first and then decreased as D increased. However, the EB kept at 5.3GHz first and then narrowed to 2.7GHz as D increased. It is confirmed that the EB kept invariant and then broadened as D increased. And the EB occurred in the low-middle frequency (f≦10GHz). Meanwhile, with D increased, the RL increased first and then decreased, however, the RF changed little. The RL at 10 GHz is -30.7 dB with EB of 5.3GHz(6.8GHz-12.1GHz) under the condition of d=8mm and D=28mm.

We investigated the RL when D is constant. As shown in Fig. 5, the RF demonstrated a strong d dependence, with the interval varying from 4 to 12 mm, the RF shifted to higher frequency and then lower frequency, meanwhile, the RL increased first and then decreased, and the EB shrinked. However, in case of d=4mm, the EB of the composites was of 2.7-9GHz, which possibly provided an idea for low-frequency composite in wide absorption band.

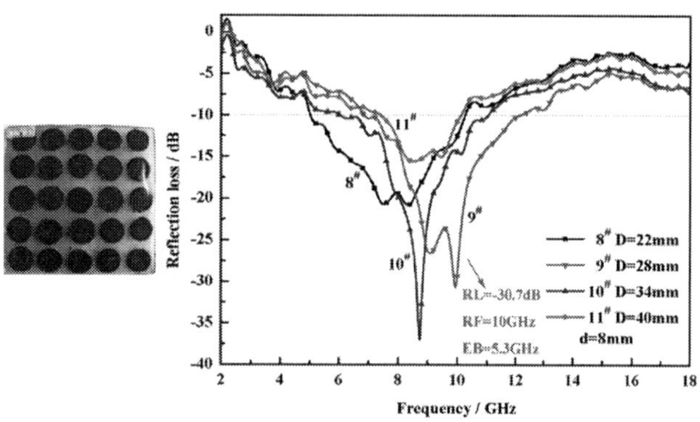

Fig. 4. The photograph of B-type FSS specimen(left) and RL characteristics with respect to D(right)

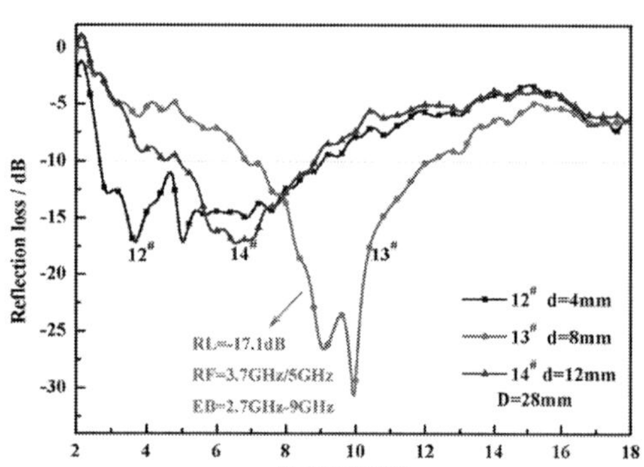

Fig. 5. The RL characteristics with respect to d

MA properties of the composites containing C-Type ACFFS

As shown in Fig. 6. The MA properties of A-type FSS(D=40mm) composites exhibited a narrow EB of 4.1GHz. In order to simultaneously achieve strong low-frequency absorption and wide band absorption, C-type FSS element with a circular patch in center of each aperture were fabricated. Specimens with different diameter of the circular patches(D_2) in circular apertures(D_1=40mm) were fabricated to measure their MA characteristics. And the the interval of the circular apertures (d) of 8mm was used.

The RL, as shown in Fig. 6, increased from -18.3dB to -11.3dB as D_2 increased. However, the RF decreased from 10.5GHz to 5.7GHz and then increased to 10.5GHz finally, as expected, the EB increased from 4.1GHz to 9.5GHz as D_2 increased. As seen from Sample 16# and 17#, the RL increased a little and double peaks appeared, nevertheless, the RF shifted to low-frequency(5.7GHz and 5.9GHz), meanwhile, the EB increased from 4.1GHz to 6.0GHz and 6.4GHz. The results showed that C-type FSS composites realize low-frequency absorption and wide band absorption.

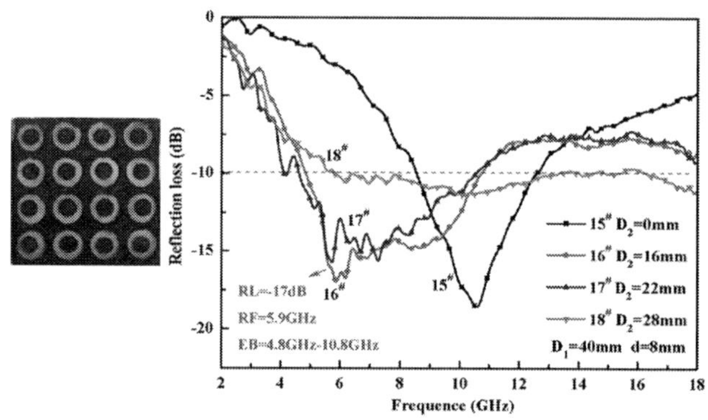

Fig. 6. The photograph of C-type FSS specimen(left) and RL characteristics with respect to D_2(right)

CONCLUSION

The composites with A-type FSS exhibited outstanding MA performances, and RL of -27.1 dB can be reached around EB of 6.1 GHz, however, RF moved to lower frequency (2-10GHz) obviously with B-type FSS. The EB were enhanced in case of the composites with C-type FSS, the EB being greatly improved from 4.1GHz to 9.5GHz when D_2=28mm.

REFERENCES

[1] Lin, H., Zhu, H., Guo, H., et al. *Materials Letters*, 16, 2007: 3547-3550.

[2] Li, X., Feng, J., Zhu, H., et al. *RSC Advances*, 63, 2014: 33619.

[3] Liu, L., He, P., Zhou, K., et al. *Journal of Materials Science*, 12, 2014: 4379-4386.

[4] Vinayasree, S., Soloman, M.A., Sunny, V., et al. *Journal of Applied Physics*, 2, 2014: 024902.

[5] Wang, X.X., Lu, M.M., Cao, W.Q., et al. *Materials Letters*, 125, 2014: 107-110.

[6] Lee, S.E., Park, K.Y., Oh, K.S., et al. *Carbon*, 8, 2009: 1896-1904.

[7] Zang, Y.J., Zhuang, A., Ou, X.H., et al. *Materials Science Forum*, 789, 2014: 163-168.

[8] Fallahi, A., Mishrikey, M., Hafner, C., et al. *Metamaterials*, 2, 2009: 63-74.

Deformation of Dispersed Phase in PP/PS Blend Fiber During Melt Spinning Process

Dan Pan, Long Chen, Houkang He, Yifei Zhou, Qin Wang
State Key Laboratory for Modification of Chemical Fibers and Polymer Materials, College of Material Science and Engineering, Donghua University, Shanghai, China
happyjack@dhu.edu.cn; chnspan@163.com

ABSTRACT
Isotactic polypropylene (PP)/atactic polystyrene (PS) blends with ratio of 92/8 were prepared at the take-up velocity of 125-1000 m/min. The quantity (N), length diameter (L), diameter (B) and shape of dispersed phase were studied. Then the deformation mechanism of dispersed phase was researched. The simulated values of L/B were compared with measured value. The simulated results show that Delaby's model can accurately predict the deformation of the dispersed phase along the evolution of the spinning process. The radial gradient of temperatures from center to the surface was calculated by using dispersed phase as temperature indicator.

INTRODUCTION
The properties of polymer blends were largely controlled by morphology structure, especially the morphology of dispersed phase in immiscible polymer blends. So the research on deformation mechanism of the dispersed phase is particularly important.

APPROACH
Rheological Evaluations
Rheological properties of polymer melt were studied by use of capillary rheometer. Cogswell's theory was used to research the tensile rheological properties of PP and PS.

Preparation of Blend Fibers
PP/PSs with the ratio of 92/8 were melt spun on a lab-scale melt spinner at the temperature of 220℃. Samples were collected at different zones of spinning line and different take-up velocities.

Morphological Evaluations
The diameter (D) of fiber samples was calculated from SEM micrographs. The cross and longitudinal sections were prepared by rotary microtome. Then the quantity (N), length diameter (L), diameter (B) and shape of dispersed phase were studied from the SEM micrographs after etching the dispersed PS domains in THF.

RESULTS AND DISCUSSION
Morphology Analysis
The SEM micrographs were evaluated with image analysis software. Image analysis was automated converting the micrographs into image displaying black droplets of PS on gray background (matrix PP). The section micrographs were divided into five zones from the center to the surface, and the dispersed phase was classified to corresponding zones was show in Figure 1. The number average Length to Diameter ratio ($\overline{L/B_n}$) for very zone was decided by equation(1).

$$\overline{L/B_n} = \frac{\sum L/B}{N_n} \qquad (1)$$

Where N_n is the total number of n-th area.

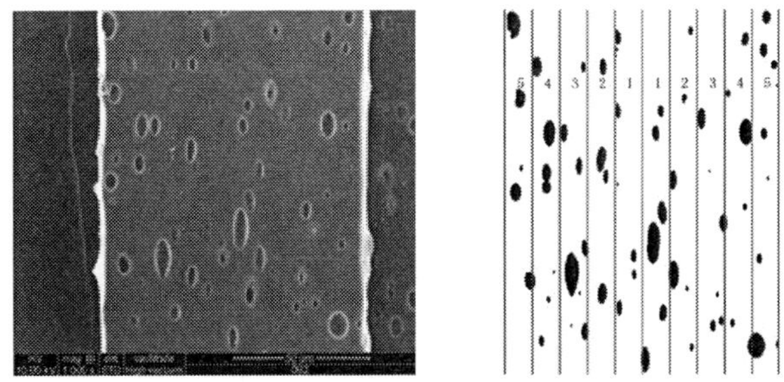

Figure 1. Approach of morphology analysis: (a) longitudinal sections SEM micrograph of sample; (b) the analyzed dispersed phase scattered in 5 radial zones.

By combining the tensile rheological properties of PP and PS with our previous research [1] about distributions of temperature in spinning line, the viscosity ratio ($P=\eta_d/\eta_m$) at vinous shear rate was calculated. The deformation ratio of matrix ($\lambda_m=D_0/D$), where D_0 and D is the diameter of sample exited from the spinneret orifice and the diameter in the different zones in spinning line.

According to Taylor [2] and Huneaults' [3] theory and Delaby's model [4-5], the deformation ratio of dispersed phase (λ_d), and the predicted values of L/B under different take-up velocities and different zones in spinning line could be calculated. As show in Figure 2, the simulated values and measured values of L/B at the take-up velocity of 125m/min are almost same. It was found that the theoretical calculation results accurately predict the deformation of the dispersed phase along the evolution of the spinning process, especially when velocity is at the range of 125-500m/min.

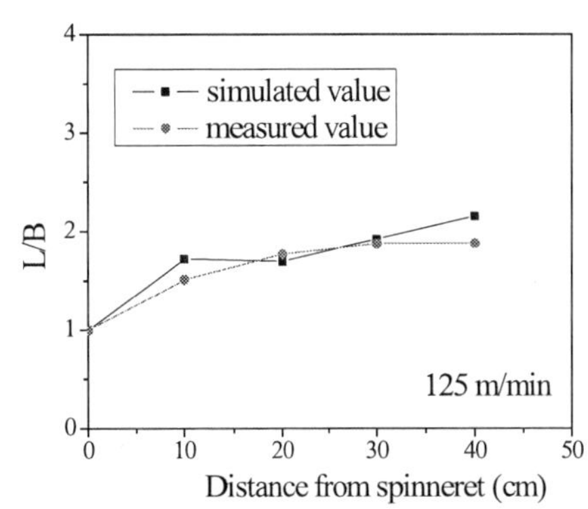

Figure 2. Simulated and measured values of L/B of dispersed phase as function of distance from spinneret.

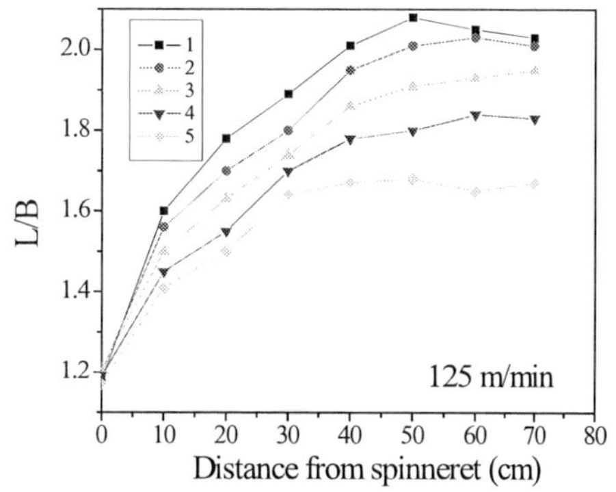

Figure 3. L/B of dispersed phase in different radial zones as a function of distance from spinneret at the take-up velocity of 125 m/min.

Deformation mechanism of the dispersed phase

As show in figure 3, there is a gradient in the deformation ratio of dispersed phase from the center to the surface of fiber. Since the matrix experience the same deformation along the radial, so the differences between different zones were caused by the differences of *P* according to Delaby's model. And the deformation ratio increase with increase of *P*. According to Arrhenius equation (equation 2), *P* is controlled by the temperature. In a word, it is the temperature which causes the radial gradient of deformation ratio.

$$\eta = A\exp\left(\frac{\Delta E_a}{RT}\right) \tag{2}$$

$$P = \left(\frac{A_d}{A_m}\right)\exp\left(\frac{\Delta E_{a,d} - \Delta E_{a,m}}{RT}\right) \tag{3}$$

Where A_d and A_m are the intrinsic viscosity of dispersed phase and matrix, R the Boltzmann constant, $\Delta E_{a,d}$ and $\Delta E_{a,m}$ the activation energy of dispersed phase and matrix , and T the absolute temperature.

Based on the measured λ_d and λ_m, *P* at different zones along radial was calculated according to Delaby's model. Then the radial distribution of temperature was calculated according to equation (3) and the distribution of temperature at the take-up velocity of 125m/min was shown in Figure 4. As polymer blend melt draw down, the radial gradients of temperatures from center with higher temperature to the surface with lower temperature was formed. The radial gradients of temperature increase with the increase of distance from the spinneret orifice and take-up velocity as show in figure 5.

CONCLUSION

It is the temperature which causes the radial gradient of deformation ratio. Delaby's model can accurately predict the deformation of the dispersed phase along the evolution of the spinning process. The radial gradient of temperatures from center to the surface was calculated by using dispersed phase as temperature indicator.

ACKNOWLEDGMENT

The authors acknowledge the Fundamental Research Funds for the Central Universities (Project No. 2232013A3-02) for financial support.

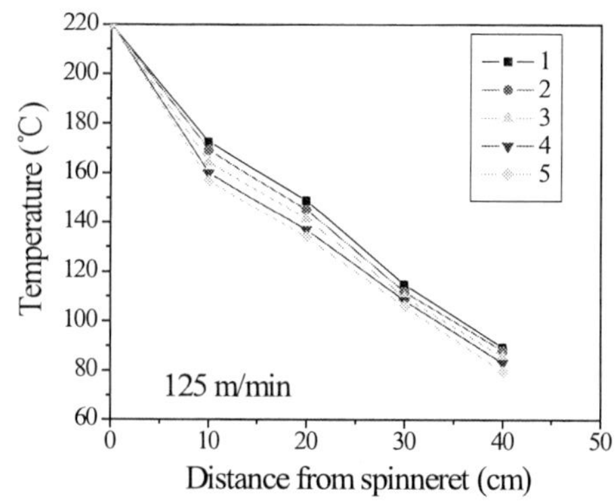

Figure 4. The radial distribution of temperature at the take-up velocityof 125 m/min.

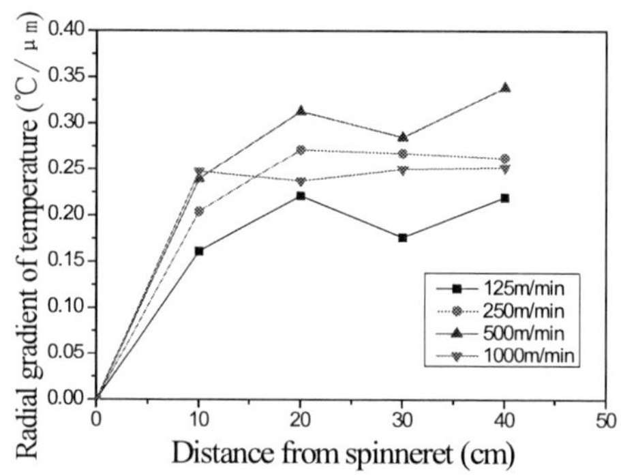

Figure 5. The radial gradient of temperature in spinning line at at different take-up velocities.

REFERENCES

[1] He, Houkang. *Morphology Evolution of PP/PS Blend Fibers along the Spinning Line: Experiments and simulations*. Donghua University, 2014.

[2] Taylor, G.I. "The Viscosity of a Fluid Containing Small Drops of Another Fluid." *Proceedings of the Royal Society A: Mathematical, Physical and Engineering Sciences*, 138.834, 1932: 41-48.

[3] Huneault, M.A., Shi, Z.H., Utracki, L.A. "Development of Polymer Blend Morphology During Compounding in a Twin-Screw Extruder. Part IV: A New Computational Model with Coalescence." *Polymer Engineering & Science*, 35.1, 1995: 115-27.

[4] Delaby, I., et al. "Droplet Deformation in Immiscible Polymer Blends During Transient Uniaxial Elongational Flow." *Polymer Engineering & Science*, 36.12, 1996: 1627-35.

[5] Delaby, I., et al. "Droplet Deformation in Polymer Blends During Uniaxial Elongational Flow: Influence of Viscosity Ratio for Large Capillary Numbers." *Journal of Rheology*, 38.6, 1994: 1705.

Cyclic Disulfide-Based Click Functionalization of Biodegradable Polyurethane Urea Elastomers

Jun Fang[1], William R. Wagner[2], Xiumei Mo[1]
[1]State Key Laboratory for Modification of Chemical Fibers and Polymer Materials, College of Materials Science and Engineering, Donghua University, Shanghai, China
[2]McGowan Institute for Regenerative Medicine, University of Pittsburgh, Pittsburgh, PA, USA
xmm@dhu.edu.cn; fangjun@mail.dhu.edu.cn

Thiol "click" chemistry is an attractive tool for post-polymerization modification of thiolmers or thiol containing substrates in the field of biomedical applications, while thiol groups are easily oxidized, limiting the potential for covalent immobilization of bioactive molecules. In this study, a series of biodegradable polyurethane elastomers incorporating stable cyclic disulfide groups was developed and characterized, via the subsequent click reaction to provide a novel facile and efficient strategy of biofunctionalization.

These poly(ester urethane)urea (PEUU-SS) polymers were based on polycaprolactone diol (PCL), oxidized DL-dithiothreitol (O-DTT), lysine diisocyanate (LDI) or butyl diisocyanate (BDI), with chain extension by putrescine. The ratio of O-DTT:PCL was altered to investigate different levels of potential functionalization. PEG acrylate was employed to study the mechanism and availability of both bulk and surface click modification of PEUU-SS polymers (Fig. 1). To evaluate the potential for incorporating a bioactive surface molecule, the peptide TPSLEQRTVYAK, which has been shown to support endothelial progenitor cell (EPC) attachment, was end-modified with maleimide, applied to an electrospun PEUU-SS surface, and this modified PEUU-SS surface was contacted with EPC to enhance acute cell adhesion.

All synthesized PEUU-SS polymers were elastic with breaking strengths of 38-45 MPa, while alternative use of LDI and BDI in polyurethane synthesis led to different crystallinity, presumably due to the interactions of the cyclic structure of O-DTT and the side chain of LDI in the hard segment, which affected the thermal and mechanical properties as well as the degradation behavior. Variable bulk click modification of PEUU-SS(LDI) polymers was achieved by controlling the amount of reduction reagent, and rapid reaction rates occurred using a one-pot, two-step process. Likewise, surface click reaction could be carried out quickly under mild, aqueous conditions. Furthermore, a maleimide-modified affinity peptide (TPS) was successfully clicked on the surface of an electrospun PEUU-SS(BDI) fibrous sheet, which improved endothelial progenitor cell adhesion versus corresponding unmodified films, suggesting the potential for this approach in cardiovascular applications.

In summary, the cyclic disulfide containing biodegradable polyurethanes described provide an option for soft tissue regenerative medicine applications where a temporary, elastic scaffold with designed biofunctionality from a relatively simple click chemistry approach is desired.

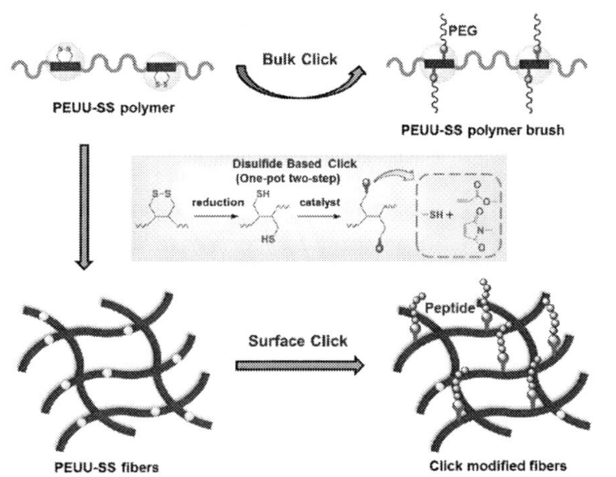

Fig. 1. Bulk click modification of PEUU-SS polymers and surface click modification on PEUU-SS films.

KEYWORDS
Biodegradable elastomer, thiol click reaction, PEGylation, bioconjugation, endothelial progenitor cell.

REFERENCES

1. Fang, J., Ye, S. H., Shankarraman, V., Huang, Y.X., Mo, X.M., Wagner, W.R. *Acta Biomater.*, 10, 2014: 639-4649.
2. Murphy, W. L., McDevitt, T.C., Engler, A.J. *Nat. Mater.*, 13, 2014: 547-557.
3. Lutolf, M. P., Hubbell, J.A. *Nat. Biotechnol.*, 23, 2005:47-55.

Simulation of Profile Degree Variation of Triangular Cross-Section Fiber

Gang Deng, Qinghua Zhang
State Key Laboratory of Modification for Chemical Fibers and Polymer Materials, College of Materials Science and Engineering, Donghua University, Shanghai, China
qhzhang@dhu.edu.cn.

INTRODUCTION

As a main method of fiber preparation, dry-spinning technology was developed in the past decades. People studied on mass transfer, heat transfer and momentum conservation of dry-spinning as well as rheological property of spinning dope, models based on highly coupled equations were established to simulate the spinning process. In previous reports, all the dry-spinning models were based on circular cross-section fiber[1-6]. When prepearing profiled fibers via dry-spinning, a more complex model, including profile degree equation of cross-section, is required.In this report, profile degree variation of triangular cross-section fiber was simulated based on cellulose acetate (CA)-acetone system. A one-dimensional model based on a viscoelastic constitutive equation combining with profile degree equation of cross-section equation was established to calculate the variation of physical quantities efficiently.

MODEL DEVELOPMENT

In order to establish the dry-spinning dynamic model and calculate more efficiently, some assumptions are proposed as follows:

a) The system is at steady state;
b) The density is a constant;
c) Filament is axial symmetry;
d) Neglect diffusion of solvent in the axial direction;
e) All force of the filament are parallel to axis direction except surface tension;
f) In spinning cabinet, residual solvent in hot air is neglected;
g) Neglect viscous dissipation; and
h) Neglect water in the solvent.

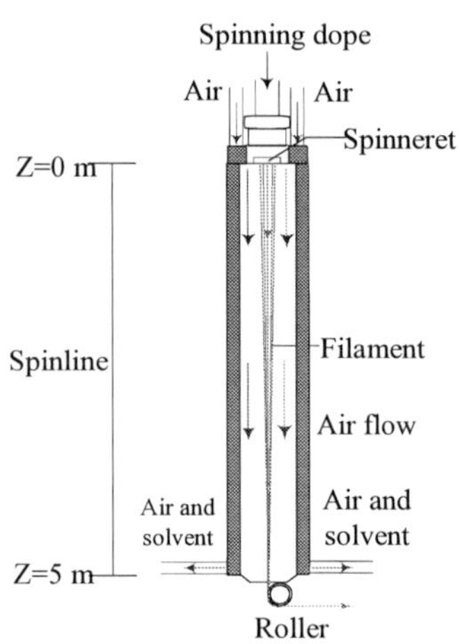

Figure 1. Schematic of dry-spinning process.

Dry spinning dynamics equations

$$\frac{dR}{dz} = \frac{R_0}{2}\sqrt{\frac{\rho_S - \rho(w_{S,N} - w_b)}{A_0 A[\rho_S - \rho(w_{S,N} - w_a)]}}\frac{dA}{dz} + \frac{Fs}{V\eta}(\frac{r}{R} - \frac{R^2}{r^2}) \quad (1)$$

$$\frac{dw_S}{dz} + \frac{2\pi \bar{R} k_y M_S (y_s - y_0)}{W_P}(1 - w_S)^2 = 0 \quad (2)$$

$$\frac{dF}{dz} = \rho A V \frac{dV}{dz} + \pi \bar{R} C_f \rho_a (V - V_{za})^2 - \rho g A \quad (3)$$

$$\rho C_p V_z \frac{dT}{dz} = \frac{2}{R}[h(T_a - T) - \Delta H_v k_y M_s (y_s - y_0)] \quad (4)$$

$$F(1 - 2\lambda \dot{\varepsilon})(1 + 2\lambda \dot{\varepsilon}) = 3\eta_0 A \dot{\varepsilon} \quad (5)$$

Numerical approach

Fourth order Runga-Kutta algorithm method is used to solve the highly coupled differential equations. A binary system is adapted for dry-spinning model in this paper. Surface tension coefficient of acetone varies with temperature in a small range and the effect is neglected here. The basic spinning parameters are listed in Table 1.

1) At z=0, considering the swell effect and neglect swelling deformation, we adopt initial values of R and r are 1.2 times as spinneret diameter. The initial value of temperature and mass fraction is the same as practice spinning parameters. The initial values of velocity and radius of filament are calculated base on mass flow rate of solution, die swell ratio and spinneret diameter. An initial filament tension of 100 dyn is assumed in this paper.

2) $(dV/dz)_{z=0}$ is determined by trial and error until filament velocity matching with take-up velocity at z=500cm, and the tolerance error is 10^{-3}.

Table 1. Spinning parameters

Parameters	Values
Initial temperature of dope(T)	60 °C
Mass fraction of solvent (w_S^0)	72%
Mass flow rate of solution (W)	0.0137 g/s
Diameter of spinneret $\Delta(d)$	60 μm
Velocity of quench air (V_{az})	50 cm/s
Length of spinline (L)	500cm
Heat capavity of polymer ($C_{p,\,p}$)	1.60J/(g ·°C)
Heat capavity of solvent ($C_{p,\,S}$)	2.16 J/(g·°C)
Real drawing ratio	1.0
Take up velocity	1000 cm/s
Desntiy of polymer (ρ_1)	1.31 g/cm^3
Desntiy of solvent (ρ_2)	0.79g/cm^3
Polymerization degree of polymer	230
Initial profile degree(D_R)	0.5

RESULTS AND DISCUSSION

Variation of profile degree and length of axis along the spinline are shown in Figure 2. Both of major axis and minor axis sharply decrease at the beginning several centimeters, then slowly declines to a platform value at z=67cm. But the decreasing percentages are different, the major axis is 35.6% and the minor axis is 57.5%, and the profile degree of filament increases from 50.0% to 67.0%. Both solvent evaporation and drawing effect lead to the thinner filament and shorter length of axis. However, the solvent evaporation rate at direction of minor axis is quicker than major axis due to small concentration gradient, which causes the decreasing percentages of the minor axis. Surface tension is another important factor effect on profile degree. It results in a radial pressure difference between the major axis and the minor axis, leading to cross-section shape approach to circular. The profiled degree is determined by both solvent evaporation and deformation of axis, a higher final profile degree can be obtained when evaporation rate is quick enough.

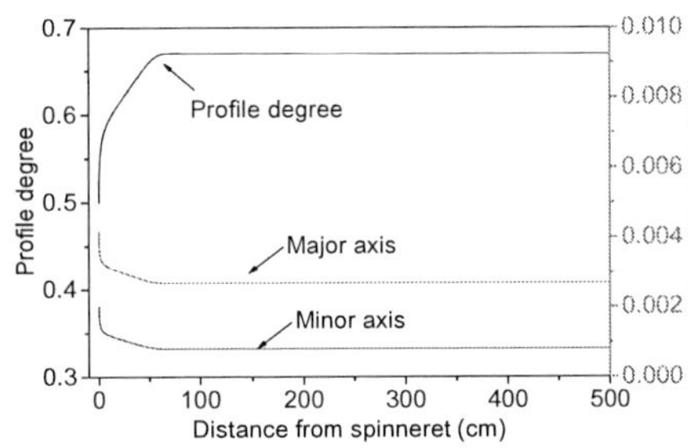

Figure 2. Variation of profile degree and length of axis.

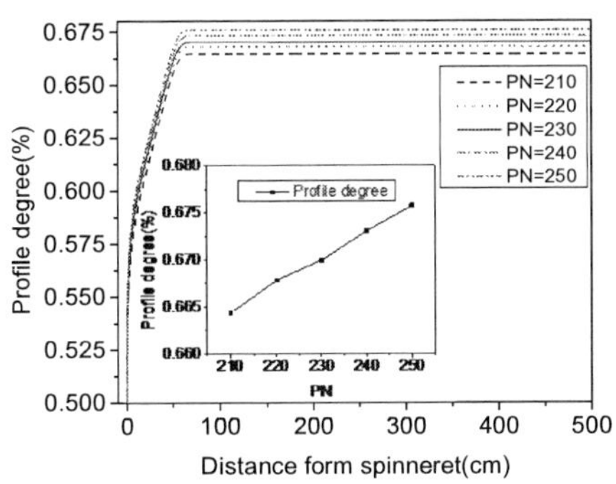

Figure 3. Effect of polymerization degree on profile degree.

Polymerization degree (PN) of polymer is an important parameter to control fiber quality. One can easily know that viscosity of spinning solution grows exponentially with PN as well as mass fraction of solvent. As shown in Figure 3, the final profile degree increased when changing PN from 210 to 250. A relative larger viscosity at the same spinning condition helps to resist deformation caused by surface tension, so the final profiled degree can be increased.

If keeping mass flow as a constant and increasing take up velocity, a larger drawing ratio (DR) is achieved. In Figure 4, DR is investigated over a range from 1 to 1.4. Apparently, an upward displacement of profile degree trend were observed as increasing DR. Higher DR results in a thiner filaments and a high specific surface area, which is beneficial to solvent evaporation. Furthermore, a higher evaporation rate means solidification point of filament shift near to spinneret. Then, the action time of surface tension before solidification is decreased and the deformation of cross-section area is more depend on solvent evaporation. Finally, a minor axis has a relative higher solvent concentration gradient, and its decrease extend is larger than major axis, so the final profile degree of filament increased with DR. As shown in Figure 5, a similar trend is observed if decreasing surface tension coefficient. When Fs=0.6 is adopted for calculation, the final profile degree increased about 1% which is not so obviously as DR. This result also illustrates surface tension is not the primary effect of final profile degree of CA/acetone system in dry-spinning process, because filament solidified in a very short time (about 0.067s) and limited deformation of cross-section area under surface tension.

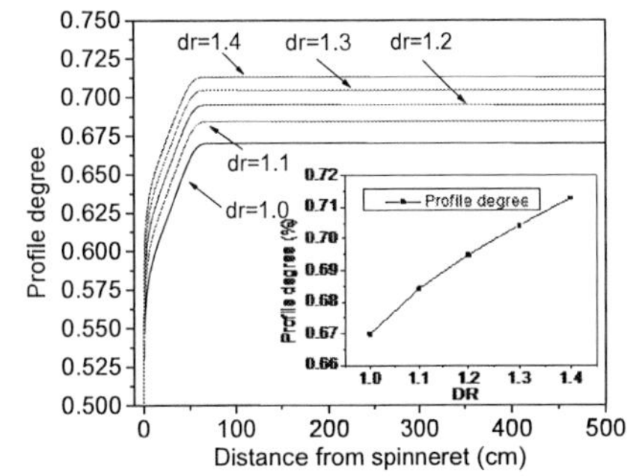

Figure 4. Effect of drawing ratio on profile degree.

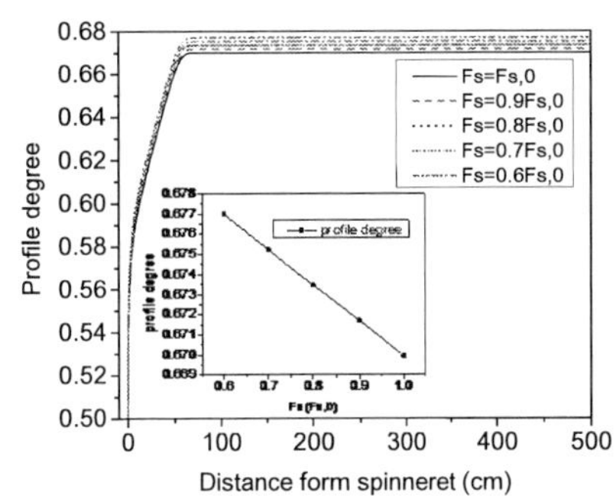

Figure 5. Effect of surface tension on profile degree.

CONCLUSIONS

A modified dry-spinning model based on viscoelastic constitutive equation combining with profile degree equation of cross-section is used to simulate the profile degree variation of triangular cross-section fiber. Control parameters that increasing viscosity of spinning solution is helpful to obtain a relative larger profile degree, but the increase amplitude is small if parameters do not obviously changed.

ACKNOWLEDGEMENT

Financial support of this work is provided by NSFC (51233001, 51173024) and 973 Plan (2014CB643603).

REFERENCES

1. Ohzawa, Y., Nagano, Y., Matsuo, T. J. *Appl. Polym. Sci.*, 13, 1969: 257.
2. Brazinsky, I., Williams, A.G., LaNieve, H.L. *Polym. Eng. Sci.*, 15, 1975: 834.
3. Sano, Y. *Drying Technol.*, 19, 2001: 1335.
4. Gou, Z., McHugh, A.J. J. *Appl. Polym. Sci.*, 87, 2003: 2136.
5. Gou, Z., McHugh, A.J. *J. Non-Newtonian Fluid*, 118, 2004: 121.
6. Gou, Z., McHugh, A.J. *Intern Polym Proc.*, 3, 2004: 244.
7. Deng, G., Xia, Q.M., Xu, Y., Zhang, Q.H. *J. Appl. Polym. Sci.*, 113, 2009: 3059.

Superhydrophobic and Transparent Paper Spray-Coated with SiO_2-Containing Alcohol Suspensions

Seon Ah Jeong, Se Young Oh, Duck Jun Hwang, Hyun Taek Jeong, Tae Jin Kang
Department of Materials and Science and Engineering, Seoul National University, Seoul, Korea
jsa121@snu.ac.kr

STATEMENT OF PURPOSE

This study fabricated a superhydrophobic and transparent paper not using the special nanomaterials and costly instruments which used in most research about superhydro - phobicity. By simply spraying alcohol suspensions containing hydrophobic SiO_2 nanoparticles, superhydro-phobic and transparent SiO_2 nanoparticle coatings were formed on paper. And the effect of the type of alcohol on the hydrophobicity of the coated paper substrate were investigated.

INTRODUCTION

Superhydrophobic surface is defined as a surface with water contact angle higher than 150°, contact angle hysteresis lower than 5°, low roll-off angle, and extreme water repellency. Superhydrophobic Surfaces are pervasive in nature (see Figures 1a-d) with various plant leaves, legs of the water strider. Inspired by natural superhydrophobic surfaces, many artificial (or synthetic) superhydrophobic surfaces have been engineered. One of the important factors which make a surface superhydrophobic is the low surface energy. The other is the surface roughness. It plays a crucial role because air trapping in the surface roughness drastically amplifies surface hydrophobicity. Therefore, hierarchical surface roughness (i.e., both nanometer- and micrometer-sized roughness) is especially effective to make surface superhydrophobic.

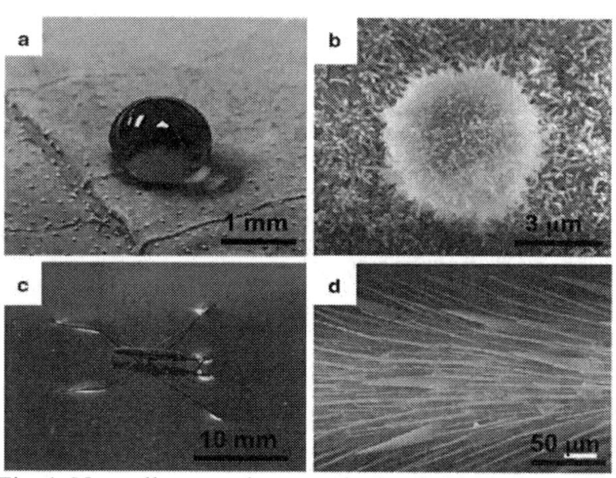

Fig. 1. Naturally occurring superhydrophobic surfaces. (a) A droplet of water (blue) displaying very high contact angle on a lotus leaf. (b) SEM image of a lotus leaf showing its rough, hierarchical texture. (c) Water strider walking on the water surface. (d) SEM image of water strider's leg, which consists of numerous oriented hairs.

EXPERIMENTAL

Preparation of SiO_2 nanoparticles

2g of SiO_2 nanoparticles (Aerosil®; the mean particle size is 16 nm) and 40 mL of toluene were placed into a Schlenk flask. After adding 1 mL of trichlorododecylsilane, they were stirred for 2 hours. The procedure imparts hydro-phobicity to the SiO_2 nanoparticles by immobilizing dodecyltrichloro groups on their surface. After filtration, the obtained SiO_2 particles were dried at 353 K and ground to a fine powder using a mortar.

Spray-coating on paper

0.3g of SiO_2 particles were added to 15mL of methanol and ethanol, 1-propanol respectively. After stirring (1h) and sonicating (30min), the suspension was manually sprayed over paper from 20cm away using an air spray gun connected to compressors. The resulting coated samples were placed horizontally and dried at room temperature overnight.

Water-repellency properties

The CAs of water droplets on the samples were measured using a contact angle meter (Attension, KSV Instruments). After dropping the 3.6±0.2 µL of water droplets onto the samples, the average contact angle was obtained by measuring at four different positions of the sample. The surface morphologies of the coated papers were examined by Scanning electron microscopy (SEM) (JSM-7600F, JEOL). The particle size distribution was measured using Particle size analyzer (Nanotrac, Microtrac™). To identify the transparency and visibility of the nanoparticle coated-substrates, pictures of a drop of water on it were taken.

RESULTS AND DISCUSSION

Table 1 summarizes left angle, right angle and their mean value; that is, contact angle of untreated sample and coated samples.

Table I. Left, Right and Contact Angles of SiO_2 Nanoparticles Coatings on Paper Substrates.

Results(°)	Left Angle	Right Angle	Contact Angle
Untreated	88.662	87.476	88.070
1-propanol	152.246	151.016	151.597
ethanol	154.942	152.225	153.534
methanol	165.032	163.076	164.054

Untreated paper substrate are hydrophobic (CA = 80-90°). But, once a paper sample was coated by SiO_2 in solvent, the substrate shows superhydrophobicity (CA > 150°). It is shown that CAs coated in 1-propanol are lower than that in ethanol and CAs in ethanol are lower than that in methanol. Among them, the CAs of the paper substrate coated in methanol are the highest values higher than 160°.

Figure 2 shows the captured images of the water droplets places on each surfaces. Figure 2(a) presents the untreated paper wetted by a water droplet. On the other hand, with decreased the length of hydrocarbon in alcohol, the paper substrate were less wetted by a water showing more spherical shape. Shown in Figure 2(d), as soon as a droplet of water was dropped, it was bounced off.

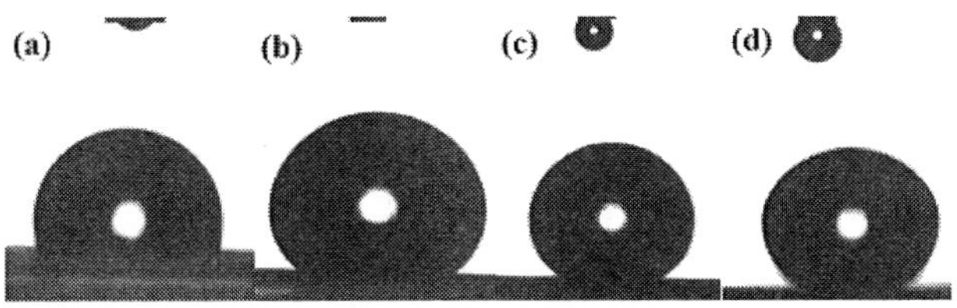

Fig. 2. Captured images of the water droplets places on each surfaces. (a) Untreated paper (b) Paper coated in 1-propanol (c) Paper coated in ethanol (d) Paper coated in methanol.

The reason why the coated sample in methanol shows higher CAs is the difference of surface roughness. As described above, the surface hydrophobicity depends on the surface energy and the surface roughness. In this study, the surface energy of the SiO_2 nanoparticles was the same; therefore, the surface roughness, which is composed of deposited SiO_2 nanoparticles, would determine the surface hydrophobicity. To discuss surface morphology, SEM images of SiO_2 nanoparticles coatings prepared using methanol, ethanol and 1-propanol were measured (Figure 3).

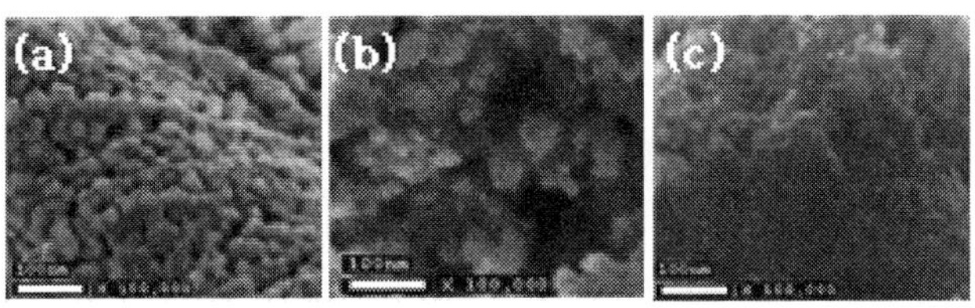

Fig. 3. SEM images of SiO_2 Nanoparticles suspensions in methanol (a) and ethanol (b), 1-propanol (c).

In high magnification images of all of suspensions, we can see the nanometer-sized SiO_2 particles. Generally, superhydrophobicity tends to be exhibited when hydrophobic surfaces have hierarchical roughness (both nanometer- and micrometer-sized roughness). Based on these SEM images, it is reasonable that superhydrophobicity was exhibited in all of them. As shown in Figure 3, the SiO_2 nanoparticles are dispersed finely in 1-propanol, and it is likely that a flat coating is formed when spraying the finely dispersed SiO_2 nanoparticles. In contrast, the SiO_2 nanoparticles in ethanol are relatively more aggregated than in 1-propanol and that in methanol are more aggregated than ethanol, causing the surface of the resulting coatings to be rougher, which is why CAs is increased with decreasing the carbon number and length of solvent.

Therefore, we assume that the morphological difference would be attributed to the difference in the aggregation states of the SiO_2 nanoparticles in the alcohols. To identify the effect of aggregation state of the nanoparticles, the particle size distribution was measured.

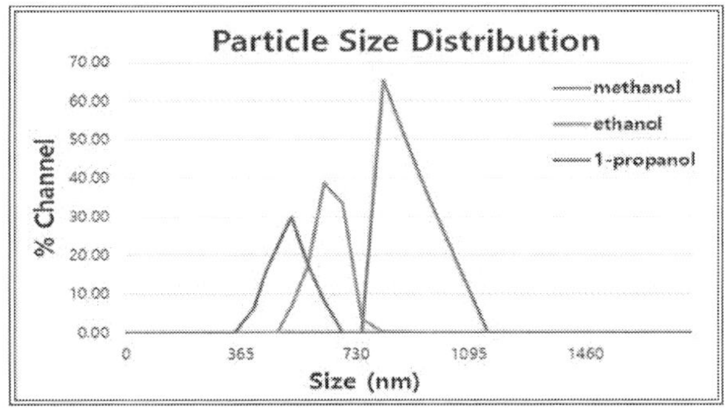

Fig. 4. Particle size distribution of suspensions containing SiO_2 nanoparticles in methanol and ethanol.

The particle size distribution data (Figure 4) suggests that the SiO_2 nanoparticles tend to form larger aggregates in lower alcohols. The average sizes of the SiO_2 nanoparticle aggregates were approximately 929 nm in methanol, 610 nm in ethanol and 530 nm in 1-propanol. The cause of these phenomena is the repulsive force between hydrophobic nanoparticles and hydrocarbon chains. It is reasonable that hydrophobic SiO_2 nanoparticles tend to aggregate more in lower alcohols (i.e., less hydrophobic alcohols) because the repulsive force between them is stronger. Therefore, the more aggregates are formed, the sedimentation rate of larger aggregates is faster.

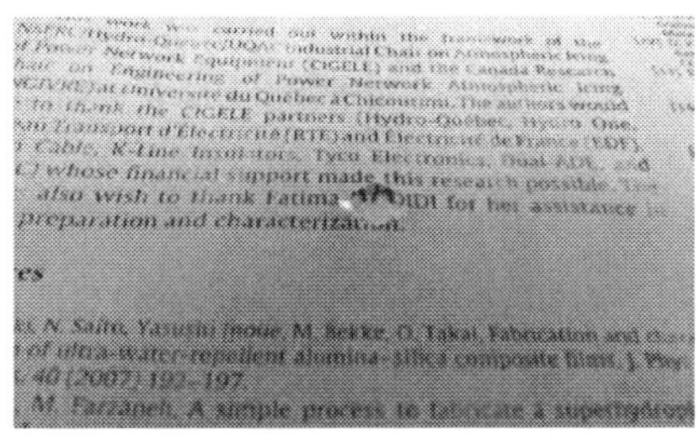

Fig. 5. Photograph the coated paper using methanol.

To identify the transparency and visibility of the nano-particle coated substrates, pictures of a drop of water on it were taken. In the coated paper sample, a water droplet on the superhydrophobic area resemble a sphere. Moreover, transparency of the coatings was confirmed and the visibility of the sample was not changed.

CONCLUSION

Superhydrophobic and transparent paper could be prepared by spraying alcohol suspension containing commercially available SiO_2 nanoparticles. However, the coated paper substrate in methanol exhibited higher CAs and water-repellency. The factor which controlled the hydrophobicity of the coatings was the aggregation state of the SiO_2 nanoparticles. The visibility of characters on the paper was maintained after coating with SiO_2 nanoparticles.

REFERENCES

1. Kota, A.K., Kwon, G., Tuteja, A. *NPG Asia Materials*, 6, 2914: e109.

2. Liu, M., Wang, S., Jiang, L. *Mrs. Bulletin*, 38, 5, 2013: 375-382.

3. Leng, B., Shao, Z., de With, G., Ming, W. *Langmuir*, 25, 4, 2009: 2456-2460.

Graphene Oxide/Polysytrene Composite Nanofibers on Quartz Crystal Microbalance Electrode for Ammonia Detection

Yongtang Jia, Hui Yu, Yumei Zhang, Lizhu Chen, Weili Nie, Fengchun Dong
The Engineering Technology Research Center for Functional Textiles in Higher Education of Guangdong Province, College of Textiles and Clothing, Wuyi University, Jiangmen, China
yongtjh@163.com

ABSTRACT

A nanostructured complex, graphene oxide (G-COOH)/polystyrene (PS) nanofiber, was fabricated as a novel sensing material coated on a quartz crystal microbalance (QCM) to realize ammonia detection in this study. Nanoporous G-COOH/PS nanofibers with an average diameter of 569 nm were fabricated via electrospinning; the fibers were composed of ultrathin nanowires with a primary diameter of 37 nm. This unique structure presented a large surface active region, making the fibers an optimal candidate for gas-sensing applications. The physical and chemical properties of the G-COOH/PS nanofibers were characterized by field emission scanning electron microscopy (FE-SEM), transmission electron microscopy (TEM), N2 physical adsorption, Fourier transform infrared spectroscopy (FT-IR), Raman spectroscopy and X-ray photoelectron spectroscopy (XPS). The results indicated that the G-COOH/PS composite nanofibers had a mesoporous structure and that G-COOH sheets were randomly dispersed in the nanofibers. A gas-sensing test showed that the G-COOH/PS nanofibers, when incorporated into a QCM sensor system, exhibited good and desirable sensing behavior, including high sensitivity, fast response and good reversibility, making them a promising candidate as an ammonia detector.

KEYWORDS

G-COOH/PS; electrospinning; nanofibrous membrane; quartz crystal microbalance; ammonia sensors

ACKNOWLEDGMENT

This work was supported by the Science Foundation for Young Teachers of Wuyi University (No.2013zk12), Foundation for distinguished young talents in higher education of Guangdong, China (No.2013LYM_0092) and the Project of Department of Education of Guangdong Province (No.2013KJCX0184).

REFERENCES

[1] Jia, Y.T., Yan, C.C., Yu, H., Chen, L.Z., Dong, F.C. "One-step fabrication of ammonia sensor by electrospinning PS-b-PMA nanofibers on quartz crystal microbalance." *Sens. Actuators*, B203, 2014: 459-464.

[2] Sauerbrey, G. "The use of quartz oscillators for weighing thin layers and for microweighing." *Z. Phys*., 155, 1959: 206-22.

[3] Ding, B., Kim, J., Miyazaki, Y., Shiratori, S. "Electrospun nanofibrous membranes coated quartz crystal microbalance as gas sensor for NH3 detection." *Sens. Actuators*, B101, 2004: 373-380.

[4] Wang, X.F., Ding, B., Yu, J.Y., Wang, M., Pan, F.K. "A highly sensitive humidity sensor based on a nanofibrous membrane coated quartz crystal

microbalance." *Nanotechnology*, 21, 2010: 1-6.
[5] Ding, B., Wang, X.F., Yu, J.Y., Wang, M. "Polyamide 6 composite nano-fiber/net functionalized by polyethyleneimine on quartz crystal microbalance for highly sensitive formaldehyde sensors." *J. Mater. Chem.*, 21, 2011: 12784-12792.
[6] Wang, X.F., Ding, B., Sun, M., Yu, J.Y., Sun, G. "Nanofibrous polyethyleneimine membranes as sensitive coatings for quartz crystal microbalance-based formaldehyde sensors." *Sens. Actuators*, B144, 2010: 11-17.
[7] Zhang, C.Y., Wang, X.F., Lin, J.Y., Ding, B., Yu, J.Y., Pan, N. "Nanoporous polystyrene fibers functionalized by polyethyleneimine for enhanced formaldehyde sensing." *Sens. Actuators*, B152, 2011: 316-323.
[8] Hua, W.L., Chen, S.Y., Liu, L.T., Ding, B., Wang, H.P. "Formaldehyde sensors based on nanofibrous polyethyleneimine/bacterial cellulose membranes coated quartz crystal microbalance." *Sens. Actuators*, B157, 2011: 554-559.
[9] Wang, X.F., Wang, J.L., Si, Y., Ding, B., Yu, J.Y., Sun, G., Luo, W.J., Zheng, G. "Nanofiber-net-binary structured membranes for highly sensitive detection of trace HCl gas." *Nanoscale*, 4, 2012: 7585-7592.

Effect of Take-Up Speed on Precursor of Polypropylene Hollow Fiber Membranes

Junfen Sun, Chuansong Xun
State Key Laboratory for Modification of Chemical Fibers and Polymer Materials, College of Material Science and Engineering, Donghua University, North People Road 2999, Shanghai, China
junfensun@dhu.edu.cn

ABSTRACT
The polypropylene (PP) with low melt flow index is applied to manufacture microporous membrane thought melt spinning, annealing, cold stretching and hot stretching method (MSCS) in this research. The effect of take-up speed on microporous membranes were characterized by using emission scanning and transmission electron microscope (SEM), X-ray diffraction (XRD) and optical microscope method. The take-up speed primarily influenced the structure and performance of hollow fiber precursor.

INTRODUCTION
Microporous membranes are commonly used in separation processes such as water and wastewater treatment. Due to the higher chemical resistance, rich resource and better mechanical properties, polypropylene (PP) is well known as the best candidates for the fabrication of microprous membranes [1].

Microporous PP hollow fiber membranes preparation can be obtained by two methods: thermally induced phase separations (TIPS) [2] and MSCS [3]. TIPS involves high consumption of water and organic solvents, which have to be disposed, so there are many advantages in using a stretching process over the other method, such as simply operation and diluent-free [4]. The first step is to prepare the precursor fibers via a melt spinning process. In this process, draw ratio controlled by take-up speed of winder is the key factor [5]. It affects the growth of lamellae of precursor. The preparation of membranes by post treatment is carried out in three main consecutive steps: annealing of the membrane to enhance the perfection of crystalline structure, cold stretching and hot stretching conducted to create voids and enlarge the pores respectively. In this work, we investigated the effect of take-up speed on PP hollow fiber precursor.

APPROACH
Materials
PP (F401) was supplied by Sinopec International Enterprise Co., Ltd. Its melt flow index and density were 2.5g/10 min and 0.9g/cm^3 respectively.

Preparation of PP hollow fiber membranes
PP particles were dried in a vacuum oven for 10h at 100 °C before melt spinning. The dried PP particles were introduced into the twin screw extruder equipped with a single-hole tube-in-orifice spinneret to fabricate hollow fibers. Nitrogen gas (N_2) was introduced into the tube. The wall thicknesses of the hollow fibers were controlled by amount of N_2. Spinning temperature was 240 °C. Melt spun hollow fiber precursors were cooled to room temperature in air, and then wound on a take-up winder. The properties of hollow fibers were investigated by using different take-up speeds of winder. Take-up speeds are 150m/min, 210m/min, 270m/min, and 330m/min respectively.

Post-treatment of the fiber precursors was conducted using four different steps which are annealing, cold stretching, hot stretching and heat setting. Annealing temperature was 130°C. The temperatures of cold stretching and hot stretching were 30°C and 100°C respectively. The drawing ratio was 2.5. Finally, the membranes were heat-set under tension at 130°C.

Geometric parameters of fiber precursors
The optical microscope was employed to investigate the influence of different take-up speeds on the geometric parameters of fiber precursors. A segment of fiber was intercepted by using blade before placed under the optical microscope. The outer diameter (Do) and inner diameter (D_i) were read from the scale of optical microscope. According equation, we can get the wall thickness (d) of the fiber.

$$d = \frac{Do - Di}{2}$$

Morphological observation
The images of hollow fiber membrane morphologies were examined by using a scanning electron microscope (Hitachi S3000N). Microscopic images of cross-sections and membranes outer surfaces were obtained under different magnifications.

X-ray diffraction (XRD)
The PP hollow fiber precursors with different take-up speeds were examined by X-ray diffraction (XRD) analysis. XRD measurements were carried out in a D2 PHASER X-ray diffractometer (Bruker, GER) using Cu Kα radiation at a scanning rate of 7°/min with an angle ranging from 0° to 60° of 2θ.

RESULTS AND DISCUSSION
Geometric parameters of fiber precursors

The hollow fiber precursor geometric parameters decrease with increasing take-up speed (Fig 1). When the N_2 rate was constant, the results showed that there was a fine linear relation between diameter and take-up speed within 150-330m/min.

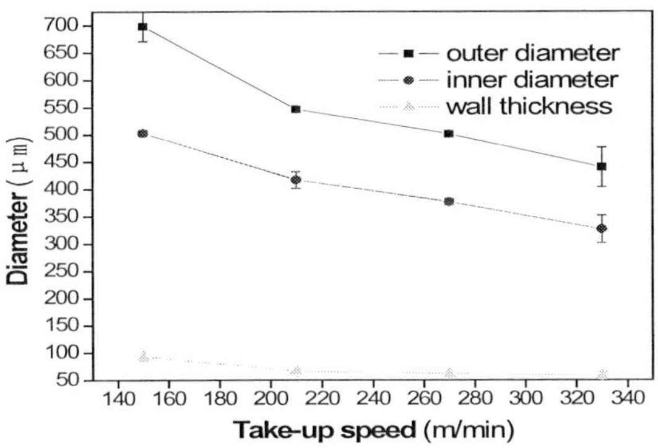

Fig. 1. Effect of take-up speed on geometric parameters of hollow fiber precursors

X-ray diffraction (XRD)

The detailed structural investigations were obtained by X-ray diffraction studies of PP hollow fiber precursors. Fig. 2 shows the X-ray patterns of hollow fiber precursors prepared with various take-up speeds. There are about four diffraction peaks corresponding to 2θ (14.0°, 16.8°, 18.6° and 21.2°), which represent α crystal form. It is seen that the diffraction angles of the different diffraction peaks remain the same for PP hollow fiber precursors prepared with take-up speeds. However their relative intensities change significantly with increasing take-up speeds. It means that high take-up speed leads to high crystallinity of PP hollow fiber precursors, as show in Table 1.

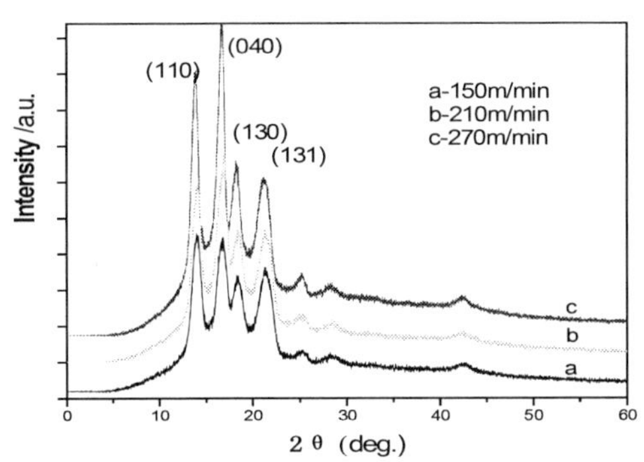

Fig. 2. XRD patters of the PPs hollow fiber precursor prepared with take-up speeds

Table 1. Crystallinity of the PP hollow fiber precursor prepared with take-up speeds

Take-up speed (m/min)	150	210	270
Crystallinity (%)	42.0	43.3	44.6

Morphological observation

Fig. 3 shows the cross section and surface of PP hollow fiber precursor prepared with take up speeds of 270m/min. As shown in Fig. 3 (a) and (b), PP hollow fiber precursor is dense and non-porous structure. It means that pores are not produced in the take up process. So post treatment is needed to promote the formation of pores in the PP hollow fiber membrane. The annealing, cold stretching and hot stretching will be conducted to create pores in the next step.

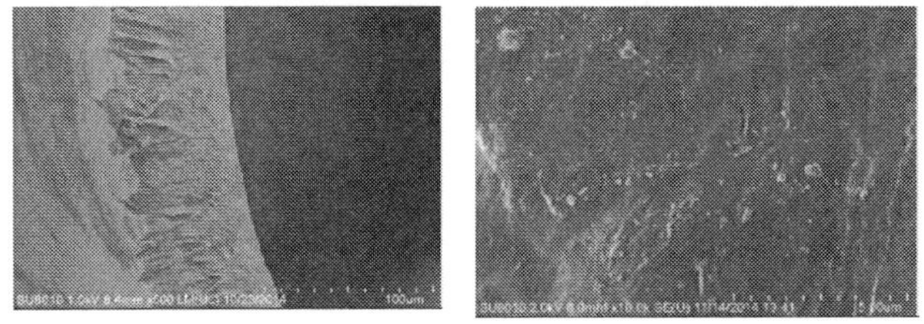

(a) (b)

Fig. 3. SEM images of PP hollow fiber precursor prepared with take up speeds of 270m/min a) cross section of precursor; b) outer surface of membrane

CONCLUSION

In this work, we investigated the structure of PP hollow fiber precusors prepared by MCSC with various take-up speeds. The hollow fiber precursor geometric parameters decrease with increasing take-up speed. High take-up speed leads to high crystallinity of PP hollow fiber precursors. In future work, we plan to study the effect of post treatment on porous PP hollow fiber membranes.

ACKNOWLEDGMENT

The authors thank National Natural Science Foundation of China (51203020); Scientific Research Starting Foundation for Returned Overseas Chinese Scholars, Ministry of Education of China.

REFERENCES

[1] Tabatabaei, S.H., Carreau, P.J., Ajji, A. "Microporous membranes obtained from polypropylene blend films by stretching." *J. Memb. Sci.*, 325,2008: 772-782.

[2] Berghmans, S., Berghmans, H., Meijer, H.E.H. "Spinning of hollow fibers via the TIPS mechanism." *J. Memb. Sci.*, 116, 1996: 171-189.

[3] Kim, J.J., Hwang, J.R., Kim, U.Y. "Operation parameters of melt spinning of polypropylene hollow fiber membranes." *J. Memb. Sci.*, 108, 1995: 25-36.

[4] Kim, J., Kim, S.S., Park, M. "Effects of precursor properties on the preparation of polyethylene hollow fiber membranes by stretching." *J. Memb. Sci.*, 318, 2008: 201-209.

[5] Han, S.W., Woo, S.M., Kim, D.J. "Effect of annealing on the morphology of porous polypropylene hollow fiber membranes." *Macromol. Res.*, 22, 2014: 618-623.

Microwave Fabrication of Hierarchical Carbon Nanotube/Carbon Fiber Nanocomposite

Yang Liu[1], Xinyu Zhang[2], John H. Xin[1]
[1]Institute of Textile and Clothing, The Hong Kong Polytechnic University
[2]Department of Polymer and Fiber Engineering, Auburn University
gerald.liu@connect.polyu.hk

INTRODUCTION

Due to its excellent mechanical strength, thermal and electrical conductivity, light weight, carbon fiber has been used extensively in the advanced engineering systems such as aerospace ships, planes, and hybrid vehicles.[1] As a result of its unique fibrous structure, carbon fiber normally serve as the reinforcing component to enhance the multi-scale properties (i.e. mechanical, thermal, electrical) of the bulk materials, such as plastics, metals, and concrete.[2] Much research effort has been devoted to enhance the reinforcing performance of the carbon fiber, where the phase separation and shearing upon stress can be reduced or even eliminated, and strong, light-weight, durable composite performance can be achieved.[3]

The growth of carbon nanotube (CNT) on the carbon fiber surface seemed to be one of the most promising solution to resolve these problems. As the nanostructured CNT forests aligned on the carbon fiber surface can provide sufficient contact surface area to enhance the adhesion between the carbon fiber and the bulk matrix.[4] Due to the distinct mechanical, thermal, and electrical properties of CNT, enhanced transverse and shear resistance can also be expected from the CNT/carbon fiber hierarchical structure, as a result of the intense interfacial interaction existed at the CNT/carbon fiber micro-nano interface.

In general, the CNT forest was grown on the carbon fiber using the bottom-up method such as chemical vapor deposition (CVD) or top-down method such as lithography.[5] However, both of these techniques requires harsh processing parameters, i.e. high vacuum, high temperature, and high temperature. And the fabrication process is usually not scalable as the size of the sample is limited and the processing time is lengthy. In order to address these obstacles and realize the practical application of the CNT/carbon fiber composite, here we proposed a novel microwave fabrication technique which can achieve a superfast aligned growth of the CNT forest over the carbon fiber surface, typically within minutes. This high quality aligned CNT/carbon fiber nanocomposite thus can have wide-range applications across the engineering fields of fiber reinforcement, supercapacitors and microelectronics.

APPROACH

pre-treatment of the carbon fiber

Before the micro-wave process, the carbon fiber mesh was re-fluxed in 6M nitric acid for 4 hours, so that more reactive sites can be created for the subsequent CNT growth.

microwave growth of CNT on carbon fibers

The pre-treated carbon fiber mesh was laid flatly at the bottom of a glass vial. Afterwards, layers of ferrocene and nickel coated graphite powder were deposited evenly on top of the carbon fiber mesh, respectively. The glass vial was then capped and put into a conventional kitchen microwave oven. The glass vial was microwaved at the maximum power in the oven and intensive reactions can be observed inside, as indicated by the sparking flames and strong gaseous carbon generation. After microwaved for 30 s, the carbon fiber mesh with the CNT forest grown was taken out and washed with copious amount of ethanol and D.I. water.

RESULTS AND DISCUSSION

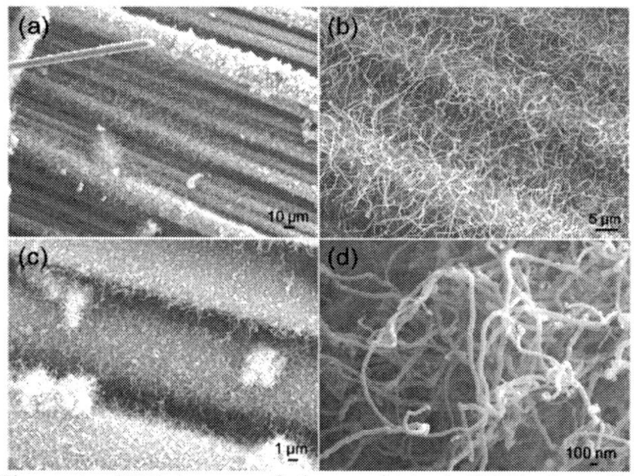

Figure 1. SEM images of the as-obtained CNT/carbon fiber nanocomposite from lower to higher magnification (a-d). The detail of the CNT forest was also shown (d).

After the microwave processing, CNT forests were grown intensively over the carbon fiber surface, covering almost entire the fiber surface with extruded, entangled nanofibrous assembling, as indicated by the SEM images (Figure 1a, b). At higher magnifications, the micro/nano interface between the CNT and carbon fiber can be clearly observed, as the small CNT nano-filaments were generally grown perpendicularly to the carbon fiber surface (Figure 1c). The surface coverage of the carbon fiber by the CNTs was high, as the growth was observed to be spanned along the fiber axis. Both long-winding and shot-rigid CNTs were existed in the forest, indicating the heterogeneous catalytic growth induced by the microwave.[6] Details of the as-obtained CNTs were also investigated, and their average diameter was calculated to be 75 nm while their length can extend to tens of microns (Figure 1d). The high aspect ratio of the CNTs thus can provide sufficient surface area to induce the enhanced interfacial interactions and elevated multi-scale functions can also be expected from the creation of the new interfaces.

It thus revealed that the microwave technique was sufficient to generate the CNT/carbon fiber micro-nano hierarchical structure in high yield and high density within a short period of time (< 30 s). This rapid CNT growth on the carbon fiber surface is probably induced by the high reaction temperature between the components upon microwaving. When receiving the microwave plasma, the nickel coated graphite would instantly combust and release large amount of heat. A chain effect, including the generation of the gaseous carbon and the decomposition of ferrocene would occur as a result of the decomposition of graphite. The iron nanoparticles generated from the ferrocene decomposition thus would serve as the catalysts for the CNT growth, and the carbon source was also provided through the decomposition of graphite, respectively. After all, the CNT forest would grow on the carbon fiber surface through this solid-vapor-liquid mechanism.

CONCLUSION

Novel carbon nanotube/carbon fiber hierarchical nano-composite with aligned micro-nano interfacial structure can be fabricated within minutes using the microwave treatment. High quality CNT forest can be obtained on the carbon fiber surface with high yield, high aspect ratio, and high coverage density. This hierarchical advanced carbon nanotube/carbon fiber nanocomposite thus can have wide-spread application potential in fillers, supercapacitors, and electronics. This high-rate, cost-effective fabrication technique also secures the industrial scale production of the advanced nanocomposite. Future work will be focused on the tuning of the CNT forest morphology to obtain the optimal material mechanical, electronic, and thermal performance, respectively.

KEYWORDS

microwave, carbon nanotube, carbon fiber

ACKNOWLEDGMENT

The financial support from Auburn University and The Hong Kong Polytechnic University was gratefully acknowledged.

REFERENCES

Chand, S., J. Mater. Sci., 35, 2000: 1303-1313.

Thostenson, E.T., Ren, Z., Chou, T. Compos. Sci. Technol., 61, 2001: 1899-1912.

(a) Delamar, M., Desarmot, G., Fagebaume, O., Hitmi, R., Pinsonc, J., Saveant, J.M. Carbon, 35, 1997: 801-807. (b) Xu, B., Wang, X., Lu, Y. Appl. Surf. Sci., 253, 2006: 2695-2701. (c) Yuan, H., Wang, C., Zhang, S., Lin, X. Appl. Surf. Sci., 259, 2012: 288-293.

(a) Thostenson, E.T., Li, W.Z., Wang, D.Z., Ren, Z.F., Chou, T.W., J. Appl. Phys., 91, 2002: 6034. (b) Samsur, R., Rangari, V.K., Jeelani, S., Zhang, L., Cheng, Z.Y. J. Appl. Phys., 113, 2013: 214903. (c) Yu, B., Jiang, Z., Tang, X.Z., Yue, C.Y., Yang, J. Compos. Sci. Technol., 99, 2014: 131-140.

(a) Zhao, Z.G., Ci, L.J., Cheng, H.M., Bai, J.B. Carbon, 43, 2005: 651-673. (b) Chen, H., Roy, A., Baek, J.B., L. Zhu, J. Qu, L. Dai, Mater. Sci. Eng. R., 70, 2010: 63-91.

Liu, Z., Wang, J., Kushvaha, V., Poyraz, S., Tippur, H., Park, S., Kim, M., Liu, Y., Bar, J., Chen, H., Zhang, X. Chem. Commun., 47, 2011: 9912-9914.

Heat and Moisture Transfer Properties of Breathable, Waterproof Fabrics Under High and Low Environmental Temperatures

Eunkyung Oh, Eunae Kim
Functional Textile System Research Lab, Department of Clothing & Textiles, Yonsei University, Korea
buti5645@naver.com; eakim@yonsei.ac.kr

INTRODUCTION

Competitiveness of breathable waterproof fabrics are determined by the two conflicting properties; breathability to remove sweat and resistance to the penetration of rain or snow. Advanced technologies have solved this conflicting problem by selecting appropriate membranes and their pore sizes. The breathability of the fabrics is often tested at the standard condition or at high temperature where the molecular movement of water is very active. But the actual condition could be much different from lab test conditions and mostly worn at low temperature or cold conditions. Therefore, molecular movement of water molecule is not active and condensation or freezing on the surface of the membrane can deter the breathability. In this study, the effect of environmental temperature on the breathability is discussed in terms of pore sizes and how it affected to the microclimate temperature and humidity.

EXPERIMENTAL

Four breathable waterproof fabrics are selected and their characteristics are shown in Table Ⅰ.

Table Ⅰ. Characteristics of the specimens

Specimen	Description	Weight (g/m^2)	Thickness (mm)
A	PU coating (dry)	140	0.22
B	PU coating (wet)	131	0.23
C	hydrophilic non-porous lamination	79	0.15
D	PU lamination	150	0.22

Water vapor transmission rate

The water vapor transmission rate was tested by vapor cup method in an environmental chamber. The air temperature was 35±0.3℃, 60±2.5%RH and 5±0.3℃, 60±2.5%RH.

$$\text{WVTR } (g/m^2\cdot 24hr) = (A_0 - A_1)/S \quad (1)$$

Where, WVTR = rate of water vapor transmission
A_0 = weight before vaporization (g)
A_1 = weight after vaporization (g)
S= test area (m^2)

Pore size and distribution

Capillary flow porometer (1100-AEHXL Porous Meida Inc., Ithaca, NY) was used to measure the pore size and distribution. Galwick with a defined surface tension of 15.9 dynes/cm (Porous Media Inc., Ithaca, NY) was used as the wetting agent.

Heat and moisture transfer properties

The microclimate within the layer was measured by the Human-Clothing-Environment simulator (U.S. Patent No. 7,680,638.). For high temperature, the chamber was set at 27±0.5℃, 60±5%RH to simulate hot summer season. For low temperature, the chamber was set at -10±0.5℃ to simulate cold winter season. The hot plate temperature was set at 35±0.5℃ to simulate the skin temperature of the human body in comfortable conditions. In order to simulate the sweat pulse triggered by body activity, 140μl of distilled water was supplied by the automatic dispenser (ISM 937, Ismatec®, Switzerland) on the surface of the plate.

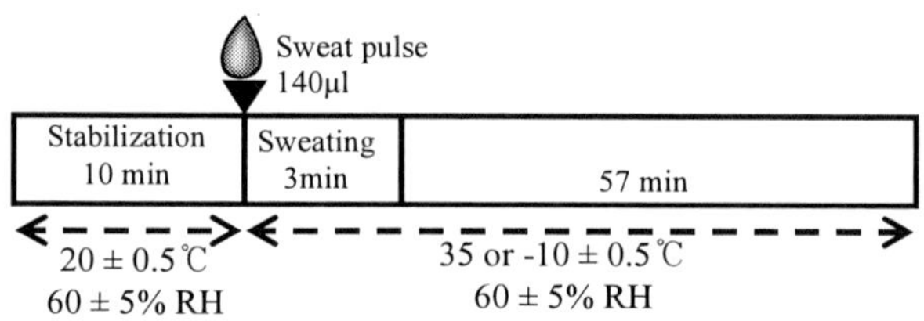

Figure 1. Test protocol

RESULTS AND DISCUSSION

Water vapor transmission rate

Table Ⅱ shows the mean pore size and WVTR at two different temperatures. At high temperature, the WVTR is proportional to the pore size except the non-porous hydrophilic membrane. At low temperature, specimen A and D showed more decrease in WVTR than specimen C of no porosity. The WVTR at cold condition was 29%, 42%, 36% and 33% of hot conditions, respectively.

Table Ⅱ. Pore size and water vapor transmission rate

Specimen	Mean Pore diameter (μ)	Water vapor transmission ($g/m^2\cdot 24hr$)	
		Hot condition	Cold condition
A	4.0192	1015	299
B	7.0623	1123	472
C	-	1051	382
D	0.0548	878	287

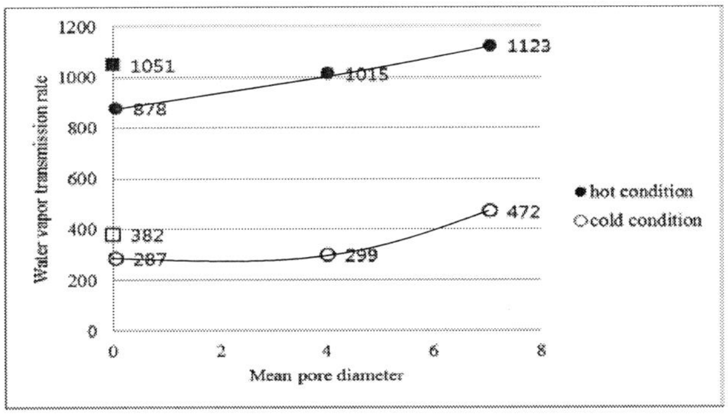

Figure 2. Correlation between mean pore diameter and water vapor transmission rate at hot condition and cold condition

Heat and moisture transfer properties

Figure 3 and Figure 4 show the results of microclimate temperature at hot and cold condition. At hot condition, there was not much difference in the microclimate temperature but at cold condition, there was a decrease in temperature in the order of A>B>D>C. This is attributed to the difference in the pore size. Even though the thickness is almost the same as there are the differences in the pore size, the thermal insulation is decreased.

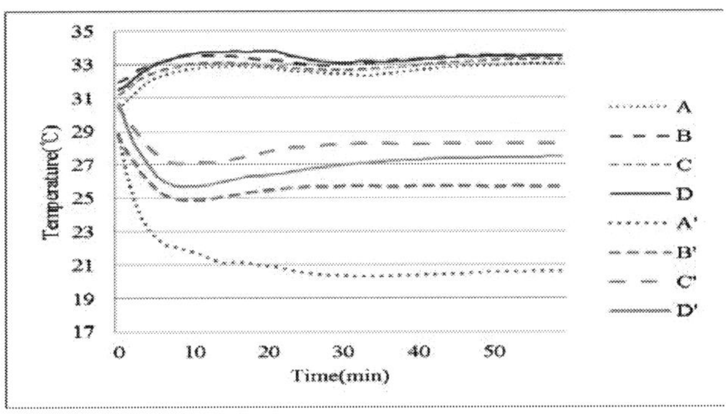

Figure 3. Microlimate temperature at hot condition (A, B, C, D) and cold condition (A', B', C', D')

Figure 4 shows the microclimate humidity at hot and cold condition. At high temperature, the relative humidity was almost same until 20 minutes for each specimen but at low temperature it was in the order of A>D>C>B. It is not just a matter of pore size of the membranes and we need further study to analyze the various factors of porosity, mean pore size, pore size distribution thickness, and WVTR at low and high temperatures. But the micro climate vapor pressure is the same order to the WVTR of the specimens in the cold condition.

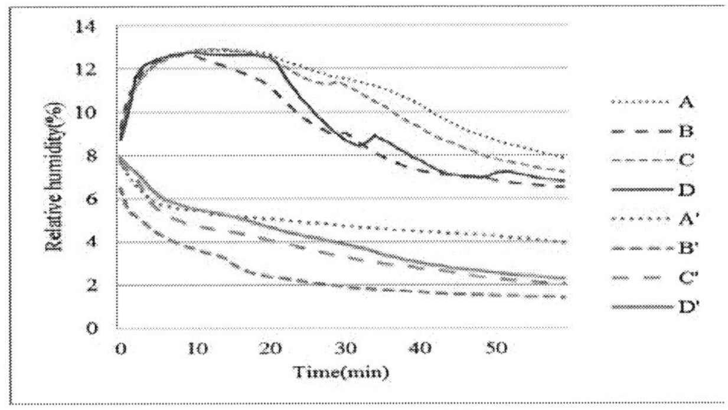

Figure 4. Vapor pressure at hot condition (A, B, C, D) and cold condition (A', B', C', D')

CONCLUSIONS

WVTR is affected by the environmental condition; at cold temperature the WVTR shows different quantity from the hot condition. To select the water breathable fabrics for different seasons, we need to compare the WVTR tested at the seasons they are going to be used instead the test results obtained at the standard condition. We need further study how this will affect to the comfort sensation on human body.

KEYWORDS

Breathable waterproof fabrics, Water vapor transmission rate, Mean pore diameter, microclimate temperature and humidity

ACKNOWLEDGMENT

This work was supported by the National Research Foundation of Korea (NRF) grant funded by the Korea government (MEST) (No. NRF-2012R1A2A2A03047265).

REFERENCES

[1] He, J., Li, J., Kim, E. "Assessment of the heat and moisture transfer in a multilayer protective fabric system under various ambient conditions." *Textile Research Journal*, 85 ,3, 2015: 227-237.

[2] Celcar, D., Meinander, H., Geršak, J. "Heat and moisture transmission properties of clothing systems evaluated by using a sweating thermal manikin under different environmental conditions." *International Journal of Clothing Science and Technology*, 20, 4, 2008: 240-252.

[3] Celcar, D., Meinander, H., Geršak, J. "Heat and moisture transmission properties of clothing systems evaluated by using a sweating thermal manikin under different environmental conditions." *International Journal of Clothing Science and Technology*, 20, 4, 2008: 240-252.

[4] Kim, Eunae, Shinjung Yoo. *Human-clothing-environment simulator.* U.S. Patent No. 7,680,638. 16 March 2010.

Structural and Dyeing Properties of Pigment Dyeing Modified with $NH_4H_2PO_4$

Shan Jiang[1], Heng Pan[1], Dan Sheng[1], Ruixue Li[1], Weilin Xu[2], Genyang Cao[2]
[1]School of Textile Science and Engineering, Wuhan Textile University, Wuhan, China
[2]State Key Laboratory of New Textile Materials and Advanced Technology, Wuhan Textile University, Wuhan, China

992069640@qq.com; 249747014@qq.com)

ABSTRACT
In order to overcome the shortcomings of poor rub fastness and being not easily dyed deep of conventional pigment dyeing, focusing on the nomex/nylon blended fabric, this paper has proposed a new approach of adding ammonium dihydrogen phosphate at the time of preparing the working dye liquor. The effect of different appending proportion on the dyeing results have been discussed. By the test of dyeing properties, it is found that the rub fastness and K/S value is much better than the original. In addition, the dyed fabric were subsequently characterized by differential scanning calorimeter (DSC), Fourier transformed-infrared spectroscopy (FT-IR), X-ray diffraction (XRD), and Transmission electron microscopy (TEM) to understand the change of the property, the crystalinity, and the particle shape of the pigment on the fabric. Thus, the experimental results obtained from the new method are analyzed microscopically. This project has provided some help to further expand the applications of pigment dyeing.

KEYWORDS
Pigment dyeing; NH4H2PO4; rub fastness; DSC; XRD

Influence of Acetone Addition on the Formation of Polyaniline Nanofibers in Water/Chloroform System

Jianyun Zhao[1], Zongyi Qin[1,2], Tao Li[1], Na Liu[2], Zhe Zhou[2], Meifang Zhu[1,2]
[1]State Key Laboratory for Modification of Chemical Fibers and Polymer Materials, Donghua University, Shanghai, China
[2]College of Material Science and Engineering, Donghua University, Shanghai, China
zjy18702928255@163.com; phqin@dhu.edu.cn

INTRODUCTION

Owing to its facile preparation, low cost, reversible acid/base doping/dedoping chemistry, and good environmental stability, polyaniline (PANI) as one of the most useful conducting polymers has been widely used in many applications [1, 2]. Generally, PANI nanofibers could be synthesized with the assistance of hard and soft templates in the early research. However, due to the limitation of templates on the product, the tedious post-treatment as well as the aggregation after the removal of templates, interfacial polymerization without the templates have been widely used to prepare PANI nanofibers. However, PANI nanofibers with amorphous structure were always prepared by conventional interfacial polymerization, which exhibited relatively weak electronic conductivity and electrochemistry properties. Recently, Nurage et al. [3] found that the extended reaction time at the interface, as well as lower concentration of monomer and oxidant were expected to produce highly crystalline products. In general, acetone was always used during the post-treatment of PANI. Therefore, an amount of acetone into water/chloroform system could affect the formation of PANI nanofibers, which is expected to generate products with high crystallinity.

APPROACH

Materials. Aniline, ammonium peroxydisulfate (APS), hydrochloric acid (HCl), chloroform ($CHCl_3$), and acetone were obtained from Guoyao Group Chemical Reagent Co. Ltd., and used as received without any further purification. The deionized water was used for all experiments.

Preparation. Aniline (0.1 M) was dissolved in 10 mL $CHCl_3$, and APS (0.025 M) in 10 mL aqueous solution of 1 M HCl with 5% acetone addition (b). For comparison, conventional interfacial polymerization without any addition of acetone (a) was prepared under the same conditions. After being mixed and ultrasonic treated for 10 min, uniform solutions were formed. Then the water phase was carefully poured into the organic solution alongside the wall to avoid any disturbance. After 24 h, the products were washed with deionized water and finally dried at 60 °C for 24 h under vacuum.

RESULTS AND DISCUSSION

The interfacial polymerization process with time was shown in Fig. 1. As can be seen, a clear products filled the interface for biphasic system without acetone addition after reaction of 5 min, and the migration of PANI into the water phase could be observed at 10 min. While in the system with acetone addition, PANI filled the boundary in 10 minutes. After 20 min, the aqueous phase was full of products whereas no apparent diffusion occurred with the acetone addition. From the point of view of formation mechanism, due to the hydrophilic property of PANI, initial products generated at the interface can diffuse into water phase. The aqueous phase was filled with products whether there are any addition of acetone in water/chloroform system after the reaction time of 24 h. Obviously, the reaction in the system with acetone addition is slower than that without any additions. It can be ascribed to the existence of acetone, which may effectively suppress the effective reaction in interfacial polymerization. In short, the system with acetone addition can slow down the reaction rate, which reversely increases the reaction time.

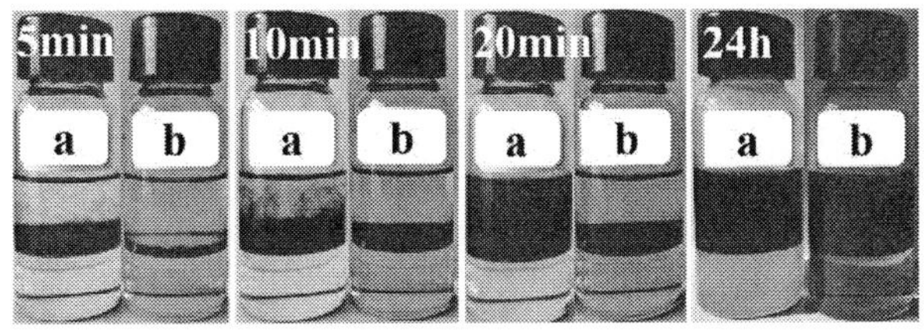

Fig. 1. Images of PANI nanofibers prepared by interfacial polymerization in water/chloroform system without (a) and with acetone addition (b).

During the interfacial polymerization process, the lack of aniline monomers in the water phase effectively restricted the further growth of initial PANI nanofibers. Therefore, the final products should be the nanofibrous morphology. Fig. 2 shows SEM images of PANI nanofibers synthesized by interfacial polymerization in water/chloroform system without and with acetone addition, respectively. Obviously, the morphology of products obtained in (b) was more uniform. With acetone addition, the diameter of PANI nanofibers is 80-120 nm and the length is 400-600 nm. The diameter of PANI nanofibers increased to 100-120 nm while its length decreased to 300-500 nm without acetone addition. In a word, PANI nanofibers with larger ratio of length to diameter and less entanglement can be produced in the water/chloroform system with acetone addition, which may be benefitted from the extended reaction time at interface. In conclusion, the acetone addition had an important effect on the morphology of PANI nanofibers.

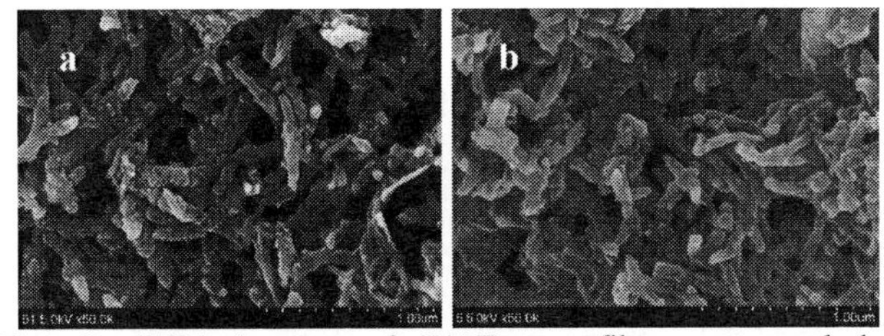

Fig. 2. SEM images of PANI nanofibers prepared by interfacial polymerization in water/chloroform system without (a) and with acetone addition (b).

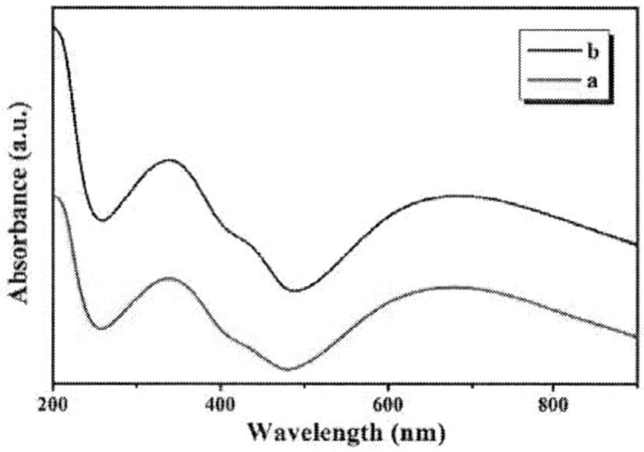

Fig. 3. UV-Vis spectra of PANI nanofibers prepared by interfacial polymerization in water/chloroform system without (a) and with acetone addition (b).

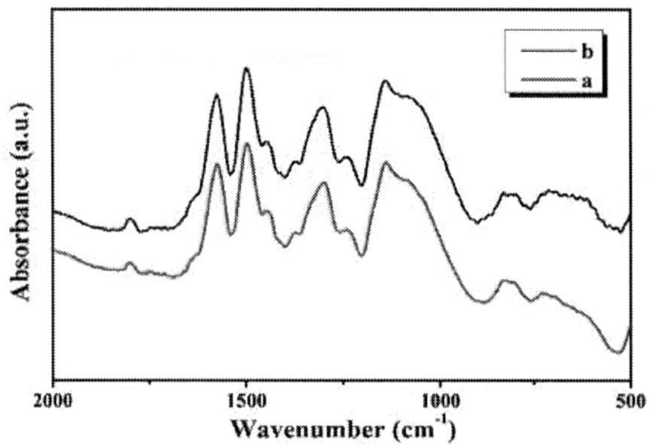

Fig. 4. FT-IR spectra of PANI nanofibers prepared by interfacial polymerization in water/chloroform system without (a) and with acetone addition (b).

The chemical structure of PANI nanofibers were studied by UV-Vis spectra and shown in Fig. 3. Clearly, the two samples exhibited similar characteristic bands. For PANI nanofibers prepared in conventional system, there occur three characteristic bands, including the bands around 338 nm, 425 nm and 667 nm, which can be assigned to π-π^*, polaron-π^* and π-polaron transition, respectively [4]. Because of the restriction of organic solvent on the doping process of polymers. A slight blue-shift at the π-π^* band appears in the spectrum of PANI obtained with acetone addition.

Fig. 4 gives the FT-IR spectra of PANI nanofibers prepared by interfacial polymerization. The characteristic band at around 1299 cm^{-1} can be attributed to C-N stretching vibration mode. The bands at 821 and 1133 cm^{-1} can be assigned to out-of-plane and in-plane bending vibration of C-H on the 1,4-desubstituted ring, respectively. Furthermore, the bands at 1560 and 1490 cm^{-1} can be ascribed to the C=C stretching vibration of quinoid and benzenoid rings, respectively [4, 5]. Generally, the intensity ratio of 1560 and 1490 cm^{-1} represents the relative content of quinone and benzene structures in polymers. The intensity ratio of PANI nanofibers obtained in the system without and with acetone addition is 0.79 and 0.78, respectively. In summary, the acetone addition in the interfacial polymerization has little influence on the chemical structure of PANI nanofibers.

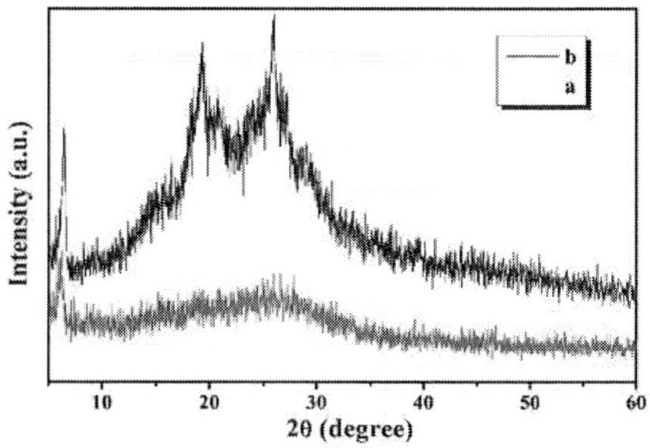

Fig. 5. XRD patterns for PANI nanofibers prepared by interfacial polymerization in water/chloroform system without (a) and with acetone addition (b).

To certify the higher crystallinity of PANI nanofibers obtained with acetone addition than that without acetone addition, XRD patterns were employed and shown in Fig. 5. For the sample synthesized in water/chloroform system with acetone, there appear three diffraction peaks centered at 6°, 19° and 26°, whereas only one broad peak can be observed at higher degree for that without acetone addition. The amorphous structure of PANI prepared by interfacial polymerization without any acetone is similar with that by conventional chemical oxidation reaction. However, the system with acetone addition can produce PANI with higher crystallinity. Therefore, it may be an effective route to improve the crystalline property of polymers.

CONCLUSION

Thin and long PANI nanofibers with less entanglement and high crystalline property were synthesized through interfacial polymerization with acetone addition in aqueous phase. Moreover, its effect on the chemical structure and crystalline property of PANI nanofibers were studied. It is found that the system with acetone addition is suitable to prepare relatively high crystalline polymers.

ACKNOWLEDGMENT

The authors gratefully acknowledge financial supports from the National Natural Science Foundation of China (21274019), and Program for Changjiang Scholars and Innovative Research Team in University (IR1221).

REFERENCES

[1} Ramya, R., et al. *Electrochimi. Acta* 101, 2013: 109-129.
[2] Ciric-Marjanovi, G. *Synth. Met.*, 177, 2013: 1-47.
[3] Nuraje, N. et al. *ACS Nano.*, 3, 2008: 502-506.
[4] Chutia, P. et al. *Physica B*, 436, 2014 : 200-207.
[5] Zhang, Z., et al. *Mater. Chem. Phys.*, 115, 2009: 275-279.

Wet Spinning of *Bletilla Striata* Polysaccharide/Polyethylene Glycol Fibers

Qian Zhang, Na Wang, Ruiqin Hu, Weilin Xu, Hongjun Yang
Key Laboratory of Green Processing and Functional New Textile Materials of Ministry of Education, Wuhan Textile University, Wuhan, China
691950977@qq.com; h_j.yang@yahoo.com

ABSTRACT

Polyethylene glycol (PEG) was used to meet the demand for weaving of polysaccharide fibers isolated from the traditional Chinese herb, *Bletilla striata* through wet spinning at room temperature using ethanol as a coagulant. The hybrid fibers were subjected to detailed analysis by scanning electron microscopy (SEM), micro-Fourier transform infrared spectroscopy (Micro-FTIR). FTIR demonstrated that the concentration of the hydroxyl groups of the hybrid fibers was added with the increasing PEG concentration. SEM results investigated that the morphology of the hybrid fibers were affected by the addition of PEG. Increasing PEG concentration caused a loosening of this structure and pore structure on the cross-section of hybrid fibers due to the different solidification conditions of BSP and PEG in ethanol. The excellent biocompatibility and mechanical properties of these hybrid fibers makes them ideally suited for use as potential wound dressings.

KEYWORDS

Bletilla striata; Polysaccharide glycol; Wet-spinning; Fiber.

Dyeing of Meta-Aramid/Nylon Blended Fabric

Dan Sheng[1], Yunli Wang[2,3], Xuefeng Lu[1], Shan Jiang[1], Ji Cai[3], Meilin Wu[3], Weilin Xu[2], Genyang Cao[2]

[1]School of Textile Science and Engineering, Wuhan Textile University, Wuhan, China
[2]State Key Laboratory of New Textile Materials and Advanced Technology, Wuhan Textile University, Wuhan
[3]School of Chemistry and Chemical Engineering, Wuhan Textile University, Wuhan, China

184430904@qq.com; 249747014@qq.com

ABSTRACT

Meta-aramid fibers are extensively used for its excellent flame retardance, thermostability, and radiation resistance. However, its extremely high glass-transition temperature (Tg 270°C) and high crystallinity also bring some bad effects, the highlighted feature is that meta-aramid is hard to be deeply colored. Nylon fibers are easy to be dyed, and has relatively complete chromatography. But its range of use has be restricted for its poor heat resistance, light fastness. This paper took advantage of the complementation of two fibers' mechanical properties and dyeing properties to study the best dyeing process of Nylon and Meta-aramid blended fabric, which meet the requirements of good mechanical properties and fastness in high temperature condition. In this article, the blended fabric was separately dyed by cationic dyes and acid dyes with two bath two step process. The influences of pH, dyeing temperature and dyeing time to fibers' crystallinity, K/S, fastness and strength were respectively discussed. XRD, tensile breaking strength tester, rubbing fastness tester and some other instrument were used for sample testing. The test results as follows: The aramid dyeing process had a far greater impact than nylon's on the fabric's mechanical properties. The tensile strength and color depth were both improved when the aramid dyeing time was 50min, dyeing temperature was 110°C and pH was 4.5-5.5. It also had good fastness, which was rated at 4-5.

KEYWORDS

Meta-aramid/Nylon blended fabric; cationic dyes; acid dyes; tensile strength; K/S.

Application Research of Flexible Manpower Line in Garment Workshop of Clothing Industry

Long Wu
Apparel & Art Design College, Xi'an Polytechnic University, Xi'an, China
wulong7398@163.com

STATEMENT OF PURPOSE

Garment industry has been a pillar one in our country. In recent years, continued rising of domestic input cost, RMB appreciation, reduction of export tax rebates, and resurgence of international trade protectionism, Chinese garment industry met up with the unprecedented challenge. In this new environment and new demands, the apparel industry of our country was executed innovation from the production process, management mechanism and industrial chain fusion etc. to change the preceding extensive mode of growth. For boosting the routing selection for the fusion of informatization and industrialization and promoting technical transformation, in this paper, the reform in the garment company is analyzed and reviewed. In the hope of renovation and upgrading, it improves core competitiveness and promotes the mode of economic development in garment manufacturing enterprise.

INTRODUCTION

China put forward the guiding ideology of "the fusion of informatization and industrialization" in 2002, which is an important turning point in the informationization advance and even the entire economic development of our country. New industrialization is insisting on informatization driving industrialization, and industrialization promoting informatization. In 2009, the Ministry of Industry and Information Technology in conjunction with the relevant association organized seven industries including textile and garment etc. to broaden and deepen the evaluation work. The apparel industry is still a labor-intensive industry of our country, but in recent years rising labor costs and labor shortages conversely forced garment factory in the direction of the industry transformation and saving labor, meeting the new production mode of diversification, high quality, short cycle and smaller batch. So-called flexible manpower line is according to how much production to increase or decrease labor.

Since 2009, the company has established the industry-study-research cooperation with our university. The company invested a total of more than 50 million RMB until 2014, and continued to purchase a large number of digital intelligent advanced equipment, for example, the intelligent pattern making CAD system, automatic flexible paving machine, CIM automatic cutter, automatically open cylinder machine with infra-red positioning, decorative sewing machine with four needles and six threads, JIT (Just In Time) automatically production line etc. To apply information technology into clothing production management as soon as possible on the software side, the company has purchased the Richpeace software system and the Kingdee ERP system, set up LAN management system of garment production, laying a solid foundation to implement ERP data management. Through the fusion of informationization and industrialization, product quality and added value was enhanced and the response speed of the market was increased. Combined with the company's technical transformation and process of saving labor, the paper systematically summarizes the experience.

APPROACH

Business requirement analysis

In the process of reengineering, the first thing to do goes to fully understand the enterprise's own needs. If the informatization investment is blind, it would incline to cause a waste of resources and the failure of the fusion of informatization and industrialization. It is a long-term investment process of garment enterprises informatization along with enterprise growth. Before carrying out the fusion of informatization and industrialization, it must be clarified what business process is suitable for the informatization. If the type is the integration of brand and process, management information system (MIS) may be adopted in sales and warehouse segment, computer aided design (CAD) system for design segment, computer aided manufacturing (CAM) system for production, computer aided process planning (CAPP) system for process planning, decision support system (DSS) for leadership decision, and so on. If the company belongs to foreign trade enterprises, it only needs to responsible for the order receiving and supply chain; If it is the part of the pure processing, regardless of sales and market; If it pertains to the brand type, it is mainly responsible for product design and marketing. Before informatization is achieved, certain foresightedness is needed. Once the business processes change, it would cause the information system to become ineffective.

Route of flexible manpower line in workshop

Planning and design of products
Process planning and management
Enterprise Resource Planning
Enterprise Resource Planning
Technological updating of enterprises

Content of flexible manpower line in workshop

A. Apparel CAD to the development of 3D field

Through 3D human body measurement, 3D body modeling, 3D garment design system, 3D garment cutting and sewing, 3D fashion virtual display etc., the development and application of systems is the ability to offer strong technical support for the apparel informatization in 3D direction.

B. Apparel computer aided process planning technology

By means of input clothing style, structure and process information, the computer automatically search and ratiocinate information and knowledge to make the decision, which can guide the practical production process route and content. To implement these functions, all kinds of materials must to have, such as, national standard, seam diagram, apparel, equipment's resources repository and complete database support.

C. Apparel CAM system extend to processing segment

Either apparel CAM is appropriate for mass production on the market, or suitable for MTM (Made To Measure). Apparel flexible manufacturing system mainly includes real-time monitoring during pieces of sewing process, on-line quality inspection and repair, etc. Compared against the traditional manual cutting, it can reflect the efficiency most.

D. System design and implementation of ERP

Production management system would decide whether need to entrust outsourcing garment factory for production, according to the sales order, sales forecast, product design and development, manufacturing, production planning, material supply, production operation and control, warehousing logistics, personnel management, financial cost and other kinds of information, balancing production capacity at the same time. Depending upon the clothing production process, production plan, and material requirements, purchasing plan of clothing materials would be made afterwards.

RESULTS AND DISCUSSION

Leadership's high attention

Before choosing to implement Flexible Manpower Line application, the leaders must first understand the business of enterprises, process flow and the demand of each department. Then they put forward existing problems in apparel production to determine which goes to undergo transformation with the aid of the informatization system. This step should be endorsed by senior leadership in the first place. If the leadership don't have a resolute attitude, some problems containing wrangling, shuffle and perfunctoriness may easily arise in the process of the Flexible Manpower Line application.

Long-term consideration and reasonable planning of CIO (Chief Information Officer)

Depending on its own production processes, design plans were personalized. By scientific studies and the examination of both sides, the customized implementation was conducted to meet the needs of enterprise quick response. Eliminating obstacles between departments and collaborating the work dispersed in various departments via the information system. CIO should regulate and standardize the action with the help of effective managing regulation.

Support from CFO (Chief Finance Officer) to avoid underfunded investment

Enterprise is necessary to adjusted and improved for the products according to customer requirements. For sustaining the growing trend, the enterprise needs to constantly innovative products and expand new markets. It is a quite incorrect opinion that some clothing enterprise thought just buying an information system can realize the informatization. After informatization, it is also needed to pay for the charge of software maintenance, upgrading and service, etc. If increasing function module or reforming the original system, the amount of the needed funding may demand a lot more.

CONCLUSION

Through the investigation of the information system over these years, it greatly reduces the difficulties of the apparel industry due to labor shortages. Judging by statistics, the number of labor decreased by 40% on the original basis, whereas output increased by 22%. This allowed us to see the huge benefits yield from advanced science and technology and equipment in the production line. If enterprises want to implement information management, they must be associated with the hardware and software environment which includes staff training of information management awareness, process reengineering of enterprise business, the corresponding soft & hardware investment, etc. At the same time, the new processes and technology should be constantly enriched in the software which should be further adjusted and enhanced according to the enterprise needs in the reengineering procedure.

KEYWORDS

Apparel manufacturing; Fusion of informatization and industrialization; Informatization; Digitization

ACKNOWLEDGMENT

The work was substantially supported by a grant from the Doctoral Scientific Research Start-up Foundation of Xi 'an Polytechnic University (310-0400608).

REFERENCES

Barnard, Malcolm. *Fashion Theory: An Introduction*. Routledge, 2014.

Chuanlei, Wang. "Research on the Fusion of Informatization and Industrialization." *Information Studies: Theory & Application*, 11, 2009: 010.

Dai, James B., Lee, Neville K.S. "Economic Feasibility Analysis of Flexible Material Handling Systems: A Case Study in the Apparel Industry." *International Journal of Production Economics*, 136, 1, 2012: 28-36.

Gan, Wen. *Several Reflections on Chinese Apparel Enterprises' Improving Employees' Qualities*, 2014.

Nabi, Farhatun, Rezwan, Mahmud, Islam, Mazedul. "Improving Sewing Section Efficiency through Utilization of Worker Capacity by Time Study Technique." *International Journal of Textile Science*, 4, 1, 2015: 1-8.

Shuying, Fu, Weidong, Yu. "Quick Response System in Textile Apparel Industry." *Journal of Donghua University, Natural Science*, 1, 2001: 026.

Wang, Wendy, Hwang, Drew. "A Case Study of a Government-Sponsored Enterprise Resource Planning Project in a Chinese Apparel Company." *International Journal of Enterprise Information Systems*, 8, 4, 2012: 63-76.

Zhu, Hai Feng, Huang, Ming Lang, Zheng, Zhe. "Research on the Technology Innovation Ability of Textile and Apparel Industry." Paper Presented at the *Advanced Materials Research Conference*, 2013.

An Improved Quantitative Method for Analyzing Performance of Flame Resistant Clothing Under Flashover

Jiazhen He[1,2], Min Wang[1,2], Jun Li[1,2]
[1]Fashion Institute , Donghua University, Shanghai, China,
[2]Protective Clothing Research Center, Donghua University, Shanghai, China,
[3]Key Laboratory of Clothing Design & Technology , Donghua University, Shanghai, China
lijun@dhu.edu.cn

INTRODUCTION

It is reported that tens of thousands of firefighter injuries occur in the line of duty every year as firefighters are the first responders to suppress unwanted fires and rescue victims. The role of protective clothing and other personal protective equipment made of heat and flame resistant textiles is fundamental for the firefighter's safety. In order to develop advanced protective ensembles and equipment for emergency responders, test methods for evaluating the ability of personal protective systems are required by research institutions and companies.

Although the thermal protective performance of materials can be evaluated using either bench-scale tests[1] or instrumented manikin tests[2], the evaluation methodology of the test approaches is different. In the manikin test, the instrumented manikin will be exposed to simulated fire for a fixed period of time.[2] Existing bench-scale test, on the other hand, expose a fabric and collect test data until a second degree burn criterion is exceeded, rather than for some fixed duration.[3]

This study proposed a new index to quantitatively assess the thermal protective performance during the bench-scale fire test and the flame engulfment test. This new index can not only be used to evaluate the thermal protective performance of fabrics with fixed exposure duration, but also could be applied to the manikin test. The aims of this study were to investigate the validity of the new index and to compare the results with the measurements obtained by the existing bench-scale test and manikin test.

APPROACH

Materials and garments

Table 1 shows the specifications of the selected fabrics. Garments used in the study were made of the materials presented below, and all of the testing garments were single layer coveralls to the same size. The details of the garment style and their configurations could be found in a previous study.[4]

Table 1. Specifications of the fabric specimens

Code	Fiber content	Weight (g/m^2)	Thickness (mm)
G1	98%Nomex® IIIA/2%CF	205	0.60
G2	98%Nomex® IIIA/2%CF	261	0.65
G3	100%m-aramid	158	0.51
G4	100%m-aramid	216	0.61
G5	98% PSA/2% CF	152	0.55
G6	98%PSA/2% CF	258	0.62
G7	100% FR cotton	325	0.73

CF is the conductive fiber; PSA is polysulfonamide fiber; FR means flame resistant.

Manikin tests

The flash fire manikin test of protective coveralls was performed on a newly developed flame manikin system, which corresponds to the technical requirements specified in ASTM F1930. A total of 135 individual calorimeter sensors are distributed over the body surface of the manikin to measure the incident heat flux. The exposure time was always 4s for these single layer coveralls. The total data acquisition time was set to 60s, so that the thermal energy transferred through and from the test specimen during and after the exposure could be measured by thermal energy sensors.

Bench-scale tests

A versatile laboratory test apparatus, widely known as the thermal protective performance (TPP) tester (Custom Scientific Instrument Corporation, USA) was modified and used in this study. The duration of exposure selected was 4s, the same length of exposure time as in the manikin test. After heat exposure, samples were separated from the heat source by inserting the blocking shutter. Sensor data continued to be recorded after terminating the heat exposure until the total data acquisition time reached 60s.

Maximum attenuation factor to second-degree burn injury (MAF)

In the light of ASTM F2703 for the bench-scale test, the response of the sensor requires to be compared with the Stoll criterial to predict the onset of the second-degree burn. In order to examine the relations between the sensor response and the Stoll criterial, the ratio between them at each time step was introduced in this study, and a new index named maximum attenuation factor to second-degree burn injury (MAF) was proposed, which was calculated as the maximum ratio of the sensor response to the Stoll criterial throughout the test:

$$MAF = max\left[\frac{Q(t)}{S(t)}\right] \quad (1)$$

The index of MAF was firstly introduced in the bench-scale test, and then it was extended to the flame manikin test. In the manikin test, the MAF of each single sensor could be determined by using equation 1. The overall MAF (OMAF) of a garment in the manikin test was obtained from the area weighted average of the index over the entire covered surface:

$$OMAF = \sum_{i=1}^{n}\{max\left[\frac{Q_i(t)}{S(t)}\right]\}\cdot\frac{A_i}{A} \quad (2)$$

RESULTS AND DISCUSSION

MAF assessment in the bench-scale test

The repeatability of the method for the determination of the MAF was firstly verified in this study. Table 2 shows the standard deviation and coefficient of variation of the MAF in the bench-scale test. The standard deviations are in the range of 0.6% to 3.6% when compared to the means of each test specimen, indicating that the proposed method has a high repeatability.

Table 2. Maximum attenuation factor to second-degree burn injury (MAF), standard deviation (SD), and coefficient of variation (CV) of the MAF in the bench-scale test

Code	G1	G2	G3	G4	G5	G6	G7
MAF	1.71	1.41	1.57	1.62	1.95	1.51	1.35
SD	0.01	0.05	0.01	0.01	0.03	0.04	0.03
CV/%	0.8	3.6	0.6	0.6	1.5	2.7	2.5

To further compare the values in MAF and the results derived from the standard test method, the correlation between the MAF and the TPE value is analyzed, as presented in Figure 1. The two indices display a strongly negative linear relationship with $R^2 = 0.961$. The fabric with a higher MAF value is found to have a smaller value in the TPE rating, which indicates the decrement in the predicted thermal protection of the fabric. The detailed comparisons between the MAF and the TPE values demonstrate that thermal protective performance assessed by the proposed method agree well with the outcomes derived from the standard test, but the proposed new method could be further used to distinguish the thermal protective properties of fabrics even though the test procedure is set with fixed exposure duration.

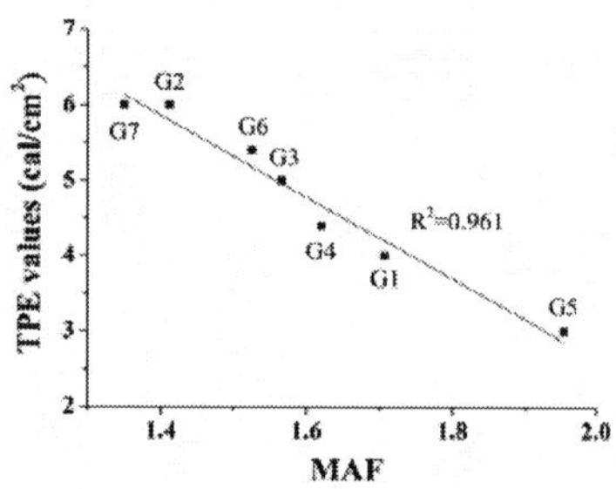

Figure 1. Relationship between the MAF and TPE values in the bench-scale test

MAF assessment in the manikin test

Similarly, the repeatability statistics of the proposed method for assessing the OMAF in the manikin test are listed in Table 3. For all the types of garments, the coefficient of variation never exceeds 4%, which indicates that the method is repeatable to be analyzed in the manikin test.

Table 3. Overall maximum attenuation factor to second degree burn injury (OMAF), standard deviation (SD), and coefficient of variation (CV) in the manikin test

Code	G1	G2	G3	G4	G5	G6	G7
OMAF	1.30	0.90	1.21	1.35	1.57	0.99	0.66
SD	0.05	0.01	0.03	0.01	0.02	0.03	0.02
CV/%	4.0	0.6	2.7	0.7	1.2	3.3	2.3

The total percentage of burn injury can be obtained from the existing manikin evaluation method. The relationship between the total percentage of burn injury and the OMAF is presented in Figure 2. The total percentage of burn injury examined here includes the percentage of second and third degree burn. A strongly positive linear relationship with R^2=0.905 is predicted between the two indices. The overall trend demonstrates that the predicted OMAF in the manikin test goes up with the increasing of the total percentage of burn injury. That is to say, the predicted thermal protective performance indicated by the OMAF is reduced as the OMAF value goes higher.

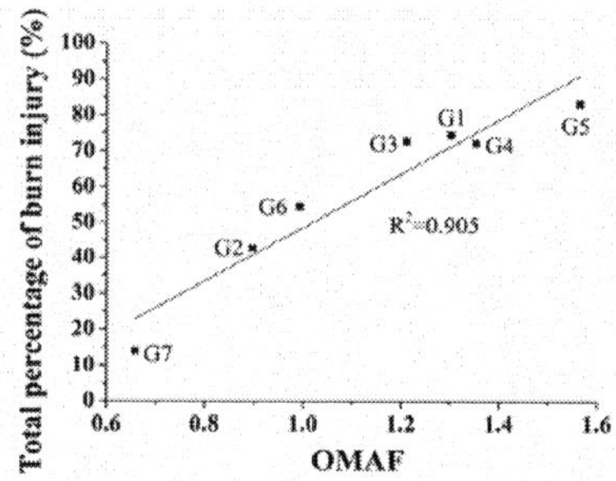

Figure 2. Relationship between values in the OMAF and the total percentage of burn injury in the manikin test

CONCLUSION

A new method to assess the performance of protective clothing upon thermal exposure was proposed in this study. This method can be used during both the bench-scale test and the manikin test. Statistical analysis of the data obtained from this study show that the proposed method has a high repeatability. In the bench-scale test, utilizing the proposed maximum attenuation factor to second-degree burn injury (MAF), the thermal protective performance of the fabric can be rated even with fixed duration of exposure. The result obtained in this study shows that the overall maximum attenuation factor (OMAF) of the garment correlates well with the total percentage of the burn injury. By using the MAF method, results derived from the bench-scale test can be further compared with the results from the manikin test as the test procedure and the evaluation index are unified.

REFERENCES

1. *Clothing for Protection Against Heat and Flame-Determination of Heat Transmission on Exposure to Both Flame and Radiant Heat.* ISO 17492:2003.
2. *Standard Test Method for Evaluation of Flame Resistant Clothing for Protection Against Fire Simulations Using an Instrumented Manikin.* ASTM F1930-13.
3. Torvi, D.A., Dale, J.D., Ackerman, M.Y., Crown, E.M. "A Study of New and Existing Bench Top Tests for Evaluating Fabrics for Flash Fire Protective Clothing." *Performance of Protective Clothing*, 6th Volume, ASTM STP 1273. West Conshohocken, PA: American Society for Testing and Materials, 1997: 108-125.
4. Li, X.H., Lu, Y.H., Zhai, L.N., et al. "Analyzing Thermal Shrinkage of Fire-Protective Clothing Exposed to Flash Fire." *Fire Technol.*, 51. 2015: 195-211.
5. *Standard Test Method for Unsteady-State Heat Transfer Evaluation of Flame Resistant Materials for Clothing with Burn Injury Prediction.* ASMT F2703-08.

Release Profile of Phytoncide and Antibacterial Properties of Core/Sheath Structured Phytoncide Oil/PVA Nanofibers Prepared via Emulsion Electrospinning

Jiyoung Shin, Seungsin Lee
Department of Clothing and Textiles, Yonsei University, Seoul, Korea
SL158@yonsei.ac.kr

INTRODUCTION

Phytoncide oil is a terpene based antimicrobial allelochemic volatile organic compound, derived from plants. It is well known for its natural characteristics of antibacterial activity, deodorization, stress relief and so on [1,2].

Emulsion electrospinning is a simple method of producing core/sheath structured nanofibers from ordinary single-nozzle electrospinning setup with emulsion as a working liquid [3]. Unlike co-axial electrospinning, which is another way of producing core/sheath structured nanofibers, emulsion electrospinning does not require a special apparatus [4,5]. Emulsion containing two immiscible liquid phases that are one oil phase and the other water phase is electrospun from an ordinary single-nozzle electrospinning setup. Emulsion phases can be elongated and converted into core/sheath structured fibers during electrospinning [6].

This study aims to propose a way of fabricating core/sheath structured phytoncide oil/PVA nanofibers via emulsion electrospinning and assess the release profile of phytoncide from the core of nanofibers. Antibacterial properties of phytoncide oil/PVA nanofibers were examined.

EXPERIMENTAL

Pure Chamaecyparis phytoncide oil was purchased from ECOMIST Co., Ltd. (Korea) and used as an oil phase. Poly(vinyl alcohol) (>99% hydrolyzed, Mw=89,000–98,000, Sigma Aldrich Co., USA) and distilled water was used as a water phase. To make homogeneous emulsions, Tween 80 (KAO, Japan) was used as a surfactant.

An oil-in-water (O/W) type emulsion, consisting of PVA solution as the water phase and phytoncide oil as the oil phase was prepared prior to electrospinning. Tween 80 was added as a surfactant to stabilize the emulsion.

Electrospinning was performed in a vertical electrospinning setup under a variety of conditions to find an optimum spinning condition. A range of electrospinning conditions was set as follows: solution feed rate 0.9-3.6 ml/h; needle gauge 23-27; voltage 21-29 kV; tip-to-collector distance 14-19.5 cm. Nanocomposite fibers were directly collected on polypropylene nonwoven substrate.

Morphology of core/sheath structured phytoncide oil/PVA nanofibers was examined using a field emission scanning electron microscope (FE-SEM) and a transmission electron microscope (TEM). Phytoncide oil/PVA nanofibers webs were heat-treated at 155 °C for 3 min to stabilize PVA structures against dissolution in water.

The release profile of phytoncide from the core of nanofibers was assessed through GC-MS analysis. Antibacterial properties of core/sheath structured phytoncide oil/PVA nanofibers were evaluated quantitatively in accordance with ASTM E 2149-01 by measuring the bacterial reductions of a Gram-positive bacterium (*Staphylococcus aureus*) and a Gram-negative bacterium (*Escherichia coli*).

RESULTS AND DISCUSSION

The SEM images of phytoncide oil/PVA nanofibers showed that at the concentrations of 13 wt% PVA, 4.64 wt% phytoncide oil, and 0.93 wt% surfactant, the emulsion yielded relatively uniform nanofibers. Nanofibers with diameters of 250-350 nm were fabricated at a feed rate of 1.8 ml/hr, a voltage of 25 kV, a tip-to-collector distance of 18 cm with a 27-gauge needle.

The nanocomposite fiber was characterized by a TEM to verify the core/sheath structure of emulsion electrospun nanofibers. TEM images illustrated a well-aligned core/sheath structure of phytoncide oil/PVA nanofibers.

The released amount of phytoncide was measured with gas chromatography after collecting the phytoncide released from the fibers. The release profile of phytoncide from the emulsion electrospun nanofiber webs showed continuous emission over 0 to 21 days through GC-MS analysis.

The effect of encapsulated phytoncide as the core of nanofibers was evaluated for its natural antibacterial property. The antibacterial activity of phytoncide/PVA nanofiber webs at 3.0 g/m^2 web area density exhibited a 99.9 % reduction rate against both *Staphylococcus aureus* and *Escherichia coli* whereas PVA nanofiber webs without phytoncide oil (control) exhibited no bacterial reduction.

CONCLUSIONS

This study demonstrated the applicability of one-step encapsulation of phytoncide oil without using an intermediate encapsulation step or an extra nozzle to fabricate core/sheath structured nanofibers. Nanofiber webs containing phytoncide oil, which is a natural antibacterial substance derived from plants, is of great potential for various applications such as medical textiles and healthcare textiles. Moreover, emulsion electrospinning can be useful in producing high-value added textile products such as functional substance releasing textile items.

ACKNOWLEDGMENT

This research was supported by Basic Science Research Program through the National Research Foundation of Korea (NRF-2013R1A1A2009730).

REFERENCES

[1] Li, Q., Otsuka, T., Kobayashi, M., Wakayama, Y., Inagaki, H., Katsumata, M., Hirata, Y., Li, Y.J., Hirata, K., Shimizu, T., Suzuki, H., Kawada, T., Kagawa, T. "Acute effects of walking in forest environments on cardiovascular and metabolic parameters." *European Journal of Applied Physiology*, 111, 11, 2011: 2845-2853.

[2] Lee, J,Y., Oh, S.J., Lee, M.S., Park, J.Y., Ryu, J.J., Lee K.H. "Preparation of phytoncide-emitting nylon/PP sheath/core fiber and the release profile of phytoncide" *Fibers and Polymers*, 13, 9, 2012: 1209-1213.

[3] Wang, C., Tong, S.N., Tse, Y.H., Wang, M. "Conventional electrospinning vs. emulsion electrospinning: a comparative study on the development of nanofibrous drug/biomolecule delivery vehicles." *Advanced Materials Research*, 410, 2012: 118-121

[4] Li, X., Su, Y., Liu, S., Tan, L., Mo, X., Ramakrishna, S. "Encapsulation of proteins in poly(l-lactide-co-caprolactone) fibers by emulsion electrospinning." *Colloids and Surfaces. B: Biointerfaces*, 75, 2, 2010: 418-424.

[5] Viry, L., Simon, E., Moulton, S.E., Romeo, T., Suhr, C., Mawad, D., Cook, M., Wallace, G.G. "Emulsion-coaxial electrospinning: designing novel architectures for sustained release of highly soluble low molecular weight drugs." *Journal of Materials Chemistry*, 22, 2012: 11347-11353.

[6] Xu, X.L., Zhuang, X.L., Chen, X.S., Wang, X.R., Yang, L.X., Jing, X.B. "Preparation of core/sheath composite nanofibers by emulsion electrospinning." *Macromolecular Rapid Communications*, 27, 19, 2006: 1637-1642.

Effect of Irradiation on PET Flame Retardant Fabric

Shifeng Zhu[1,3], Meiwu Shi[2], Lijunqu[1,3], Tianmingwei[1,3], Guoxiaoqing[1], Xixianglong[4]
[1]College of Textiles, Qingdao University, Shangdong, China
[2]Quartermaster Institute of General Logistic Department, CPLA, Beijing, China
[3]Laboratory of New Fiber Materials and Modern Textile, the Growing Base for State Key Laboratory, Qingdao University, Qingdao, Shandong, China
[4]Binzhou Technical College, Shandong, China
qduzhu@163.com

ABSTRACT

Irradiation crosslinking of PET flame-retardant fabric with the presence of trimethylolpropane triacrylate (TMPTA) was studied. TG, DSC and SEM were used to analyze the effects of irradiation crosslinking on structure and properties of PET flame retardant fabric with TMPTA. There were no melt drips of PET flame-retardant fabric after irradiation crosslinking while the flammability was weakened. The melting temperature of irradiated fabric decreased and thermal gravimetric analysis showed that the onset temperature of degradation of PA6 fibers changed a little, the amount of non-volatile residue at 800°C increased as the irradiation dosage increased, the rate of maximum mass loss decreased after irradiation. SEM photographs showed that the residue from irradiated PET fabrics is different from that of the untreated which remained the intrinsic crossed structure of PET fabric and enlarged graphs showed that there was uniformly distributed and tight honeycombs char structure.

The Preparation and Properties of PU Films with Discarded Coffee Grounds

Ruixue Li[1], Xiaolin Shen[1,], Hao Wang[1], Shan Jiang[1], Weilin Xu[2]
[1]School of Textile Science and Engineering, Wuhan Textile University, Wuhan, China
[2]State Key Laboratory of New Textile Materials and Advanced Technology, Wuhan Textile University, Wuhan, China

550730089@qq.com; xiaolin_shen527@126.com

ABSTRACT

With the development of people's living standard, the public's environmental awareness is strengthened. As the environmental pollutant, the discard coffee grounds which was useless for coffee shop, has increased the burden of environment. Herein a method of using discard coffee grounds is discovered and it endows the textiles more functions. The discard coffee grounds was dried and crushed into micron-sized coffee powders using physical means. Thus, the coffee powders could be used to prepared polyurethane (PU) films with different contents (0wt%, 1wt%, 2wt%, 3wt%, 4wt%, 5wt% and 6wt%). The anti-ultraviolet properties and thermal comfort performances of difference PU films were measured. The results show that: The ultraviolet property factor (UPF) value of the film is increased with the increase of coffee powders, and it can achieve to 174.5 when the content of coffee powders is 6wt%. The CLO value of the film increases from 0.0101 to 0.0950 while the warmth retention significantly increases from 1.83% to 17.27%. These indicate that the coffee grounds could be applied to summer clothing and sunscreen products in which the anti-ultraviolet property and thermal comfort performance are all essential and determinative. This study also provides an effective measure to reuse discard coffee grounds, which is consistent with the environmental-friendly society principle.

KEYWORDS

Coffee grounds; environmental-friendly; anti-ultraviolet; thermal comfort

Fabrication of Honey-Loaded Silk Fibroin Nanofibrous Mats by Green Electrospinning

Si Lin, Xing-Xing Yang, Yun-Yi Wang, Fan Yu, Hong-Sheng Wang
Key Laboratory of Textile Science and Technology, Ministry of Education, Donghua University, Shanghai, China
whs@dhu.edu.cn

ABSTRACT REQUIREMENTS

A lot of studies have demonstrated that honey has many beneficial effects on wound sites[1], such as providing moist environment which is good for wound healing, anti-bacterial and anti-inflammatory effect, promoteing angiogenesis, reducing scarring and pain of patients and so on. Silk fibroin (SF) nanofibrous mats has multiple unique properties, including excellent biocompatibility, low inflammatory response[2], good permeability of water and gas, and good adhesive on wound sites[3]. Hence, we hypothesized that the incorporation of honey into SF nanofibrous mats may result in good effect on wound healing.

Here, a novel honey-loaded SF nanofibrous mats was fabricated successfully via green electrospinning different content of honey.The existence of honey in the composite nanofibrous mats was confirmed by Fourier Transform Infrared (FTIR) spectra (Fig.1). The morphology of the SF composite nanofibrous mats loading different content of honey (10%, 30%, 50%, 70% w/v) was round and smooth as shown by the scanning electron microscopy (SEM). The diameter of the composite nanofibers increased with the increase of the honey content (Fig.2), which is due to the sugar in the honey increase the viscosity of the spinning solution. Our preliminary data showed that this composite nanofibrous mats has the ability to inhibit bacteria. Further study need to evaluate the beneficial effects of this composite nanofibrous mats on wound healing.

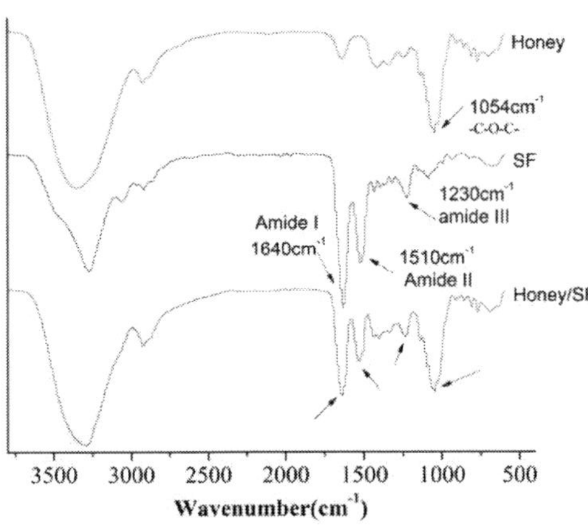

Fig. 1. FTIR spectra of the fibrous mats.

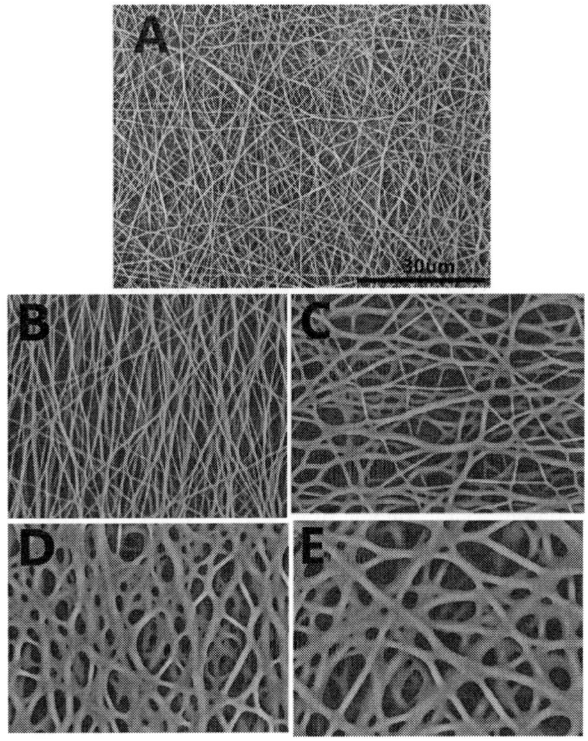

Fig. 2. SEM image of the fibrous mats. A. pure SF; B. 10% Honey/SF; C. 30% Honey/SF; D. 50% Honey/SF; E. 70% Honey/SF.

KEYWORDS
Honey; Silk Fibroin; Anti-bacterial.

REFERENCES

[1] Gethin, G.T., Cowman, S, Conroy, R M. The impact of Manuka honey dressings on the surface pH of chronic wounds" *International Wound Journal*, 5, 2: 2008: 185-94.

[2] Fan, L., Cai, Z., Zhang, K., Han, F., Li, J., He, C., Mo, X., Wang, X., Wang, H. "Green electrospun pantothenic acid/silk fibroin composite nanofibers: Fabrication, characterization and biological activity." *Colloids and Surfaces B: Biointerfaces*, 117, 2014: 14-20.

[3] Horan, RL, Antle, K., Collette, A.L., et al. "In vitro degradation of silk fibroin." *Biomaterials*, 26, 17, 2005: 3385-93.

Preparation of Alginate Gel Fibers Based on Coordination with Lanthanide

Lingang Wen, Jiefu Li, Zhiliang Wang, Lijuan Shao, Jing Nie, Shuguang Yang
State Key Laboratory for Modification of Chemical Fibers and Polymer Materials, College of Material Science and Engineering, Donghua University, Shanghai, China
shgyang@dhu.edu.cn

ABSTRACT
Alginate fibers were prepared through the coordination of sodium alginate with lanthanide ions. Alginate-La fiber has stronger tensile strength than alginate-Ca fiber. FTIR spectroscopy and conductivity measurement were used to investigate polymer-metal coordination.

KEYWORDS: polysaccharide; coordination; lanthanide ions; gel fiber.

INTRODUCTION
Alginate is a natural polysaccharide, which composed of (1→4) α-L-guluronate (G) and β-D-mannuronate (M) (Figure 1). This polysaccharide has an inherently rigid chain due to the repeated units of six-membered pyranoid rings [1]. And the mechanical properties, gelation and swelling behavior of alginate gel are strictly affected by the rigidity of the alginate chain [2, 3]. As a water-soluble polymer, alginate is an excellent gel forming material due to capable of holding a large amount of water, which is used for would dressing. The alginate can coordinate with various metal ions, such as calcium divalent cations, because of abundant hydroxyl groups and carboxyl groups. Alginates are extensively used for the biologically active materials for applications in the controlled release of drugs, food, biological catalysis, and enzyme transport in detergents [4]. Hybrid materials containing trivalent lanthanide ions have been widely used as photoluminescent materials [5,6]. In this work, we prepare alginate-lanthanide fiber, and study the coordination between Lanthanide ion and repeat units of alginate.

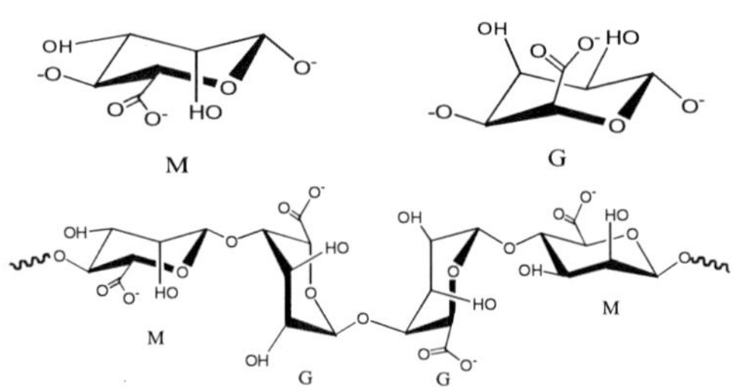

Figure 1. Schematic Presentation of Linked α-L-Guluronate (G) and β-D-Mannuronate (M) isomers

APPROACH
Alginate-lanthanide complex fibers were prepared by wet spinning. Alginate power is dissolved in water to form a homogenous spinning solution, and the coagulation bath consisted of an aqueous solution of lanthanide ions. The formation of complex fiber is due to the coordination between the carboxyl and lanthanide ions.

RESULTS AND DISCUSSION
The formation and the mechanic properties of the alginate fiber may be depend on coordination between metal ion and the structure unit of alginate. The repeated units of six-membered pyranoid rings of alginate give a sort of rigidity for alginate fiber, and lanthanide ions have a higher coordination number than calcium ion. By comparison, alginate-La complex fiber has stronger tensile strength and modulus but a lower break elongation. The tensile strength of alginate-La complex fiber increases 36% compared to alginate-Ca complex fiber.

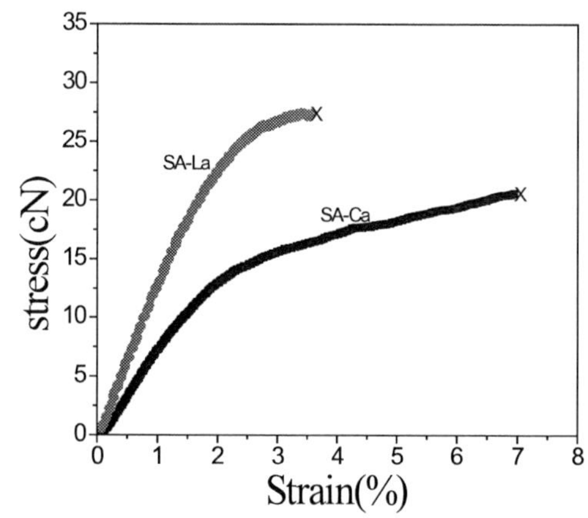

Figure 2. The stress-strain curves of alginate-La and alginate-Ca complex fibers

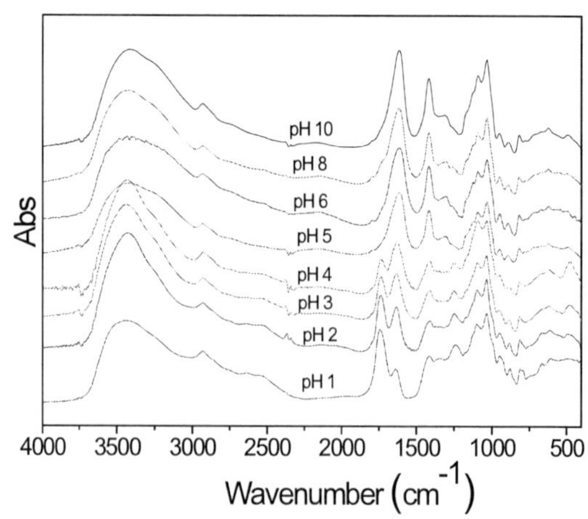

Figure 3. The FTIR spectra of alginate in various pH values. (For clarity, the spectra were intentionally overlaid with arbitrary offset.)

Alginate is a weak polyelectrolyte. It was reported that $pK_{a,guluronic} = 3.20$ and $pK_{a,mannuronic} = 3.38$ [7]. Ionization degree will affect coordination between alginate and lanthanide ion. Carboxylate show stronger interaction with lanthanide than non-ionized COOH.

The FTIR spectra of alginate in various pH values were shown in Figure 2. For the FTIR analysis of alginate, three distinct adsorption bands of the carboxylic acid functional groups of alginate were considered: the $1636cm^{-1}$ (asymmetric stretching band of COO^-),$1409cm^{-1}$ (symmetric stretching band of COO^-) and 1735 cm^{-1} (C=O stretching of COOH). The degree of ionization of alginate at a given pH can be calculated as $\alpha = [\nu(COO^-)]/[\nu(COOH) + \nu(COO^-)] \times 100\%$. FT-IR spectra indicate that above pH 5.0, alginate can be considered to be fully ionized.

Conductivity measurement was used to investigate the interaction between alginate and lanthanide ion. Increasing the concentration of $LaCl_3$ solution, the change of mixed solution conductivity is nearly constant when the mole ratio of rare earth ions with the carboxyl was 1:2 (Table 1). So we conclude that a La^{3+} coordinates with two carboxylate groups when coordinated with alginate.

Table 1. The conductivity of different mole ratios of La^{3+} with the carboxyl group.

	C ($LaCl_3$) mol/L	K_a($LaCl_3$) us/cm	K_b(SA) us/cm	nLa^{3+}:nCOO-	$K_0=K_a+K_b$ us/cm	K_1 (mixed) us/cm	$\Delta K=K_1-K_0$ us/cm
1	2.5×10^{-3}	701	507	1:4	1208	846	362
2	3.3×10^{-3}	928	507	1:3	1435	992	443
3	5.0×10^{-3}	1356	507	1:2	1863	1343	520
4	6.0×10^{-3}	1730	507	1:1.67	2237	1720	517
5	7.0×10^{-3}	1941	507	1:1.43	2448	1930	518
6	8.0×10^{-3}	2250	507	1:1.25	1757	2238	519
7	9.0×10^{-3}	2480	507	1:1.11	2987	2469	518
8	10×10^{-3}	2670	507	1:1	3177	2660	517

FTIR spectroscopy was also used to investigate the interactions of metal-carboxylate. FT-IR spectra of the alginate fibers prepared with rare earth ions are shown in Figure 4. The separation between asymmetric and symmetric carboxyl vibration peak can reflect the interaction mode of metal ion and carboxylate. The separation between asymmetric and symmetric carboxylate of the alginate-La fiber is 192 cm^{-1}, which is smaller than alginate sodium, the coordination mode is suggested as unidentate mode.

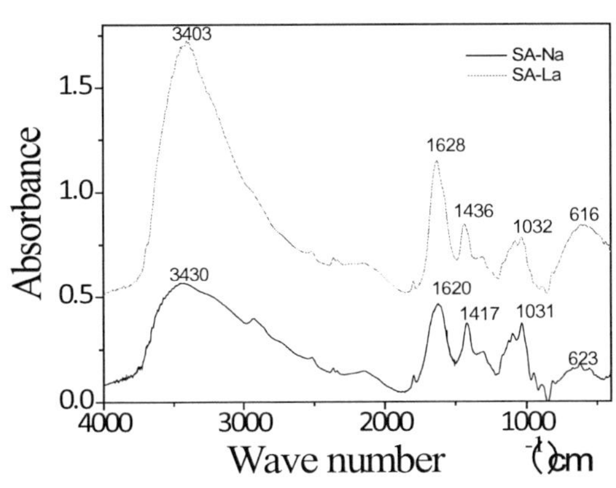

Figure 4. FT-IR spectra of sodium alginate and alginate-La fiber.

TG was used in order to obtain information about the water content of the alginate complex fiber which dried at ambient temperature. The weight of the alginate-La fiber reduced 20% when the temperature increased to 100 ℃. So there is loosely bound water in the fiber. Thus, basing on the experimental results of conductivity, FTIR and TG, the coordination structure in alginate-La fiber is suggested as shown in Figure 5.

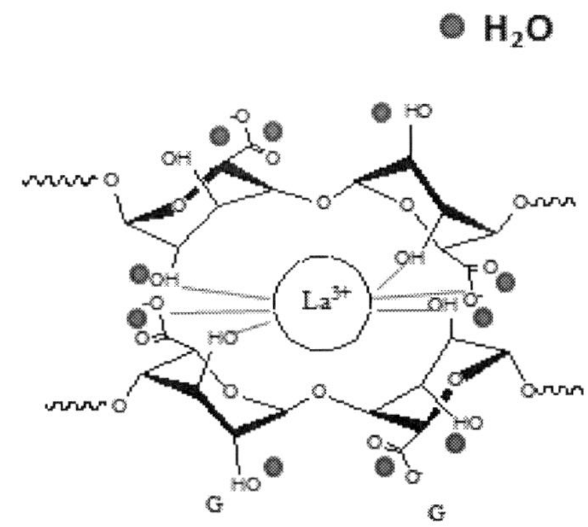

Figure 5. The coordination structure of alginate-La fiber

CONCLUSIONS

Alginate fibers were prepared through the coordination of sodium alginate with lanthanide ions. In the alginate fiber, carboxylate group and La^{3+} take unidentate coordination. One La^{3+} interacted with four guluronate units and two carboxylate groups involved coordination.

ACKNOWLEDGEMENT

The work is supported from National Natural Science Foundation of China (NSFC, Grant No. 51373032), Innovation Program of Shanghai Municipal Education Commission, Fundamental Research Funds for the Central University and DHU Distinguished Young Professor Program.

REFERENCES

[1] Lee, K.Y., Bouhadir, K.H., Mooney, D.J. *Biomacromolecules*, 3, 6, 2002: 1129-1134.
[2] Qin, Y. *Polymer International*, 57, 2, 2008: 171-180.
[3] Augst, A.D., Kong, H.J., Mooney, D.J. *Macromolecular Bioscience*, 6, 8, 2006: 623-633.
[4] Robitzer, M., David, L., Rochas, C., et al. *Langmuir*, 24, 21, 2008: 12547-12552.
[5] Liu, F., Carlos, L.D., Ferreira, R.A.S., et al. *Biomacromolecules*, 9, 7, 2008: 1945-1950.
[6] Sun, L.N., Zhang, H.J., Peng, C.Y., et al. *The Journal of Physical Chemistry B*, 110, 14, 2006: 7249-7258.
[7] Martinsen, A., Storrø, I., Skjårk-Bræk, G. *Biotechnology and Bioengineering*, 39, 2, 992: 186-194.

In situ SAXS and WAXD Characterization of Poly(ethylene terephthalate) Industrial Yarns During Tensile Deformation Above Glass Transition Temperature

Yatao Liu[1], Lixin Yin[2], Huirong Zhao[2], Guangkun Song[2], Fangming Tang[2], Lili Wang[2], Huili Shao[1], Yaopeng Zhang[1]
[1]State Key Laboratory for Modification of Chemical Fibers and Polymer Materials, College of Materials Science and Engineering, Donghua University, Shanghai, China; [2]Jiangsu Hengli Chemical Fibre Co.,Ltd., Jiangsu Province, China
zyp@dhu.edu.cn; liuyataojay@163.com

ABSTRACT

Time-resolved synchrotron radiation small-angle X-ray scattering (SAXS) and wide-angle X-ray diffraction (WAXD) measurements were performed to study structure evolution of poly(ethylene terephthalate) (PET) industrial yarns during tensile deformation at 125 °C. We found that both crystalline and amorphous phase were involved during the tensile process. WAXD data indicated that the crystallinity of the PET yarns decreased with increasing strain up to the fracture. The estimated crystallite size associated with the (010) reflection remained unchanged at strains under 15% and decreased quickly at strains from 15% to 32% (breaking point). The crystallite size associated with (-110) reflection decreased, while the crystallite size of (100) reflection nearly remained unchanged during the whole tensile process. SAXS data showed that the long period, crystalline and amorphous thickness increased with the increasing strain during tensile deformation. The enlargement of long period indicated that the lamellar structures of the yarns became loosely packed during the tensile deformation.

KEYWORDS

Poly(ethylene terephthalate) industrial yarns; WAXD; SAXS; Structure evolution

INTRODUCTION

Poly(ethylene terephthalate) (PET) is widely used in the form of plastic, film and fiber due to its excellent mechanical properties and low cost. Extensive studies of PET during tensile deformation have been carried out in the past few years [1-7]. Kawakami et al. [5] studied the lamellar structure development during uniaxial stretching of amorphous PET above glass transition temperature. They found that all structures evolution were mainly induced by strain and the load-strain curve could be categorized into three stages based on the mechanical behavior and structure characterstics. Shioya et al. [6] investigated the tensile fracture process of PET fiber at room temperature. It was found that the strain of the long period structure increased in accordance with the external fiber at small strains while it deviated from the fiber strain at lager fiber strain. Microvoids much larger than the long period structures were formed at a larger fiber strain and the number of the microviods continued to increase with increasing fiber strain up to fiber fracture.

Most of the previous studies [1-7] focused on the structure evolution of amorphous PET above the glass transition temperatures and PET yarns at room temperature. In this paper, time-resolved synchrotron radiation small-angle X-ray scattering (SAXS) and wide-angle X-ray diffraction (WAXD) measurements were performed to investigate the structure evolution of PET industrial yarns during tensile deformation at 125 °C.

METHODS

The tensile deformation were carried out in a custom-built stretching instrument which can be heated by hot air. The sample was equilibrated in the heated chamber for 2 min at 125 °C before the engagement of the tensile deformation. The stretching rate was 4 mm/min and the X-ray beam always illuminated on the same position of the specimen during tensile deformation. The load-strain curve was recorded simultaneously during the collection of SAXS and WAXD signals.

Time-resolved synchrotron radiation SAXS and WAXD measurements were carried out at BL16B1 in Shanghai Synchrotron Radiation Facility (SSRF). The energy of the X-ray was 10 keV, resulting in a X-ray wavelength of 1.24 Å. The two-dimensional SAXS and WAXD patterns were collected with a X-ray MarCCD detector with a acquisition time of 10 s and 5 s, respectively. All data were corrected for background scattering before analysis and treated with the software Fit2D. The SAXS and WAXD measurements were performed separately with different speciments under the same preparation and experimental conditions.

RESULTS AND DISCUSSION

Structure evolution by WAXD

The stress-strain curve is shown in Fig. 1(a). It is necessary to point out that the experimental yarns produced at high spinning speed, low drawing ratio and high heat-setting temperatures have a stable structure with high crystallinity. We found that the crystallinity of the PET yarns decreased with increasing strain up to the fracture during tensile deformation, as shown in Fig. 1(b). The estimated crystallite size associated with the crystalline (010) reflection remained unchanged at strains under 15% and decreased quickly at strains from 15% to 32% (breaking point), shown in Fig. 1(c). However, the crystallite size along the plane normal of the (-110) reflection decreased from the beginning to the end (Fig. 1(c)). In the case of (100) reflection, the crystallite size nearly remained unchanged during the tensile deformation (Fig. 1(c)), indicating that the (100) plane is more stable than (010) and (-110) planes. It can be inferred that under the external force, the equatorial crystalline dimensions of the PET industrial yarns decreased, which may relate to the decrease of the crystallinity during tensile deformation.

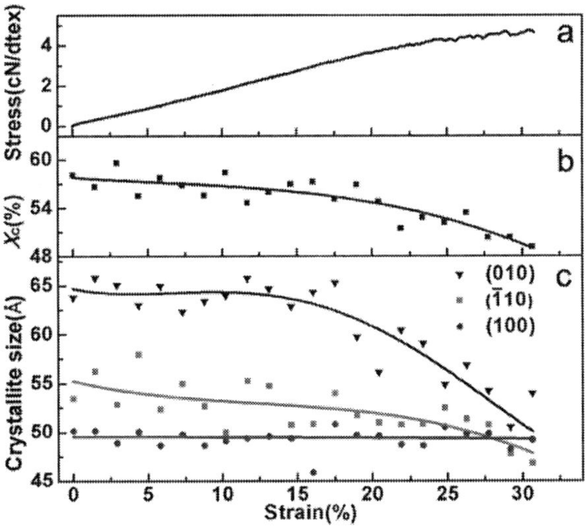

Fig. 1. (a) Stress-strain curve and the evolution of (b) crystallinity and (c) crystallite size during tensile deformation in *in-situ* WAXD characterization. X_c represents crystallinity.

Structure evolution by SAXS

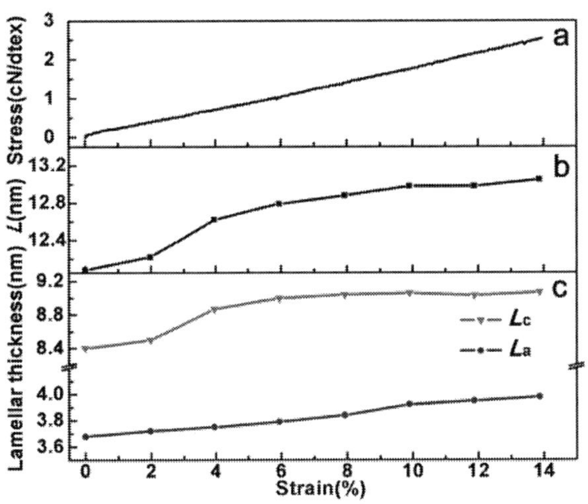

Fig. 2. (a) Stress-strain curve and the evolution of (b) long period and (c) lamellar thickness during tensile deformation in *in-situ* SAXS characterization. L is long period. L_c and L_a represent crystalline and amorphous thickness, respectively.

The one-dimensional electron intensity correlation function was adopted to analyze the detailed information of the lamellar structures, such as long period (L), crystalline thickness (L_c) and amorphous thickness (L_a) [8]. Fig. 2(a) shows the stress-strain curve, of which the strains are only from 0% to 14% due to some experimental conditions. The lamellar structure evolutions during tensile deformation are illustrated in Fig. 2(b) and Fig. 2(c). It is known that long period implies the repeat unit distances of crystalline and amorphous phase in the fiber. A small value of long period means compactly packed lamellar stacks. As shown in Fig. 2(b), the long period of the yarns increased with the increasing strain during tensile deformation. The enlargement of long period indicates that the lamellar structures of the yarns became loosely packed during the tensile deformation. Similar to the long period, crystalline thickness and amorphous thickness show a trend of rising with the strain during tensile deformation (Fig. 2(c)), which implies that both crystalline and amorphous molecules were involved during tensile deformation above the glass transition temperature (i.e. 125 °C) of PET.

CONCLUSIONS

Both crystalline and amorphous regions of PET yarns were involved during tensile deformation at 125 °C. In this procedure, the fiber structure became losely packed. The crystallinity and crystalline equatorial dimensions of PET yarns decreased from the beginning to fracture, while the long period, crystalline and amorphous thickness increased at strains from 0% to 14%.

ACKNOWLEDGMENT

This work was supported by the China Postdoctoral Science Foundation (2014M55129).

REFERENCES

1. Kawakami, D., Ran, S.F., Burger, C., Fu, B., Sics, I., Chu, B.,Hsiao, B.S. *Macromolecules*, 36, 2003: 275-9280.
2. Kawakami, D., Hsiao, B.S., Ran, S.F., Burger, C., Fu, B., Sics, I., Chu, B., Kikutani, T. *Polymer*, 45, 2004: 905-918.
3. Kawakami, D., Hsiao, B.S., Burger, C., Ran, S. F., Avila-Orta, C., Sics, I., Kikutani, T., Jacob, K. I., Chu, B. *Macromolecules*, 38, 2005: 91-103.
4. Kawakami, D., Ran, S.F., Burger, C., Avila-Orta, C., Sics, I., Chu, B., Hsiao, B.S., Kikutani, T. *Macromolecules*, 39, 2006: 2909-2920.
5. Kawakami, D., Burger, C., Ran, S., Avila-Orta, C., Sics, I., Chu, B., Chiao, S.-M., Hsiao, B.S., Kikutani, T. *Macromolecules*, 41, 2008: 2859-2867.
6. Shioya, M., Kawazoe, T., Okazaki, R., Suei, T., Sakurai, S., Yamamoto, K., Kikutani, T. *Macromolecules*, 41, 2008: 4758-4765.
7. Okada, K., Higashioji, T., Nakagawa, T., Uchida, H., Takahashi, K., Inoue, R., Nishida, K., Kanaya, T. *Polymer Journal*, 45, 2013: 50-56.
8. Yujing, T., Zhiyong, J., Yongfeng, M., Lijia, A., Enderle, H.F., Lilge, D., Roth, S.V. Gehrke, R., Rieger, *J. Polymer*, 48, 2007: 5125-5132.

Low-Temperature Preparation of Rice Husk Char/TiO_2 Composite and Its Properties

Cheng Xu, Yueming Li, Biao Wang, Jianming Jiang
State Key Laboratory for Modification of Chemical Fibers and Polymer Materials,
College of Material Science and Engineering, Donghua University, Shanghai, China
1129025@mail.dhu.edu.cn; wbiao2000@dhu.edu.cn

STATEMENT OF PURPOSE

In this paper, a simple low-temperature preparation method of rice husk char/TiO_2 composite was developed by utilizing agriculture waste. Nanosized TiO_2 particles were loaded to rice husk char by in-situ thermal treatment of impregnated rice husk. The composite combined porous rice husk char matrix with nanosized TiO_2 particles offered high specific surface area and good photocatalytic property.

INTRODUCTION

In China, rice is one of the most widely cultivated agriculture crops. As a by-product of rice, the rice husk is produced approximately 54 million tons per year, but only a small part of it is reused.[1] Rice husk char is the solid product from rice husk pyrolysis, and it can be used in various fields because its properties such as emitting far-infrared ray, releasing anion, absorbing electromagnetic wave and possessing rich minerals.[2] Rice husk char prepared by conventional pyrolysis shows low specific surface area, and usually conduct a activation process at high temperature with activating agent before use.[3-4]

Photocatalysis is an important process in wastewater treatment, air purification, heavy metal remediation, etc.[5] Among all photocatalyst, nanosized titanium dioxide is the most widely used for its low-cost, chemical stability and efficiency. There are still some problems that limit the use of nanosized titanium dioxide, as it is hard to recovery and to disperse.[6] In order to overcome these shortcomings, titanium dioxide is loaded to bamboo charcoal or graphene by dip-dry, sol-gel or spray coating method to solve these problems.[7-9]

In this study, we proposed rice husk as a precursor to synthesize rice husk char/TiO_2 composite through a low-temperature preparation method. This work was a combination of activation of rice husk char and in-situ preparation of nanosized TiO_2 particles, and would develop a different method to protect the environment by utilizing the agriculture waste in a energy-saving route.

APPROACH

Preparation method

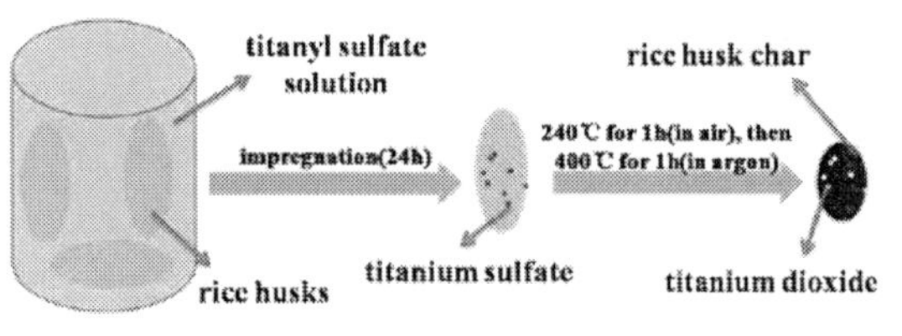

Fig. 1. Illustration of rice husk char/TiO_2 composite preparation

Rice husk was washed by 1M HCl and dried at 105°C before use. The composites were produced by using a method as showed in Figure 1. The titanyl sulfate solution concentrations were 0, 0.125, 0.25 and 0.5M, and the corresponding prepared samples were referred as C0, C1, C2 and C3, respectively.

Characterization

The microscopic features of the samples were characterized by scanning electron microscopy (SEM, Quanta-250, FEI) and field emission transmission electron microscopy (TEM, JEM-2100F, JEOL). The specific surface areas of the samples were characterized by nitrogen-adsorption measurements (Autosorb-1 MP, QUANTACHROME). The photocatalytic activity of the samples was characterized by UV-vis spectroscopy (UV-1100, MAPADA).

RESULTS AND DISCUSSION

Titanium dioxide contents of the samples

The TiO_2 content was calculated as follows:

$$Y_T = \frac{m_{C,ash} - m_{C0,ash}}{m_c}$$

Where Y_T, $m_{C,ash}$, $m_{C0,ash}$ and m_C referred to TiO_2 content(wt%), ash weight of composite (gram), ash weight of C0 sample (gram) and weight of corresponding composite (gram), respectively.

Table I. TiO_2 contents of samples

sample	C1	C2	C3
TiO_2 content (wt%)	4.6	18.5	23.6

Table I represented the TiO_2 contents of samples. As the concentration of impregnation solution increased, the TiO_2 content of the composites elevated. There was only 4.5wt% TiO_2 of C1 sample, then sharply increased by 13.9wt% of C2 as the solution concentration rise to 0.25M. When further elevating the solution concentration to 0.5M, the increase of TiO_2 content decreased to 5.1wt%.

Morphologies of the samples

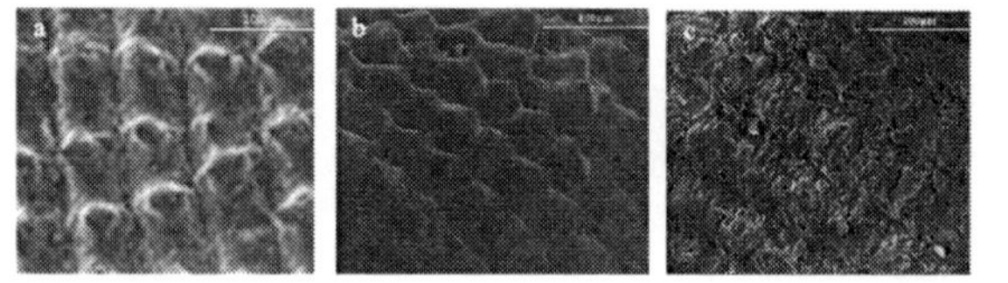

Fig. 2. SEM images of rice husk (a), C0 (b) and C2 (c) samples

The SEM photographs of rice husk, C0 and C2 samplesin Figure 2 showed huge differences. The C0 sample maintained the structure of original rice husk with smooth surface morphology. Compare to C0 sample, C2 sample observed broken fragments over the surface. This morphology changes took a positive effect to enlarge the specific surface area of the composite.

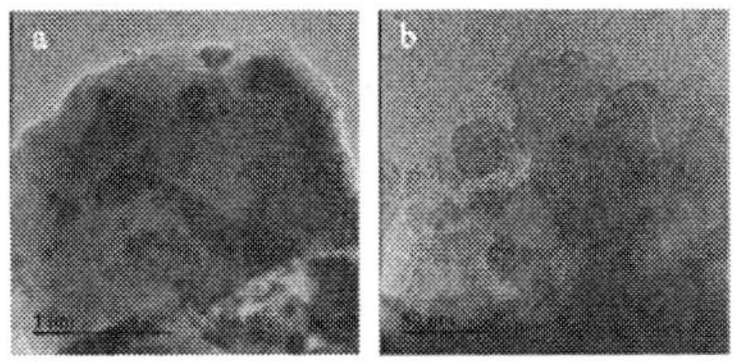

Fig. 3. TEM images of C2 sample

The TEM images in Figure 3 presented the morphology of TiO_2 particles. No significant clusters were observed in Figure 3a, indicating that TiO_2 particles could have good dispersion in the rice husk char matrix. In Figure 3b, the TiO_2 particle sizes were less than 100nm.

Specific surface areas (SSA) of the samples

Table II. Specific surface area of samples

sample	C0	C1	C2	C3
SSA(m^2/g)	258.8	380.4	365.4	343.6

Table II showed the specific surface areas of the samples. In this work, the specific surface areas of the composites were 380.4, 365.4 and 343.6 m^2/g, respectively. The specific surface areas of all three composites were much higher than the C0 sample. This might be the TiO_2 particles worked as activating agent that created more porous structure during the thermal treatment process. The results corresponded well to the SEM images of Figure 2. However, increased TiO_2 content had a negative effect on the specific surface area.

Photocatalytic activities the samples

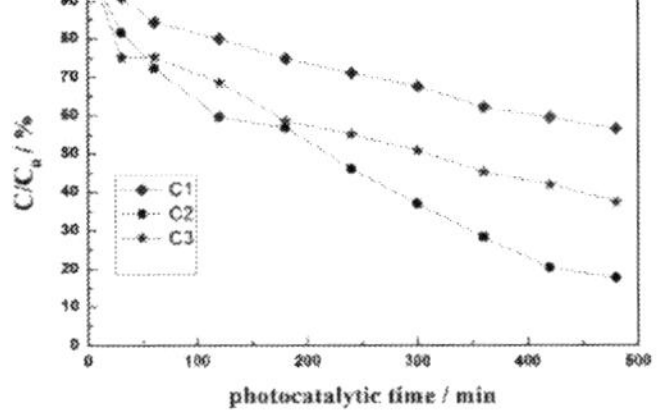

Fig. 4. Photocatalytic degradation over MB of the composites under UV light irradiation

In Figure 4, photocatalytic activity of the composites was characterized by using 10mg/L methylene blue (MB) as the model pollutant. Among all three samples, C2 presented best photocatalytic activity by decomposed over 80% MB in 8h. The sample with higher specific surface area did not possess higher activity because of its low TiO_2 content. Inversely, the C3 sample had the highest TiO_2 content, but showed a poorer photocatalytic activity than C2 sample. High TiO_2 content could promote aggregation and blocking of pores, leading to a poor photocatalytic property. Above all, the photocatalytic activity of the sample was affected by both TiO_2 content and specific surface area. In order to achieve good photocatalytic activity, the TiO_2 content and specific surface area should be carefully adjusted.

CONCLUSION

In this study, the rice husk char/TiO_2 composites were successfully prepared by a low-temperature method. The prepared C2 composite of a final temperature at 400℃ had a specific surface area of 365.4m^2/g and a good photocatalytic property that decomposed over 80% MB in 8h. The low-temperature preparation method provided an energy-saving route that might offer a great potential in environment protection or other areas.

REFERENCES

[1] Liu, Y., Yuan, X.Z., Huang, H.J., et al. "Thermochemical liquefaction of rice husk for bio-oil production in mixed solvent (ethanol-water)." *Fuel Processing Technology*, 112, 2013: 93-99.

[2] Li, Q.S., Xu, M.S., Zhou, G.J., et al. "Preparation and characterization of white bamboo charcoal PET fiber." *Chinese Chemical Letters*, 21, 2010: 995-998.

[3] Claoston, N., Samsuri, A., Ahmad Husni, M., et al. "Effects of pyrolysis temperature on the physicochemical properties of empty fruit bunch and rice husk biochars." *Waste Management & Research: the Journal of the International Solid Wastes and Public Cleansing Association*, 32, 2014: 331-339.

[4] Li, D., Ma, T., Zhang, R., et al. "Preparation of porous carbons with high low-pressure CO_2 uptake by KOH activation of rice husk char." *Fuel*, 139, 2015: 68-70.

[5] Han, T., Fan, T.X., Chow, S.K., et al. "Biogenic N-P-codoped TiO_2: Synthesis, characterization and photocatalytic properties." *Bioresource Technology*, 101, 2010: 6829-6835.

[6] Gaya, U.I., Abdullah, A.H. "Heterogeneous photocatalytic degradation of organic contaminants over titanium dioxide: A review of fundamentals, progress and problems." *Journal of Photochemistry and Photobiology C-Photochemistry Reviews*, 9, 2008: 1-12.

[7] Zhang, J., Zhao, D.S., Wang, J.L., et al. "Photocatalytic oxidation of dibenzothiophene using TiO_2/bamboo charcoal." *Journal of Materials Science*, 44, 2009: 3112-3117.

[8] Cheng, D.L., Jiang, S.X., Zhang, Q.S. "Analysis and characterization of microstructures of a TiO_2/bamboo charcoal composite and modified poplar veneer." *Forestry Studies in China*, 14, 2012: 240-245.

[9] Yang, J., Zhang, X., Li, B., et al. "Photocatalytic activities of heterostructured TiO_2-graphene porous microspheres prepared by ultrasonic spray pyrolysis." *Journal of Alloys and Compounds*, 584, 2014: 180-184.

The Influence of Monomer Ratio on the Properties of Liquid Crystal Poly(oxybenzoate co-oxynaphthoate) Copolyester

Peng Wei, Yimin Wang
College of Materials Science and Engineering, State Key Laboratory for Modification of Chemical Fibers and Polymer Materials, Donghua University
Wang Yimin@dhu.edu.cn; appletree0322@163.com

ABSTRACT
Thermotropic liquid crystal copolyester (TLCP) has been widely studied and well developed in past three decades, due to its many unique properties, such as excellent mechanical properties and thermal properties, etc[1-3].

Though fully aromatic poly(hydroxybenzoic acid) (PHBA) exhibits excellent mechanical properties, high melting temperature (above 600°C) of this copolyester limits its further application[8]. Structure modification has been proved as an effective method in lowering melting temperature of this polyester[4-6]. Based on above principles, poly(oxybenzoate co-oxynaphthoate) with suitable melting temperature were obtained by copolymerizing HBA(hydroxybenzoic acid) with HNA(6-hydroxy-2-nanthplene acid). Subsequently, copolyester prepared from 73mol% HBA and 27mol% HNA had been commercialized by Celanese under trademark Vectra in 1980s[7]. Researchers focused on melt processing, rheology, thermal properties and structure of Vectra copolyester[8-10], however, no work has been reported on systematic study of the influence of compositions on the properties of this copolyester. Therefore, the purpose of this study is to synthesize a series of HBA/HNA copolyesters with HBA molar ratios ranged from 20-85mol% by melt polymerization method and investigate the effect of monomer ratios on the thermal behaviors, structure and UV absorption properties.

EXPERIMENTAL

All of the copolyesters were synthesized via melt transesterification polymerization reaction of the respective acetylated monomers according to previous research[3]. The HBA/HNA copolyester is abbreviated as P-HBAx thereafter, where x stands for the molar ratio of HBA.

CHARACTERIZATION

Thermal Analysis

Thermal properties were measured using a TA Q50 Thermogravimetry analyzer (TGA) at a heating rate of 10 °C/min from 30 to 800 °C under the N_2 atmosphere. Thermal behavior was observed using a TA Q200 DSC instrument under a N_2 atmosphere. The heating rate was set as 20 or 10 °C/min for the first and second heating circles respectively.

X-Ray Diffraction

X-ray diffraction patterns of powder samples were obtained on a Bruker AXS Generator using the nickel-filtered CuK_α radiation (λ= 0.154 nm, 40 KV, 40 mA) and scanned from 5.0 to 40.0°.

UV Spectrum

The UV absorption properties were evaluated by TU-1901 Lambda 950 solid state UV spectrometer scan rate was 200nm/min.

RESULTS AND DISCUSSION

The DSC heating curves of P-HBAx are shown in Figure 1 and 2.It is found that melting temperature, crystallization temperature of copolyester P-HBAx ranges from 240 to 330 °C and 180 to 265 °C respectively, when the HBA composition varies from 20 to 85mol%.

The thermal stability of the copolyesters was evaluated by TGA under N_2 atmosphere. The temperature at 5% and 10% weight loss ranged from 442-505℃ and 475-515℃, respectively.

Figure 3 shows the WAXD graph of copolyesters. The copolyester has 3diffraction peaks .Peaks located at around 18.6, 21.7 and 26.6° on pattern of P-HBA20 were attributed to (110), (200) and (211), respectively, which were coincided with P-HNA homopolymer, indicating P-HBA20 was orthorhombic. With increasing of HBA content, the crystal form of copolyesters changed in following: orthorhombic(HBA<50%), pseudohexgonal (55-75%), orthorhombic(HBA>80%).

Figure 4 shows the UV graph of P-HBA20,P-HBA30 and P-HBA40. All of copolyesters show absorption at 215, 295, 340 and 398 nm. With increasing HBA contents, absorption peak of carbonyl group red shifts from 340nm to 345nm.

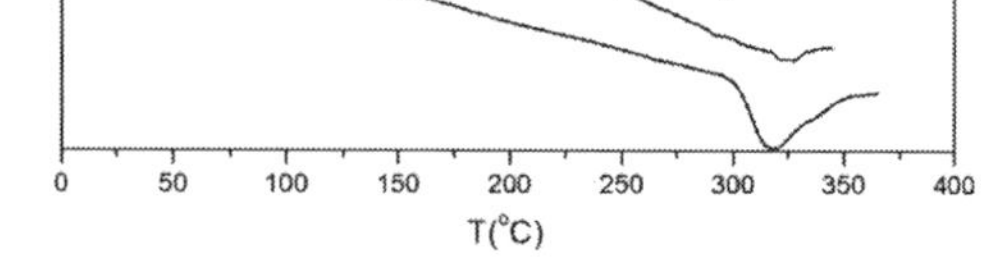

Figure 1. The first DSC heating curve of HBA/HNA copolymer

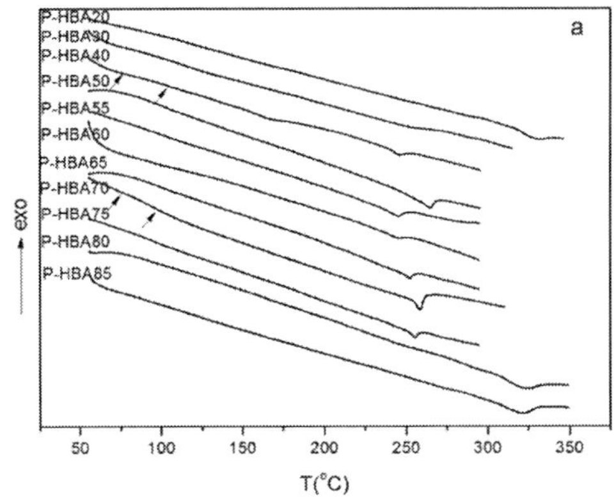

Figure 2. The second DSC heating curve of HBA/HNA copolymer

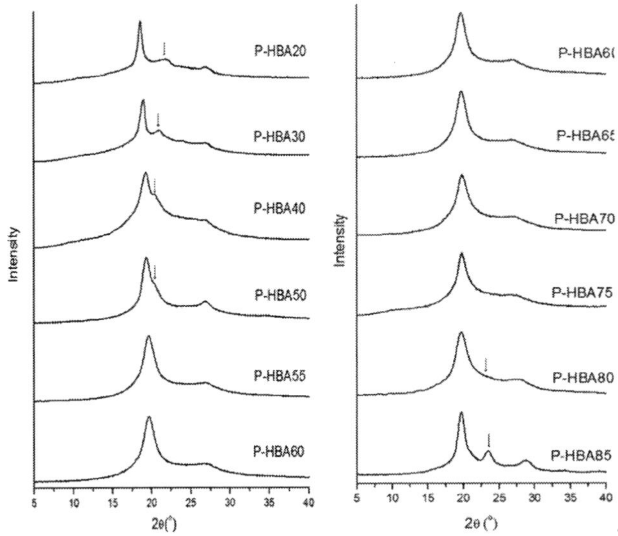

Figure 3. The WAXD patterns of HBA/HNA copolyester

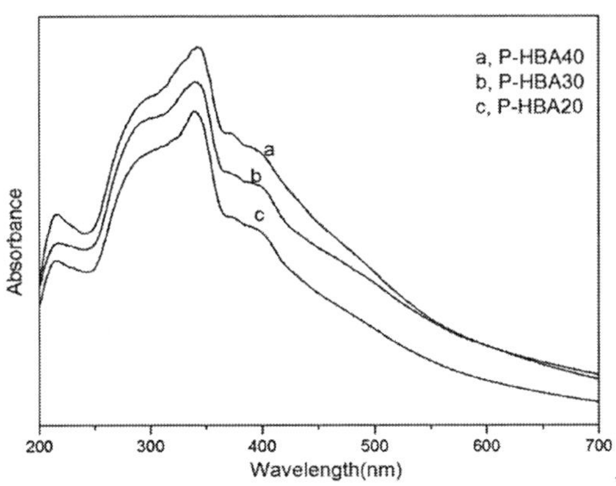

Figure 4. The solid state UV spectrum of HBA/HNA copolymers

CONCLUSION

The molar ratio has apparent impact on the thermal properties, structure and UV absorption properties, the reason caused these phenomena was discussed in detail.

ACKNOWLEDGMENT

We gratefully acknowledge the valuable help and great support from Innovation program of Shanghai Municipal Education No. 12ZZ062, the Fundamental Research Funds for the Central Universities, the Shanghai Natural Science Founcation and Programme of Introducing Talents of Discipline to Universities No. 111-2-04.

REFERENCES

1. Wilsens, C.H.R.M., Verhoeven, J.M.G.A., Noordover, B.A.J., Hansen, M.R., Auhl, D., Rastogi, S. *Macromolecules*, 2014.
2. Jang, K.-S., Johnson, J.C., Hegmann, T., Hegmann, E., Korley, L.T.J. *Liq. Cryst.*, 41, 2014: 1473.
3. Wei, P., Cakmak, M., Chen, Y., Wang, X., Wang, Y., Wang, Y. *J. Appl. Polym. Sci.*, 10.1002, 2014: 40487.
4. Mitra, S., Pillai, C.K.S. *J. Appl. Polym. Sci.*, 107, 2008: 778.
5. Yerlikaya, Z.Y., Aksoy, S.I., Bayramli, E. *J. Polym. Sci., Part A : Polym. Chem.*, 39, 2001: 3263.
6. Vijayanathan, V., Prasad, V.S., Pillai, C.K.S. *J. Appl. Polym. Sci.*, 82, 2001: 1021.
7. Calundann, G.W. U.S. Patents, 4,161,670: 1979.
8. Blackwell, J., Gutierrez, G.A., Chivers, R.A., Ruland, W. *Journal of Polymer Science: Polymer Physics Edition*, 22, 1984: 1343.
9. Han, C.D., Chang, S. J. *Rheol.*, 38, 1994: 241.
10. Muramatsu, H., Krigbaum, W.R. *Macromolecules*, 19, 1986: 2850.

Electrospun Composites of PHBV/Pearl Powder for Bone Repairing

Jingjing Bai, Jiamu Dai, Guang Li
College of Material Science and Engineering, Donghua University, Shanghai, China
lig@dhu.edu.cn; jingjing-bai@outlook.com

INTRODUCTION

Electrospinning is a simple method to fabricate nanofiber. Recently many synthetic polymers have been used in electrospinning. Poly(3-hydroxybutyrate-co-3-hydroxyvalerate) (PHBV) produced by microbial fermentation is one of the biomaterials applied in bone tissue engineering. Using Pearl powder to enhance biological and mechanical performances has been studied recently [1]. Pearls contain growth factors that are able to increase osteoblast proliferation [2]. Furthermore, it is approved that pearl shows great biocompatible and biodegrade ability [3]. In this study, we selected PHBV as the main matrix and pearl powder as inorganic additive.

APPROACH

Before added in PHBV solution, pearl powder was pretreated in polyethylene glycol (PEG) solution with the aim at raising compatibility. First, suspensions were obtained by dispersing pearl powder in deionized water by stirring, then PEG powder was added to the suspensions combing ultrasonic and stirring. Each process lasted for 1 hour. The suspensions were frozen overnight, and then were freeze-dried for two days. Polymer solution was obtained by dissolving PHBV in HFIP at room temperature, and then pearl powder was added to the solution with sufficient stirring.

RESULTS AND DISCUSSION

Electrospinning

To evaluate the effect of concentration and voltage, the electrospinning was undertook with different concentration solutions (4, 5 wt %) and different voltage (11kV, 17kV, 22kV). The Morphology of electrospun PHBV nanofiber scaffolds prepared from different concentration solutions were shown in Fig. 1. The nanofiber resulted from 4wt% solution showed a better morphology than that from 5wt% and more uniform diameter distribution of fibers. Thus, the solution of 4 wt% was fixed for further experiments. While Fig. 2 showed PHBV nanofiber scaffolds fabricated on different voltage (11kV, 17kV, 22kV). The fiber diameter became smaller and more uniform with the voltage increasing, while the fiber displayed a collapsing type under low voltage. Fig. 3 and Fig. 4 were PHBV/pearl composite nanofibers with the voltage of 11kV and 22kV respectively. As shown in these pictures, the addition of pearl powder improves the spinnability of the PHBV solution. The diameter of nanofibers was thinner because of the addition of pearl powder. The reason maybe that calcium carbonate in pearl powder plays a role as a lubricant, which may lead to the viscosity reduction of the blend solutions.

The ATR-FTIR spectrum of PHBV/pearl (Fig.5) showed absorption peaks at around 1433, 863 and 713 cm^{-1}.

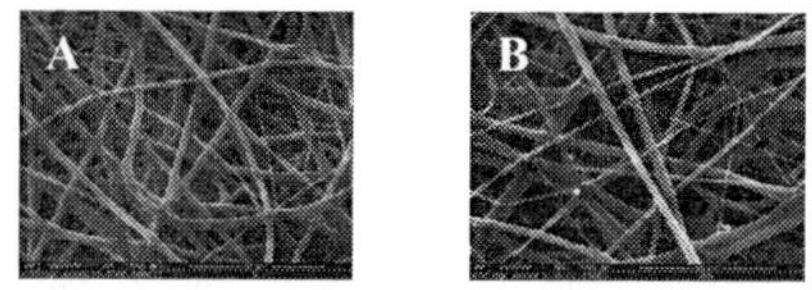

Fig. 1. SEM micrographs of electrospun fibers using PHBVsolution in HFIP.(A) 4 wt%, (B) 5wt%.

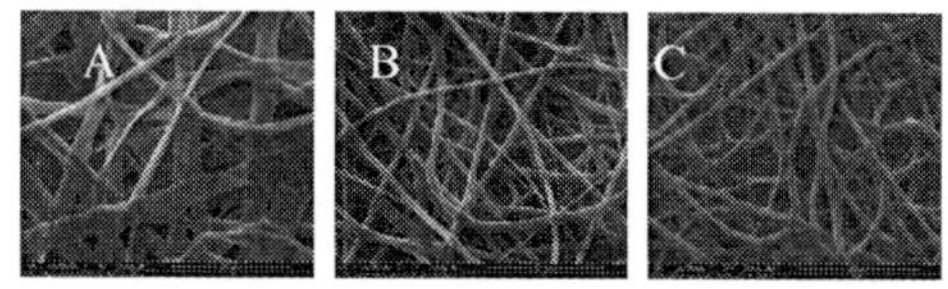

Fig. 2. SEM micrographs of electrospun fibers using different voltage.(A) 11kV, (B) 17kV. (C) 22kV.

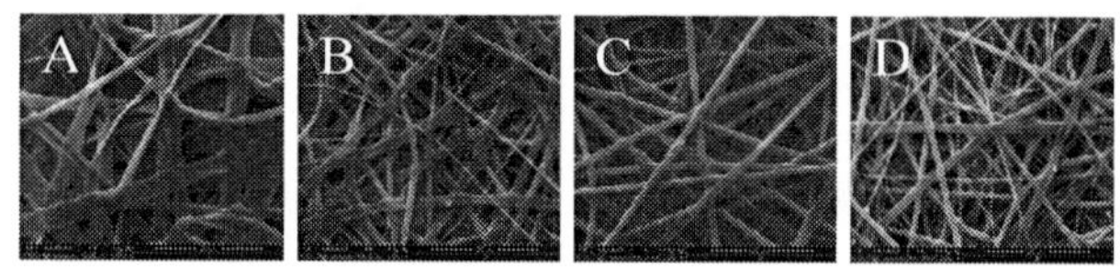

Fig. 3. SEM micrographs of electrospun fibers with different content of pearl powder using voltage of 11kV.(A) 0 wt%, (B) 2 wt%, (C) 4 wt%,(D) 6 wt%.

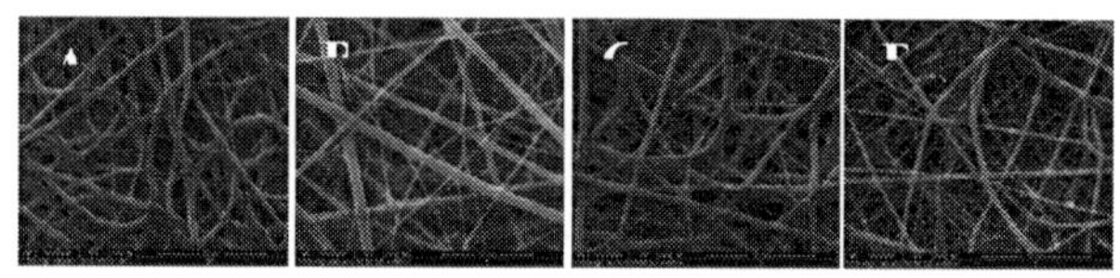

Fig. 4. SEM micrographs of electrospun fibers with different content of pearl powder using voltage of 22kV.(A) 0 wt%, (B) 2 wt%, (C) 4 wt%,(D) 6 wt%.

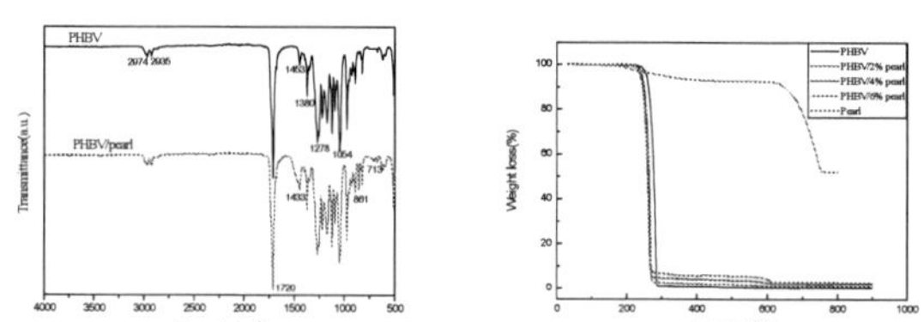

Fig.5 ATR-FT IR spectra and TGA curves of PHBV and PHBV/pearl nanofiber.

When electrospinning of organic–inorganic compounds, special attention should be paid in solution preparation since the composite solution brings about dispersion problems [4, 5]. The TG analysis was used to evaluate the content of pearl powder in composite nanofibers. As shown in Fig. 5, the thermogram of PHBV nanofiber showed a one-step thermal degradation between 260 and 285 °C, whereas the thermogram of PHBV/ pearl nanofiber showed two steps, with the second step of degradation at about 600 °C, which can be attributed to the decomposition of pearl. In a short summary, it was showed that the pearl powder evenly dispersed in the viscous polymer solution.

Mineralization studies

After soaked in 1.5SBF for a few days, the weight increasing rate of the composite nanofiber scaffolds were calculated as shown in Fig. 6. The HA deposited on the surface of pristine PHBV nanofiber scaffolds was less than that on composite nanofiber scaffolds. This indicates that pearl performed as a crystal nucleus for HA particle depositing. For HA deposition, crystal nucleus is needed. However, there's no nucleus existing on PHBV nanofiber, therefore a long nucleus inducing period is acquired; then the first formed HA particles work as crystal nucleus for further mineralization. Furthermore, as shown in Fig. 7, formation of HA particles were observed throughout the PHBV nanofiber scaffolds as expected.

HA structure analysis

The FTIR spectra of nanofiber scaffolds before and after mineralization showed that new peaks appeared for mineralized scaffold due to the HA particle deposition (Fig. 8). XRD diffraction patterns of PHBV and PHBV/pearl nanofiber scaffold before and after mineralization for 30 days were given in Fig. 9. The deposited HA on the surface of different nanofiber scaffolds were almost the same both in constitution and structure.

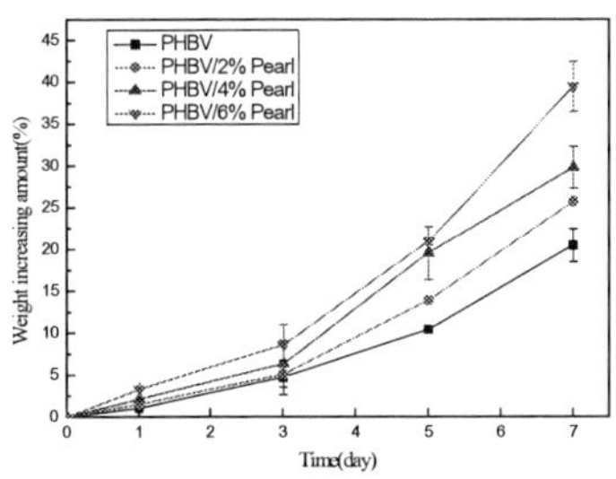

Fig. 6. The influence of pearl concentration and soaking time on HA deposition.

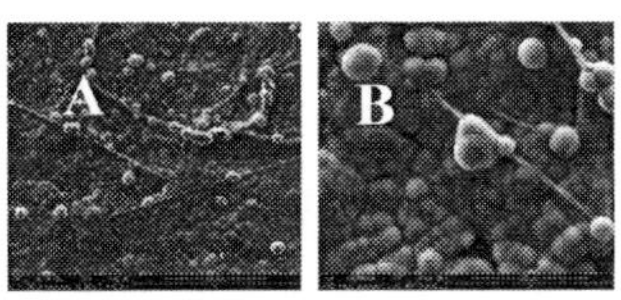

Fig. 7. SEM micrographs of HA deposited on PHBV scaffold after soaked in SBF for 7 days.

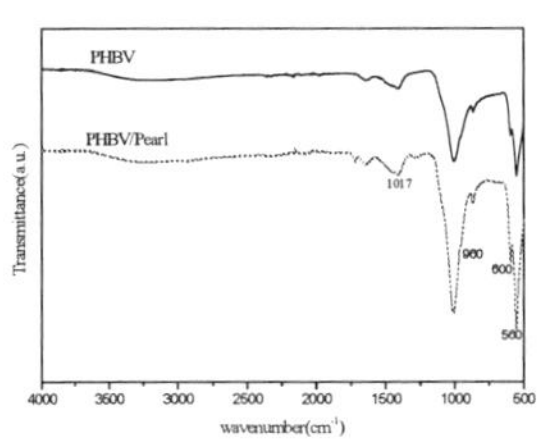

Fig. 8. ATR-FTIR spectra of PHBV and PHBV/pearl after mineralization in SBF for 30 days.

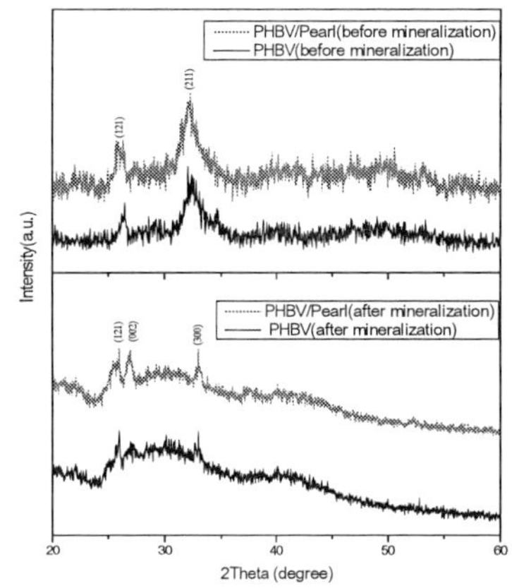

Fig. 9. XRD diffraction patterns of PHBV and PHBV/pearl nanofiber scaffold before and after mineralization for 30 days.

CONCLUSION

In this work, we successfully fabricated PHBV/pearl composite nanofiber scaffolds. The incorporation of pearl powder is in favor of spinning of PHBV. Mineralization ability of these composite nanofiber scaffolds was greatly enhanced due to naturally non-toxic pearl powder, which act as crystal nucleus for HA particle depositing. The weight increase of HA on scaffolds is proportional to the content of pearl powder. Therefore, the prepared PHBV/pearl composite nanofiber scaffolds would be a promising candidate as an osteoconductive composite material for bone repairing.

KEYWORDS

Hydroxyapatite; Pearl; PHBV; Mineralization.

REFERENCES

[1] Flausse, Alicia, et al. "Osteogenic differentiation of human bone marrow mesenchymal stem cells in hydrogel containing nacre powder." *Journal of Biomedical Materials Research Part A*, 101.11, 2013: 3211-3218.

[2] Shen, Yutian, et al. "In vitro osteogenetic activity of pearl." *Biomaterials*, 27.2, 2006: 281-287.

[3] Yang, Yueh-Lung, et al. "Osteogenic activity of nanonized pearl powder/poly (lactide-co-glycolide) composite scaffolds for bone tissue engineering." *Bio-medical Materials and Engineering*, 24.1, 2014: 979-985.

[4] Rezwan, Kurosh, et al. "Biodegradable and bioactive porous polymer/inorganic composite scaffolds for bone tissue engineering." *Biomaterials*, 27.18, 2006: 3413-3431.

[5] Armentano, I., et al. "Biodegradable polymer matrix nanocomposites for tissue engineering: a review." *Polymer Degradation and Stability*, 95.11. 2010: 2126-2146.

Flammable Behavior of Nylon 6 Fabrics With and Without Spandex Using Cone Calorimeter Test and Vertical Burning Test

Shengli Luo[1], Xin Chen[2], Ruitian Zhu[1], Zhijie Jiang[3], Hailiang Zhang[3], Zhicheng Zhan[3], Yurong Yan[3]
[1]Guangzhou Fiber Product Testing Institute, Guangzhou, China
[2]Guangdong Xinhui Meida Nylon Co.,Ltd, Jiangmen,Guangdong, China
[3]School of Material Science and Engineering, South China University of Technology, Guangzhou, China
yryan@scut.edu.cn

INTRODUCTION

Nylon is one of common used fiber materials in textiles industry, especially in undergarment, sportswear, stockings and in automobile parts. In order to get some elasticity, nylon is weaved or knitted with spandex. As an organic polymer material, Nylon is considered to have a lower degree of flammability although it is not as combustible as other fabric materials. But for spandex (polyurethane), it is extremely easy to burn, the limiting oxygen index (LOI) of polyurethane is only 17.3 [1]. Whereas the limiting oxygen index (LOI) of polyamide 6 is 21 [2]. How the flammability of Nylon/Spandex was not reported clearly.In this study, vertical burning test[3] and cone calorimeter test[4] were used for assessing the flammability of flammable behavior of nylon 6 fabrics with and without spandex.

EXPERIMENTAL

Materials

All samples were prepared by Guangdong Xinhui Meida Nylon Co., Ltd. Two kinds of samples were used in this study: (1) 100% nylon 6 knit fabric; (2) nylon 6 knit fabrics containing 5 percent spandex.

Vertical burning test

The Vertical test was carried out on a YG(B)815D-I vertical burning test instrument (Wenzhou Darong Textile Instruments Co., Ltd.) according to GB/T 5455-97. The dimension of all the fabric samples was $(300\times76)mm^2$.

Cone calorimeter test

The cone calorimeter tests (CTT) were finished on a FTT0007 cone calorimeter (Fire Testing Technology Ltd.,UK) according to ISO 5660. The dimension of all samples was $(100\times100\times50)$ mm^3. All the specimens were exposed horizontally to an external heat flux of 35kw/m2 (corresponding to the temperature of 690°C), and an air flow of 24 l/s.

RESULTS AND DISCUSSION

Vertical burning test

Vertical burning test is used to measure and describe the response of materials, products, or assemblies to a heat source and flame under a controlled laboratory conditions. We found that no afterflame and afterglow could be found after the flame was removed for Nylon 6 fabrics. But for nylon 6 fabrics containing spandex burned sorely after the flame was removed and had an afterflame time of 25.8s. Besides, dripping and setting fire to the absorbent cotton were also happened. Furthermore, two nylon 6 fabric samples in our study were burned completely. So, spandex improved the flammability of nylon 6.

Cone calorimeter test

Cone calorimeter can provide useful flammability information to predict the combustion behavior of materials in a large scale through its simulation in real fire. The HRR, THR and RSR curves of the two samples were shown in Figure 1 to Figure 3 respectively. According to Figure 1, the time of main peaks of nylon 6 containing spandex was earlier than that of the pure nylon 6 fabric. And the peak area of nylon 6 fabric sample containing spandex was smaller than that of the pure nylon 6 fabrics.

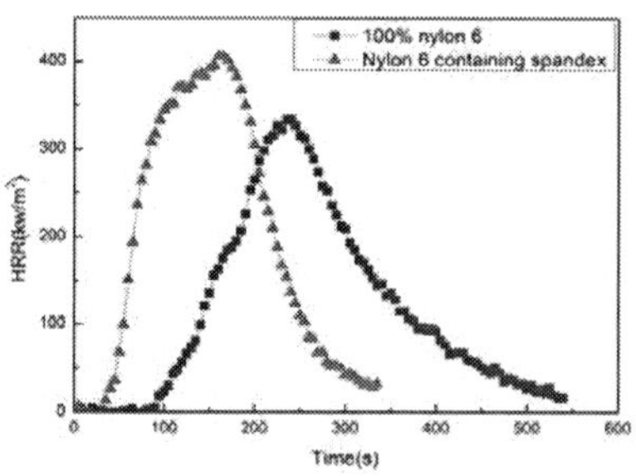

Figure 1. HRR curves of different samples.

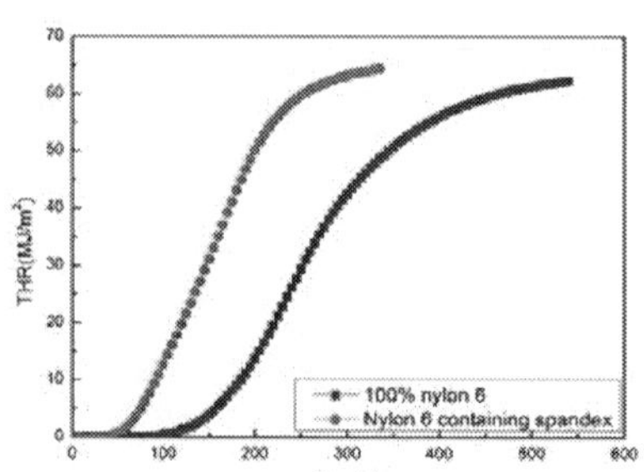

Figure 2. THR curves of different samples.

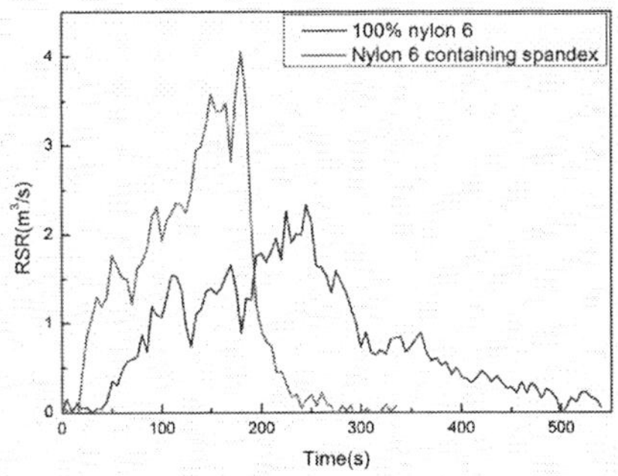

Figure 3. RSR curves of different fabric samples.

CONCLUSIONS

To study the influence of spandex on combustion behavior of textile fabrics, two kinds of nylon 6 fabrics were tested using cone calorimeter test and vertical burning test. Combustion parameters such as heat release rate (HRR), total heat release (THR) and rate of smoke release (RSR) were obtained from cone calorimeter test and afterflame time and damaged length from vertical burning test. Results indicated that the nylon 6 fabric containing spandex showed 21% higher HRR value. But THRs of the nylon 6 fabric containing spandex and the pure nylon 6 fabric were close. RSR curves indicated that RSR value of the nylon 6 fabric containing spandex was 73% higher than that of the pure nylon 6 fabric. The afterflame time and damaged length of the nylon 6 fabric containing spandex were much longer than that of the pure nylon 6 fabric. Spandex does accelerate the combustion behavior on Nylon fiber.

KEYWORDS

Nylon 6, Flammable Behavior, Spandex, Cone Calorimeter Test, Vertical Burning Test

REFERENCES

[1] Wang, J.C., Chen, Y.H. *Polym. Mater. Sci. Eng.*, 20, 2004: 168-172.

[2] Ma, S.Y., Yang, W., Jiang, S., Bai, Y., et al. *Chin. Plast. Ind.*, 35, 2007: 25-28.

[3] Shekar, R.I., Kotresh, T.M., Damodhara Rao, P.M., Satheesh Kumar, M.N., et al. *J. Appl. Polym. Sci.*, 122, 2011: 2295-2301.

[4] Tata, J., Alongi, J., Carosio, F., et al. *Fire Mater.*, 35, 2011: 397-409.

Reinforced and UV Resistant Silks from Silkworms Fed with Titanium Dioxide Nanoparticles

Lingyue Cai, Huili Shao, Yaopeng Zhang
State Key Laboratory for Modification of Chemical Fibers and Polymer Materials,
College of Materials Science and Engineering, Donghua University, Shanghai, China
zyp@dhu.edu.cn; yuefavorever@163.com

ABSTRACT

As the perfect combination of strength and luster, silkworm silks have been widely used in many fields. Numerous researches have been conducted to improve its properties for years. This paper demonstrates an *in vivo* uptake of titanium dioxide (TiO_2) nanoparticles by the silkworms, leading to the direct production of intrinsically modified silk. The method easily incorporated the nanoparticles into the inner structure of silks. Results suggest that both the strength and the anti-UV ability of the modified silks were well improved at a low additive amount.

INTRODUCTION

Silkworm silks, produced by silkworms from an aqueous circumstance, are fibrous proteins composed of fibroin and sericin.[1] Lots of researches focused on the improvement of silkworm silk's performance to make it rival or even exceed the best natural silk-spider silk. Recently, a report demonstrated that TiO_2 improved the toughness of artificial silk spun from regenerated silk fibroin solution.[2] Li found that TiO_2 with anti-UV ability can attach to silks along with the oxide's chemical assembly onto a silk substrate.[3]

To find out a proper way to incorporate some xenobiotics into silkworm silks, some researchers came up with a new method. Tansil et al. approved that dyes can show up in silk when silkworms was fed with dye-containing diet.[4] Subsequently, Nisal et al. attained colored cocoons by adding azo dyes into feed of silkworms.[5]

In this work, we fed silkworms by artificial diet mixed with TiO_2 nanoparticles to improve the mechanical properties and the UV resistant properties of the silk. The effects of concentration gradient of the nanoparticles on the silk properties were also investigated.

METHODS

Preparation of artificial diet and silkworm raising

Nano-TiO_2 solution (20% mass fraction) with different TiO_2 contents of 2.5 g, 5 g, 7.5 g and 10 g was mixed with 100 mL of water respectively. Fifty grams dry powder of artificial silkworm diet was mixed with the diluted solution uniformly and microwaved for 3 min. The mass fraction ratio of TiO_2/dry diet powder was calculated as 1%, 2%, 3% and 4%. Normal feed was heated without adding TiO_2 solution. Cooked artificial food was pressed to be wafer for further feeding.

Silkworms were raised starting with eggs in a constant temperature and humidity chamber. The silkworm larvae were fed with normal feed from their hatch to forth instar. In the fifth instar, they were separated into five groups. Four groups were fed with modified diet and the other was fed with normal diet as control until fiber spinning.

Characterization

Synchrotron radiation wide angle X-ray diffraction (SR-WAXD) patterns were obtained at the BL15U1 beamline at the Shanghai Synchrotron Radiation Facility (SSRF). FIT2D (V12.077) and Peakfit (V4.12) software were applied to process the WAXD patterns to get the crystallinity and crystallite size of the degummed silk. The mechanical properties of the fibers (sample number 20) were measured with an Instron 5565 material testing instrument.

RESULTS AND DISCUSSION

Table I shows the crystallinity and crystallite size of all degummed silks. The crystallinity decreased with the increase of TiO_2 content with an exception of TiO_2-1%. Compared with control, the TiO_2 modified silks exhibit reduced crystallite size in *c* direction (direction along the fiber axis) and increased crystallite size in *a* direction (interchain direction). However, it keeps almost constant in *b* direction (intersheet direction).

Table I. Crystallinity and crystallite size of degummed silks

Sample	X_c (%)	Crystallite size (nm)		
		L_a	L_b	L_c
Control	49.27	3.6	3.8	10.7
TiO_2-1%	54.72	4.3	3.7	7.4
TiO_2-2%	48.12	4.6	3.9	8.0
TiO_2-3%	46.80	5.7	3.8	8.6
TiO_2-4%	42.34	5.1	3.7	8.9

The stress-strain curves of the degummed silks are shown in Fig. 1(a). The breaking strength of the control silk ranges from 360 to 430 MPa, and the elongation at break is 13-17%. In the case of TiO_2-1%, the breaking strength of the obtained silk exhibits obviously increases to 520-580 MPa, while the elongation at break is 15-18%. In the case of TiO_2-2%, the breaking strength of the obtained silk is similar to the control, but the elongation at break is extended to 19-23%. However, the mechanical properties of the modified silk become poorer when the TiO_2 content further increases to 3% and 4%.

UV resistant properties of the TiO_2 modified silks were investigated by comparing the loss of the mechanical properties of the fiber after irradiation with ultraviolet light. TiO_2-3% and TiO_2-4% samples were not tested as the poorer mechanical properties than the control group. The degummed silks of control, TiO_2-1% and TiO_2-2% silk were exposed to UV light at a wavelength of 390 nm and an intensity of 1245 $\mu W/cm^2$ for 3 h in a Rayven UV Oven before tensile testing. The stress-strain curves are plotted in Fig. 1(b). As expected, the breaking strength and the elongation at break of the UV irradiated silks is lower than those of the untreated silks because the UV light may break protein chains. The control group was weakened most while TiO_2-1% kept most of the mechanical properties.

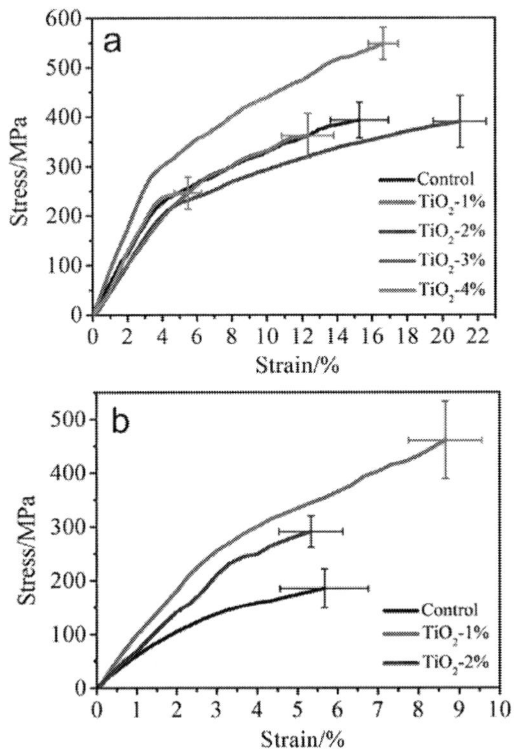

Fig. 1. Stress-strain curves of (a) degummed silks and (b) the silks after exposure to UV for 3 h.

At the low content of 1%, TiO_2 firstly incorporated into amorphous region due to the looser structure than crystalline region. The nanoparticles with less probability to agglomerate at a low concentration could be well absorbed by silkworms and evenly dispersed into silks. Zhou et al.[6] found that metallic ion, such as Mg^{2+}, induces the conformation transition of silk fibroin to β-sheet. As Ti is also a multivalent metallic ion, it may has similar effect on silk fibroin. Therefore, the crystallization process is promoted at TiO_2 content of 1%. Lee[7] proved that Ti interacts with protein molecules through coordinate bonds and hydrogen bonds, so TiO_2 may link protein molecules in amorphous region. The increased crystallinity and improved links in amorphous region result in the reinforcement of the TiO_2-1% fiber. However, nano-TiO_2 particles at higher concentration began to enter crystalline region, which lead to an opposite effect on the performance of the silks. The nano-confined crystallization[2] of silk fibroin in crystalline region leads to the decrease of crystallinity of the fiber. The larger elongation of the TiO_2-2% silk may result from the higher content of amorphous region. Moreover, the TiO_2 nanoparticles with higher concentration are easier to form inhomogenous agglomerates with larger size and broader distribution, which may act as defects in the fibers. Consequently, the strength and toughness of the silk with 3% and 4% TiO_2 are both smaller than those of the control silk. It should be noted that the food intake and growth of silkworms can be affected unavoidably when the concentration of nano-TiO_2 is relatively high. This may be another reason why TiO_2-4% silk has very poor mechanical properties.

As shown in Fig. 1(b), the breaking strength of control, TiO_2-1% and TiO_2-2% group decreased 52.9%, 15.9% and 25.4% (reduction of the average strength/original average strength), respectively. Therefore, the addition of the TiO_2 certainly improved the ultraviolet resistance of silkworm silk. In general, the ultraviolet resistance of the material may be enhanced with the increase content of TiO_2, but the tensile properties of TiO_2-2% silk decreased more than TiO_2-1% silk. This is because TiO_2-2% silk has more amorphous region which is easier to be damaged than crystalline region under UV light.

CONCLUSION

Silks were modified by raising silkworms with TiO_2-containing diet. At the TiO_2 content of 1%, the breaking strength of the modified silk is 520-580 MPa, while the elongation at break is 15-18%, which increased 39.2% and 8.9% on average respectively compared with those of control silk. The breaking strength of TiO_2-1% silk just decreased 15.9% while that of control silk decreased 52.9% after exposing to UV light. However, both the breaking strength and UV resistant property reduced when the content of TiO_2 was further increased. This feeding method can potentially apply to other metallic compounds for improving the performance of silks or effectively combining some new functions with silks at a low-cost. The functional silks with exceptional mechanical properties can be used for non-conventional textiles as well as tissue engineering.

ACKNOWLEDGMENT

This work is supported by the National Natural Science Foundation of China (21274018), the Innovation Program of the Shanghai Municipal Education Commission (12ZZ065), DHU Distinguished Young Professor Program (A201302), and the Fundamental Research Funds for the Central Universities (2232013A3-11).

REFERENCES

[1] Omenetto, F.G., Kaplan, D.L. *Science*, 329, 5991, 2010: 528-531.

[2] Pan, H., Zhang, Y.P., Shao, H.L., Hu, X.C., Li, X.H., Tian, F., Wang, J. *J. Mat. Chem. B*, 2, 10, 2014: 1408-1414.

[3] Li, G.H., Liu, H., Zhao, H.S., Gao, Y.Q., Wang, J.Y., Jiang, H.D., Boughton, R.I. *J. Colloid Interface Sci.*, 358, 1, 2011: 307-315.

[4] Tansil, N.C., Koh, L.D., Han, M.Y. *Adv. Mater.*, 24, 11, 2012: 1388-1397.

[5] Nisal, A., Trivedy, K., Mohammad, H., Panneri, S., Gupta, S. Sen, Lele, A., Manchala, R., Kumar, N.S., Gadgil, M., Khandewal, H., More, S., Laxman, R.S. *ACS Sustain. Chem. Eng.*, 2, 2, 2014: 312-317.

[6] Zhou, L., Chen, X., Shao, Z.Z., Huang. Y.F, Knight, D.P. *Journal of Physical Chemistry B*, 109. 35, 2005: 16937-16945.

[7] Lee, S.M., Pippel, E., Gosele, U., Dresbach, C., Qin, Y., Chandran, C.V., Brauniger, T., Hause, G., Knez, M. *Science*, 324, 5926, 2009: 488-492.

PLA/SA Emulsion Electrospinning and Its Use as Oral Cell Scaffold Materials

Weihong Xu[1], Renzhe Shen, Jie Gao, Yurong Yan
Department of Polymer Materials and Engineering, South China University of Technology, Guangzhou, China
yryan@scut.edu.cn

INTRODUCTION

Polylactic acid (PLA) is a kind of cell scaffold because of its biodegradable property and non-toxic degradation production. However, the surface of PLA scaffold is hydrophobic which limits its use in actually applications[1]. Sodium alginate (SA) is a kind of widely used dressing materials because they tend to dissolve or gel in the wound bed and form a kind of swelled calcium alginates after the ion-change reaction between sodium alginate and calcium in blood or tissue fluid, and block the further bleeding of wound. Also because of its good hydrophilicity and biocompatibility, it have been used in dental impression materials[2,3]. In our research before, nanofibers with sodium alginate as skin and polylactic acid as core was designed by using w/o emulsion electrospinning. Here, sodium alginate can gel and block the pore of the dressing then provide a micro-environment for the wound bed, and PLA can provide the tensile strength for the whole dressing. In this paper we will discussion if this kind of special "composite fiber" can be used as a kind of Oral Cell Scaffold Materials.

EXPERIMENTS

PLA/SA emulsion electrospinning and characteristics of PLA/SA nanofibers

Emulsion solution preparation and PLA/SA emulsion electrospun nanofiber mats were prepared and characterized according to Ref [3-4].

Preparation and biological properties of PLA/CA mats

$CaCl_2$ solution was electrostatic spraying on the PLA/SA membrane to obtain PLA/CA membrane[5]. MTT method was used to test Cell proliferation bioassay[6]. The cell mineralization was tested by Alizarin red staining method. In order to improve the antibacterial properties of the PLA/SA mats, some nano-Ag were added.

RESULTS AND DISCUSSION

In order to identify the sodium alginate forming the skin part of the fiber, we dyed fiber with Rhodamine B, and tested it on Leica TCS SPE Confocal Laser Scanning Microscope, and the result was shown in Figure. It was shown that the dyed sodium alginate formed the skin part of fibers. So we can use CA to improve the cell cultivation of PLA by using ion-exchange reaction between SA and $CaCl_2$.

\

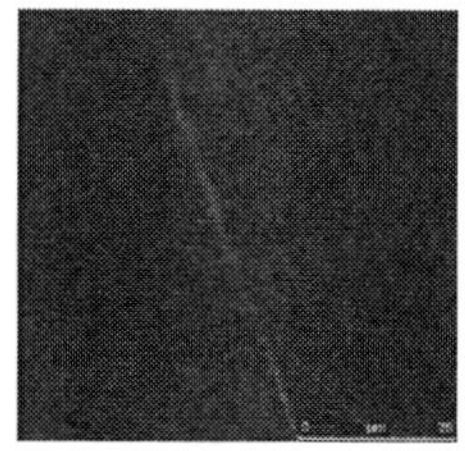

Figure. 1. CLSM image of nanofibers with sodium alginate content of 10% comparing to PLA content.

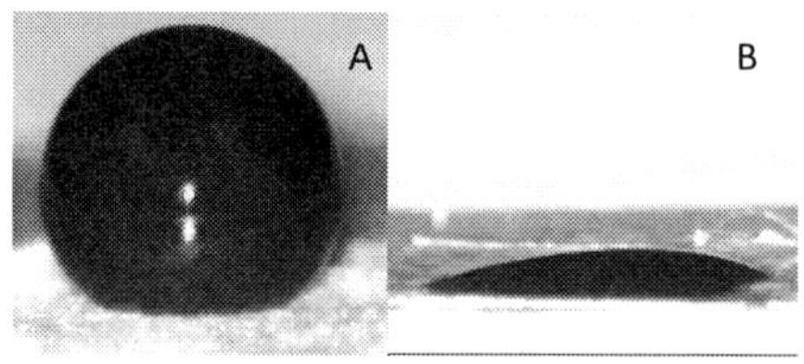

Figure. 2 Hydrophilic test photos of PLA mat(A) and PLA/SA mat(B).

According to Figure 2, we can find that the surface properties of PLA/SA mats were hydrophilic due to the SA formed the rich phase on the surface of the PLA/SA nanofiber and thus favorite the cell adhesion onto the mats

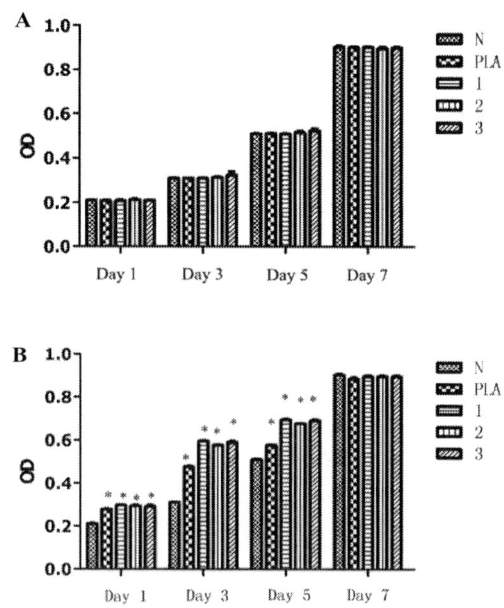

Figure. 3. Influence of PLA/CA/50ugnAg extract and PLA/CA/50ug nAg nanofiber mats on cell proliferation. A: PLA/CA/50ugnAg extract; B: PLA/CA/50ug nano-Ag nanofiber mats.

According to Figure 3, we can find that the PLA/CA/50ug nano-Ag extract could not promote the proliferation of cells (Figure.3A), but PLA/CA/50ug nano-Ag nanofiber mats promote the proliferation of cells at beginning of the period(from1to 5days (Figure.3B). That means CA can not be looked as a good cell scaffold materials unless some nanofiber 3-Demention structure formed in the scaffold.

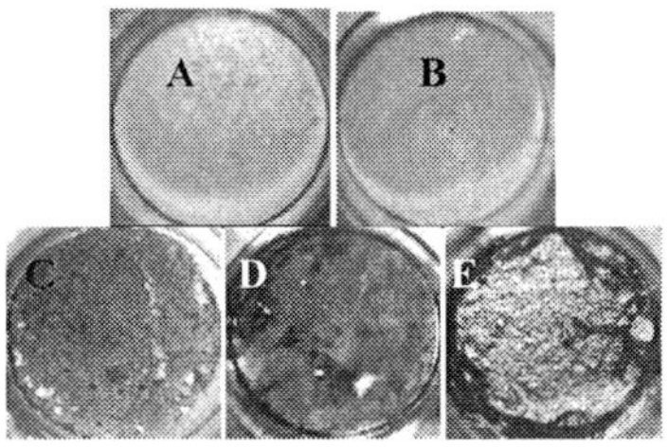

Figure. 4. Mineralized nodules staining (A: NC, B: PLA, C: PLA/CA (0.5ml 0.2g/ml); D: PLA/CA (1ml 0.2g/ml); E: PLA/CA (100ml0.01g/ml).

Another proof for the promotion behavior of fiber structure and CA on the cell formation is shown in Figure 4. We can also find that the PLA/50ugnAg could not promote the formation of mineral node of cells, but PLA/CA/50ug nano-Ag nanofiber mats promote the formation of mineral node of cells, and this tendency is improved with the proportion of CA in PLA/CA fiber mats.

CONCLUSIONS

Emulsion electrospinning can be used as a widely used method and as a promising technique to produce polymer scafford for tissue regeneration applications. Both Polylactic acid and Sodium alginate are biological material. Though emulsion electrospinning, hydrophilic SA was added on the surface of PLA nanofibers mostly, which make PLA/SA with a hydrophilic surface and favorite the adhesion of cells and provide a good environment for cells to grow. PLA and nAg can not prompt the formation of mineral node of cells. CA can provide cells mineralization. with Ca^{2+}. But only PLA/CA fiber mates can promote this progress with some special fiber scaffold structure.

KEYWORDS

PLA; Sodium alginate; Cell scaffold; Emulsion electrospinning.

ACKNOWLEDGMENT

Many thanks to the supports of National Nature Science Funds for Young Scholar of China (Grant No. 51003033),Fundamental research Funds for the Central Universities of South China University of Technology(Grant No. 2009ZM0225),Opening Project of the Key Laboratory of Polymer Processing Engineering, Ministry of Education, China (No.20111006).

REFERENCES

[1] Smith, LA, Ma, PX. "Nano-fibrous scaffolds for tissue engineering." *Colloids and Surfaces B: Biointerfaces*, 39, 3, 2004: 125-131.

[2] Rubio, M.R., Ghaly, E.S. "In-vitro release of acetaminophen from sodium alginate controlled release pellets." *Drug Dev. Ind. Pharm.*, ,20, 1994: 1239-51.

[3] Yan, Y.R., Zhao, J.Y, Xie, J.F, Hong, S.H, Li, C.Y. "Emulsion Electrospinning of Sodium Alginate and Polylactic Acid Fiber with Skin-Core Structure." *Fiber Society 2013 Spring Conference Proceedings*, Geelong, Australia, 2013: 22-24

[4] Li, X.Q, Su, Y., Liu, S.P. "Encapsulation of proteins in poly(l-lactide-co-caprolactone) fibers by emulsion electrospinning." *Colloids and Surfaces B: Biointerfaces*, 75, 2, 2010: 418-424.

[5] Chan, L.W., Lee, H.Y., Heng, P.W.S. "Production of alginate microspheres by internal gelation using an emulsification method." *International Journal of Pharmaceutics*, 242, 1-2, 2002: 259-62.

[6] Mosmann, T. "Rapid Colorimetric Assay for Cellular Growth and Survival: Application to Proliferation and Cytotoxicity Assays." *Journal of Immunology Methods*, 65, 1-2, 1983: 55-63.

Investigation of the State and Distribution of Water in Poly(ethylene terephthalate)-Poly(ethylene glycol) Block Copolymer Matrix

Peng Ji, Chaosheng Wang, Jin Jin, Xianglin Chen, Huaping Wang
State Key Laboratory for Modification of Chemical Fibers and Polymer Materials, College of Materials Science and Engineering, Donghua University, Shanghai, China
cswang@dhu.edu.cn

ABSTRACT

In this paper, the interaction between water molecules and PETG copolymers was investigated by low-field nuclear magnetic resonance (LF-NMR) and Fourier transform infrared spectroscopy Attenuated total reflection-Fourier transform infrared spectroscopy (ATR-FTIR). The water state and distribution of Poly (ethylene terephthalate)-Poly (ethylene glycol) (PETG) copolymer during water adsorption were analyzed. A model of the interaction between water and PETG copolymers was proposed that there are three categories of water in PETG copolymers during water adsorption process. It is believed the method developed could be also applied to investigate various interactions between macromolecules and other small molecules without using deuterium samples.

INTRODUCTION

Water is one of the simplest and most significant molecules. The interactions between polymer and water have been considered to represent a subject of primary interest both from a fundamental and a technological perspective, for the desired properties of polymer materials can be greatly affected by the water absorbed [1]. Copolymers generally have complex morphological structures due to phase separation and segregation, and water diffusion in copolymers complicated. Recently poly (ethylene terephthalate)-poly (ethylene glycol) (PETG) copolymer has been drawn growing attention in numerous fields especially in textile [2] for their good water adsorption properties.

To better understand the interaction between water molecules and polymer materials, many efforts has been made to investigate water diffusion into polymers, including gravimetric sorption; nuclear magnetic resonance spectroscopy (NMR) [3], dielectric relaxation spectroscopy, and FTIR [4]. Among these methods, LF-NMR spectroscopy is a powerful technique for investigating the structure, mobility and properties of various polymeric systems, which is the most direct approach to evaluate the mobility changes of water molecules on polymer without destructive analytical tools and deuterated reagent [5].

In the present study, LF-NMR and ATR-FTIR methods were used to investigate the interaction between water and PETG copolymers at the molecular level. The states and distribution of water in PETG copolymers were analyzed.

EXPERIMENTAL APPROACH

Materials

PETG copolymers were synthesized by macromolecular interesterification method between poly (ethylene terephthalate) and Poly (ethylene glycol) (Mn=6000g/mol). Deuterated water (D_2O) and manganese chloride used in experiments was purchased from J&K Chemical. All materials were used without further purification.

Sample Preparation

The PETG copolymers films used for LF-NMR and ATR measurements were cast from the PETG solution at room temperature, which was maintained under vacuum at 40°C for 24 h to remove residual CF_3COOH. The thickness of the film was about 20μm.

Methods

A low field pulsed NMI 20-Analyst ((Resonance Instruments, Shanghai Niumag Corporation China) with 18.4 MHz was used in the experiment. The maximum point of every second echo was accumulated using a 90° pulse of 17 μs and a 180° pulse of 34 μs. The delay between the 180° pulses, τ, was 500 μs. The amplitude of every second echo was measured, a total of 18000 points were collected, and scans were coded using a repetition time (TR) of 4s. The relaxation measurements were performed at 32°C.

ATR-FTIR measurements were carried out on a Bruker Vertex-70 spectrometer equipped with a horizontal ATR accessory (Zinc selenide crystal, 45°). All spectra were taken by 16 scans at a resolution of 4 cm^{-1}. For the water absorption experiments, a PETG film was sandwiched between the Zinc selenide crystal and a piece of filter paper, and the assembly was mounted in the ATR cell. Ultrapure water was then injected into the filter paper, and data acquisition was started simultaneously.

RESULTS AND DISCUSSION

It has been confirmed that the protons of pure water are observed in only one status since it is a homogeneous single content [6]. In order to detect the mobility of protons in PETG copolymers, D_2O was used in our experiment. It is clear that there is no clear peak appearing in the copolymers mixture with D_2O, which means that the protons component of PETG is too short to be detected in LF-NMR measurement.

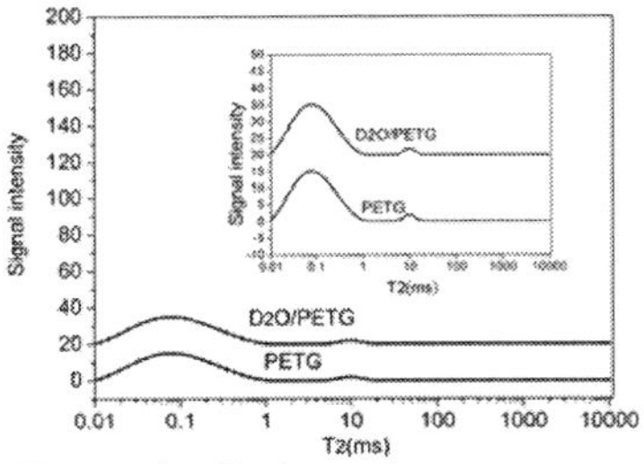

Figure 1. T_2 inversion spectra of PETG/D_2O by LF-NM method.

Compared with the inversion spectra of PETG/D_2O, the inversion spectra of PETG/H_2O disappears three new peaks (shown in Fig2), and the new peaks belongs to the water components. There are three categories of water in PETG copolymers during water adsorption process. The water in PETG copolymers were designed as S_2, S_1 and S_0. The integrated signal intensity of water peak in the inversion spectrum reflects the amount of water in PETG copolymers. The results of ratio of water in copolymers are shown in Table. 1.

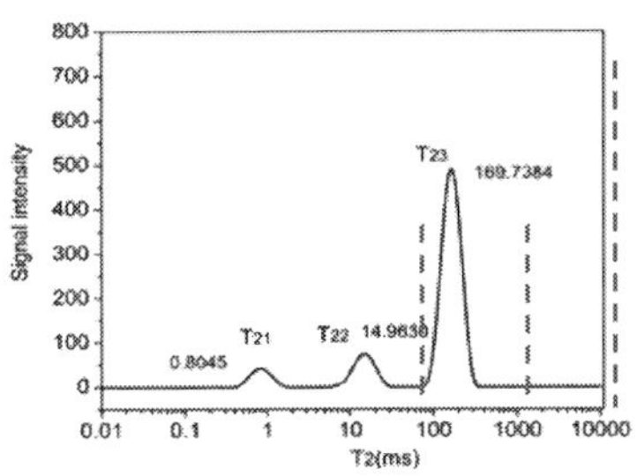

Figure 2. T_2 inversion spectra for PETG/H_2O by LF-NM method.

To further verify the state of water in PETG copolymers during water absorption process, ATR-FTIR method was used. The synchronous and asynchronous 2D correlation spectra in the region of 3800-3000 cm^{-1} are shown in Fig 3. There is only one strong auto peak located at 3450 cm^{-1} in the synchronous spectrum, which is assigned to the O-H stretching of water. On the other hand, there are one broad positive cross-peak (3530/3120 cm^{-1}) and one negative cross-peak (3700/3530cm^{-1}) in the asynchronous spectrum, indicating that the broad band in the region of 3800-3000 cm^{-1} consist of three bands located at 3700, 3530 and 3120cm^{-1}. The result of analysis is consistent with that of LF-NMR method.

Table 1. The water state and distribution of PETG copolymer.

Water categories	S_2	S_1	S_0
Ratio (%)	4.5	9.3	60.1

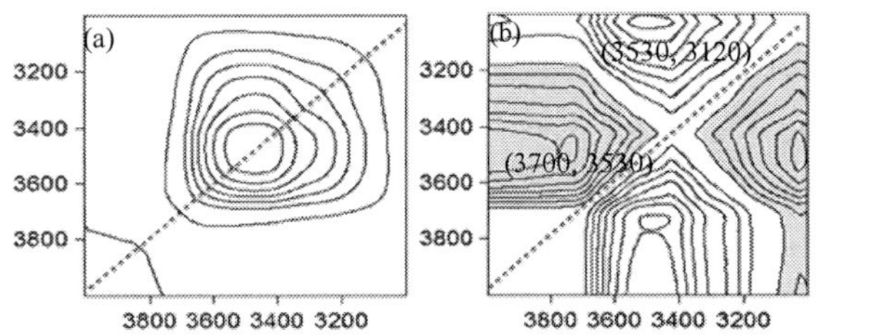

Figure 3. 2D correlation FTIR spectra of the OH stretching band in the range of 3800-3000 cm^{-1}: (a) Synchronous and (b) asynchronous contour maps.

Based on the results of LF-NMR and ATR-FTIR methods, a model of the interaction between water and PETG copolymers is proposed (shown in Fig. 4). Water absorbed to PETG copolymers can form different kinds of hydrogen bonds due to the polarity of ether bond in chains.

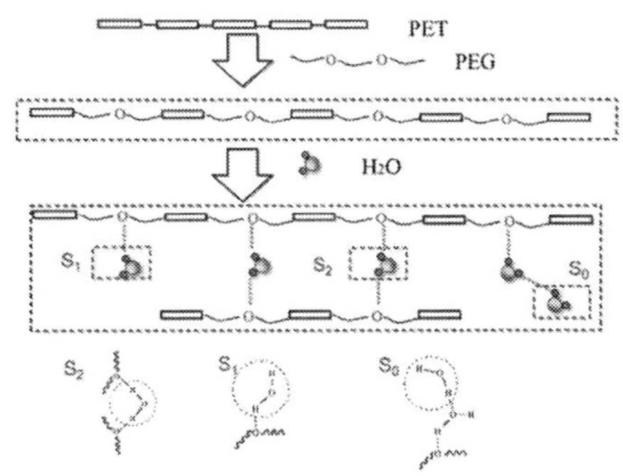

Figure 4. A model of the interaction between water and PETG copolymers.

CONCLUSIONS

The methods of LF-NMR and ATR-FTIR can be successfully applied to investigate water state and distribution of PETG copolymer during water adsorption. The analysis of T_2 component with different states showed that there are three categories of water in PETG copolymers during water adsorption process.

REFERENCES

[1] Hossain, K.M.Z, Parsons, A.J., Rudd, C.D., Thielemans, I.A.W. *Eur. Polym. J.*, 53, 2014: 270-281.

[2] Meng, Q., Hu, J. *Sol. Energ. Mat. Sol. C*, 92, 2008: 1245-1252.

[3] Volke, F., Eisenblatter, St., Klose, G. *Biophys. J.*, 67, 1994: 1882-1887.

[4] Kitano, H., Nagaoka, K., Tada, S., Gemmei-Ide, M., Tanaka, M. *Macromol. Biosci.*, 8, 2008: 77-85.

[5] Wu, J., Wang, Z., Lin, W., Chen, S. *Acta. Biomater.*, 9, 2013: 6414-6420.

[6] Wu, J., Chen, S.F. *Langmuir*, 28, 2012: 2137-2144.

Stress Distribution in Micro-Region of Carbon Fiber Composites by Carbon Nanotube Raman Strain Media

Shi Chen[1], Yanni Shi[1], Shaoxu Xia[1], Haiyu Quan[1], Qilin Wu[1,2]
[1]State Key Laboratory for Modification of Chemical Fiber and Polymer Materials, Donghua University, Shanghai, China
[2]College of Material Science and Engineering, Donghua University, Shanghai, China
wql@dhu.edu.cn; chensswork@163.com

OBJECTIVE

Tiny addition of carbon nanotubes into polymer matrix enables Raman mapping investigation of micro-region feasible via carbon nanotubes strain media. Raman mapping investigations reveal the stress profiles of micro-region nearby a single carbon fiber.

INTRODUCTION

Growing interests have been focused on carbon nanotubes(CNTs) due to excellent electrical, mechanical, and chemical performances[1]. As a new technique, Raman spectroscopy has been an important tool in investigating CNTs because their spectra are particularly sensitive to the microstructure of CNTs[2; 3]. For instance, Raman G' band (2500-2700cm^{-1}) of CNTs is very sensitive to axial deformation: it shifts significantly from its original position to a lower wavenumber position when CNTs are under tension and to a higher wavenumber position when under compression. Based on the mechanical sensitivity of Raman spectra of CNTs, Wagner et al. systematically studied the behavior of single-walled CNTs under uniaxial strain via resonance Raman scattering and sensed matrix stresses around a single glass fiber by the use of CNTs[4-6]. Especially, Qiu et al. proposed Raman strain rosette, an experimental study on a new technique of strain measurement on the basis of a theoretical model of a CNTs strain sensor[7; 8]. The usage of a carbon nanotube sensor acts as a wireless sensor for measuring strain at the microscale was displayed.

Most research focused on the methodology of carbon materials working as mechanical media or sensor for strain effect by Raman Spectroscopy[9; 10]. Nevertheless, less efforts have been made to explore micro-regions within composite materials via Raman mapping. As a matter of fact, a good understanding of micro-regions' profile plays an important role in the architecture of polymer-fiber composites.

APPROACH

Specimen

Carbon fiber was manufactured in our laboratory: Polyacrylonitrile precursors have been oxidized at 280 ℃ in air, and carbonized at 1300 ℃ in nitrogen atmosphere. The products taken were further heated at graphitization temperature of 2200 ℃ in argon atmosphere so as to yield resulting fibers, a single carbon fiber being of a diameter of 7 μm. Multi-walled CNTs were provided by Chengdu Organic Chemicals Co.Ltd., Chinese Academy of Sciences. The epoxy resin (E-44) with the curing agent (T-31), as the matrix of the composites, was provided by Shanghai Resin Factory Co. Ltd. CNTs were dispersed uniformly in E-44 by sonicating for 20 min and the curing agent (0.5 wt%) was then added and mixed mechanically. After the curing of the composites, samples of sizes 2×1×0.5 cm^3 were prepared.

Raman testing

The sample was placed in a home-made tensile device and Raman spectra were obtained with a Renishaw inVia microscopy. 632.8 nm line of a 2 mW He-Ne laser was applied. The laser beam was focused on the specimen through a ×50 objective lens. Laser spot of 2 μm diameter was focused on surface of specimen. The schematic diagram of Raman mapping area is shown in Fig.1.

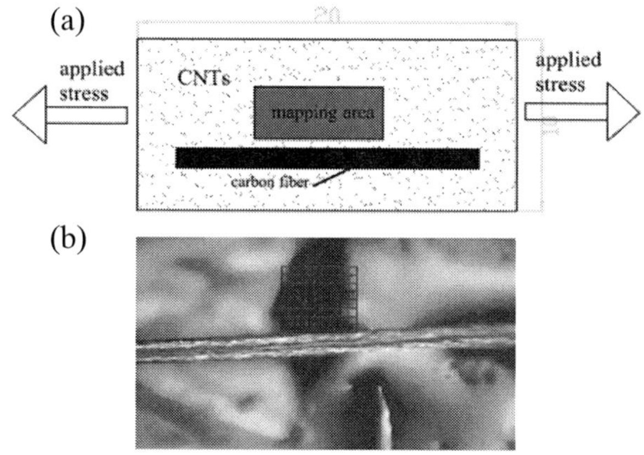

Fig. 1. (a) Schematic diagram of Raman detection (Unit:mm); (b) Specimen photograph (red grids stand for mapping area).

RESULTS AND DISCUSSION

A linear relationship exists between the G' band wavenumber of MWCNTs embedded in the matrix and the applied strain, as shown in Fig.2. The inset is CNTs raman spectra under different applied strains. The calibration relationship implies that for every percent of positive strain change, the G' band reduced in wavenumber has a constant value of 25.41cm^{-1}.

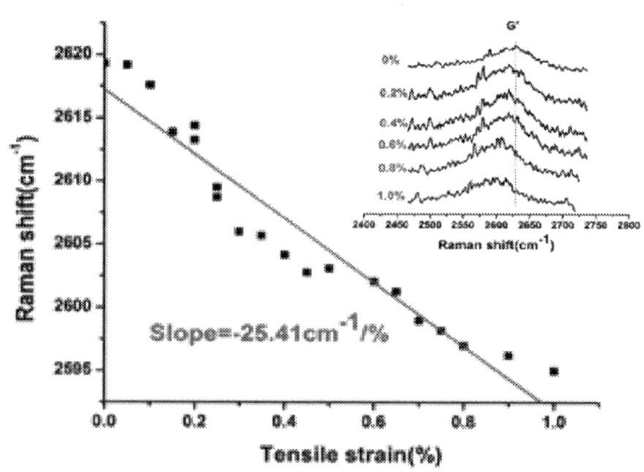

Fig. 2. Calibration of G' band of MWCNTs under different strains.

$$\sigma = \frac{\Delta\omega * E}{-25.41 * 100} \quad (1)$$

The practical stress in MWCNTs is quantified by equation (1). Where, σ is the stress in the matrix, E is the matrix Young's modulus and Δωis the measured w
minus the wavenumber at zero applied strain.

(a)

(b)

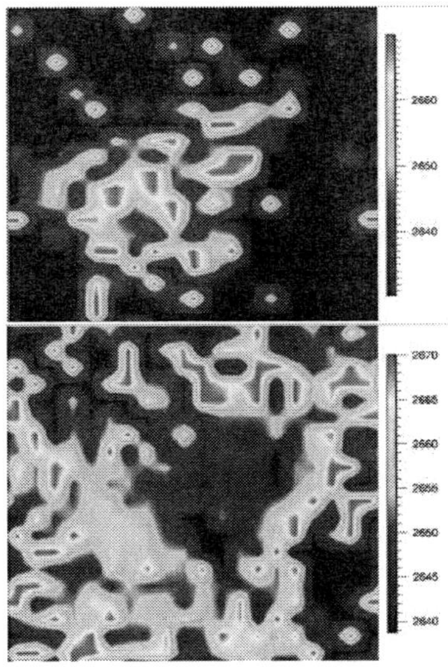

Fig. 3. Distribution profiles (25×25 μm^2) of G' band offsets by Raman mapping under (a) 0% and (b) 0.4% external strain.

In order to deeply detect the nearby micro-region of carbon fiber within composites, Raman mappings with equivalent mapping step length were conducted. Two dimensional mapping of G' band offsets are shown in Fig.3, which presents and indirectly displays the unstrained and strained micro-region nearby carbon fiber. The range of contrast scale of both Raman images has been fixed within range of c.a. 2640(dark) ~ 2670 cm^{-1}(bright). Raman G' band shifts show a universal red-shift over the center region, as reflected by the increment of darkness.

The pratical stress distribution near the single carbon fiber can be caculated from equation 1. As shown in Fig.4, tensile stress and compressive stress are identified by color labeling. The positive values stand for tensile stress, while the negative values mean compressive stress.

The non-uniform of stress distribution profiles reveal the real situation of micro-region within composites. The reason for nonuniformity might be due to the contact among carbon fiber, MWCNTs and epoxy resin, consequently, the van der Waals force differs in the micro-region. Such ununiformity calls for caution on preparing composite samples, especially how carbon fibers are placed within composites.

CONCLUSION

CNTs added into polymer matrix as strain medias enable Raman mapping investigation of micro-region feasible. The G' band of MWCNTs shifts proportionally under different external strain, and the micro-region Raman mapping nearby a single carbon fiber is displayed. Stress distributions of strained and unstrained micro-regions reveal the mechanical profiles of composites.

(a)

(b)

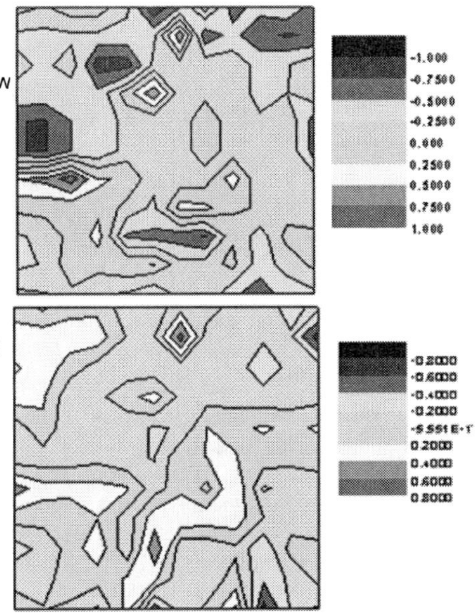

Fig. 4. Distributions profiles of practical stress of mapping area under (a) 0% (b) 0.4% external strain.

KEYWORDS

Raman mapping; carbon nanotubes; micro-region

ACKNOWLEDGEMENTS

This work was funded by grants from Chinese NSFC (No.60975059), the Research and Innovation Project of Shanghai Municipal Education Commission (No. 14zz069) and the Key Laboratory of Advanced Civil Engineering Materials of Ministry of Education, Tongji University (No.201301).

REFERENCES

[1] De Volder, M.F., Tawfick, S.H., Baughman, R.H., Hart, A.J. *Science*, 339, 6119, 2013: 535-9.

[2] Qiu, W., Kang, Y.L., Lei, Z.K., Qin, Q.H., Li, Q., Wang, Q. *J. Raman Spectrosc.*, 41, 10, 2010: 1216-1220.

[3] Schadler, L.S, Giannaris, S.C, Ajayan, P.M. *Appl. Phys. Lett.*, 73, 26,1998: 3842-3844.

[4] Barber, A.H., Zhao, Q., Wagner, H.D., Baillie, C.A. *Compos. Sci. Technol.*, 64, 13-14, 2004: 1915-1919.

[5] Zhao, Q., Wagner, H.D. *Compos. Part A: Appl. S.*, 34, 12, 2003: 1219-1225.

[6] Zhao, Q., Wagner H.D. *Phil. Trans.*, 362, 1824, 2004: 2407-2424.

[7] Qiu, W., Li, Q., Lei, Z.K., Qin, Q.H., Deng, W.L., Kang, Y.L. *Carbon*, 53, 2013: 161-168.

[8] Qiu, W., Kang, Y.L., Lei, Z.K., Qin, Q.H., Li, Q. *Chinese Phys. Lett.*, 26, 8, 2009: 080701.

[9] Frank, O., Tsoukleri, G., Parthenios, J., Papagelis, K., Riaz, I., Jalil, R., Novoselov, K.S., Galiotis, C. *ACS Nano.*, 4, 6, 2010: 3131-3138.

[10] Bissett, M.A., Izumida, W., Saito, R., Ago, H. *ACS Nano.*, 6, 11, 2012: 10229-10238.

The Microdeformation Behaviors in the Vicinity of V-Shaped Defects in CNTs/Epoxy Composites

Guizhi Ren[1], Shi Chen[1], Shaoxu Xia[1], Yanni Shi[1], Qilin Wu[1,2]
[1]State Key Laboratory for Modification of Chemical Fiber and Polymer Materials, Donghua University, Shanghai, China
[2]College of Material Science and Engineering, Donghua University, Shanghai, China
wql@dhu.edu.cn

OBJECTIVE

This work investigates the different effects (external strain value and position of defect) on the micro-deformation behaviors in the vicinity of V-shaped defects in CNTs/epoxy composites by confocal Raman Spectroscopy.

INTRODUCTION

Material failure induced by stress concentration is a sophisticated situation, which always takes place in materials containing wispy defects. The subject about acquiring the stress distribution information in the vicinity of defects to predict or detect the onset of failure in materials is concerned by researchers all the time.

As a new developed technique, Raman spectroscopy is non-contact, effective and precise in micro-scale spatial solution for micro-deformation measurement[1], such as the interfacial shear stress distribution along the fiber length in single fiber fragmentation test[2], and adhesive property between fiber and matrix[3]. Unfortunately, most of the materials applied in engineering are Raman-insensitive. Interestingly, CNTs could become sensitive strain sensors in Raman-insensitive materials, which firstly proposed by Wagner H D can break the limitation[4]. The G and G' Raman bands of CNTs shift linearly with the external strain. Based on the theory, Dally and Raley established the empirical relationship between Raman shifts and the external strain by using CNTs embedded in the vicinity of circular hole as sensing media. However, the research was on the premise that all CNTs were parallel to the tensile stress. Considering CNTs' random directions, an alternative is to use polarized Raman spectroscopy. Qiu, et al,[5; 6] acquired the distributions of stress around the circular hole by the use of polarized Raman spectroscopy.

APPROACH

Specimen

An amount (5 wt%) of pristine multi-walled CNTs (L-1020, Shenzhen Nano-Tech Port Co.) were dispersed into a liquid epoxy (6002, Shanghai Resin Factory Co., Ltd) and mixed for 12h in a ultrasonic bath, and the curing agent (T31, Shanghai Resin Factory Co., Ltd, mass ratio 5:1) and acetone (3 wt%) as the diluent was mixed in, in the mean time. The mixture was stirred at room temperature until all the trapped air was eliminated and the mixture became homogeneous. The mixture was dropped on a glass sheet under a spin coater, resulting in a film of about 100 μm thickness. After curing at room temperature for 2h, the incomplete curing film was peeled from the glass sheet carefully and then was cut into a 8.5×2.5 cm^2 strip with a V-shaped defect, punched inner or at the edge of the strip. Finally, CNTs/epoxy films with V-shaped defects were prepared after continuously curing at room temperature for 72h.

Raman testing

The CNTs/epoxy composite films were tensiled using a self-made stretching device, and mapping areas are shown in Fig. 1. A confocal microRaman spectroscope (Invia Reflex, Renishaw, UK) with a He-Ne laser source (633nm, ~2mW power) was employed in the backscattering. The incident laser beam was focused on the specimen through a 50× objective lens with a 1μm^2 sampling spot. Each spectrum was acquired in 1s exposure time and 10 times accumulation. The data were fitted by a mixed Gauss/Lorentz function.

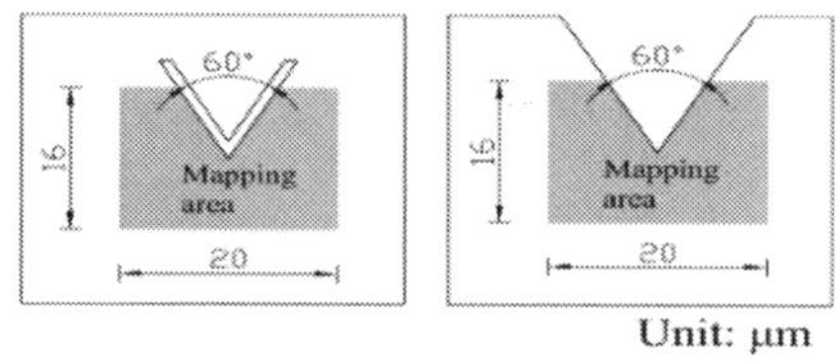

Fig. 1. Geometrical shapes and Raman mapping areas

RESULTS AND DISCUSSION

Calibration procedure

The Raman spectra of pristine CNTs and CNTs/epoxy composite are shown in Fig. 2 (inset). The Raman G' band, a sensitive peak located in 2500~2700 cm^{-1}, with no epoxy peak overlap in the region.

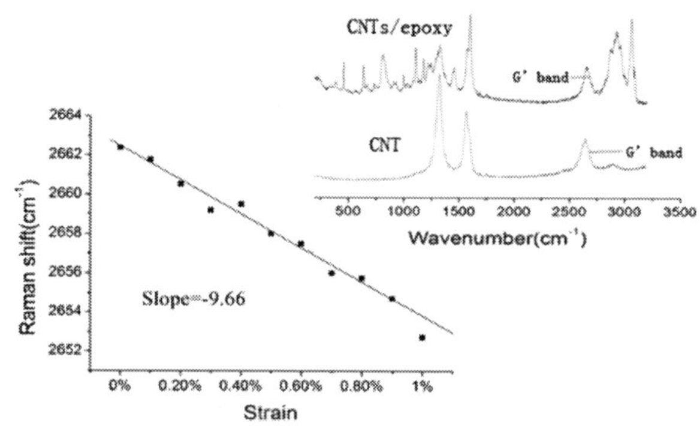

Fig. 2. Relationship of Raman shift and applied strain of CNTs/epoxy composite film and Raman spectra(inset) of CNTs and CNTs/epoxy composite.

The G' Raman wave number downshifts when CNTs are under tension and up shifts when under compression. Such shifts could be correlated with applied stress through a calibration procedure. According to Fig. 2, a linear relationship exists between the G' Raman wave number of CNTs and the applied strain. On the basis of Hooke's law and the linear relationship, the local stress σ (in MPa), and the G' Raman wave number shift $\Delta\omega$ (in cm^{-1}) relative to the wave number at zero applied strain could be related by equation (1):

$$\sigma = -\frac{15\Delta\omega}{9.66} \qquad (1)$$

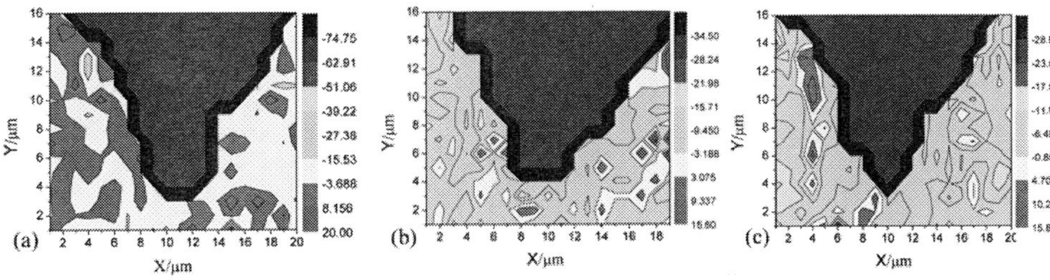

Fig. 3. Stress distributions of specimen containing inner V-shaped defects under (a) 0 (b) 0.5% and (c) 1% strain.

The external strain effect

The G' Raman wave number data on mapping area in the vicinity of V-shaped defects are substituted into Equation 1, leading to a contour plot of the stress which is shown in Fig. 3, where the X-axis and Y-axis are the distance around the defects. The darkest color area is the V-shaped defect, and the positive number shown in the legend means tensile stress, while, the negative number means compressive stress.

As is well known that there exists residual stress in internal polymer during the whole curing process. As seen in Fig. 3a, the film was probed under no tension (0 strain), and stress values are mostly negative number, which illustrates that the film is mainly affected by compressive stress and the values are almost about 15.53 MPa (the yellow area). At a strain of 0.5% (Fig. 3b), the stress distribution is similar to the previous one (0 strain), whereas the stress values almost decrease to 9.45 MPa (the green area), and the stress values further decrease to 6.40 MPa (the green area) at 1.0% strain (Fig. 3c). So overall, stress redistribution in the vicinity of V-shaped defect is obviously observed and the residual compressive stress has remarkable reduction with the increasing applied strain.

In addition, the stress concentration zone was changed under the applied strain. Stress is relatively well-distributed in the mapping area without tension at the beginning, and then the highest stress concentration occurs at the both sides of the V-shaped defect, and by increasing the strain, the stress concentration transfers to the lower-left corner of the defect finally. The process implies the floating stress concentration resulting from stress redistribution has significant effect on the onset of material failure.

The position of defect effect

As shown in Fig. 3(b) and Fig. 4, we can easily observe the stress concentration zones are different from each other when the V-shaped defect is within or at the edge of the films under 0.5% strain. In detail, the edge defect shows an apparent stress concentration near the rim of the V-shaped defect and is more sensitive to the external stress. Furthermore, the obtained stress mainly run the range of values from -9.45 MPa to -3.18 MPa for inner defect, which is lower than the result of edge defect, between -5.47 MPa and -1.35 MPa.

Comparing with the inner defect, the edge defect is more sensitive to the external stress or strain, and easily leads to stress concentration and material failure. Also found in daily production, the fatal defects occurring in the materials tend to aggregate at the edge of materials. Therefore, we should pay more attention to defects in the surface or at the edge of materials.

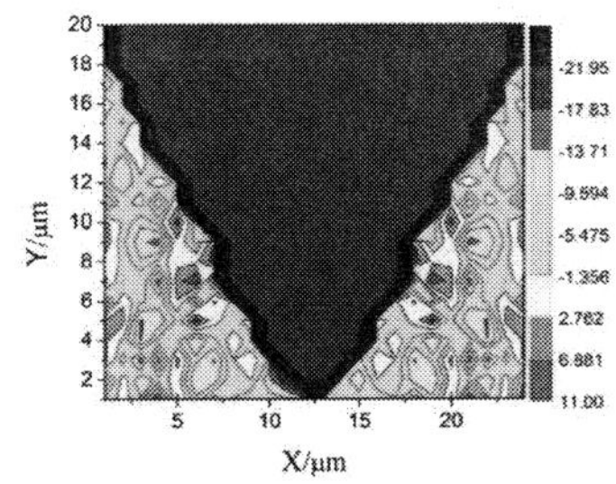

Fig. 4. Stress distribution near V-shaped defect at the edge of specimen under 0.5% strain.

CONCLUSION

Carbon nanotubes are used as strain or stress sensors to achieve stress distribution in the vicinity of V-shaped defect via Raman spectroscopy. The external stress can easily eliminate the residual stress, and with the increment of the applied stress, the stress concentration occurs in the vicinity of V-shaped defect and the zone is close to the defect gradually. The different stress sensitivities are observed in the defects in different position. Under the same tension, for edge defect, the residual stress is more offset and the stress concentration zone is closer to V-shaped defect than the inner defect.

ACKNOWLEGEMENTS

This work was funded by grants from Chinese NSFC (No.60975059), the Research and Innovation Project of Shanghai Municipal Education Commission (No. 14zz069) and the Key Laboratory of Advanced Civil Engineering Materials of Ministry of Education, Tongji University (No.201301).

REFERENCES

[1] Bounos, G., Andrikopoulos, K.S., Karachalios, T.K., Voyiatzis, G.A. *Carbon*, 76, 2014: 301-309.

[2] Liu, L., Li, L., Gao, Y., Tang, L., Zhang, Z. *Compos. Sci.Technol.*, 77, 2013: 101-109.

[3] Barber, A.H., Zhao, Q., Wagner, H.D., Baillie, C.A. *Compos. Sci. Technol.*, 64, 13-14, 2004: 1915-1919.

[4] Zhao, Q., Frogley, M.D., Wagner, H.D. *Compos. Sci. Technol.*, 61, 14, 2001: 2139-2143.

[5] Qiu, W., Kang, Y.L., Lei, Z.K., Qin, Q.H., Li, Q., Wang, Q. *J. Raman Spectrosc.*, 41, 10, 2010: 1216-1220.

[6] Qiu, W., Li, Q., Lei, Z.K., Qin, Q.H., Deng, W.L., Kang, Y.L. *Carbon*, 53, 2013: 161-168.

Study on the Application of Ammonium Dihydrogen Phosphate in Textile Size

Xuefeng Lu[1], Dan Sheng[1], Hao Liu[1], Rui Wang[3], Weilin Xu[2], Genyang Cao[2]

[1]School of Textile Science and Engineering, Wuhan Textile University, Wuhan, China
[2]State Key Laboratory of New Textile Materials and Advanced Technology, Wuhan Textile University, Wuhan, China
[3]School of Chemistry and Chemical Engineering, Wuhan Textile University, Wuhan, China

443744886@qq.com; 249747014@qq.com

ABSTRACT

Modification effect of silane coupling agent on the mixed size including phosphate starch and acrylate was researched. Effect of silane coupling agent (KH580) usage on the mixed slurry, size film, and its adhesion. Studying the composition and the crystalline state of size film by FTIR, XRD and tensile fracture test characterization methods. The test result shows adhesion of starch to cotton fiber can be improved, mechanical properties of mixed size film can be largely improved and size film wear can be reduced when adding silane coupling agent in mixed size. Silane coupling agent did not change the chemical composition of mixed size and the crystallization peak of the size film became sharper, the crystal type got better. As the increase of silane coupling agent usage, size film breaking elongation rate and the strength was increased, while size viscosity was reduced. Wear resistance and break work of starch size film can be improved when silane coupling agent usage in the range of 0.1% to 0.5% (Weight ratio), adhesion of size film to cotton fiber can be increased.

KEYWORDS: Size Mixture, Silane Coupling Agent, Size Film, Adhesion

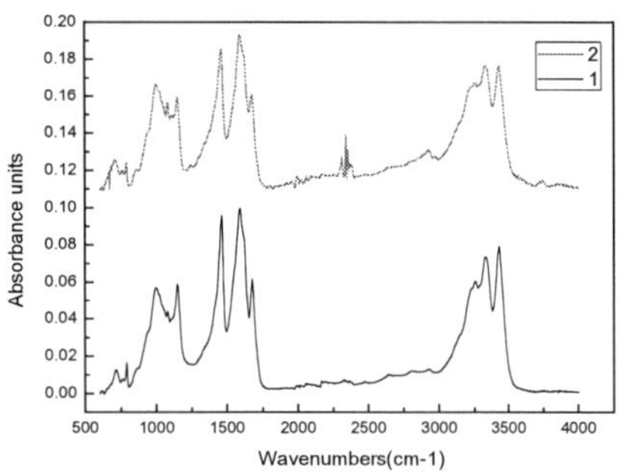

Fig. 1. The FTIR picture of the mixed size

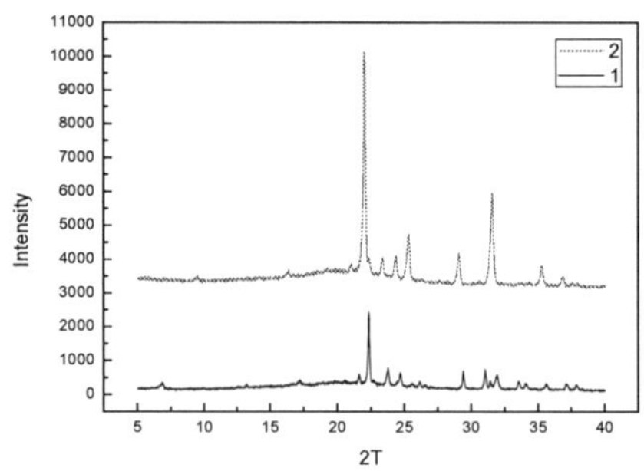

Fig. 2. The XRD picture of the mixed size

Centrifugal Spinning of Viscoelastic Liquids: Flow Visualization and Discretized Modeling of the Jetting Behavior

Yevgen Zhmayev[1], An-Cheng Ruo[1], Maunica Jyothi[1], Tao Huang[2], Yong Lak Joo[1]
[1]School of Chemical & Biomolecular Engineering, Cornell University, Ithaca, NY, USA
[2]DuPont Central Research & Development, Experimental Station, Wilmington, DE, USA
ylj2@cornell.edu

ABSTRACT
We present an experimental visualization and discretized modeling study of centrifugal spinning, which is a novel and scalable method of producing nanofibers. The investigation was conducted using Newtonian and viscoelastic liquids to study the effect of viscoelasticity, flowrate and driving force on the initial thinning behavior, jet contour shapes and radii. Newtonian polybutelene (PB) and viscoelastic poly-isobutelene (PIB)-based Boger fluids were utilized as test fluids in the study. Our results reveal that increasing the viscoelasticity leads to a faster initial thinning of the polymer jet. However, the effect is strongly coupled with the rotation speed, and due to a faster increase in extensional viscosity for highly viscoelastic fluids, the thinning slows down with the increase in angular velocity. Initial thinning is shown to be faster for the lower flow rates. Viscoelasticity and centrifugal force have a significant influence on the jet contour radii. The maximum radius decreases for more viscoelastic fluids, and with the increase in angular velocity, in a certain range of rotation speeds. The comparison with the discretized element modeling is reported to confirm the model predictive potential. Results from the centrifugal spinning experiments are compared to electrospinning in order to observe a qualitative similarity.

INTRODCUTION
Centrifugal spinning is a novel process for producing nanoscale or submicron fibers, where the centrifugal force is utilized to accelerate and elongate a fluid jet. Similarly to electrospinning process, which uses electric force to spin the fluid jet, centrifugal spinning has two main stages. The first stage ("stable jet") starts when the fiber leaves the edge of the rotating disk and ends at the onset of non-spiral motion. The main advantage of the needless centrifugal spinning process is the high production rate and low process cost, which gives a potential for future industrial utilization. Productivity was an issue for well-established spinning process, such as electrospinning (0.01-0.1 mL/min per nozzle), however centrifugal spinning can offer a much higher production rate (10-20 mL/min per nozzle in current study). In this paper, we focus on the experimental visualization of the fluid flow and examine the effects of viscoelasticity, driving force and flow rate on the initial thinning of the polymer jet and the contour shapes and radii. The initial thinning behavior can elucidate the interaction between forces acting on the polymer jet, and it is a good starting point for the comparison with simulation, due to the absence of instability. The effect of the fluid properties and experimental conditions on the maximum contour radii is beneficial for the optimization of fiber collection. Experimental results from centrifugal spinning experiments of Boger fluids are compared with discretized modeling predictions. Generally, theoretical predictions are in a good agreement with the experimental results, but there are still some discrepancies in the case of viscoelastic fluids. The model is sensitive to driving force and the prediction is generally better for the higher values of angular velocity. After some modifications it can be further utilized for the analysis of the emanation and propagation of instability. The effect of viscoelasticity, driving force and flow rate on the initial thinning behavior are then compared to the results from electrospinning process [1] and repeated in this work.

EXPERIMENTAL
The fluids used in this study are viscoelastic PIB/PB Boger fluids of different compositions (2000, 4000 and 8000 ppm), and a low molecular weight PB that was used to examine the effects of driving force and flow rate on Newtonian fluid. The density and surface tension of the fluids are 850 kg/m^3, 1.46 x 10^{-2} N/m, respectively, and the viscosity, and relaxation time of the fluids are summarized in Table 1.

Table I. Viscosity and relaxation time of test fluids.

PIB Conc.	Viscosity, [Pa s]	Relaxation time, [s]
0	0.028	-
2000 ppm	0.072	2.50×10^{-3}
4000 ppm	0.102	3.11×10^{-3}
8000 ppm	0.175	6.25×10^{-3}

The centrifugal spinning setup is depicted in Fig. 1. The fluid is pumped to a cavity inside of the disk, it travels radially outward to a thin circular channel (to prevent die swell) and then on to the orifice of a certain diameter.

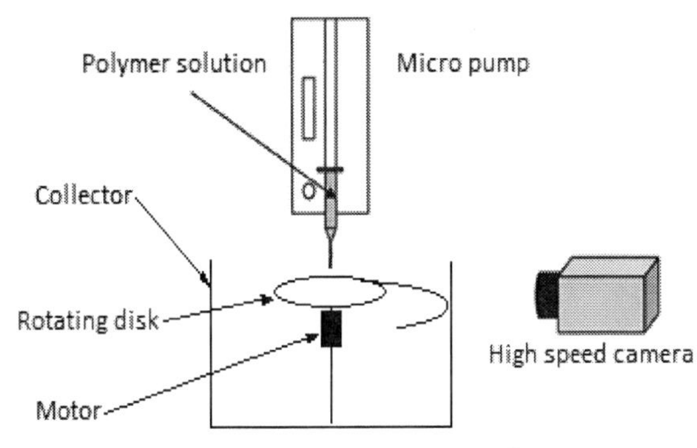

Fig. 1. Centrifugal spinning setup.

RESULTS AND DISCUSSION
The effect of viscoelasticity was probed by changing the polymer relaxation time through the concentrations of Boger fluid. Deborah number, De, along with other parameters is summarized in Table II. a_0 is the orifice radius, R_{disk} is the disk radius, λ is the relaxation time, t_0 is the characteristic

time, ρ and μ is the density and viscosity, Ω is the angular velocity, ε_D is the permittivity of the free space, E_0 is the electric field strength and U_0 is the initial velocity.

Table II. Characteristic scales and dimensionless groups.

Characteristic scales			Dimensionless groups
Radius	Time	Length	$De = \lambda U_0/a_0$; $Re = \rho U_0 a_0/\mu$
a_0	$t_0 = a_0/U_0$	R_{disk}	$Rb = 1/\Omega t_0$; $\varepsilon_E = \varepsilon_D E_0^2/\rho U_0^2$

It can be seen from Fig. 2, that increasing the viscoelasticity will result in faster thinning of the polymer jet in the early stage of centrifugal spinning, and the thinning slows down as the jet moves farther from the spinneret, resulting in thicker fibers for more viscoelastic fluids. The fluid viscoelasticity has also an effect on the jet contour radius, which is the maximum distance from the spinning origin to the polymer jet at a single plane. The maximum radius of the contour for the same angular velocity, decreases with increasing De (Fig. 3). This behavior can be explained by the hoop stress, $\tau_{hoop}=(\tau_{zz}-\tau_{rr})/(R/d)$, which arises from the curvilinear trajectory and viscoelasticity of the fluid.[1] The hoop stress is more significant for higher De, due to the increase in the axial stress.

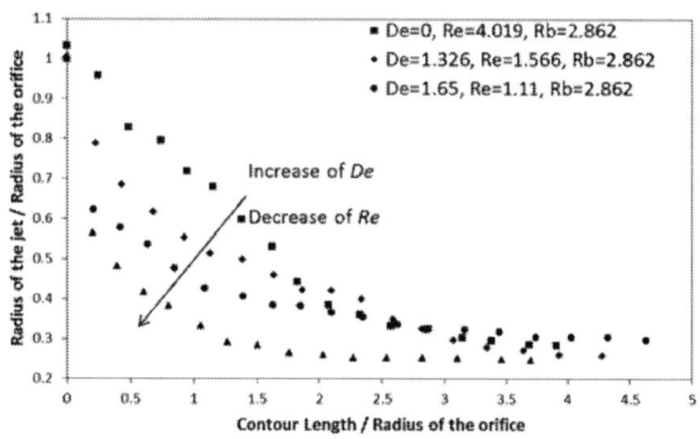

Fig 2. Radius evolution of the polymer jet during centrifugal spinning (12.5 ml/min, 1770 rpm) for different De.

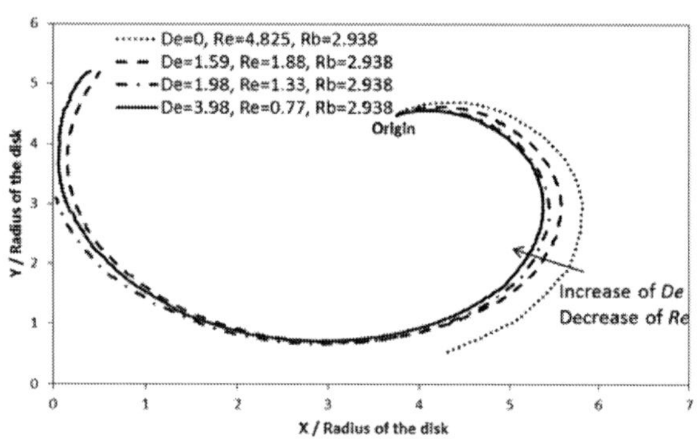

Fig 3. Contour radii profiles for centrifugal spinning (15 mL/min, 2070 rpm) of four test fluids.

The effect of flowrate can be seen from Fig. 4. The initial thinning is faster for slower flow rates (small Re), which matches the earlier observations from electrospinning studies. The effect of the flow rate is coupled with the effects of viscoelasticity and driving force, which can be observed from the change of De and Rb (Rossby number, which is inversely proportional to centrifugal force) with the change of the flow rate.

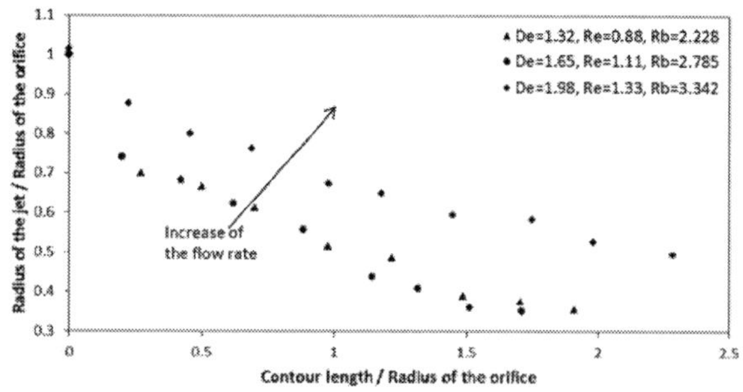

Fig 4. Flowrate effect on initial thinning during centrifugal spinning of the 4000 ppm Boger fluid at 1820 rpm.

The effect of driving force, *i.e.* centrifugal force on contour radii is shown in Fig. 5. As we increase the rotation speed during the spinning of viscoelastic Boger fluid – the jet tends to move closer to the disk.

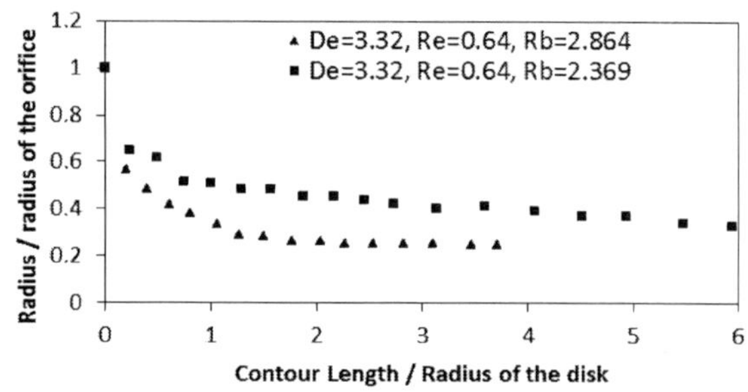

Fig. 5. Influence of the rotation speed on the thinning behavior for 8000 ppm Boger fluid (1770 and 2140 rpm).

The comparison with discretized modeling for the initial thinning behavior is shown in Fig. 6. In discretized modeling, the jet is considered to be a series of discrete beads connected by viscoelastic springs. [2, 3] Each bead is pulled by its neighboring connector elements and a smooth trajectory can be formed. It is confirmed that the modeling prediction is generally in good agreement with experimental observations, and can correctly predict faster initial thinning for more viscoelastic fluids

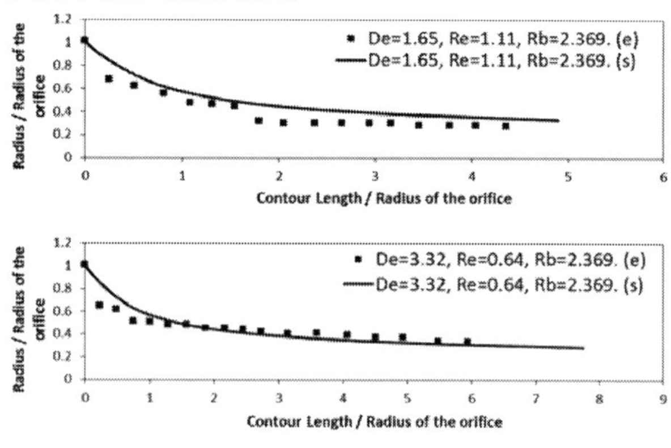

Fig. 6. Comparison between the simulation prediction ("s") and experiments ("e") for 4000 and 8000 ppm Boger fluids.

REFERENCES

[1] Caroll, C.P., Joo, Y.L. *Phys. Fluids*, 18, 053102, 2006.
[2] Caroll, C.P., Joo, Y.L. *J. Appl. Phys.*, 109, 094315, 2011.
[3] Ruo, A., Jyothi, M., Zhmayev, Y., Joo, Y.L. *Phys. Fluids*, 2015.

Relative Contribution of Stereoregularity and Molecular Weight on the Dissolution of Isotactic Poly(acrylonitrile) into Organic Polar Solvents

Masatomo Minagawa,[1] Yuji Hashimoto[2], Makoto Shimizu[2], Nobuhiro Sato3, Tomochika Matsuyama[3]

[1]NPO, Dream-Create-Laboratories, Naoe-sekitei 3800-1, Yonezawa, Japan
[2] Department of Polymer Science & Engineering, Yamagata University, Yonezawa, Japan
[3]Research Reactor Institute, Kyoto University, Kumatori, Japan
gakusai-minagawa@memoad.jp

ABSTRACT

The effect of stereoregularity and molecular weight (MW) on the dissolution of isotactic poly(acrylonitrile) (PAN) under elevated temperatures was quantitatively studied. Dissolution temperature (T_{sol}) determined by a laser-optical method was a good measure of this parameter. From systematic experiments, it was concluded that the former was predominantly larger than the latter, and the relative contribution was roughly estimated to be (8:2~7:3) as follows. When the value of T_{sol} was plotted against the stereoregularity, a simple saturation type of curve was obtained. The results were explained in terms of the configurational entropy (Fig.1). In the inverse plots of T_{sol} against the stereoregularity, a linear decreasing line was obtained. For example, in homologous dinitrile solvents, CN-(CH2)n-CN (n=1~3), the strongest solvent was MaN (n=1), where these two factors were obtained in two different areas in the plots. Stereoregularity layed on a decreasing line, whereas the MW scattered in a square region crossed over the line (Fig. 2). In poor solvent (n=2, 3), the straight line shifted toward upward direction with milder slopes and the latter-region depressed and disappeared (Figs. 3,4). Under such extreme conditions, the former was only one remaining factor.

From the chain scission experiments by US irradiation, the relative contribution was estimated to be (8:2~7:3). Outstandingly predominant effect of the former over the latter was verified in this way.

INTRODUCTION

PAN can be used as a starting material of carbon fiber. PAN fiber is generally prepared by wet spinning from conc. solution, since it has no melting point (T_m) [1]. Selection of solvent is important not only from the cost but also from the high performance of the fiber. That is, polymer chain is generally extended in a good solvent, whereas in a poor solvent, it is rather contracted [2].

RESULTS AND DISCUSSION

Isotactic PAN was obtained in a form of fine white powder. It did not dissolve in any solvent at RT. Dissolution occurred only when the temperature was increased and exceeded some critical threshold. This process was well characterized by a laser-optical instrument. The T_{sol} was determined uniquely as a cross-over point between two lines in the turbidity data.

When T_{sol} was plotted against MW, a large scatter of the data resulted. When it was plotted against the stereoregularity, a simple saturation type of curve was obtained. It is obvious that (1) T_{sol} is mainly dedfined by the extent of stereoregularity, and (2) the higher the stereoregularity the higher the T_{sol}. The results were reasonably explained in terms of the concept of configurational entropy (ΔS) (Fig.1).

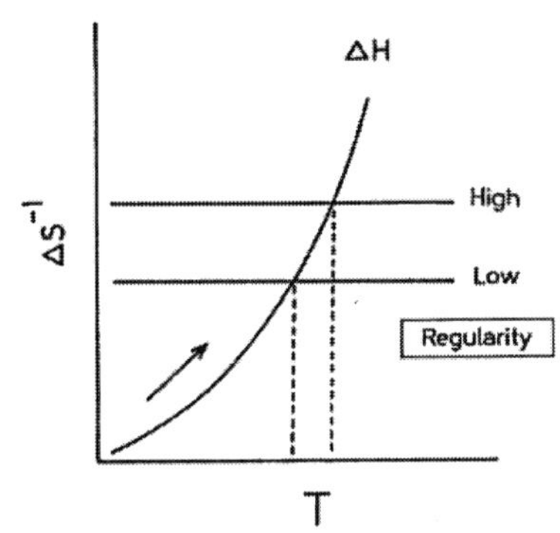

Figure 1. Configulational entropy in T_{sol} of PAN [3].

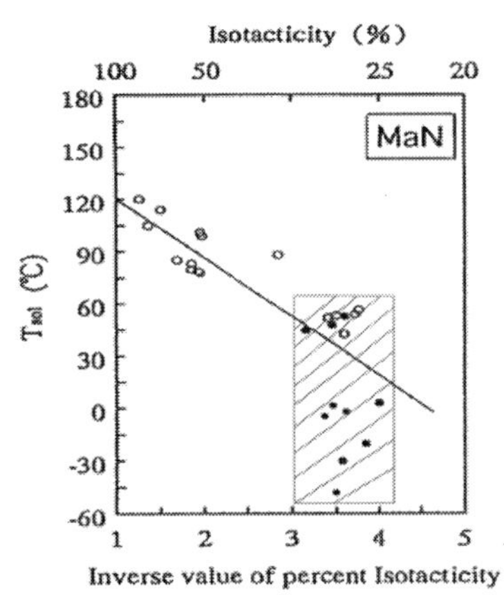

Figure 2. T_{sol} of PAN into MaN in the inverse plots.

sample was evaluated critically. These results are comprehensively discussed.

CONCLUSION

We can evaluate the relative contribution of these factors as follows: (1) The effect of stereoregularity was exceedingly larger than that of MW, and (2) the relative contribution of both factors on T_{sol} was about 8:2 in PAN.

REFERENCES

[1] Kriegbaum, W.R., Tokita, N. *J. Polym. Sci.*, 18, 1956: 467-488.

[2] Gedde, U.W. *Polymer Physics*. Berlin: Springer, 1995: 23.

[3] Minagawa, M., Miyano, K., Morita, T., Yoshii, F. *Macromolecules*, 22, 1989: 2054-2058.

[4] Minagawa, M., Hashimoto, K., Shirai, H., Morita, T., Yoshii, F. *Colloid & Polym. Sci.*, 278, 2000: 352-357.

[5] Minagawa, M., Agatsuma, K., Shimo, Y., Sato, N., Matsuyama, T. *Preceedings of the Fiber Society 2010 Spring Conference*, Turkey, May 12-14, 2010: 183-184.

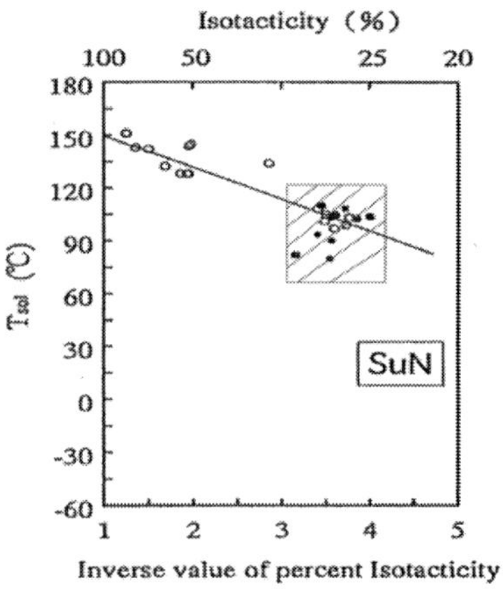

Figure 3. T_{sol} of PAN into SuN in the inverse plots.

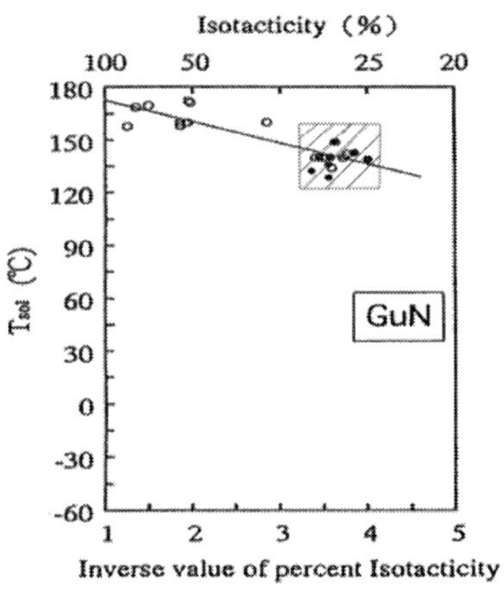

Figure 4. T_{sol} of PAN into GuN in the inverse plots.

This T_{sol} experment was made for dinitrile solvents. When these T_{sol} were plotted against the inverse value of isotacticity $(mm)^{-1}$, a linear decreasing line was obtained (Figs.2~4). In a good solvent (n=1; MaN), MW dependent region was represented by a red hatched square region, which crossed-over the solid line (Fig.2). In a poor solvent (n=2~3; SuN and GuN), however, solid line shifted to a higher temperature with milder slopes, and a square region was depressed and diminished (Figs.3,4). It is possible to summarize that MW dependence of T_{sol} is evident only in a good solvent, while it is not do in a poor solvent, and stereoregularity is only one remaining factor under extreme conditions [4]. It is worthwhile noting that this linear relationship holds good for all the organic polar solvents (11 kinds) investigated previously. From the chain scission experiments using ultra-sonic apparatus [5], one can evaluate the relative contribution of microtacticity and MW on T_{sol} quantitatively. That is, each PAN with different isotacticities was chain fragmented with sufficiently long time (~ a few months), and covered

Preparation of Polyamide 6 Modified with Sulphur-Containing Flame Retardant and Its Continuous Spinning Properties

Lili Li, Yinping Tao, Tianze Tian, Shouyang Lu, Bin Sun, Wenping Chen, Meifang Zhu
State Key Laboratory for Modification of Chemical Fiber and Polymer Materials, College of Materials Science and Engineering, Donghua University, Shanghai, China
zmf@dhu.edu.cn

ABSTRACT

A PA6/sulphur-containing flame retardant (AS, 10 wt%) master batch was prepared, as well as PA6 composites with different AS contents. Limiting oxygen index (LOI), mechanical test demonstrated that the LOI of PA6 increased from 24.4% to 33%, the tensile strength and the elongation at break decreased with the increase of AS content. Moreover，thermal gravimetric analysis (TGA) pointed that the effect of AS increased the char residue. Differential scanning calorimetry (DSC) showed that AS can promote the crystallization of PA6 and increased the melting endotherm. The flame retardant PA6 fibers were prepared by melt spinning of dried PA6 composites chips and master batches. The mechanical properties of PA6/AS fibers containing different AS content were tested by the multifilament yarn strength tester. The results showed that AS makes the fiber strength and elongation at break of composite fiber decrease while the master batch technology can improve the performance and appearance of the fiber effectively.

INTRODUCTION

At present, the main flame retardant additives for PA6 vary from a broad range, and the halogen, phosphorus and nitrogen containing additives are included. For example, the halogen-containing flame retardant is highly efficient, thus, its usage is limited for releasing many harmful gases when burning, which brings about a huge problem [1]. Besides, high content of phosphorus and nitrogen-containing flame retardant required to achieve acceptable level are always results in the deterioration of the processibility and mechanical property of the PA6 fibers [2,3].

Sulphur-containing flame retardant additives such as sulfamic acid and sulfonate are initially found applied in the flame retardant of cotton, paper and wood [4,5] . Lewin, et al., found that low content of ammonium sulfamate made the PA6 composite possess excellent flame retardancy and mechanical properties, in which the sulfation reaction occurring produced crosslinks and networks creating a hard and compact residue char [6]. Basing on the former conclusions, ammonium sulfamate(AS) was applied in the flame retardant PA6 fibers and the influence of AS content on the flame retardant, mechanical and continuous spinning properties was studied.

EXPERIMENTAL

First, PA6/AS composites (1.0 wt%, 1.5 wt%, 2.0 wt%, 2.5 wt%, 3.0 wt%) and master batch was prepared by twin-screw extruder. PA6, PA6 composites and PA6/AS master batchs were pre-crystallized at 90℃ and then dried at 120℃ for 36 hours. Used melt spinning machine (Ube, Japan) to prepare PA6/1.5AS fibers, PA6/2.0AS fibers, and PA6/2.0AS fiber through master batch technology. Finally, the prepared PA6 fibers and flame retardant PA6 fibers were woven into a stocking using groups of each four cylinder roller.

RESULTS AND DISCUSSION

The effect of the content of AS on the LOI of the PA6 was shown in Figure 1. It indicated that the LOI increased to 30% and 31.1% with 1.5wt% and 2.0 wt% in PA6 composites respectively. Moreover, the LOI was up to 33% 3 wt% AS. it showed that the LOI increased remarkably with the addition of AS, which can improve the flame retardancy of PA6. It may because that the expansion foaming of PA6 composites after ignition can form intumescent char layer rapidly to insulate the transfer of material and energy.

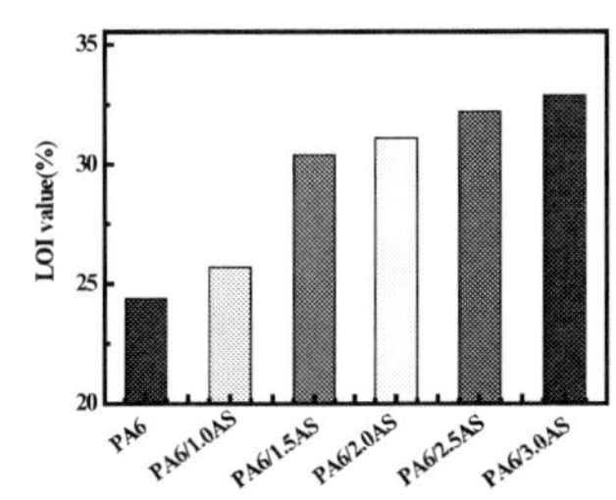

Figure 1. The limiting oxygen index of the composite of PA6 treated with different content of AS.

The tensile strength of the composite decreased from 59.2MPa to 25.6MPa when the amount of AS increase to 3.0 wt%. Besides, the elongation at break of the composite decreased remarkably when the amount of AS was over 2.0 wt% and then the brittle fracture took place, which may because of the decomposition of AS during the process with releasing of the NH3, causing the decomposition of PA6.

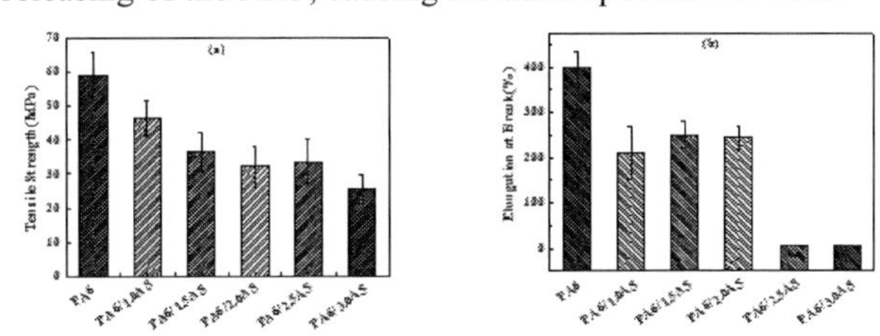

Figure 2. (a) The tensile strength and (b)the elongation at break of PA6 composites with different content of AS.

The addition of AS decreased the initial decomposition temperature ($T_{5\%}$, the temperature when the weight loss was 5%) as well as the temperature of the maximum thermal decomposition for the first time (peak 1of DTG), but had little influence on the second one, which was relative to the decomposition of AS and its catalytic degradation of the chain breaking of PA6. From the analysis of the char residue at 700℃, AS could promote the char forming of PA6.

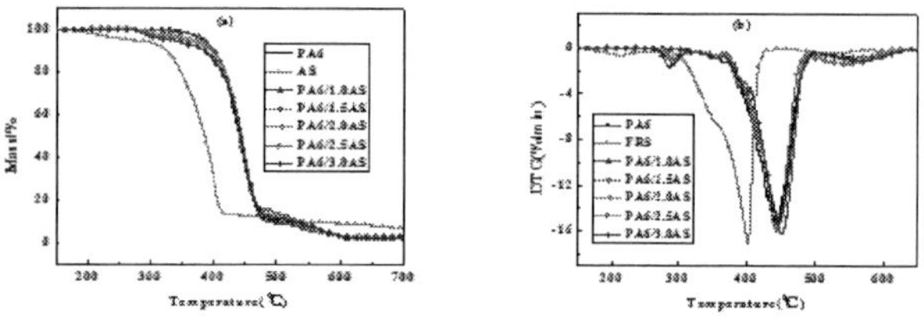

Figure 3. TG curves (a) and DTG curves (b) of PA6, AS and PA6 composites with different contents of AS.

In Figure 4, compared with pure PA6, the PA6/AS composites had a lower melting temperature (Tm) and a broader melting endothermic peaks, which contributed to the improvement of flame retardancy. Besides, the PA6/AS composites will crystallize at a higher temperature with the existence of AS and the crystallinity increased with the rising content of AS, which indicated AS can enhance the crystallization of PA6 to improve the processability of PA6.

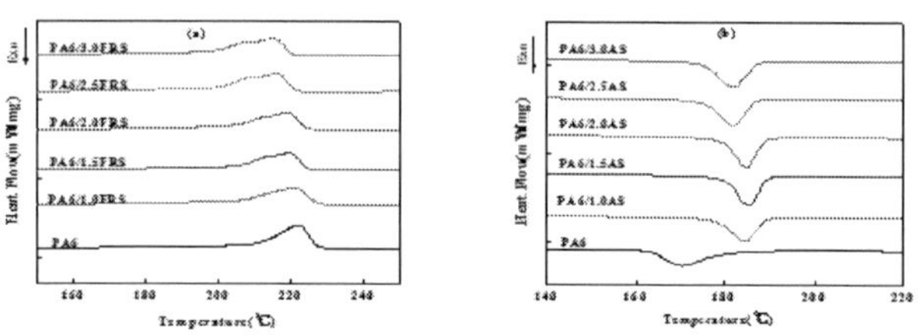

Figure 4. The heating curve (a) and the cooling curve (b) of PA6 and PA6 composites with different content of AS.

Two kinds of PA6 composite fiber with different content of AS were prepared. Besides, master batch technology is taken for the preparation of the PA6/AS composite fiber of with 2.0 wt% of AS. The spinnability of PA6/AS composites were good and the process is consecutive without broken filament. The photos of the prepared PA6 fiber and composite fibers were shown in Figure 5. It was seen that the pure PA6 was white and had the best luster, whereas the composite fibers produced by chip method showed a little yellow stain. Meanwhile, the color of the PA6/AS produced by materbatch process was sharply improved, which implied a good dispersion of AS and decreasing in thermal oxidization of molecular chain.

Figure 5. The photos of PA6 and flame retardant PA6 fibers (a) PA6, (b) PA6/AS (1.5%), (c) PA6/AS (2.0%), (d) PA6/AS (2.0%) by master batch process.

The stress-strain curves of PA6 and PA6/AS composite fibers were shown in Figure 6. The fiber strength and the restrain the orientation of PA6 chain during the drafting and defects were generated because of the stress concentration spots' appearance. However, the tensile properties of the composite fiber prepared by master batch technology was dramatically improved. Therefore, it proved that it's really a good way to achieve a good dispersion of fillers by master batch process.

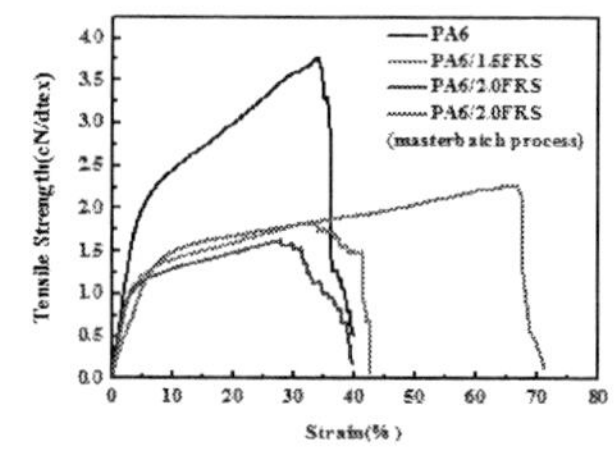

Figure 6. The stress-strain curves of PA6 and PA6 composite fibers.

CONCLUSIONS

PA6/AS composites were prepared by melt blending in two-screw extruder , and the flame retardant PA6/AS fibers with different contents of AS were also prepared by melt spinning. LOI of PA6/AS composites equaled to 30% when the amount of the AS was 1.5wt%, and LOI increased to 33% when the amount of AS up to 3wt%. As the increasing of AS to 3.0%, the tensile strength decreased from 59.2MPa to 25.6MPa. And the elongation at break dropped sharply as the content of AS over 2.0%. AS can promote the char forming and lead to a decrease of $T_{5\%}$. Besides, PA6/AS had a lower T_m and a broader melting endothermic peak compared to PA6, which is beneficial to the flame retardant property. The spinnability of PA6/AS composites were good and the process of melt spinning was consecutive. The tensile properties and the color of the prepared fibers by master batch process can be improved.

ACKNOWLEDGMENT

This work is supported by the Program for Changjiang Scholars and Innovative Research Team in University (IRT1221), Research Program of Shanghai Science and Technology Commission(12nm0500200).

REFERENCES

1. Lu, S.Y., Hamerton, I. *Prog. Polym. Sci.*, 27, 2002: 1661.
2. Dogan, M., Bayramli, E. *Fiber. Polym.*, 14, 2013: 1595.
3. Gijsman, P., Steenbakkers, R., Furst, C., Kersjeset, J. *Polym. Degrad. Stab.*, 78, 2002: 219.
4. Aarons, R. U.S. Patent, 2935471-1960.
5. Morton, T. British Patent, 634690-1950.
6. Lewin, M., Brozek, J., Martens, M.M. *Polym. Advan. Technol.*, 13, 2002: 1091.

Mechanical Properties and Transparency of Corn Straw Cellulose Film Prepared at Different Casting Solution Temperatures

Junfen Sun, Wenjuan Niu

State Key Laboratory for Modification of Chemical Fibers and Polymer Materials, College of Material Science and Engineering, Donghua University, Shanghai, China
junfensun@dhu.edu.cn

ABSTRACT

In this work, 6.5wt% corn straw pulp is dissolved in paraformaldehyde/dimethyl sulfoxide (PF/DMSO) to prepare the film by changing the temperature of casting solution from 20℃ to 80℃. The properties of film are characterized by mechanical properties and transparency.

KEYWORDS: corn straw cellulose, film, preparation, mechanical properties, transparency

INTRODUCTION

The preparation of cellulose products has attracted much attention due to the increasing demand for biocompatible and environmentally friendly products recently [1]. However, cellulose is difficult to be dissolved due to the strong hydrogen bonds. In recent years, new solvents have been developed to dissolve cellulose such as PF/DMSO, ionic liquids, NMMO and alkali/urea aqueous systems, providing great opportunities for the development of cellulose products. Meanwhile corn straw as an abundant resource mainly consists of cellulose (45%), hemicelluloses and lignin [2]. A few corn straws are commonly returned to field, made into biogas and fed to the animals. However most of corn straws are abandoned or burned, which make serious waste of resources and the pollution of environment. The corn straw cellulose films have good biodegradability, reproducing property and mechanical properties, unique physical and chemical properties [3, 4] and can be used widely as separation membrane [5, 6], packing film [7] and mulch film. In this study, the corn straw cellulose is dissolved in PF/DMSO to prepare the film, and the mechanical properties and transparency were studied.

APPROACH

Materials

Corn straw pulp sheets were provided by science and technology consulting service center of Henan province. Paraformaldehyde (PF) was purchased from Shanghai Lingfeng Chemical Reagent Co., Ltd. Dimethyl sulfoxide (DMSO) was purchased from China Pharmaceutical Chemical Reagent.

Preparation of casting solution

Initially, corn straw pulp sheets were torn into small pieces and thoroughly smashed to tiny pieces by a disintegrator. Then the smashed cellulose was dried by vacuum drying oven. The dried corn straw cellulose (19.5g, 6.5wt% based on weights of the solution), PF (19.5g) and DMSO (237ml) were placed in three-necked flask. The mixture was stirred in an oil bath for 1h at 60℃, then the temperature of oil bath was increased to 120℃ and stirred for 2h. Then the solution was filtered to exclude the impurities. The casting solution was obtained and then placed in oven in order to remove bubbles.

Preparation of film

The film was prepared by immersion precipitation. The casting solution was cast on a glass plate at 20℃, 40℃, 60℃, 80℃ respectively, then immersed into water bath at room temperature immediately and the film formed a few minutes after immersion. The films were washed with water repeatedly to remove residual solvent and dried in air.

Tensile strength measurement

The films were cut into rectangle-shaped samples with a length of 70mm and a width of 15mm. At least five samples were performed for every film using a universal testing machine WDW3020 which was produced by Changchun Kexin test instrument co., LTD at room temperature and the speed of stretching is 5mm/min.

Transparency

Optical transmittance of the films was measured with a UV-1800 spectrophotometer which was produced by SHIMADZU Company at the wavelength from 190 to 900nm.

RESULTS AND DISCUSSION

Mechanical properties of the corn straw cellulose films

Fig.1 shows the mechanical properties of the films prepared at different casting solution. The tensile strength and elongation decrease with rising temperature of casting solution. High temperature of casting solution leads to low viscosity and the double diffusion rate between solvent and coagulant increases. The pores with large size form with rising temperature of casting solution. It means that the film is inclined to form a dense structure at low temperature of casting solution and form a porous structure at high temperature. The mechanical properties of the films are the best at 20℃.

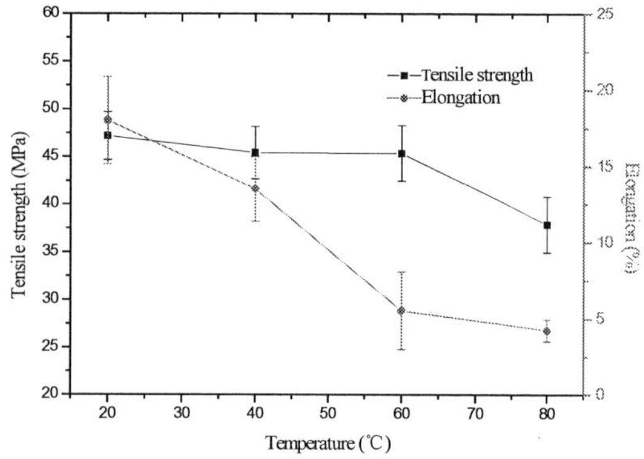

Fig. 1. Tensile strength and elongation of films prepared at different temperature of casting solution.

Transparency of the corn straw cellulose films

Fig.2 shows the transparency of the films prepared at different temperature of casting solution. Transmittance of films was measured in the UV and visible light range. The transmittance of UV light is very low at 200 nm for all films. Therefore, the corn straw cellulose films have excellent barrier properties against UV. The transmittance of visible light is more than 85% at 800 nm for all films. It is suggested that the transparency of films has little change with rising casting solution temperature.

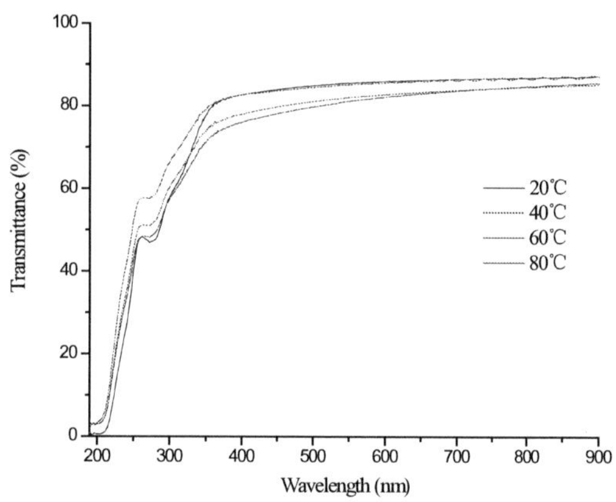

Fig. 2. Transmittance of films prepared at different temperature of casting solution.

CONCLUSION

The corn straw cellulose films were prepared at different casting solution temperature. The mechanical measurement shows that the tensile strength and elongation decrease with rising the temperature of casting solution. But the transparency has little change with rising the temperature of casting solution. The properties of the corn straw cellulose film are the best at 20℃.

ACKNOWLEDGMENT

The authors thank National Natural Science Foundation of China (51203020); Scientific Research Starting Foundation for Returned Overseas Chinese Scholars, Ministry of Education of China.

REFERENCES

[1] Chang, C.Y., Zhang, L.N. "Cellulose-based hydrogels: Present status and application prospects." *Carbohyd. Polym.*, 84, 2011: 40-53.

[2] Prsad, S., Singh, A., Joshi, H.C. "Ethanol as an alternative fuel from agricultural, industrial and urban residues." *Resour. Conserv. Recy.*, 50, 1, 2007: 1-39.

[3] Qiu, X.Y., Tao, S.M., Ren, X.Q., Hu, S.W. "Modified cellulose films with controlled permeatability and biodegradability by crosslinking with toluene diisocyanate under homogeneous conditions." *Carbohyd. Polym.*, 88, 2012: 1272-1280.

[4] Mao, Y., Zhou, J.P., Cai, J., Zhang, L.N. "Effects of coagulants on porous structure of membranes prepared from cellulose in NaOH/urea aqueous solution." *J. Memb. Sci.*, 279, 2006: 246-255.

[5] Wu, J., Liu, J.H., Yuan, Q. "Gas permeability of novella-cellulose membrane." *J. Chem. Ind. Eng.*, 54, 3, 2003: 333-338.

[6] Madaeni, S.S., Heidary, F. "Improving separation capability of regenerated cellulose ultrafiltration membrane by surface modification." *Appl. Surf. Sci.*, 257, 2011: 4870-4876.

[7] Ghaderi, M., Mousavi, M., Yousefi, H., Labbafi, M. "All-cellulose nanocomposite film made from bagasse cellulose nanofibers for food packaging application." *Carbohyd. Polym.*, 104, 15, 2014: 59-65.

Research on the Preparation and Adsorption of Thermoplastic Polymer Functional Nanofibers

Dandan Xu, Ru Xiao
State Key Laboratory for Modification of Chemical Fibers and Polymer Materials, College of Materials Science and Engineering, Donghua University, Shanghai, China
xudandan23@163.com; xiaoru@dhu.edu.cn

INTRODUCTION

The Cr ion is one of the most commonly present heavy metal pollutants in wastewaters. A variety of techniques, such as chemical reduction and precipitation, ion exchange, membrane separation and adsorption have been developed for the removal of Cr (VI) from aqueous solution[1]. Among these, adsorption has been considered as an effective and promising method due to its advantages of simplicity, low cost, high selectivity and environmental appeal. Bhaumik reported a process for that Cr (VI) uptake that included ion-exchange followed by reduction of Cr (VI) by polypyrrole-polyaniline (PPy-PANI) nanofibers to form Cr (III)[2]. Some researchers have reported the use of PPy coated on a variety of materials such as Fe_3O_4[3], sepiolite[4], graphene[5] for the removal of Cr (VI).

Polymer nanofiber membranes were used as the base material of the adsorbents because of its high gas permeability, high porosity, small interfibrous pore size, and, above all, high specific surface area per unit mass, which affords high adsorption capacity. It is relatively easy to modify these membranes with functional groups. For example, poly(vinyl alcohol-co-ethylene) (PVA-co-PE) prepared by melt extrusion, functionalized with iminodiacetic acid were used for the removal Cu^{2+} [6]. The electrospun polyacrylonitrile (PAN) nanofibers were modified by diethylenetriamine to be used as the absorbents of Cu^{2+} [7]. Polyacrylonitrile (PAN) nanofibers functionalized with amine groups were prepared for the removal of Cr (VI) from aqueous solutions[8].The melt blending extrusion method for preparing various kinds of thermoplastic polymer nanofibers is environment friendly, high efficiency, versatility and continuous.

In this reported study, the CAB was the matrix used to prepare the EVOH nanofibers by melt blending extrusion. Then nanofiber membranes were prepared by layer deposition technology to remove Cr (VI) from aqueous solution. This was accomplished by functionalized using in situ oxidative polymerization from an aqueous solution of pyrrole (Py) at room temperature, employing $FeCl_3$ as a catalyst. Kinetic data for Cr (VI) removal at four different temperatures were analyzed using pseudo-first and pseudo-second order kinetic models, and the linearized Langmuir, Freundlich and Tempkin equations were used to fit the equilibrium isotherms for Cr adsorption systems.

APPROACH

The blends were prepared using a co-rotating twin screw extruder (D = 16 mm, L/D = 40, EUROLAB16, Thermo-Haake Co.). The morphology of EVOH/PPy nanofiber functionalized membranes was characterized by scanning electron microscopy (SEM, S-3000N, Hitachi, Ltd., Japan). (SEM).The functional groups on the surface of synthesized adsorbents were detected by Nicolet Nexus 8700 FT-IR model. Cr (VI) concentration was measured with an inductively coupled plasma mass spectrometry (ICP-AES prodigy Leeman American). The surface chemical composition of functionalized membranes before and after adsorbtion were analyzed by X-ray photoelectron spectroscopy (XPS). The EVOH resin and CAB were dried in vacuum prior for 24 h at 80°C before melt blending. EVOH andCAB with the weight ratios of 25/75 were fed into the co-rotating twin-screw extruder with a screw speed of 80 rpm and a melting zone temperature ranging from 190 to 220°C. The blends were extruded by a take-up device and water-cooled to room temperature. The extrudants were immersed in acetone via a Soxhlet extractor at room temperature for 15 minutes to remove CAB from the blends and deposited onto a supporting PP nonwoven mat. After natural drying, the EVOH nanofiber membrane with a controlled thickness was released from the supporting PP nonwoven mat.

The EVOH/PPy nanofiber functionalized membrane was synthesized via in situ chemical oxidative polymerization technique. In a typical polymerization procedure, a piece of EVOH nanofiber membrane was added to the $FeCl_3 \cdot 6H_2O$ (0.002 mol) solution with 10 mL deionized water, then pyrrole (0.02 mol) in 50mL of ethanol was mixed into the solution a conical to start the polymerization. The mixed reaction solution was shaken in a constant temperature shaking bath, maintained at 25°C and 100rpm. Finally, the EVOH nanofiber membranes was removed , washed repeatedly with ethanol, followed by deionized water to remove the residual pyrrole monomers and dried in a vacuum oven at 60°C overnight.

The adsorption behavior of for Cr (VI) was established using batch adsorption experiments. The various concentrations of Cr (VI) ions were prepared using potassium dichromate ($K_2Cr_2O_7$). During this process, the beaker was shaken in a constant temperature shaking bath.

RESULTS AND DISCUSSION

SEM showed that PPy well-distributed on the nanofiber membranes and the polymerization process occurred both at the surface and in the bulk of the EVOH nanofibers.

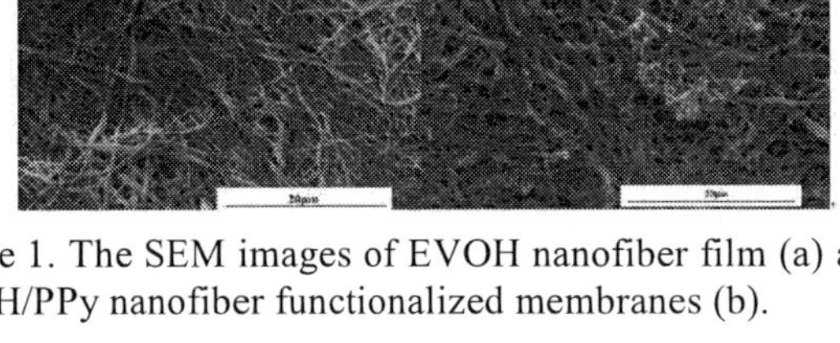

Figure 1. The SEM images of EVOH nanofiber film (a) and EVOH/PPy nanofiber functionalized membranes (b).

It is observed that after pyrrole polymerization on the surface of EVOH nanofiber film, the absorption bands assigned to pyrrole ring stretching, C-N stretching vibrations, C-H stretching vibration and C-H deformation appeared at 1545 cm^{-1}, 1164 cm^{-1}, 1036 cm^{-1} and 878 cm^{-1}, respectively. These results indicate that PPy was coated on the surface of PAN nanofibers.

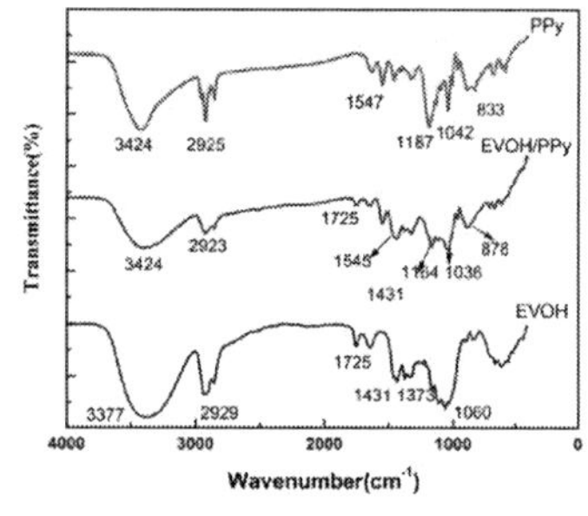

Figure 2. FTIR Spectra of EVOH nanofiber film, EVOH/PPy nanofiber functionalized membranes and Ppy.

After the adsorption process, energy band at about 576 eV and 586 eV appeared, which is attributed to Cr $2p_{3/2}$ and Cr $2p_{1/2}$ orbital. The Cr $2p_{3/2}$ spectrum can be fitted into two peak components with BEs of about 575.5and 576.8 attributable to Cr III) and Cr (VI), and so does 585.35 and 586.5 for the Cr $2p_{1/2}$ spectrum. The presence of the Cr (VI) on the surface of functionalized membranes was due to the anion exchange of PPy by replacing the doped Cl^{-}. The presence of Cr (III) was due to the reduction property of PPy for the Cr (VI), and it can be observed that the PPy has a strong reduction ability for Cr (VI) (Fig. 3c).

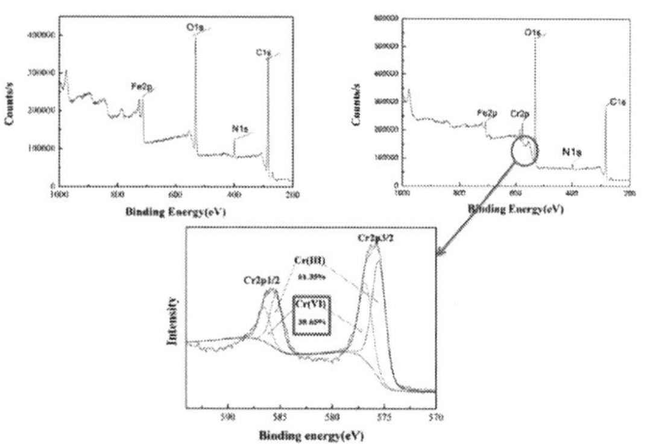

Figure 3. XPS full scan of EVOH/PPy nanofiber functionalized membranes before(a) and after (b) and adsorption and Cr $2p_{1/2}$ and Cr $2p_{3/2}$core level spectrar.

As shown in Fig. 4, when the pH of the initial solution changed from 6.0 to 2.0, the adsorption efficiency increased from 55% to nearly 86.1%. Therefore the following adsorption isotherms and kinetic properties were performed at the condition of initial pH 2.0 of the solution (Fig. 4a). The adsorption capacity of Cr (VI) increases with an increase in adsorption time until equilibrium is reached between the adsorbents and Cr (VI) solution (Fig. 4b). The equilibrium time increased from 30 to 100 min as the initial solution concentration increased from 50 to 150 mg/L. The equilibrium capacity is 12.9, 19.17, 25.28, and 36.41 mg/g for the50, 75, 100 and 150 mg/L initial Cr (VI) solution, respectively. The process can be described using the pseudo-second-order model (Fig. 4c). Isotherm data fitted well to the Langmuir isotherm model (Fig. 4d).

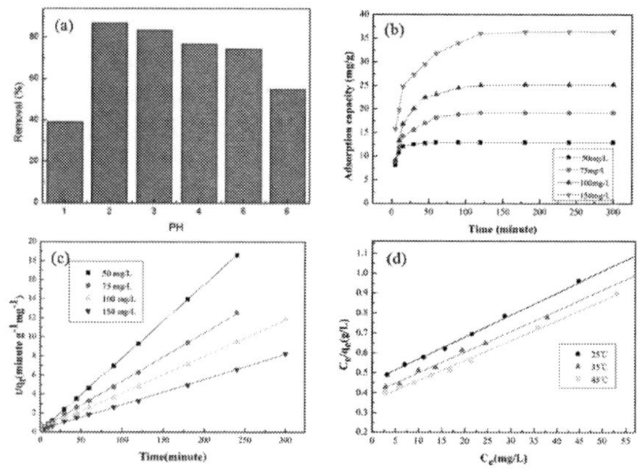

Figure 4. Effect of pH on the removal of Cr (VI) (a): Effect of contact time and initial concentration on the adsorption of Cr (VI) (Temperature = 25°C and initial solution pH 2.0); (b): pseudo-second-order kinetic model (c) and Langmuir isotherm model (d) for adsorption for by EVOH/PPy nanofiber functionalized membranes.

KEYWORDS: Poly (ethylene-co-vinyl alcohol); Polypyrrole; Nanofiber membrane; Adsorption; Hexavalent chromium.

ACKNOWLEDGMENT

This research was financially supported by the National Natural Science Foundation of China (No.20874010), the program of Introducing Talents of Discipline to Universities (No.111-2-04).

REFERENCES

[1] Golbaz, S., Jafari, A.J., Rafiee, M., Kalantary, R.R. "Separate and simultaneous removal of phenol, chromium, and cyanide from aqueous solution by coagulation/precipitation: Mechanisms and theory." *Chem. Eng. J.*, 253, 2014: 251-257.

[2] Madhumita Bhaumik, A.M., Srinivasu, V.V., Onyango, Maurice S. "Removal of hexavalent chromium from aqueous solution using polypyrrole-polyaniline nanofibers." *Chem. Eng. J.*, 2012: 323-333.

[3] Bhaumik, M., Maity, A., Srinivasu, V.V., Onyango, M.S. "Enhanced removal of Cr (VI) from aqueous solution using polypyrrole/Fe3O4 magnetic nanocomposite." *J. Hazard. Mater.*, 190, 2011: 381-390.

[4] Chen, J., Hong, X., Xie, Q., Li, D., Zhang, Q. "Sepiolite fiber oriented-polypyrrole nanofibers for efficient chromium(VI) removal from aqueous solution." *J. Chem. Eng. Data*, 59, 2014: 275-2282.

[5] Yao, W., Ni, T., Chen, S., Li, H., Lu, Y. "Graphene/Fe3O4@polypyrrole nanocomposites as a synergistic adsorbent for Cr (VI) ion removal." *Compos. Sci. Technol.*, 99, 2014: 15-22.

[6] Lu, Y., Wu, Z., Li, M., Liu, Q., Wang, D. "Hydrophilic PVA-co-PE nanofiber membrane functionalized with iminodiacetic acid by solid-phase synthesis for heavy metal ions removal." *React. Funct. Polym.*, 82, 2014: 98-102.

[7] Neghlani, P.K., Rafizadeh, M., Taromi, F.A. "Preparation of aminated-polyacrylonitrile nanofiber membranes for the adsorption of metal ions: comparison with microfibers." *J. Hazard. Mater.*, 186, 2011: 182-189.

[8] Avila, M., Burks, T., Akhtar, F., Göthelid, M., Lansåker, P.C., Toprak, M.S., Muhammed, M., Uheida, A. "Surface functionalized nanofibers for the removal of chromium(VI) from aqueous solutions." *Chem. Eng. J.*, 245, 2014: 201-209.

Electric-Optic Response of Nanowhiskers Smart Material

Yuwei Chen[1,2], Xinhang Wang[1], Yanping Wang[1], Miko Cakmak[2], Yimin Wang[1,3]
[1]Donghua University; [2]University of Akron; [3]State Key Laboratory for Modification of Chemical Fibers and Polymer Materials
jslyjy@163.com

KEYWORDS: Electric-optic response, cellulose nanowhiskers, smart material

ABSTRACT

In this study, cellulose microwhiskers (CMWs) and nanowhiskers (CNWs) were prepared by acid hydrolysis method, both CMWs and CNWs show outstanding optic anisotropic properties which observed by polarizing microscope. We dispersed CNWs in DMF and Norland 65, then we found the CNWs can be aligned or release by apply or remove electric field, and this alignment and disalignment endow this material show electric-color response under polarizing microscope. We suppose this new discovery has potential application in sensors, displays, etc.

INTRODUCTION

The utilization of cellulose nanowhiskers (CNWs) in nanocomposites has attracted significant attention during the last decade. The size of the CNWs depends on the source, and is for example around 5nm in width and 200 nm in length for whiskers from. It has been reported that CNWs in suspension can be oriented by superconducting magnets, shearing forces and by an electric field.[1] In this study, we firstly reported a novel electric-optical response phenomenon that happens in CNWs solution.

APPROACH

Materials

Microcrystalline cellulose (MCC, particle size about 50 mm), commercially available as CEOLUSTM, was kindly provided by Asahi Kasei. Sulfuric acid (98%), acetone (99%), and DMF (98%) were purchased from Sigma-Aldrich and used as received.

Preparation of CMWs and CNWs

CMWs and CNWs were prepared using the following procedure: 8.7 g MCC was mixed with 100 mL 64 wt % sulfuric acid solution. The MCC/sulfuric acid suspension was then put in a water bath and vigorously stirred at 50C for 90 min under ultrasonic treatment. The resultant suspension was repeatedly rinsed with deionized water and centrifuged until the supernatant became turbid.[2] Figure 1(a, b) shows the morphology of CMWs and CNWs. Fig. 1c shows the optical anisotropic of CNWs between two polarizers. 10 wt% CNWs were added to DMF and Norland 65 to get electric-optic response solution.

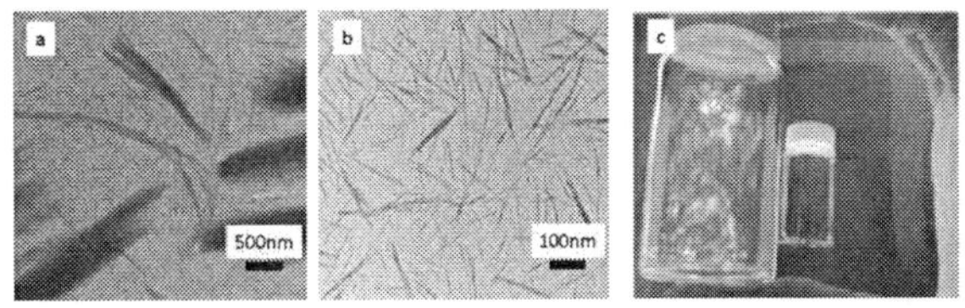

Figure 1. Morphlogy of CMWs and CNWs synthesized by acid hydrolysis method: a. CMWs, b. CNWs, c. Optical anisotropic of CNWs solution.

RESULTS AND DISCUSSION

The most sensitive area of the Michel-Levy Birefringence Chart is first-order red (550 nanometers), because even a slight change in retardation causes the color to shift dramatically either up in wavelength to cyan or down to yellow. We take advantage of this sensitivity by applying a first-order red compensator with polarizing microscopes to determine the properties of birefringent property of CMWs. As shown in Fig. 2, CNWs shows cyan when they parallel to the slow axis of red compensator. Although we cannot observe a single CNW by microscope, we can judge the color by the same way if all the CNWs are aligned parallel or perpendicular to the red compensator. As shown in Fig. 3, without electric field , the solution shows the red color which caused by the red compensator, while it turns to cyan after applied electric field, after remove the electric field, the color change back to red. It is because the CNWs aligned to electric field direction when applied electric field which is parallel to the slow axis of the red compensator. Once the electric field removed, those CNWs released to disorder direction.

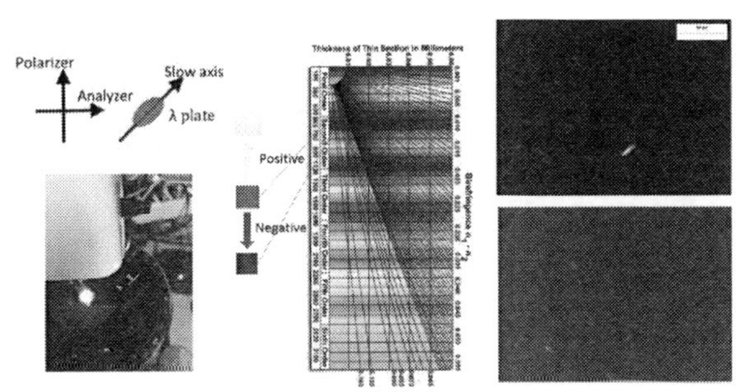

Figure 2. Optical anisotropic of cellulose microwhiskers.

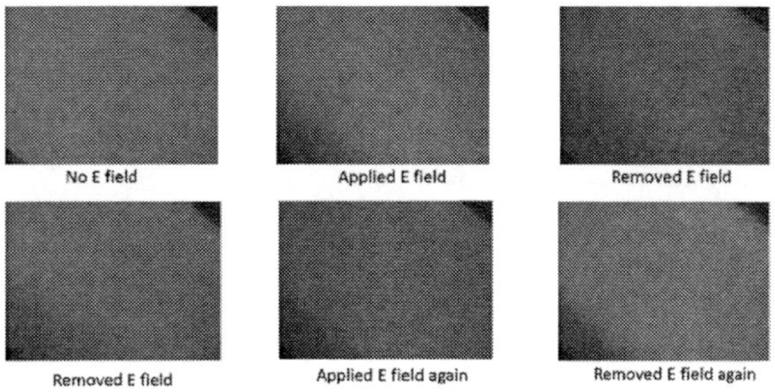

Figure 3. Electric-Optical response.

CONCLUSION

The electric field assisted alignment of CNWs can lead to optical birefringence of CNWs solution, which shows optical-electric response between two polarizers.

FUTURE WORK

In order to get a better understanding of the electric-optical response phenomenon, we will continue this study by measuring the birefringence on real time during alignment by customer made device.[3]

ACKNOWLEDGMENT

This work was financially supported by Third frontier of state of Ohio that provided funding to develop the roll to roll manufacturing line. In addition the authors gratefully acknowledge the doctoral innovation fund and scholarship of Donghua University and China scholarship council.

REFERENCES

1. Kvien, I., Oksman, K. *Applied Physics A*, 87, 4, 2007: 641-643.

2. Wang, X., Xia, Y., Wei, P., Chen, Y., Wang, Y., Wang, Y. *Journal of Applied Polymer Science*, 131, 2014: 19.

3. Unsal, E., Drum, J., Yucel, O., Nugay, I., Yalcin, B., Cakmak, M. *Review of Scientific Instruments*, 83, 2, 2012: 025114.

Polymerization Kinetics of Acrylonitrile, Acrylamide and 2-Acrylamido-2-Methylpropanesulfonic and the Thermal Characterization of Their Copolymer

Zhaopo Zeng , Ru Xiao , Yonggen Lu
State Key Laboratory for Modification of Chemical Fibers and Polymer Materials & College of Materials Science and Engineering, Donghua University, Shanghai, China
xiaoru@dhu.edu.cn

ABSTRACT

The effect of comonomers acrylamide (AM) and 2-acrylamido-2-methylpropaneesulfonic (AMPS) on the pre-oxidation and spinnability of polyacrylonitrile (PAN) at the same time were investigated using differential scanning calorimetry (DSC) method and the reactivity ratios of AN/AM and AN/AMPS in suspension polymerization were calculated by Fineman-Ross method and Kelen-TÜdÖs method. The apparent activation energies (E_a) of cyclization reactions were calculated by Ozawa method and Kissinger method. The results indicated that the stabilization and spinnability of P(AN-AM) has been remarkably improved with the incorporation of AMPS monomer compared with PAN homopolymer, such as broadened exothermic peak and smaller E_a of cyclization, which is attributed to the initiation of comonomer through ionic mechanism.

INTRODUCTION

Carbon fibers are dominantly manufactured from polyacrylonitrile (PAN), rayon, and mesophase pitch precursor fibers, etc. PAN is a majority precursor material for preparing high performance carbon fiber due to its excellent specific strength, stiffness and lightweight. In general, the polymerization methods of fiber-forming copolymer and the spinning conditions are involved in the determining the quality of high performance carbon fibers [1]. Homo-polyacrylonitrile (homo-PAN) has hardly been used for the production of carbon fibers because of its poor properties and processibility. Flexibility in the design of these polymers may be enhanced by the incorporation of a low percentage of suitable acidic comonomers, such as methylacrylic acid (MAA), acrylic acid (AA), itaconic acid (IA), and so on. It has been documented that PAN polymerization is a homo- or/and a heterophase free-radical process. Homophase synthesis, mainly including solution polymerization, is generally considered to be difficult to obtain high molecular weight PAN as a result of the higher chain transfer coefficients of some organic solvents like dimethylsulfoxide (DMSO), dimethylformamide (DMF) as reaction mediums [2]. Mixed-solvent deposited polymerization and aqueous deposited copolymerization belongs to heterophase polymerization system using deionized water (H_2O) or solvent-water mixture as reaction medium, respectively [3]. The latter is the preferred method due to zero chain transfer coefficient of H_2O, which plays an important role in increasing the molecular weight of precursor polymer.

APPROACH

Polymerization of AN with AM as a comonomer was conducted using a 250ml three-necked round bottom flask at 60°C for 25 minutes under a nitrogen atmosphere. In the reaction system, the total monomer concentration was 22wt.%, the monomer ratio was 98/2 (wt./wt.), and the initiator concentration was 0.6wt.% (on the basis of total monomers). The copolymerization was terminated at a low conversion of ≤10%; then the product was precipitated by pouring the mixture into 500 ml water. Obtained polymer was washed many times to eliminate the residual monomers and solvent, and dried under vacuum at 60°C for 24 hours. Similar polymerization conditions were used to prepare the polyacrylonitrile homopolymer, P(AN/AM), P(AN/AMPS) and P(AN/AM/AMPS) copolymers, respectively.

RESULTS AND DISCUSSION

Apparent reactivity ratios of AN and AMPS in aqueous suspension polymerization system was also calculated by Kelen-TÜdÖs method. The experimental results are presented in Figure 1. As shown from Figure 1, the reactivity ratios of AN reaches a maximum at ~2% polymerization conversion and has a minimum value at ~26% conversion in the AN-AMPS system. With an increase of polymerization conversion, reactivity ratio for AN rises rapidly and that for AMPS reduces quickly. The investigated data indicates that, there is an interaction between the apparent reactivity ratios of AN & AMPS and polymerization conversion in the copolymer of AN-AMPS system which may be due to oligomeric radicals precipitate out after attaining a certain critical weight and then act as primary particles. Propagation would then occur either in water phase or the oligomeric radicals phase. When the polymerization conversion is lower than 6%, homogeneous phase solution polymerization played a key role in the early stage of the polymerization reaction system and the reactivity ratios of the two monomers have no significant changes. With the proceeding of the polymerization, the free radicals attached to the macroinitiator backbone polymerized with AN molecules dissolved in the monomer phase. The resultant macroinitiator with grafted oligoacrylonitrile chains exhibited some surface activity and facilitated the consumption of AM units in the continuous aqueous phase. As a result of these factors, the monomer reactivity ratio of AN was reduced significantly, while that of AMPS increases.

The DSC curves of PAN, P(AN-co-AM) and P(AN-co-AMPS) heated at 5°C/min from ambient temperature to 300°C under N_2 (50ml/min) are shown in Figure 2. Under the condition of N_2 and no oxidative reactions occurred during this process, the DSC exothermic peaks of PAN homopolymer and PAN copolymers can be attributed to the cyclization reactions of nitrile (C≡N) groups. As shown in

Figure 2, there is only one sharp exothermic peak from 248°C to 263°C attributed to intramolecular reactions in PAN homopolymer. It is well known that the cyclization reactions are initiated by a free radical mechanism, causing the release of a significant amount of heat and gaseous products at the same time, which results in the breakage of molecular chains and structural defects of resultant carbon fiber. What should be noted is the changes in the cyclisation exotherm of the AN units, associated with incorporation of AM or AMPS into AN copolymers (b, c). The exotherm is broadened considerably for the appearance of two exothermic peaks which implies the occurrence of two distinct reactions, namely the existence of two different cyclization pathways. The lower temperature peak is usually assigned to the cyclization, whilst the higher temperature peak is usually assigned to the dehydro-oxy. The temperatures of two peak maxima for P(AN-AM) (236 and 257°C, respectively) are higher than that for P(AN-AMPS) (232 and 252°C, respectively), correspondingly. In addition, ΔT for P(AN-AM) is smaller than that for P(AN-AMPS). This indicates that, in the case of the close initiating exothermic temperatures for two precursor, critical reactions in P(AN-AMPS) release more uniformly with wider exothermic peaks in DSC curves.

We also studies the DSC curves of PAN homopolymer, P(AN-co-AM) copolymer and P(AN-co-AMPS) copolymer heated at different rates (5,10,15,20,30°C/min) from ambient temperature to 300°C under N_2 (50ml/min). The exotherms of all PAN polymers wholly shift to high temperature and the exothermic peaks become stronger. Ozawa method and Kissinger method have been widely employed for determining the apparent activation energies (Ea) of cyclization reaction for a long time in the literatures, because no prior knowledge of reaction mechanism is required to quantify Ea, but just requiring a series of DSC curves heated at different rates. According to the results, the Ea calculated by two methods show good agreement. The Ea of cyclization reactions in PAN homopolymer is about 164 kJ/mol, while in all PAN copolymers the Ea has been splitted into two parts. It is widely accepted that the first exothermic peak is assigned to the cyclization reactions initiated by comonomers through ionic mechanism and the second to radical mechanism. The latter value is more considerably to reflect the degree of difficulty or easiness to the pre-oxidation. Obviously, the Ea of PAN copolymers (~141 and ~120 kJ/mol) have been reduced significantly compared with PAN homopolymer, which is beneficial to the improvement of re-oxidation. It also should be noted that due to the accelerated rate of stabilization, broadening of the exothermic peak does not permit us to follow the peak positions accurately in advanced stages of stabilization.

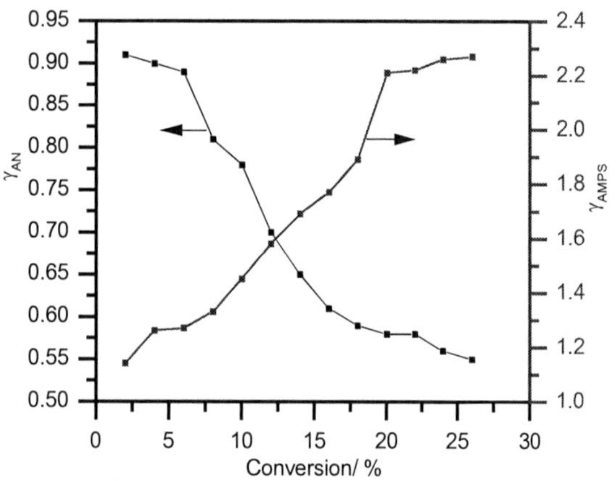

Figure 1. Relationship between reactivity ratios and conversion

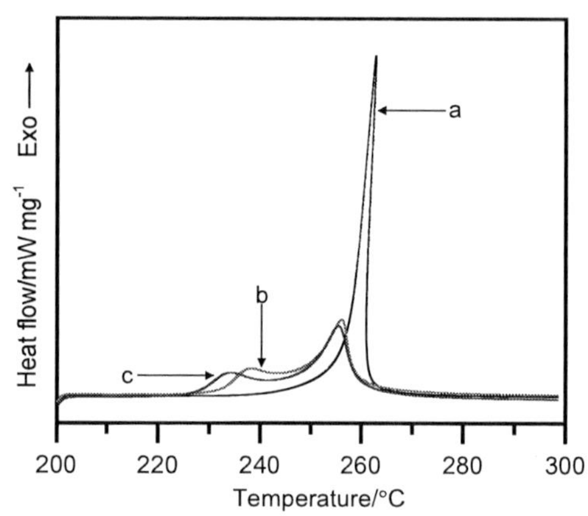

Figure 2. DSC curves of PAN homopolymer, P(AN-co-AM) and P(AN-co-AMPS) copoloymers

KEYWORDS
Polyacrylonitrile; acrylamide; 2-acrylamido-2-methylpropaneesulfonic; stabilization reaction; spinniability

ACKNOWLEDGMENT
We acknowledge the National 973 Project (2011CB605602 and 2011CB605603) for providing financial support.

REFERENCES

[1] Shcherbina, L.A., Alekhina, E.A., Chirtulov, V.G., Gellerm V.E. "Effect of thiourea dioxide on the dynamics of continuous homophase synthesis of fiber-forming copolymers of acrylonitrile, methyl acrylate, and 2-acrylamido-2-methylpropanesulfonic acid."

[2] Zhang, W.X., Liu, J., Wu, G. "Evolution of structure and properties of PAN precursors during their conversion to carbon fibers." *Carbon*, 41, 2003: 2805-2812.

[3] Rwei, Syang-Peng, Way, Tun-Fun, Hsu, Yuan-Shuin. "Kinetics of cyclization reaction in poly(acrylonitrile/methyl acrylate/dimethyl itaconate) copolymer determined by a thermal analysis." *Polym. Degrad. Stab.*, 98, 2013: 2072-2080.

Phosphorous-Contained Flame Retardant, Thermotropic Crystalline Polyarylate: Synthesis and Properties

Jiangwei Wang[2], Guanghong Pang[3], Hongjie Fan[3], Yimin Wang[1], Yanping Wang[1]

[1]State Key Laboratory for Modification of Chemical Fibers and Polymer Materials, Donghua University, Shanghai, China
[2]College of Material Science and Engineering, Donghua University, Shanghai, China
[3]Zhejiang Xinglunkai New Material Technology Co., Ltd. Zhejiang Province, China

Contact Author: wyp@dhu.edu.cn; Speaker: jiangwei2114@163.com

ABSTRACT: A series of Phosphorous-contained Flame-retardant Thermo-tropic Liquid Crystalline Polyarylate (P-F-TLCP) was synthesized by melt polycondensation from p-acetoxybenzoic acid (p-ABA), 6 - acetoxy - 2 - naphthalene carboxylic acid (ANA), terephthalic acid (TPA), 2-(6-oxido-6H-dibenz <c,e> <1,2> oxaphosphorin-6-yl)-1,4-diacetoxy phenylene (DOPO-AHQ). The chemical structure and properties were characterized by Fourier transform infrared (FT-IR) spectroscopy, ^{13}C nuclear magnetic resonance (^{13}C NMR) spectroscopy, hot-stage polarizing microscope (POM), X-ray diffraction analysis (XRD), differential scanning calorimetry (DSC), thermo-gravimetric analysis (TGA) and limiting oxygen index (LOI) analysis. P-T-TLCP exhibited a wide melting temperature range from 244 to 280°C, which showed excellent liquid crystalline texture and a high flame retardancy with a limiting oxygen index value of 61.6%.

KEYWORDS: Thermo-tropic Liquid Crystalline Polyarylate, Flame-Retardant, Phosphorous-Contained

EXPERIMENT

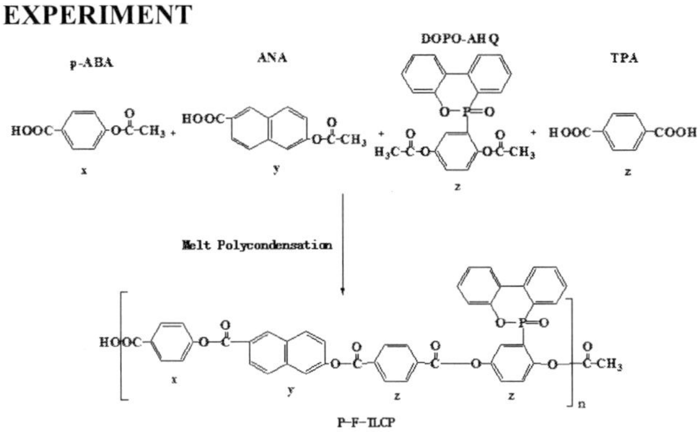

Figure 1. Synthesis of P-F-TLCPs

Table 1. Molar ratio of monomers

NO.	x	y	z
P-0	70	30	0
P-25	70	25	2.5
P-50	70	20	5
P-75	70	15	7.5
P-100	70	10	10
P-150	70	0	15

RESULT AND DISCUSSION

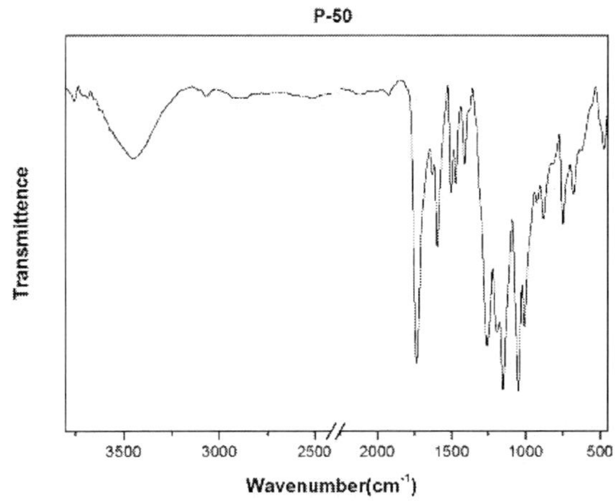

Figure 2. FT-IR patterns of P-50

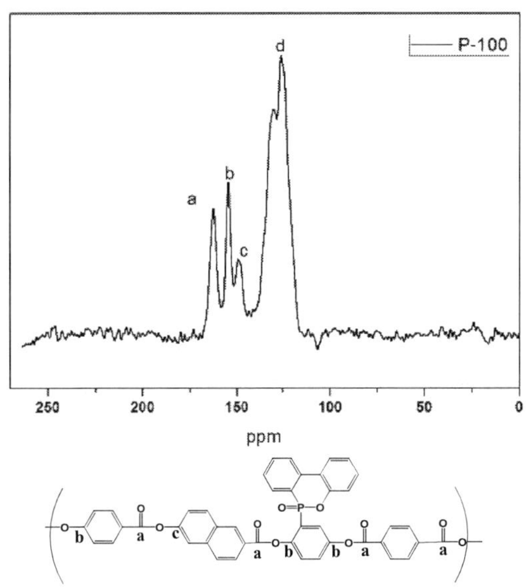

Figure 3. ^{13}C NMR patterns of P-50

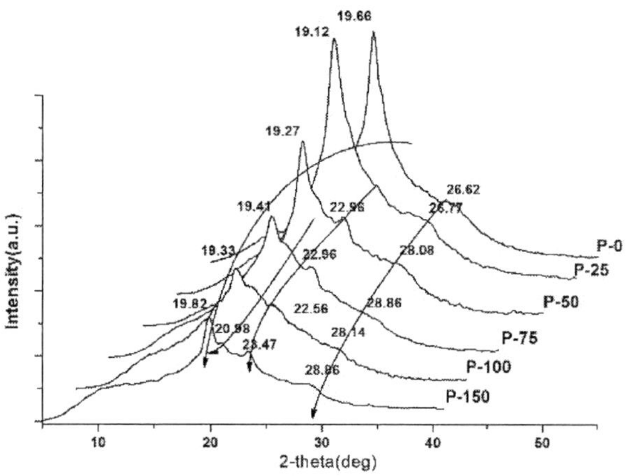

Figure 4. XRD analysis of P-F-TLCPs

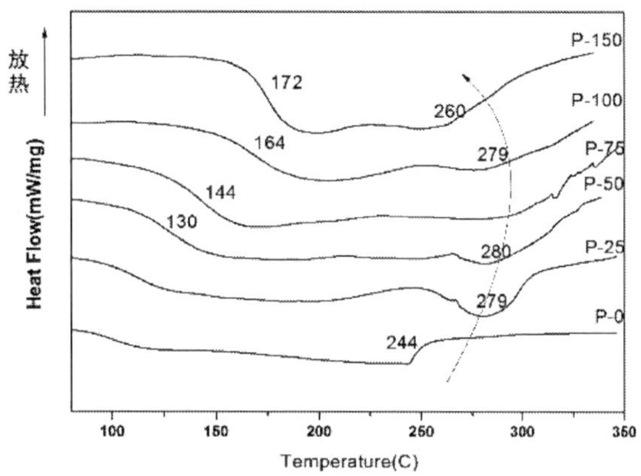

Figure 5. DSC thermogram of P-F-TLCP

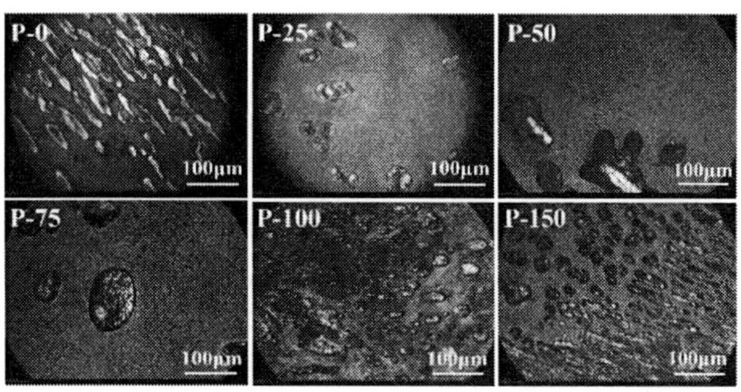

Figure 6. Optical micrographs of P-F-TLCPs: all of these patterns obtained under 320°C

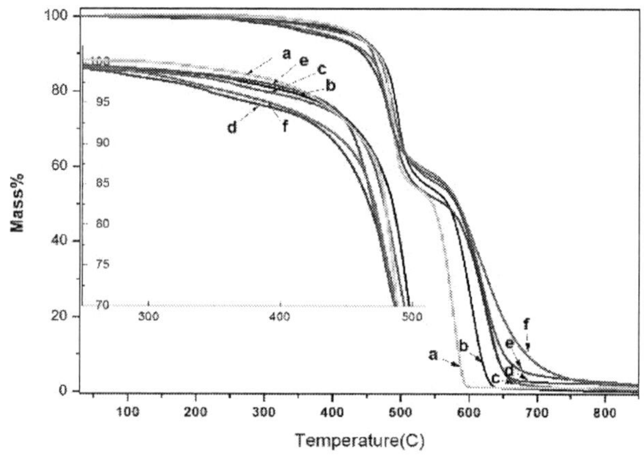

Figure 6. TGA curves of P-F-TLCPs: a:P-0; b:P-25; c:P-50; d:P-75; e:P-100; f:P-150

Table 2. LOI of P-F-TLCPs

NO.	P-0	P-25	P-50	P-75	P-100	P-150
LOI%	41.1	47.0	51.8	57.2	59.0	61.6

CONCLUSION

The chemical structure of the P-F-TLCPs was confirmed. The crystal form was found to be similar with hexagonal packing. It has been studied that P-F-TLCPs has high mesophase temperature and excellent thermal decomposition property in air with the increasing ratio of DOPO-AHQ. The flame retardancy enhanced by the molar ratio of DOPO-AHQ, which showed maximum LOI with 61.6%. The P-F-TLCPs with high performance in thermal and liquid crystal properties has a vast potential for future apply in fiber manufacturing.

ACKNOWLEDGMENT

We gratefully acknowledge the (valuable help and great) support from Innovation Program of Shanghai Municipal Education No. 12ZZ062, the Fundamental Research Funds for the Central Universities, the Shanghai Natural Science Foundation and Programme of Introducing Talents of Discipline to Universities No. 111-2-04.

REFERENCES

[1] Serbezeanu, D., et al.. Macromolecular Chem. and Phys., 211, 13, 2010: 460-1471.

[2] Wei, P., et al. *Journal of Appl. Polym. Sci.*,131, 13, 2014.

[3] Vlad-Bubulac, T. et al. *Polymer*, 50, 10, 2009: 2220-2227.

[4] Donald, A.M., et al. *Liquid Crystalline Polymers*. UK: Cambridge University Press, 2005: 402-404.

[5] Senthil, S. et al. *Polymer*, 45, 11, 2004: 3609-3614.

[6] Wang, Y.Z. et al. *Journal of Appl. Polym. Sci.*, 86, 5, 2002: 1278-1284.

Structural and Properties of Thermotropic Liquid Crystal Polyarylate Modified by Grapheneoxide

Shuohan Huang[1], Hong Fan[1], Peng Wei[1], Yimin Wang[2]
[1]College of Materials Science and Engineering
[2]State Key Laboratory for Modification of Chemical Fibers and Polymer MaterialsDonghua University, Shanghai, China
hsh199477@163.com

ABSTRACT

In this work, HBA/HNA-GO composite was prepared by melt polycondensation via in situ polymerization, and the effects of GO contents on the structure and properties of composites were studied. The results proved that the thermal stability of HBA/HNA copolyester was enhanced with the increase of GO content. For example, the 1.2 wt% HBA/HNA-GO composite shows a 17.2°C improving in the 5% weight loss temperature in comparison with the pure TLCP. The work provides a good direction to further the development of high performance TLCPs.

KEY WORDS

HBA/HNA, liquid crystal polyarylate, in situ polymerization, grapheneoxide, composite

EXPERIMENT

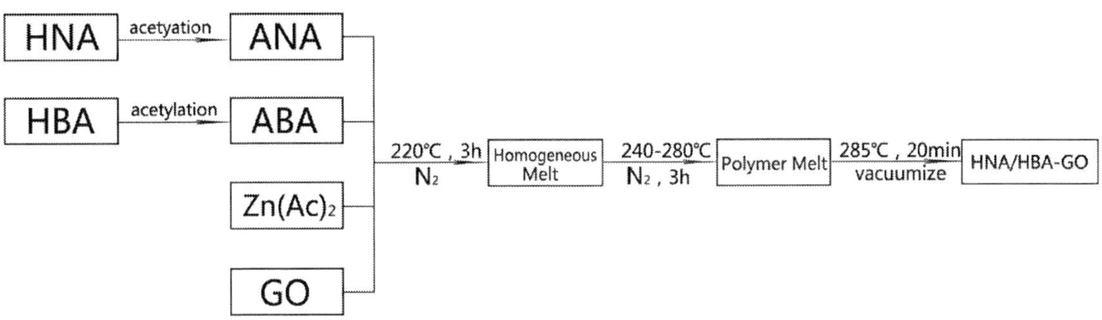

RESULTS AND DISCUSSION

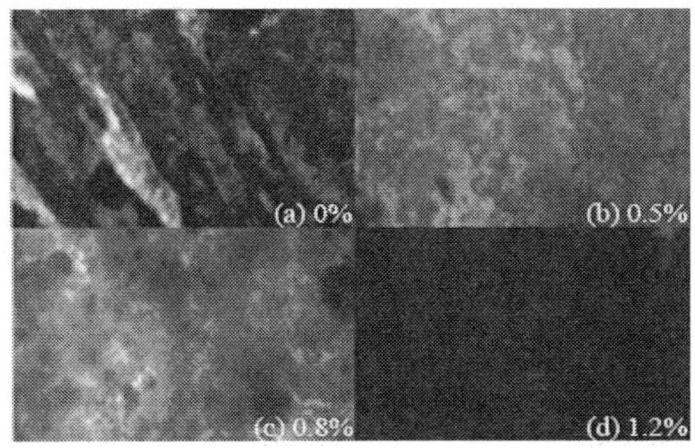

Polarized Optical Micrograph of HNA/HBA-GO

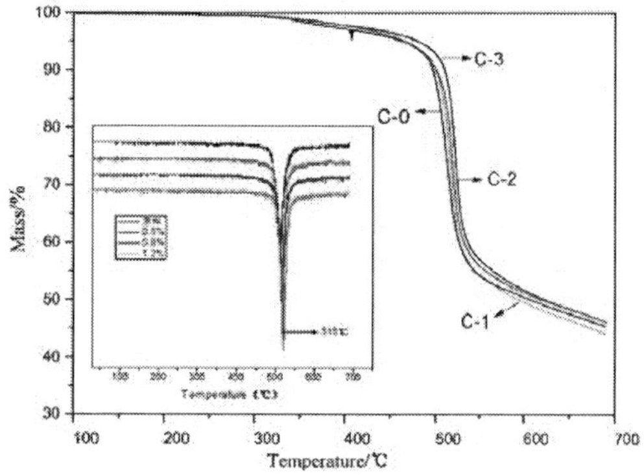

TG and DTG curve of HNA/HBA-GO

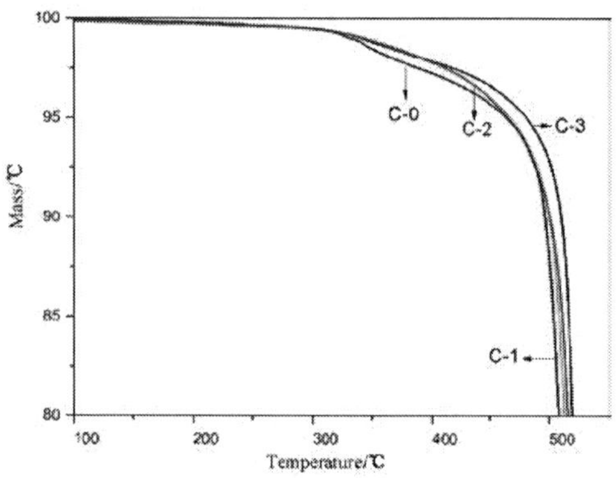

TG curve of HNA/HBA-GO

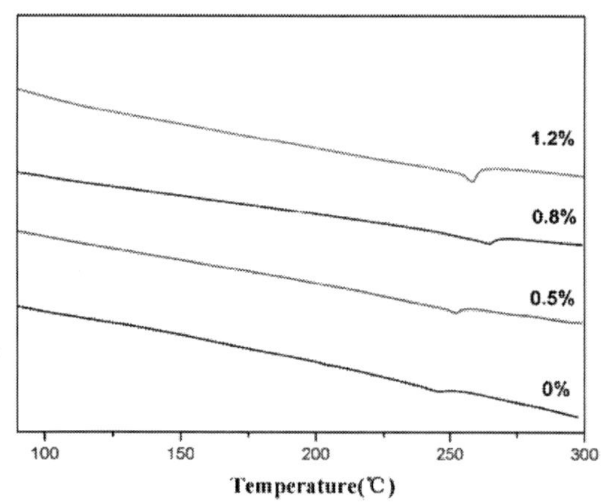

Drilling String Compensator curve of HNA/HBA-GO

CONCLUSION

(1) The melting point of the TLCPs are all around 250°C, while the TLCPs modified by grapheneoxide is higher. (2) The thermostability of the composite material shows the trend of increasing, for the 1.2 wt% HBA/HNA-GO composite shows a 17.2°C improving in the 5% weight loss temperature in comparison with the pure TLCP

ACKNOWLEDGMENT

We gratefully acknowledge the valuable help and great support from Innovation Program of Shanghai Municipal Education No. 12ZZ062, the Fundamental Research Funds for the Central Universities, the Shanghai Natural Science Foundation and Programme of Introducing Talents of Discipline to Universities No. 111-2-04.

REFERENCES

[1] Hou, J., et al. *Polymer*, 35, 4, 1994: 699-705.

[2] Bedford, S.E. et al. *Polymer*, 31, 4, 1990: 616-620.

[3] Wei, P. et al. *Journal of Applied Polymer Science*, 2014: 10.1002; app. 40487.

[4] Wei,.P. et al. *Thermochim. Acta.*, 586, 2014: 45.

[5] Jin, J.-I., et al. *Prog. Polym. Sci.*, 22, 1997: 937.

Effect of SiO_2@TiO_2 Nanoparticles on the Mechanical and UV-Resistance Properties of Polyphenylene Sulfide Fibers

Zexu Hu, Si Meng, Zhe Zhou, Long Chen, Meifang Zhu
State Key Lab for Modification of Chemical Fiber &Polymer Materials,College of Materials Science and Engineering, Donghua University, Songjiang, Shanghai, China
zmf@dhu.edu.cn

ABSTRACT

The polyphenylene sulfide (PPS) nanocomposites fiber was prepared via melt spinning, which display different crystallization behavior on variation of the diameters, as confirmed by differential scanning calorimetry (DSC). The spinnability, breaking strength and UV-Resistance Properties of PPS nanocomposites, as measured by Capillary rheometry, Melt spinning machine, Xenon-lamp Weather Resistance Test Chamber and Yarn Tensile Tester, manifest a dependence on the diameters. The addition of nanoparticle with the diameter of 25nm improves the spinability and the mechanical performance of PPS with an increase in breaking strength of 0.4cN/dtex. The UV-resistance properties of the PPS nanocomposites were remarkably improved by the addition of SiO_2@TiO_2 nanoparticles.

KEYWORDS: polyphenylene sulfide, SiO_2@TiO_2, UV-resistance properties

INTRODUCTION

In last decades, nanomaterial/polymer nanocomposites have attracted considerable attention [1,2]. Most of nanoparticles used for mechanical reinforcements, as wellas the flame resistance, thermal stability, electrical properties enhancements suggest use for these fibers in a broad spectrum of applications [3-6] The mechanical properties of PPS seriously degrade as exposed to sunlight, due to the poor UV-resistance properties[7]. To improve the UV-stability, SiO_2@TiO_2 Nanoparticle with various diameters were populated into the PPS fibers by melt spinning.

EXPERIMENTAL

The PPS resin and SiO_2@TiO_2 with three kinds of diameters were melt-blended respectively by twin-screw extruder to produce the PPS nanocomposites. The contents of nanoparticles in PPS nanocomposites were about 1.5wt%..The nanocomposite fibers were spun by using homemade Melt spinning machine, under a constant spinning temperature.

RESULTS AND DISCUSSION

To evaluate this behavior of PPS nanocomposite, the spinneret draw ratio (V_L/V_0) is the maximum extensibility of the fiber. The PPS/SiO_2@TiO_2 (25nm) nanocomposite fiber shows remarkably higher value (91.7) of the spinneret draw ratio with respect to the pure PPS (75). However, the PPS/SiO_2@TiO_2 (60nm) samples show the lower values (66.7). Because of the excessive module pressure, the properties of PPS/SiO_2@TiO_2 (100-200nm) fiber were not discussed in this work.

Table 1. The mechanical performance of PPS nanocomposite fiber

Samples	Specification (dtex/f)	Breaking strength (cN/dtex)	Elongation at break (%)
PurePPS	75/24	3.1	23.7
PPS/SiO_2@TiO_2 (25nm)	75/24	3.5	22.8
PPS/SiO_2@TiO_2 (100nm)	75/24	3.0	24

The addition of nanoparticles with the diameter of 25nm improves the breaking strength of PPS with an increase in breaking strength of 0.4cN/dtex. To understand the experimental data obtained for the spinnability and mechanical performance of PPS nanocomposite fiber, the crystallization behavior of PPS nanocomposites and fiber were characterized by DSC and XRD.

Figure 1 and Table 2 show the crystallization behavior of PPS/SiO_2@TiO_2 nanocomposites with various diameters of particle. The addition of SiO_2@TiO_2 (25nm) as the nucleation promote the crystallization of PPS. On the contrary, the larger diameters of SiO_2@TiO_2 retard the nucleation of PPS, which was inferred from the DSC curves. The crystallization rate and crystallinity may be the key factor of the spinnability and mechanical performance of PPS fiber.

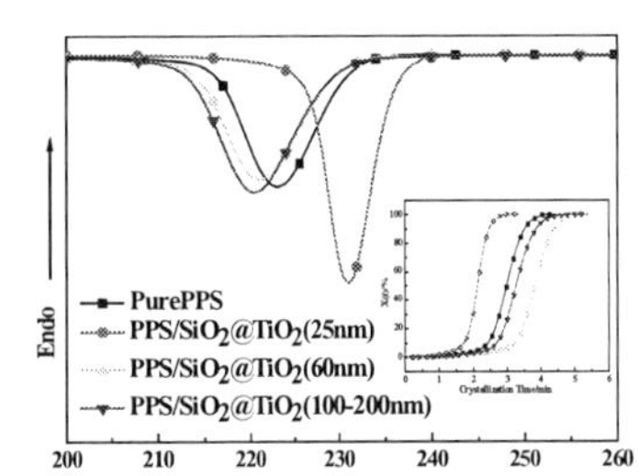

Figure 1. DSC thermograms of dynamic crystallization of PPS nanocomposite resins; the inset is the plot of degree of conversion (x) versus time

Table 2. Crystallization values of PPS nanocomposite resins

Sample	Tc_f/℃	Tc_p/℃	$t_{1/2}$/min	Tm/℃	$X_c(t)$/%	$\Delta H_m(J\cdot g^{-1})$
PurePPS	231.4	223.1	2.98	279.7	38.61	31.18
PPS/SiO_2@TiO_2 (25nm)	236.0	231.1	2.13	280.4	42.22	34.09
PPS/SiO_2@TiO_2 (60nm)	230.1	221.0	3.82	280.2	36.07	29.13
PPS/SiO_2@TiO_2 (100-200nm)	229.4	220.5	3.28	279.9	38.91	31.42

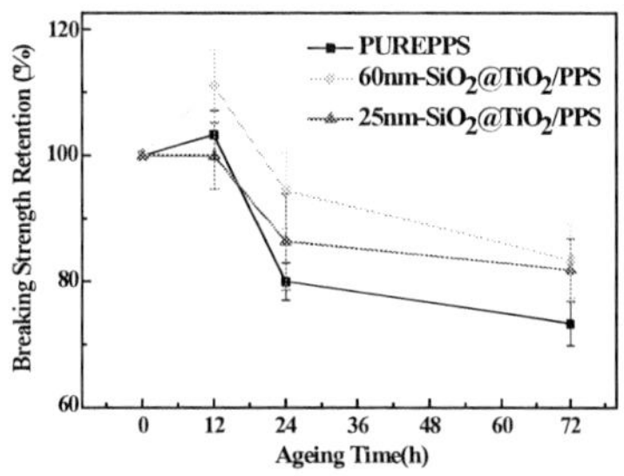

Figure 2. Breaking strength retention of PPS/SiO_2@TiO_2 fibers at 60°C

Figure 2 shows the breaking strength retention PPS/SiO_2@TiO_2 (25nm, 60nm) are about 81.8% and 83.6% respectively, which are higher than pure PPS fiber. The UV-resistance property of PPS/SiO_2@TiO_2 (60nm) fiber is the best, which may imply that more UV light would be scattered by the SiO_2@TiO_2 of larger dimension.

CONCLUSIONS

Various spinning, UV-resistance and mechanical properties of PPS nanocomposite fiber with SiO_2@TiO_2 of different diameters were studies in this work. It was found that the addition of SiO_2@TiO_2 (25nm) promoted the crystallization of PPS, which was the key factor to improve the spinning and mechanical properties. The breaking strength of PPS/SiO_2@TiO_2 (25nm) fiber was 0.4cN/dtex higher than purePPS fiber (3.1cN/dtex). The addition of SiO_2@TiO_2 (60nm) damaged the mechanical of PPS fiber. However, the breaking strength retention of PPS/SiO_2 @TiO_2 (60nm) fiber under the ageing treatment was 83.6%, which was higher than PPS/SiO_2 @TiO_2 (25nm) fiber (81.8%) and pure PPS fiber (73.3%).

ACKNOWLEDGMENT

This work is financially supported by Program for Changjiang Scholars and Innovative Research Team in University (IRT1221, T2011079); 111 Project; Research Program of Shanghai Science and Technology Commission (12nm0500200).

REFERENCES

1. Rahman, A., Ali, I., et al. *Nano.*, 6, 3, 2011: 185.
2. Okada, Akane, Usuki, Arimitsu, et al. *Macromolecular Materials and Engineering*, 91, 12, 2006: 1449
3. Litchfield, David W., Baird, Donald G., et al. *Polymer Engineering and Science*, 50, 11, 2010: 2205.
4. Zhang, S, Horrocks, A.R. *Progress in Polymer Science*, 28, 11, 2003: 1517.
5. Chiu, Chih-Wei, Lin, Chin-An, Hong, Po-Da. J. *Polym. Res.*, 18, 2011: 367.
6. Mazinani, Saeedeh, Ajji, Abdellah, Dubois, Charles. *Polymer Engineering and Science*, 50, 10, 2010: 1956.
7. Das, P.K. *Polymer Degradation and Stability*, 48, 1995: 11.

Transmutation of Structure and Properties in the Preparation of Chestnut Shell Biochar

Yanni Shi, Shaoxu Xia, Shi Chen, Guangxue Wang, Qilin Wu
State Key Laboratory for Modification of Chemical Fibers and Polymer Materials and the College of Materials Science and Engineering, Donghua University, Shanghai, China
wql@dhu.edu.cn; yannishi@outlook.com

ABSTRACT

The study is well-established to develop a set of technical routes on the preparation of biochar based on chestnut shells. Chestnut shells based biochar with calorific value has been prepared successfully through pre-oxidation and carbonization according to the innate advantages of chestnut shells at carbon content and calorific value. The pyrolysis behaviors and the transmutation of both the structure and properties during the carbonization were investigated, especially including the effect of microstructure transmutation on the properties.

INTRODUCTION

The directly burning of a large number of chestnut shells causes environmental pollution and resource waste. Several studies on the carbonization of different biomass sources have been reported, including chestnut shells, grape seeds and pistachio shells[1-3]. The chemical structure and major organic components in biomass are extremely important in the development of processes for producing derived fuels and chemicals[4, 5].

According to the innate advantages of chestnut shell at carbon content and calorific value, chestnut shell based biochar with high calorific value had been prepared through pre-oxidation and carbonization. The pyrolysis behaviors and the transmutation of both the structure and properties during the carbonization were investigated.

APPROACH

Carbonization process

The raw chestnut shells were first immersed in the catalyst solution containing sodium chloride and phosphoric acid. Shells were taken out after reaching equilibrium, and were dried in the oven at 60℃ for 24 hrs. Afterwards, the pre-treated shells were carbonized from room temperature escalating to 250℃ at a rate of 5℃/min in the carbonizing furnace in air. Nitrogen was then infused into the furnace to prevent oxidization reaction, and the pre-oxidized shells were further carbonized until 750℃.

Characterization methods

The pyrolysis of chestnut shell was carried out in a thermo-gravimetric(TG) analysis (NETZSCH STA 409C, Germany). Samples were analyzed at heating rates of 5 and 10℃/min from room temperature to 900℃, respectively, in the atmosphere of nitrogen (99.999%) flowing at 30mL/min. The total mass was 10mg for each run. Fourier transform infrared (FTIR) spectra of the sample powders were recorded on a Nicolet NEXUS-670 spectrometer. Elemental analysis (C, H and O) was investigated on a Vario EL III Microanalyzer (Germany). All specimens for SEM observation were coated with a thin layer of gold in a sputter-coating unit prior to observation to avoid charging in JSM-5600LV microscope. The calorific values of samples were measured by the ASTM bomb calorimeter method using an IKA Calorimeter C 7000 model instrument.

RESULTS AND DISCUSSION

Decomposition

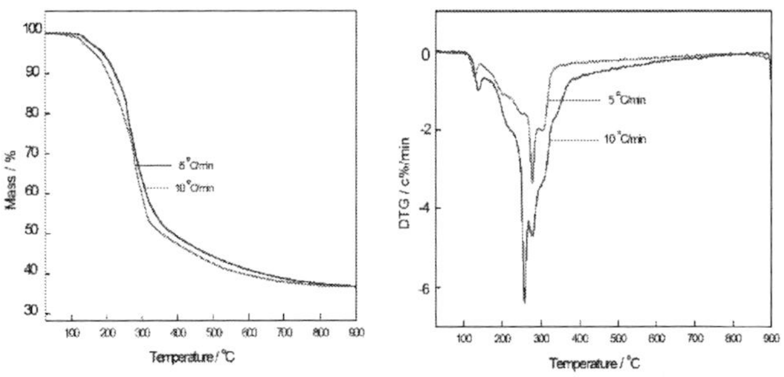

Fig. 1. (a) TG and (b) DTG curves of chestnut shells.

The TG curves for chestnut shell present reversing S-shaped. The weight loss began as soon as shells experienced heat-treatment with increasing temperature. In case of heat rate of 5℃/min, weight loss continued from 100~200℃ at a relatively slow speed, then a considerable loss, almost 50%, appeared at the range from 200~340℃, followed by a 20% loss from 340~900℃. The maximum loss rate appeared around 260°C, according to the accompanying DTG curves. The total loss reached 75%, leaving a solid residue of 25%. In this work, with proper arrangement of the catalyst and temperature, the total yield of chestnut shell biochar reached 44.31%.The analysis of the change in weight loss (Fig. 1a) and the rate of weight loss (Fig. 1b) with temperature increasing showed that the decomposition of this material took place mainly in the range of 200–340°C.

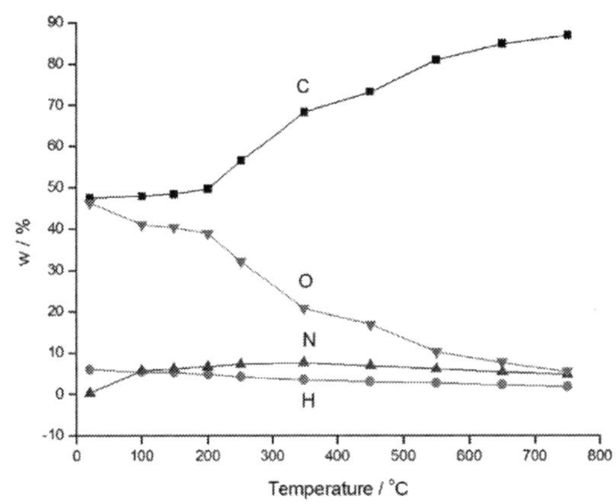

Fig. 2. Element content changing trend.

The chemical transformations during raw materials converting into biochar products by means of hydrothermal carbonization were examined. Along with the elimination of CO_2 and other carbon-containing compounds, a large amount of atom C is removed from the chestnut shell structure. At the same time, there was an obvious reduction in the oxygen content and hydrogen (shown in Fig. 2). It was also observed that temperature increasing (i.e., from 250℃ to 750℃) caused a decrease in the O/C and H/C atomic ratios, suggesting an increase in the carbon yield. Thus, the carbonization led to carbon content increasing from 44.4 wt.% (raw shell) to 90 wt.% in the biochar samples(750℃ treated) in the solid residue.

FTIR spectra of chestnut shells of different periods are shown in Fig. 3. It can be seen that the sample obtained at 750℃ exhibited completely different FTIR trace compared to the raw material. The decrease (in relation to raw shell) in the intensity of the bands at 1000-1500cm^{-1} and the broad band at 3000-3600cm^{-1} suggested that dehydration occurred during the carbonization of the chestnut shells.

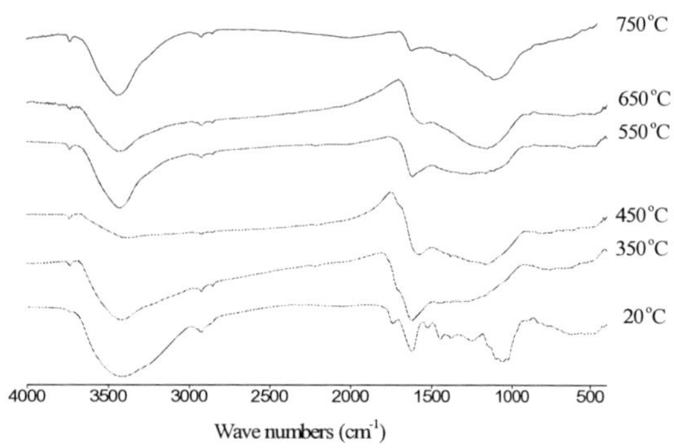

Fig. 3. FTIR spectra for chestnut shells treated at different temperatures.

Structural properties

The microstructure of samples in pre-oxidation and carbonization had been studied by SEM (shown in Fig. 4). As the process temperature increased, the holes on surface shrank and the porosity of cross section became large, while network structure grew to be regular. The catalyst made the network structure of cross section emerged and the surface became turtle shell shape.

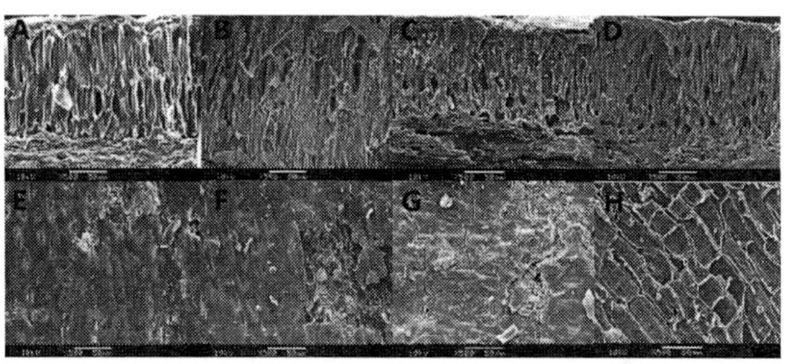

Fig. 4. SEM images: cross section of (A) raw chestnut shells and (B–D) heat treated shell at: (B) 250℃, (C) 550℃, and (D) 750℃, out surface of (E) raw chestnut shell and (F-H) heat treated shell at 7 (F) 250℃, (G) 550℃, and (H) 750℃.

Combustion performance

The gross calorific value of chestnut shell was found as 17.46 MJ kg−1. In the stage of pre-oxidation (below 250°C) the gross calorific value increased slowly, mainly associated with the removal of moisture; in carbonization stage, it had grown rapidly, reaching 35.48MJKg^{-1} (Fig. 5). Due to its pore structure and high fixed carbon content, chestnut shell biochar had a higher calorific value than the raw material.

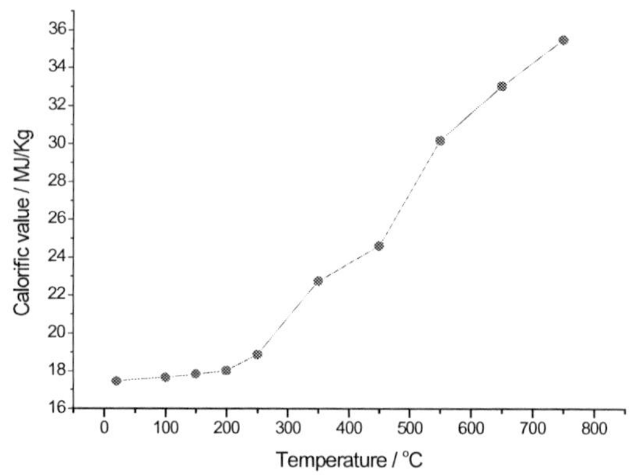

Fig. 5. The gross calorific values of samples at different temperatures.

CONCLUSIONS

We have presented a procedure for obtaining a biochar product by carbonization of chestnut shell. The pyrolysis behaviors and the transmutation of both the structure and properties during the carbonization were investigated. With the catalyst, the chestnut shell based biochar with yield of 44.31% and gross calorific value of 35.48MJKg^{-1} can be obtained at 750℃.

KEYWORDS

Chestnut shell, Biochar, Carbonization

ACKNOWLEDGMENT

This work was funded by grants from the Research and Innovation Project of Shanghai Municipal Education Commission (No. 14zz069) and the Key Laboratory of Advanced Civil Engineering Materials of Ministry of Education, Tongji University (No.201301).

REFERENCES

[1] Sevilla, M., Fuertes, A.B. “The production of carbon materials by hydrothermal carbonization of cellulose.” *Carbon*, 47, 2009: 2281-89.

[2] Okutucua, Cagdas, Dumana, Gozde, Ucarb, Suat. “Production of fungicidal oil and activated carbon from pistachio shell.” *J.* Anal. Appl. *Pyrolysis*, 91, 2011: 140-6.

[3] Ozçimen, Didem, Ersoy-Meriçboyu, Aysegul. “Characterization of biochar and bio-oil samples obtained from carbonization of various biomass materials.” *Renew. Energ.*, 35, 2010: 1319-24.

[4] Qu, Yaohui, Zhang, Zhian, Zhang, Xiahui, et al. “Highly ordered nitrogen-rich mesoporous carbon derived from biomass waste for high-performance lithium-sulfur batteries.” *Carbon*, 84, 2015: 399-408.

[5] Wu, Xilin, Wen, Tao, Guo, Hongli, et al. “Biomass-Derived Sponge-like Carbonaceous Hydrogels and Aerogels for Supercapacitors.” *ACS Nano.*, 7, 2013: 3589-3597.

Preparation and Properties of Recycled PET Fiber-Filled Polyethylene Composites

Zhigang Wang, Wenhao Xi, Jingbo Zhou, Jiaming Xu, Guang Li
College of Material Science and Engineering, Donghua University, Shanghai, China
lig@dhu.edu.cn; hpuwang@126.com

INTRODUCTION

Nowadays, the regeneration research of Polyethylene Terephthalate (PET) fibers are mainly recycled for raw materials of spinning and plastic by physical/chemical methods[1]. However, lots of problems exist with the recycling processes, such as Complex recovery system, immature technologies and high cost, therefore a new way for high-efficiency and environmental friendly recycling is quite essential. Furthermore, it is worth to find new applications for recycled PET fibers in fiber form, and there have been several studies performed on PET fiber reinforced composites[2-4]. In this study, we firstly deal with the waste PET textiles through PFI mill, and the recycled PET fibers were used to prepare fiber reinforced composites by melting blend methods.

APPROACH

Recycled PET fibers were obtained from waste PET textiles using a PFI mill. High density polyethylene (HDPE)-based composites reinforced with recycled PET fibers were manufactured by melting blend. The mechanical properties of the composites were investigated by mechanical property test. The thermal stability and crystallinity were analyzed by Thermogravimetric Analysis (TGA) and Differential Scanning Calorimetry (DSC), and their microstructures were characterized by Scanning Electron Microscopy (SEM).

RESULTS AND DISCUSSION

Preparation of recycled PET fibers with PFI mill

The optical micrograph of the recycled fibers exhibited in Fig. 1c indicates well dispersed of PET fibers after beating, and the length of fibers is kept well during milling process, it is roughly 8 mm in this study.

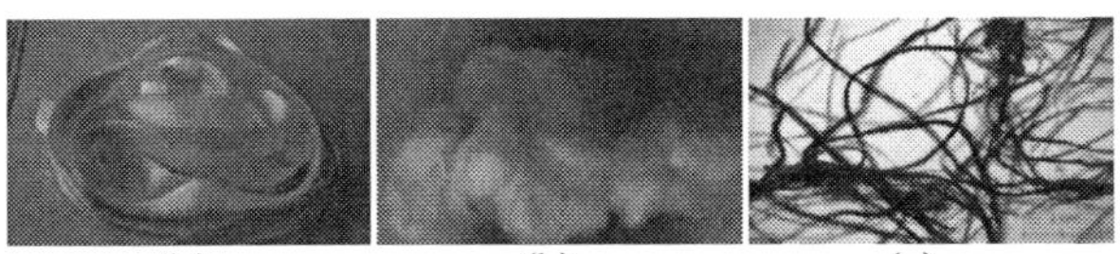

(a) (b) (c)

Fig. 1. The photos of appearance of start PET textile and recycled PET fibers (a) start PET textile, (b-c) recycled PET fibers.

Effects of PET fiber content on mechanical properties

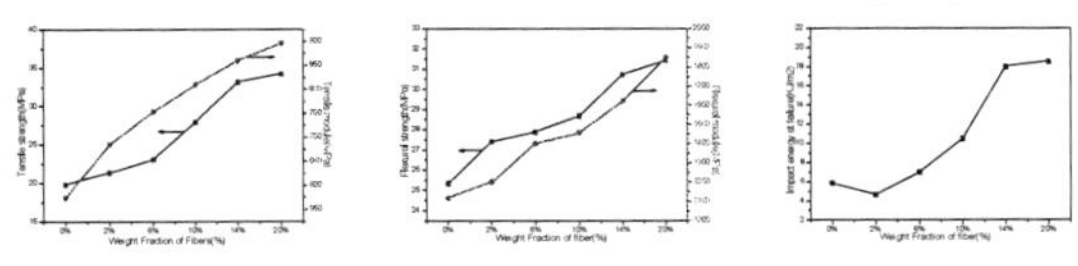

(a) (b) (c)

Fig. 2. Effect of PET fiber content on mechanical properties of composites, (a) tensile properties (b) flexural properties (c) impact properties.

From Fig. 2a, with increasing the content of PET fibers, the tensile strength and tensile modulus of the composites increased significantly, which suggests the effective reinforcement of recycled PET fibers. From Fig. 2b, the flexural strength and modulus reached 31.41MPa and 1850.3MPa when the fiber content is 20wt%, increased by 24.2% and 65.8% respectively. From Fig. 2c, the notch impact strength became high with increasing content of the PET fibers except for the content of 2wt%, which might be due to individual factor. The impact strength exhibited an increase of 218% from 5.81KJ/m^2 to 18.52 KJ/m^2 (20wt% PET), which may be related to the fiber-matrix adhesion where a strong network or bonding between fibers and HDPE matrix was built.

Differential scanning calorimetry

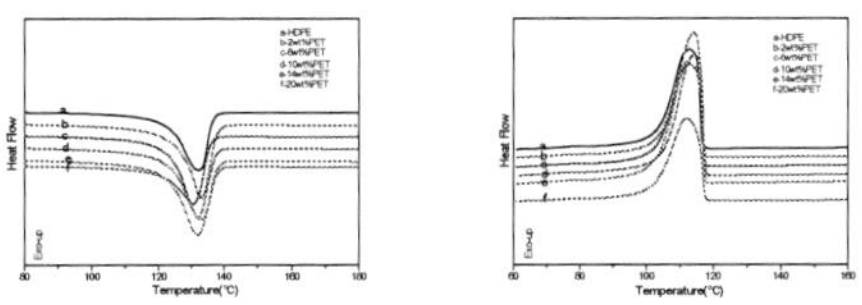

(a) (b)

Fig. 3. DSC curves of HDPE composites reinforced with different content of PET fibers, (a) melting curves, (b) crystallization curves.

The crystallization temperature (T_c), melting temperature (T_m), super-cooling degree (ΔT), melting enthalpy (ΔH_m) and degree of crystallinity (X_c) obtained from DSC analysis are summarized in Table I. From Table I the addition of recycled PET fibers slightly shifted the crystallization temperature towards higher temperature as compared to pristine HDPE, while the melting temperature did not show change. Super-cooling degree of the composites was also increased, which indicates that the addition of PET fibers may act as nucleus and improve the crystallization rate of HDPE. In addition, the degree of crystallinity increased with the addition of PET fibers.

Table I. The DSC characteristic value of the composites.

sample	T_m (℃)	T_c(℃)	ΔT(℃)	ΔH_m (J/g)	X_c
HDPE	132.2	112.6	19.6	155.4	53.6%
2wt%PET	133.1	114.3	18.8	143.2	50.4%
6wt%PET	130.4	113.2	17.2	147.2	54.0%
10wt%PET	131.6	114.5	17.1	144.7	55.4%
14wt%PET	131.4	112.7	18.7	149.1	59.8%
20wt%PET	132.1	113.1	19.0	145.6	62.8%

Thermal stability

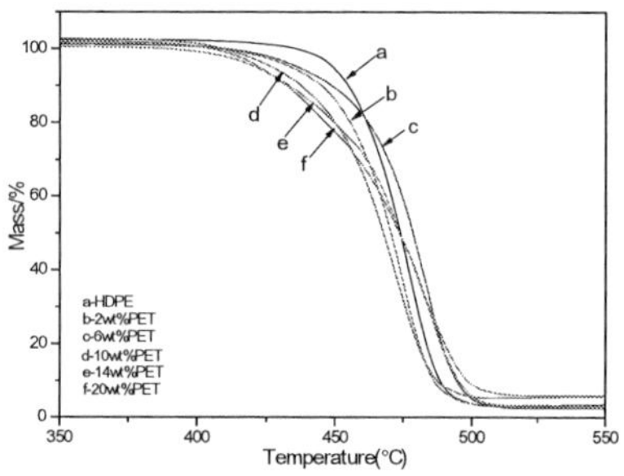

Fig . 4. TGA curves of the composites with different content of PET fibers.

The characteristic thermal parameters of the thermal degradation are listed in Table II, including the onset temperature (T_0) and the final decomposition temperature (T_d). For pristine HDPE and the obtained composites, there was single step decomposition process. The onset of thermal decomposition of the composites was slightly lower than that of pristine HDPE. This is because PET fibers have moisture absorption in air, easily hydrolyze at low temperatures.

Table II. The TGA characteristic value of the composites.

sample	HDPE	2wt% PET	6wt% PET	10wt% PET	14wt% PET	20wt% PET
T_0(°C)	383.0	372.1	371.7	368.4	364.8	366.9
T_d(°C)	511.8	512.0	528.3	512.4	523.8	524.1

Morphology characterization

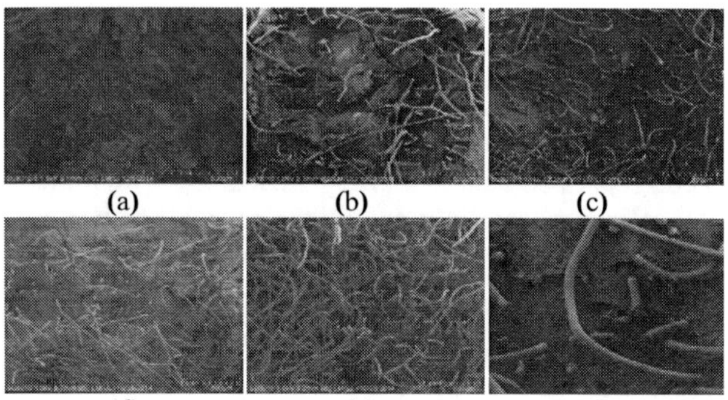

Fig. 5. SEM micrographs of fracture section of the composites, (a) HDPE, (b) 6wt%PET, (c) 10wt%PET, (d) 14wt%PET, (e) 20wt%PET, (f) 10wt%PET high magnification.

It is observed a good distribution and dispersion of PET fibers in HDPE matrix (Fig. 5a-e). Many fibers were pulled out from the matrix before breakages, leaving some holes there (Fig. f). In this case, interface adhesion of fibers and HDPE consisted of interface friction and static adsorption, and the main model of interface failure was the fiber pull-out and rupture, which was confirmed by the strengthening mechanism: PET fiber has a greater strength than the matrix, when composites are under external load, shear stresses between the matrix and fibers would be transferred to the fibers, in the meanwhile deformations of the matrix and fibers are confined by each other.

CONCLUSION

The mechanical properties, including tensile, flexural and impact behavior, of the composites showed improved performance with increasing content of the recycled PET fibers up to 20wt%. The degree of crystallinity and crystallization rate were enhanced to certain extent, whereas the thermal stability was almost unchanged. The recycled PET fibers could be dispersed in HDPE matrix with the help of 10wt% PE-g-MAH as a compatibilizer, but the interface bonding function between fibers and HDPE needs to be further improved.

KEYWORDS

Recycled PET fibers, HDPE, PFI, Composites

REFERENCES

[1] Shi, Yuan-Yuan, Chen, Yan-Xia, Shi, Yi-Dong, et al. "Exploitation of recycling of waste PET for textile product." *Progress in Textile Science & Technology*, 1, 2008: 29-31.

[2] Kayaisang, Supattra, Amornsakchai, Taweechai, Saikrasun, Sunan. "Influence of liquid crystalline polymer and recycled PET as minor blending components on rheological behavior, morphology, and thermal properties of thermoplastic blends." *Polym. Adv. Technol.*, 20, 2009: 1136-1145.

[3] Satapathy, Sukanya, Nando, Golok Bihari, Jose, Jobin, et al. "Mechanical properties and fracture behavior of short PET fiber-waste polyethylene composites." *Journal of Reinforced Plastics and Composites*, 27, 9, 2008: 967-984.

[4] Chen, San-Yue, Liang, Wen-Miin, Yen, Pau-Su. "Reinforcement of Acrylic Denture Base Resin by Incorporation of Various Fibers." *Journal of Biomedical Materials Research*, 58, 2, 2001: 203-208.

Isothermal Crystallization Behaviors of Cellulose Acetate Butylene/Poly(butylene succinate)

Xinhang Wang, Yanping Wang, Yimin Wang
State Key Laboratory for Modification of Chemical Fibers and Polymer Materials, College of Materials Science and Engineering, Donghua University, Shanghai, China
xh_w2012@163.com

INTRODUCTION

Cellulose acetate butyrate (CAB) is an abundant and promising cellulose derivative with renewable and potentially biodegradable characteristics. Though having melts processability compared with the parent cellulose, CAB still suffers relatively high melt viscosity and a narrow window between the melt flow temperature and the decomposition temperature compared to typical commodity plastics or fibers. Moreover, the rigid nature of cellulose backbone contributes to the stiffness and brittleness of CAB which limits its application. A convenient method that blending CAB with biodegradable polymers such as poly((ε-caprolactone), polylactide, poly(butylene succinate) (PBS)[1] was used to improve its melt flowing and mechanical properties. In general, it is of great significance to study the crystallization behaviors of polymer blends because this have a profound effect on the material performance. To further understanding the promising CAB/PBS blend, the isothermal crystallization behaviors were investigated in detail using differential scanning calorimetry (DSC).

KEYWORDS

Cellulose acetate butyrate, poly(butylene succinate), isothermal crystallization behaviors, the equilibrium melting point.

APPROACH

CAB/PBS blends with 0, 5, 10 and 20 wt% of CAB component were obtained through the solution casting method.

RESULTS AND DISCUSSION

Isothermal crystallization behaviors

During this work, it was found that the CAB/PBS blends with more than 30 wt % of CAB crystallized very slowly or did not crystallize, so the isothermal crystallization behaviors of only pure PBS and the blends with less than 20 wt % of CAB were studied.

The Avrami equation, which is widely used to analyze the isothermal crystallization kinetics, can be expressed as a double logarithmic form as following equation (1):

$$\log_{10}\left[-\ln\left(1 - X(t)\right)\right] = n\log_{10} t + \log_{10} K \qquad (1)$$

Where t and $\boldsymbol{K}$ denote the duration of the crystallization process and the overall crystallization rate constant, n is an Avrami exponent determined by the crystal nucleation mechanism and geometrical shape of crystallite. The relative crystallinity, (t), is defined as the proportion of formed crystallinity until time (t) to the total formed crystallinity during the whole isothermal crystallization process. Figure 1 show (t) of neat PBS and the CAB/PBS composites as a function of time at different temperature.

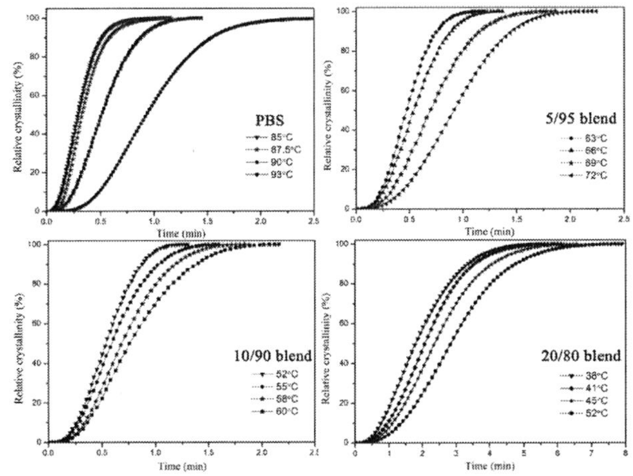

Figure 1. Relative crystallinity of pure PBS and its blend with CAB.

The Avrami plots of $\log_{10}\left[-\ln\left(1 - X(t)\right)\right]$ versus $\log_{10} K$ of the pure PBS and the CAB/PBS blends are shown in Figure 2. The value of n and $\boldsymbol{K}$ can be obtained from the slope of the starting linear portion and the intercept, respectively. Another important parameter of isothermal crystallization kinetics, the half-time of crystallization ($t_{1/2}$), which is defined as the time when (t) is 50%, can be calculated as following equation (2):

$$t_{1/2} = \left(\frac{\ln 2}{k}\right)^{1/n} \qquad (2)$$

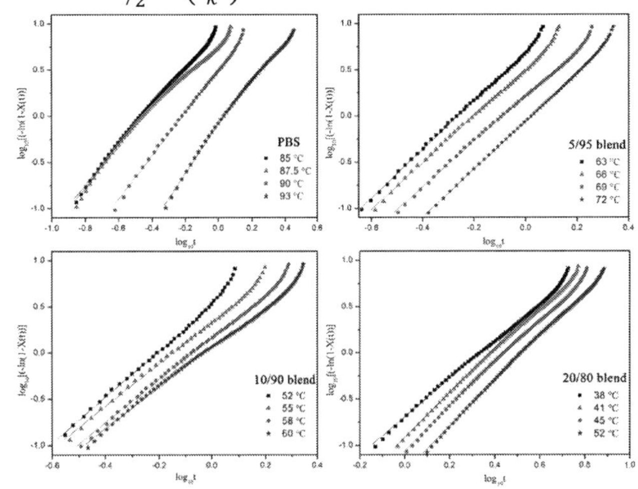

Figure 2. Avrami plots of pure PBS and its blends with CAB isothermally crystallized at predetermined temperature.

The isothermal crystallization parameters are summarized in Table 1. In this study, the n values of the pure PBS and its blends with CAB are in the same range of the 2.0 - 3.0, indicating they almost share the same crystal nucleation mechanism and geometrical shape of crystallites. The value of K and $t_{1/2}$ decreased and increased respectively for all the samples, indicating that the isothermal crystallization rate increase with decreasing the crystallization temperature (Tc). In general, after incorporating with CAB component, the K values for PBS component decreased, whereas the $t_{1/2}$ values showed the opposite trend, even though the Tc values in the CAB/PBS blends were smaller than those of pure PBS. This demonstrates that the isothermal crystallization of PBS is severely hindered by CAB. This may be attributed to the restricted molecular mobility of PBS during crystallization process by the rigid CAB molecular with high glass transition temperature. This phenomenon was also reported by T. Tatsushima.[1] Similarly, the crystallization rate of polylactide was confirmed to be significantly reduced.[2]

Melting behavior after isothermal and equilibrium melting point

The equilibrium melting point (T_m^o), which may be defined as the melting point of perfect crystals with infinite large lamella, can separate the morphological effect from the thermodynamic effect and thus is frequently employed to estimate the miscibility of polymer blends. T_m^o was determined by the following Hoffman -Weeks equation (3):

$$T_m = \eta\, T_c + (1 - \eta)\, T_m^o \qquad (3)$$

where Tm is the observed melting point, η may be regarded as a measure of the stability. In this work, the melting behaviors after isothermal crystallization were measured using DSC. As an example, Figure 3(a) showed the melting thermograms of the blend containing 5 wt% of PBS. Double melting peaks are observed which can be contributed to different size and perfection or re-crystallization.[3] T_{m1} corresponded to the melting of crystals formed during existed prior to the subsequent heating process, while the isothermal crystallization process which alreadyT_{m2} corresponded to the melting of crystals formed through recrystallization during the heating process. This phenomenon was also observed for other blends in this study. Therefore, T_{m1} should be adopted for the analysis of the Hoffman – Weeks equation. The Hoffman –Weeks plots were shown in Figure 3(b), from which the T_m^os were estimated to be around 139.0 °C 138.0 °C, 136.6 °C, 134.4 °C for pure PBS and its composites with 5, 10 and 20 wt% CAB, respectively. It is clear that T_m^os of PBS decrease with the addition of CAB. The depression of T_m^o indicates that CAB and PBS are thermodynamically miscible.

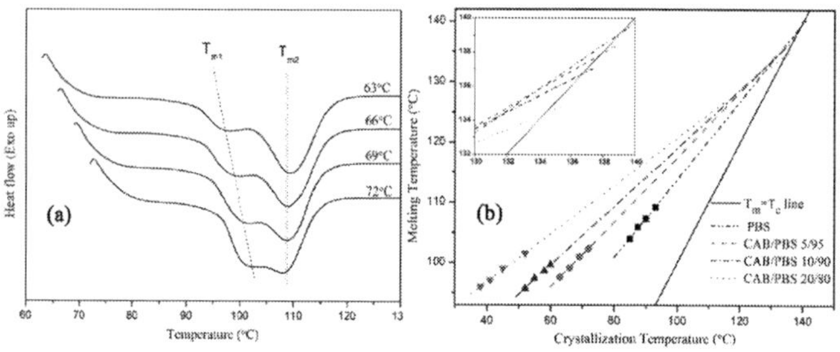

Figure 3. (a) DSC scan for CAB/PBS with 5wt% CAB after isothermal crystallization, (b) application of Hoffmn-Weeks analysis to samples for determining T_m^o.

CONCLUSION

The isothermal crystallization kinetics is analyzed using the Avrami equation in this work, and the result show that the crystallization of PBS is hindered after compositing CAB. Furthermore, the equilibrium melting point of pure PBS and CAB/PBS blends were determined by the melting point after isothermal crystallization using Hoffman-Weeks equation, the depression of Tm° expressed CAB and PBS are thermodynamically miscible.

REFERENCES

1. Tatsushima, T., Ogata, N., Nakaneand, K., Ogihara, T. *Journal of Applied Polymer Science*, 96, 2005: 400-406.
2. Wang, Y., Xu, Y., He, D., Yao, W., Liuand, C., Shen, C. *Materials Letters*, 128, 2014: 85-88.
3 Lee, Y, Porter, R.S. *Macromolecules*, 22, 1989: 1754.

Table 1. Crystallization kinetics parameters based on the Avrami analysis.

Sample	T_c / °C	n	K / min^{-1}	$t_{1/2}$ / min
Pure PBS	85	2.20	10.33	0.29
	87.5	2.18	9.27	0.30
	90	2.35	3.22	0.52
	93	2.65	0.80	0.95
CAB/PBS 5/95 blend	63	2.62	4.85	0.48
	66	2.52	3.14	0.55
	69	2.47	1.64	0.71
	72	2.59	0.87	0.92
CAB/PBS 10/90 blend	52	2.47	3.23	0.54
	55	2.28	2.19	0.60
	58	2.30	1.50	0.71
	60	2.20	1.20	0.78
CAB/PBS 20/80 blend	38	2.05	0.20	1.82
	41	2.33	0.12	2.10
	45	2.41	0.09	2.38
	52	2.47	0.05	2.87

Preparation and Characterization of Electro-Responsive Core/Shell Type Fibrous Mats as Smart Materials

Se Young Oh, Deok Jun Whang, Jeong Seon Ah, Jung Hyun Taek, Tae Jin Kang
Department of Materials Science and Engineering, Seoul National University, Seoul, Republic of Korea
teakang@snu.ac.kr; osy530182@snu.ac.kr

INTRODUCTION

Recently there has been a great deal of progress in the production of one-dimensional nanostructures with core-sheath structures from a variety of materials. Among the other nano/micro-size fiber fabrication methods, electrospinning has gained growing interest in the past decade because this technique is quite versatile and cost-effective for producing functional nano/micro-size fibers from variety of materials including polymers, polymer blends, emulsions, suspensions, sol-gels, metal oxides, composite structures as well as non-polymeric systems. Electrospun fibers and their webs have remarkable characteristics including a very high specific surface area, pore sizes within the nanoscale and very lightweight. Moreover, the control of the fiber surface morphology, fiber orientation, and cross-sectional configuration, quite feasible for obtaining multi-functional electrospun nanofibers. Nano/micro-size fibers with core-sheath, hollow, or porous structures have many potential applications in microfluidics, photonics, sensor, biomedical and energy storage.

ER fluids have been recognized as a smart and intelligent material. Their structural and rheological changes under the external electric fields are typically reversible and have swift transition between the liquid state and the solid state. ER fluids can potentially be used as an interface that translates electrical signals into mechanical signals, opening the possibility of actively controllable clutches, dampers, valves, locks, and impact absorption.

In this study, we have successfully achieved the core/shell micro-size fibers containing ER fluids by coaxial electrospinning processes; the micro-size fibers composed the PET (polyethylene terephthalate (shell)) and inorganic-coated MWCNTs based on oil medium (core). The core/shell micro-size fibers including different ER particle contents (1, 2, 4 and 8 wt%) were produced through controlling applied voltage, tip-correction distance and flow rate of core and shell. Micro-size fibers including ER fluids have shown uniform morphology and enhanced mechanical strength under the external electric fields. Our findings suggest that these core/shell fiber mats can be promising instantly hardening materials under external electric field.

EXPERIMENTAL

Preparation of inorganic-coated MWCNTs and ER fluids

Raw MWCNTs were treated with HNO3/H2SO4 mixture (3:1) at 100 °C for 24h and then washed with deionized water. NH4OH (28~30 vol.%) and distilled water are added in the F-MWNTs/ethanol mixtures and then sonicated to produce a stable and homogeneous suspension. Then, silica-coated MWNCTs were formed by the hydrolysis and condensation reactions of TEOS. Finally, dry silica-coated MWCNTs were ultrasonically added in ethanol, distilled water and acetic acid. TBT was quickly injected into the flask and stirred vigorously for 4 h. After the reaction is complete, centrifuging, washing and drying procedures of the particles were performed sequentially. ER fluids were prepared by dispersion of the dry silica-titania coated MWCNTs in silicon oil (Dielectric constant of 2.7, density of 0.96 g/cm3 at 25 oC) using mechanical stirring and ultrasonic, respectively. Viscosity of core fluid is controlled by mixture of 50 and 1000 cs silicon oil. The weight fraction of silica-titania coated MWNTs within suspension was adjusted to 1, 2, 4 and 8 wt%. The PET solution of 12 wt% was prepared by dissolving PET chip into TFA and stirring vigorously for 5 h.

Coaxial electrospinning setup

Coaxial electrospinning was performed with varying flow rates of inner and outer dope to determine appropriate levels. The basic setup was composed of two syringe pumps, a high power voltage supply, a grounded collector and a coaxial spinneret. The inner and outer size of dual capillary is 16 gauge (1.660 mm) and 21 gauge (0.8 mm). The shell flow rate fixed to 2.0 ml/h and the core flow rate varied from 0.3 to 0.6 ml/h. The applied voltages ranged from 12 to 14 kV. Distance between spinneret and collector was kept at 15 cm. A colored paper above a flat metal plate is used as a collector.

RESULTS AND DISCUSSION

Characterization of inorganic-coated MWCNTs and core/shell electrospun micro-size fibers

The morphologies of inorganic-coated MWCNTs and electrospun micro-size fibers were studied by the transmission electron microscope (TEM) and the scanning electron microscope (SEM). We confirmed that the morphology of F-MWCNTs was well preserved as well as the silica/titania shells were formed uniformly after the sol-gel reaction. EDS was conducted to characterize the elemental composition of the inorganic-coated MWCNTs and the composition of silicon (37.2 %), titanium (31.15 %), oxygen (27.7 %), and carbon (3.49 %) elements were confirmed by the line profiles recorded perpendicular to the long axis of carbon nanotubes.

Continuous micro-size fibers with bead-free morphology could be formed under the following parameters: outer (PET) solution concentration was 12 wt% as optimal condition, voltage was set as 11-13 kV and the feeding rate of inner and outer solution were set as 0.3 mL/h and 2.0 mL/h, respectively. The surface of the electrospun micro-size fibers containing 1 and 2 wt% of inorganic-coated MWCNTs is very smooth, however, low surface roughness and unevenness in the samples containing 4 wt% were observed. Considerable aggregation and local irregularities were observed as the inorganic-coated MWCNTs content increased to 8 wt%.

In the electrospinning process, the viscosity and conductivity of the core solution are very important parameters, which can affect the composite micro-size fibers diameters and uniformity of fiber surface. As described in the literature, mostly with the viscosity and conductivity increasing in the particle content, the average diameter of micro-size fibers increased.

The ER fluid of 4 wt% was fully incorporated in the core part. However, as the inorganic-coated MWCNT content in the electrospun fibers increased to 8 wt%, the micro-size fibers was irregular surface and bead, which can be related to inorganic-coated MWCNT aggregation and bundling during the electrospinning process. Also, the irregular surface and bead of the fibers can influence the mechanical properties of the composite samples. Generally, to improve the morphological properties of the composite fibers containing inorganic-coated MWCNTs, the dispersion condition is very critical and plays a significant role in achieving effective properties.

The mechanical properties of the electrospun core/shell micro-size fiber mat were measured on the universal testing machine (Instron, 5566) at room temperature. The specimens of micro-size fiber mat were cut into 1 cm x 4 cm rectangular shapes. Tensile testing was performed using a 10 N load cell at extension rate of 50 mm/min. The strength of the fiber mat was calculated from Eq. 1. The modulus and elongation at break were obtained from the stress-strain curves. The mechanical data were acquired with five independent experiments, and all data were displayed as mean values and standard deviation:

$$\sigma(Mpa) = \frac{\rho(g/cm^3) \times F(cN)}{T(cm) \times A(g/m^2)} \quad (1)$$

where σ is tensile strength, ρ is density, F is tensile force, T is the thickness of the fiber sample, and A is cross-sectional area density.

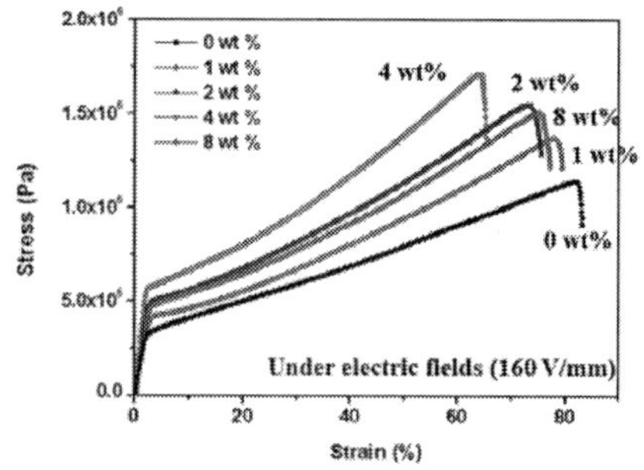

Figure 1. Stress–strain curves of core/shell nanofiber mats as various concentrations of core fluids

With the inorganic-coated MWCNTs in core solution increasing from 1 wt% to 4 wt%, the tensile strength and modulus increased. Even so, when inorganic-coated MWCNTs were increased to 8 wt%, the tensile strength and modulus decreased. It is because the increase of core particles can cause the collapse of core/shell structure as well as form unstable fiber structure.

When an external electric field was applied, the ERF/PET core/shell nanowebs showed the increase of tensile strength and modulus, but showed the decrease of strain. A dipole-dipole interaction arises between the inorganic-coated MWCNTs when the micro-size fiber containing ER fluids is exposed to an electric field and the tensile axis component of such dipole-dipole interactions can act as a resistance force against the tensile stress. As the dipole-dipole interaction is inversely proportional to the square of distance between the nanoparticles, this instant reinforcement effect of inorganic-coated MWCNT nanoparticles under an electric field is more prominent in the initial stage of tensile deformation. Consequently, further improvement was observed in modulus values than tensile strength values. With the tensile deformation progressed, the distance between the nanoparticles increases along the tensile axis, while decreases along the direction perpendicular to the tensile axis. Upon further tensile deformation, the distance between the nanoparticles oriented in the direction perpendicular to the tensile axis become close enough to form aggregations due to the dipole-dipole interaction. Such aggregations act as stress concentrators, decreasing the strain.

ACKNOWLEDGMENT

This work was supported by DAPA (Defense Acquisition Program Administration) and ADD (Agency for Defense Development, South Korea).

REFERENCES

[1] Chuangchote, Surawut, Sirivat, Anuvat, Supaphol, Pitt. “Mechanical and electro-rheological properties of electrospun poly(vinyl alcohol) nanofibre mats filled with carbon black nanoparticles.” *Nanotechnology*, 18, 2007: 145705.

[2] Nasouri, Komeil, Shoushtari, Ahmad Mousavi, Ali Kaflou, Bahrambeygi, Hossein, Rabbi, Amir. “Single-Wall Carbon Nanotubes Dispersion Behavior and Its Effects on the Morphological and Mechanical Properties of the Electrospun Nanofibers.” *Polymer Composites*, 33, 2012: 1951.

[3] Yin, Jianbo, Zhao, Xiaopeng. “Electrorheol-ogy of nanofiber suspensions.” *Nanoscale Research Letters*, 6, 2011: 256.

Coordination Effect of Plasma Pre-Treatment on Alkaline Etching and Deep Coloring of Polyethylene Terephthalate Fabrics

Amel E. A. Elabid[1], Jing Zhang[2]

[1]College of Material Science and Engineering, Donghua University, Shanghai, China
[2]College of Science, Donghua University, Shanghai, China
jingzh@dhu.cdu.cn; amelkanzi2014@gmail.com

ABSTRACT

The conventional textile dyeing process involves a high consumption of water, chemicals and energy. Therefore, the textile industry has a great interest in novel technologies with energy and resources-saving. Plasma technology received enormous attention as an environmental-friendly thenology. In this work, synergistic effect of atmospheric pressure plasma on alkaline etching and deep coloring of dyeing properties on polyethylene terephthalate (PET) fabrics was investigated. The weight loss percentage of PET fabric treated by alkaline and O_2/Ar plasma has been more than the total amount of the fabric by the two treatments separately. The K/S value and color strength of PET fabrics by various plasma pre-treatment times was increased. It indicated that plasma pre-treatment has synergistic effect on alkaline treatment of PET fabrics, which could be obtained at low temperature.

KEYWORDS: atmospheric pressure plasma, synergistic effect, alkaline treatment, polyethylene terephthalate fabric, weight loss, K/S value.

INTRODUCTION

PET fabric is one of the largest synthetic polymer commodity in the world and usually requires harsh alkaline and dyeing processes [1], which is causing environmental problems for its hydrophobicity and poor dyeing performance etc.. The demands for environment friendly process led to the search for a new plasma pre-treatment technique to improve the alkaline etching and deep coloring of PET materials with less energy, chemicals and cost. Surface properties of materials such as wettability, chemistry, charge and topography [2] are of great importance in textile applications in fabric handness or dyeing etc.. The plasma treatment is highly efficient for increasing the surface energy and modifying the uppermost molecular layers without affecting the bulk properties [3,4,5].

OBJECTIVE

To decrease energy, chemicals and time consume in the traditional alkaline etching and dyeing process of PET fabric materials by atmospheric pressure plasma pre-treatment. To understand the mechanism of the process through the relationship of the plasma process and the structure and properties of the fabric surface.

APPROACH

A dielectric barrier discharge (DBD) atmospheric pressure plasma system is used for the surface treatment in this study. AC power used is about 20 kHz. Then alkaline etching proceeds for only 5 minutes with a solution of 4% w/w at 85°C and the liquor to fabric ratio was 50:1. Dyeing was carried out in a dispersive dye bath containing 2% o.w.f shade in distilled water of 100 °C without any auxiliaries.

RESULTS AND DISCUSSION

Figure 1 (a) shows the weight loss of PET fabrics with variation of alkaline treatment time under close ventilation conditions using alkaline solution (4M) and constant temperature of 85°C. The weight loss became noticeable with alkaline treatment time of more than 30 minutes. However, after 3 minutes plasma pre-treatment, the weight loss rate was noticeable for only 5 minutes alkaline treatment as shown in Figure 1 (b). The weight loss rate of PET fabric treated by alkaline and O_2/Ar plasma has been more than the total amount of the fabric by the two treatments separately, which shows the synergistic effect of plasma pre-treatment on the alkaline etching treatment.

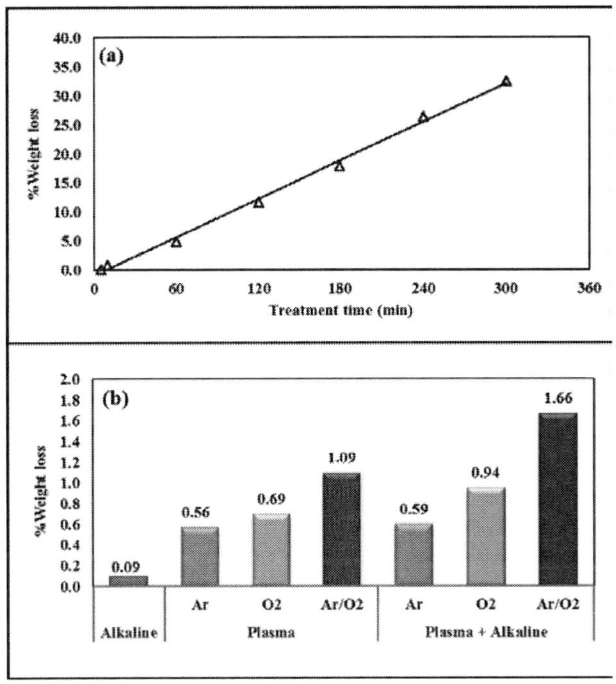

Figure 1. Weight loss rate of PET fabrics at alkaline etching temperature of 85°C (a) with variation of alkaline treatment time (b) with various plasma pre-treatments of 3 minutes and alkaline etching time of 5 minutes.

Moreover AFM analysis gave information on the plasma induced surface chemical and topographical modifications.
Figure 2 shows AFM images of PET fiber surface before and after alkaline and plasma treatment. As shown in Figure 2 (a) and (b), the untreated sample shows relatively flat surface and the plasma treated one shows small edges of hill,

uniformly distributed on the fiber surface. PET fiber surface in Figure 2 (c) the alkaline treated shows a few rise to pits and craters. While sample in Figure 2 (d) through plasma pre-treatment and alkaline etching at 4M 85°C shows the broad and smooth wavy surface originating from alkali treatment reaction. The number of ridges on the surfaces is influenced by plasma pre-treatment and alkaline etching time. Figure 3 shows the root mean square roughness (RMS) of the PET fabrics of untreated, the alkaline treatment, the plasma treatment, the plasma/alkaline pre-treatment. The RMS roughness value increases with adding another treatment, which shows the superposition effect of these treatment process on the roughness of the PET fiber surface.

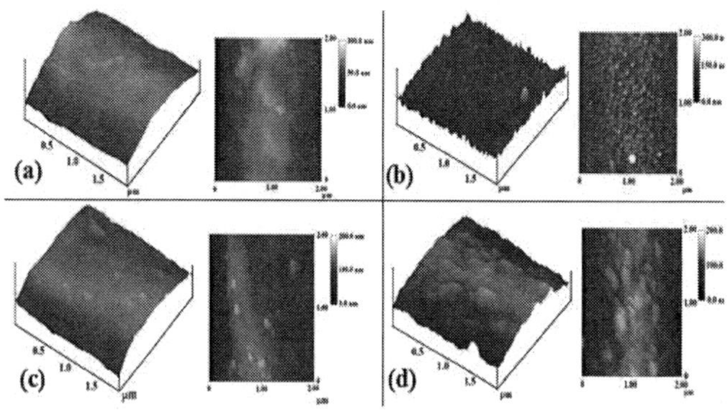

Figure 2. AFM images of PET (a) untreated (b) treated in O_2/Ar plasma treatment (c) treated by alkaline 4M at 85 °C (d) treated by O_2/Ar plasma pre-treatment on alkaline etching at 4M, 85 °C.

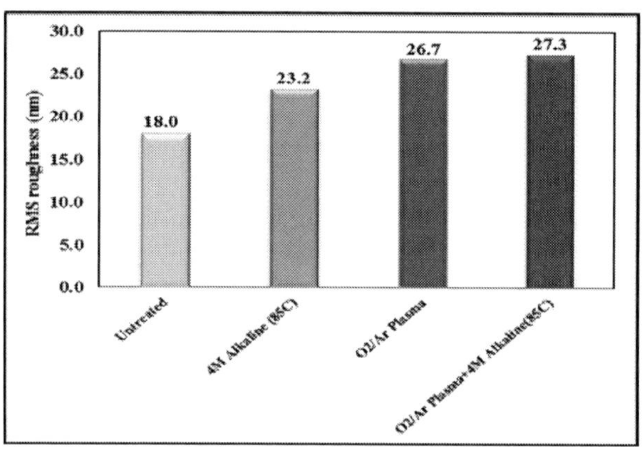

Figure 3. RMS roughness of PET fiber surface of untreated and treated at different conditions.

Figure 4 shows the dyed PET fabric samples and their K /S values. K /S values and color strength increases obviously after plasma only or alkaline treatment together.

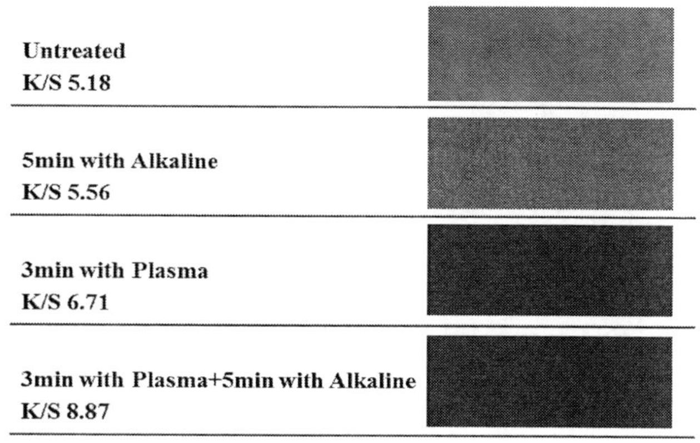

Figure 4. Photographs of dyed PET fabric samples.

CONCLUSION

Synergistic effects of plasma pre-treatment on alkaline etching was found and the weight loss rate, surface roughness, K/S value and color strength was found obviously increased after plasma pre-treatment. As shown in Figure 5, it is believed that the number of active sites and bond ability on the PET surface was changed due to the roughness increases and the introduction of additional C=O and O-C=O groups.

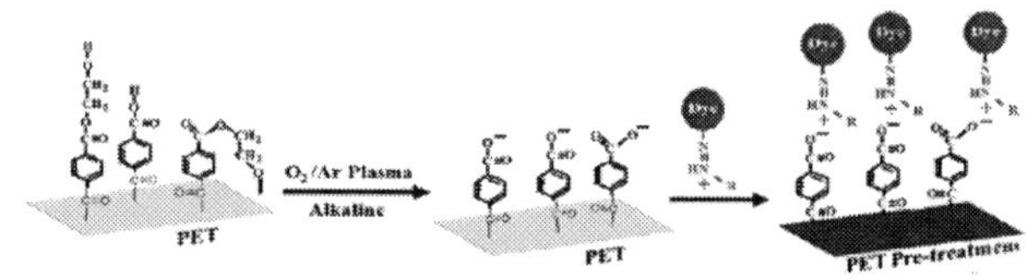

Figure 5. The mechanism of blue disperse dye with the plasma pre-treated PET fabric surface.

The increase of fiber surface energy and the surface roughness led to achieve increased coloration of the dispersive dyes.

REFERENCES

1. Parvinzadeh, M., Ebrahimi, I. "Atmospheric air-plasma treatment of polyester fiber to improve the performance of nanoemulsion silicone." *Applied Surface Science*, 257, 9, 2011: 4062-4068.

2. Khang, G., et al. "Controlled Release of Nerve Growth Factor from Sandwiched Poly(L-lactide-co-glycolide) Films for the Application in Neural Tissue Engineering." *Macromolecular Research*, 11, 5, 2003: 334-340.

3. Shah, J.N., Shah, S.R. "Innovative Plasma Technology in Textile Processing: A Step Towards Green Environment." *Research Journal of Engineering Sciences*, 2, 4, 2013: 34-39.

4. Shishoo, R. *Plasma Technologies for Textiles*. Boca Raton: CRC Press. 2007: 6.

5. Oh, K.W., et al. "Improved adhesion property and electromagnetic interference shielding effectiveness of electroless Cu-plated layer on poly (ethylene terephthalate) by plasma treatment." *Applied Polymer Science*, 84, 7, 2002: 1369-379.